Re-engineering for Sustainable Industrial Production

IFIP – The International Federation for Information Processing

IFIP was founded in 1960 under the auspices of UNESCO, following the First World Computer Congress held in Paris the previous year. An umbrella organization for societies working in information processing, IFIP's aim is two-fold: to support information processing within its member countries and to encourage technology transfer to developing nations. As its mission statement clearly states,

> IFIP's mission is to be the leading, truly international, apolitical organization which encourages and assists in the development, exploitation and application of information technology for the benefit of all people.

IFIP is a non-profitmaking organization, run almost solely by 2500 volunteers. It operates through a number of technical committees, which organize events and publications. IFIP's events range from an international congress to local seminars, but the most important are:

- the IFIP World Computer Congress, held every second year;
- open conferences;
- working conferences.

The flagship event is the IFIP World Computer Congress, at which both invited and contributed papers are presented. Contributed papers are rigorously refereed and the rejection rate is high.

As with the Congress, participation in the open conferences is open to all and papers may be invited or submitted. Again, submitted papers are stringently refereed.

The working conferences are structured differently. They are usually run by a working group and attendance is small and by invitation only. Their purpose is to create an atmosphere conducive to innovation and development. Refereeing is less rigorous and papers are subjected to extensive group discussion.

Publications arising from IFIP events vary. The papers presented at the IFIP World Computer Congress and at open conferences are published as conference proceedings, while the results of the working conferences are often published as collections of selected and edited papers.

Any national society whose primary activity is in information may apply to become a full member of IFIP, although full membership is restricted to one society per country. Full members are entitled to vote at the annual General Assembly, National societies preferring a less committed involvement may apply for associate or corresponding membership. Associate members enjoy the same benefits as full members, but without voting rights. Corresponding members are not represented in IFIP bodies. Affiliated membership is open to non-national societies, and individual and honorary membership schemes are also offered.

Re-engineering for Sustainable Industrial Production

Proceedings of the OE/IFIP/IEEE International Conference on Integrated and Sustainable Industrial Production Lisbon, Portugal, May 1997

Edited by

Luis M. Camarinha-Matos
New University of Lisbon
Lisbon, Portugal

Published by Chapman & Hall on behalf of the
International Federation for Information Processing (IFIP)

CHAPMAN & HALL
London · Weinheim · New York · Tokyo · Melbourne · Madras

Published by Chapman & Hall, 2–6 Boundary Row, London SE1 8HN, UK

Chapman & Hall, 2–6 Boundary Row, London SE1 8HN, UK

Chapman & Hall GmbH, Pappelallee 3, 69469 Weinheim, Germany

Chapman & Hall USA, 115 Fifth Avenue, New York, NY 10003, USA

Chapman & Hall Japan, ITP-Japan, Kyowa Building, 3F, 2-2-1 Hirakawacho, Chiyoda-ku, Tokyo 102, Japan

Chapman & Hall Australia, 102 Dodds Street, South Melbourne, Victoria 3205, Australia

Chapman & Hall India, R. Seshadri, 32 Second Main Road, CIT East, Madras 600 035, India

First edition 1997

Printed in Great Britain by Athenæum Press Ltd, Tyne & Wear

ISBN 0 412 79950 2

A catalogue record for this book is available from the British Library

∞ Printed on permanent acid-free text paper, manufactured in accordance with ANSI/NISO Z39.48-1992 and ANSI/NISO Z39.48-1984 (Permanence of Paper).

CONTENTS

INTERNATIONAL PROGRAMME COMMITTEE

Chairman: Luis M. Camarinha-Matos

Co-chairmen:
Forum industrial experiences: Eduardo Prata
Tutorials: J. M. Mendonça
Industrial Exhibition: J. Morais Martins
Posters & students works: L. Gomes

Afsarmanesh, H. , Univ. Amsterdam, NL
Almeida, L. , Citeve, P
Alonso Romero, L. , Univ. Valladolid, E
Barata, M. , ISEL/Estec, P
Basañez, L. , Univ. Catalonya, E
Bernhardt, R., IPK, D
Bjorner, D., United Nations Univ., MO
Borne, P. , EC, Lille, F
Browne, J. , Univ. College Galway, IR
Calheiros, J. , Setcom, P
Carelli, R. , Univ. Nac. San Juan, AR
Carmo Silva, S. , Univ. Minho, P
Doumeingts, G. , Univ. Bordeaux, F
Duarte Ramos, H. , New Univ. Lisboa, P
Encarnação, J. , Univ. Darmstadt, D
Fellipe de Souza, J.A.M. , Univ. Beira Interior, P
Fernandes, C. , Autosil, P
Fernandes, R. L., EDP, P
Ferreira, A. C. , Univ. F. S. Catarina, BR
Ferreira, F. , Schlumberger, P
Filip, F. G. , Res. Inst. Informatics, RO
Fukuda, T., Nagoya Univ., J
Gindy, N. , Univ. Nottingham, UK
Goldberg, K. Y., Univ. Berkley California, USA
Gomes, M. R. , IST / Inesc, P
Gruver, W. , Simon Fraser Univ., CA
Guimarães, T. , Tennessee Tech. Univ. USA
Haendler Mas, R., EC DG III, B
Jaccuci, G. , I
Kimura, F. , Univ. Tokyo, J
Kovacs, G.L. , Univ. Budapest, HU
Kovacs, I. , ISEG, P
Kusiak, A. , Univ. Iowa, USA
Lyons, D. , Philips Labs, USA
Mandado Perez, E. , Univ. Vigo, E
Mäntylä, M., Helsinki Univ. Tech., FI
Marik, V. , Techn. Univ. Prague, CZ
Marques dos Santos, J. , Univ. Porto, P
Mazón, I. , Univ. Costa Rica, CR
Melo, L. , CTC, P
Mertins, K. , IPK, D
Mesquita, R. , INETI, P
Mielnik, J-C., Sextant, F
Miranda Lemos, J. , IST/INESC, P
Miyagi, P.E. , Univ. S. Paulo, BR

Negreiros, F. , Univ. F. Espirito Santo, BR
Nemes, L. , CSIRO, AU
Nunes, D. , EID, P
Oliveira, E. C. , Univ. Porto, P
Pinoteau, J., Sema-Group, F
Pita, H. , Univ. Nova de Lisboa, P
Putnik, G. , Univ. Minho, P
Rembold, U. , Univ. Karlsruhe, D
Requicha, A. , Univ. South California, USA
Restivo, F. , CCP, P
Ribeiro Teixeira , Atecnic, P
Rudas, I. , Banki Donat Institute, HU
Sá da Costa, J. , IST, P
Santos Silva, V. , CSIN, P
Secca, E. , SMD, P
Silva, M. , Univ. Zaragoza, E
Singh, M. G. , UMIST, UK
Spong, M. W. , Univ. Illinois at U-C, USA
Steiger-Garção, A. , Univ. Nova de Lisboa, P
Tadeu, L. , Fordesi, P
Taia, J. , EID, P
Távora, J. , Tecnotron, P
Tenreiro Machado, J. , Univ. Porto, P
Toscano Rico, A. , ABB, P
Traça-Almeida, A. , Univ. Coimbra, P
Travassos, J.M.C., INDEP, P
Van Brussel, H. , K. Univ. Leuven, B
Van Puymbroeck, W. , EC, B
Verissimo, P. , Fac. Ciencias, P
Vernadat, F. , Univ. Metz, F
Williams, T. , Purdue Univ., USA

STEERING COMMITTEE

A.J. Simões Monteiro - Chairman
João Morais Martins
José Távora
Luis Gomes
Luis M. Camarinha-Matos
Mário Rui Gomes

HONORARY COMMITTEE

E. Maranha das Neves (OE Chairman)
J. A. Simões Cortez (Former OE Chairman)
A. Torres Lopes (OE Mining Engineering)
A. Augusto Fernandes (OE Mechanical Engineering)
A. Adão da Fonseca (OE Civil Engineering)
A. S. Carvalho Fernandes (OE Electrical Engineering)
C. Guedes Soares (OE Naval Architecture and Marine Engineering)
C. Pedro Nunes (OE Chemical Engineering)
Toru Mikami (IFIP TC 5)
George Bekey (IEEE Robotics and Automation Society)
Donald E. Brown (IEEE Systems, Man and Cybernetics Society)

ORGANIZING COMMITTEE

Chairman: A.J. Simões Monteiro
Helder Pita
João Miranda Lemos
João Morais Martins
Jorge Calheiros
Luis Gomes

TECHNICAL CO-SPONSORS:

Ordem dos Engenheiros [Portuguese Engineers Association]

IFIP WG 5.3

IEEE Robotics and Automation Society
IEEE Systems, Man and Cybernetics Society

Towards sustainable industrial production

Motivation. The development and evaluation of new manufacturing paradigms is a very active scientific and technologic area nowadays. The large number of ongoing international collaborative projects addressing topics such as Integration in Manufacturing, Intelligent Manufacturing Systems, Logistics and Supply Chain Management, Virtual Enterprises, or Electronic Commerce, clearly shows the intensive efforts being put on this sector worldwide.

On the other hand, the proliferation of international conferences and workshops on IT in manufacturing and related topics opens new opportunities for the exchange of ideas and discussion of experiences, therefore contributing to the rapid advance of new manufacturing approaches and technologies.

However, most of these projects and events are still too biased by the "large enterprise" scenario or by the context of highly industrialized regions. The issues of sustainability, environmental impacts, anthropocentric and balanced automation approaches, or re-engineering and migration from legacy shop floor systems, are not receiving an appropriate level of attention.

In spite of this situation, in most geographical regions, the industrial scenario is mainly composed of small and medium sized enterprises (SMEs). For instance, in a recent survey reported in the Green Book of Innovation, promoted by the European Commission, it is shown that more than 99% of the European companies are small enterprises with less than 100 employees!

Most of these enterprises are facing tough challenges due to the globalization of the economy, the rapid technologic changes provoked by the Information Technologies and the increasing requirements for quality and environment protection. On the other hand, the engineering staff of such enterprises is quite limited. In many companies there is only one engineer that has to take care of not only the technical aspects but also management and even marketing problems.

In our opinion, the search for new solutions and approaches in the industrial production R&D should strongly take this scenario into account.

History. The Portuguese Association of Engineers organizes, on a regular basis (every two years) a conference on Computer Aided Design, Planning and Manufacturing. This is the major scientific event in Portugal in the area of Computer Integrated Manufacturing that attracts a large number of scientists and engineers.

In order to face the challenges of the globalization of the economy and the consequent internationalization of the engineering, and also because we believe that most of the problems the Portuguese engineering faces are common to many other regions in the world, it was decided to transform this event into an international conference, the ISIP'97 - International Conference on Integrated and Sustainable Industrial Production.

Objectives of ISIP'97. A major goal was the discussion of tendencies and experiences regarding:

-Industrial strategies for peripheral regions;

-Engineering and penetration of new markets;

-Engineering and industrial surviving in open markets;

-Globalization of the economy and subcontracting;

-Achievement of appropriate combination of automation and human centered manufacturing (balanced automation systems).

The focus is on SMEs from small geographical regions.

This proceedings book, including selected and invited papers, addresses some important topics, such as virtual enterprises, systems modeling and integration, re-engineering, performance indicators, time based competition, new management approaches, mechatronics, intelligent manufacturing, machine learning, virtual reality and multi-agent systems.

As it is the first time ISIP is organized, not all the stated objectives are clearly addressed in the presented papers. Some aspects, being still at a less structured status are more appropriately discussed in a panel than in a formal paper. Therefore a number of panels with industry experts were organized within ISIP'97, but their conclusions will be published elsewhere. Nevertheless, we hope that the contributions included in this book can contribute to some re-focusing of the research and engineering efforts in industrial production systems.

The Chairman of the
International Programme Committee of ISIP'97

Luis M. Camarinha-Matos
New University of Lisbon, Portugal

PART ONE

Invited Talks

1

Business Process and shop floor Re-Engineering

DOUMEINGTS. MALHENE
LAP / GRAI - University Bordeaux 1
351 cours de la Libération 33405 TALENCE CEDEX - FRANCE
Tel : (33) (0) 5 56 84 65 30 - Fax : (33) (0) 5 56 84 66 44
E-mail : doumeingts@lap.u-bordeaux.fr

Abstract

For companies, Business Process Re-engineering (BPR) is becoming a technique to attaining a competitive advantage. But BPR is too often related to specific process while it should ensure a global performance by covering multi-processes. By combining local view with global one, and moreover, by combining physical view of the enterprise with its associated decisional part, GRAI approach allows to introduce the Business System Re-engineering (BSR) concepts.

The approach has been successfully applied in a lot of industrial experiments. One of the most recent one has contribute to demonstrate the efficiency of the approach in the domain of Shop-floor.

Key words

Business System Re-engineering, Methodology, GRAI approach, Shop-floor.

1 INTRODUCTION

Companies are under increased pressure today to improve performance while being customer oriented. The high competition in a global international market and the decrease of product life cycles have led manufacturing enterprises to increase their capability, to react rapidly, in coherence with the objective focusing on their core-business.

BPR is an approach to re-design one or several processes of a company in order to improve its performances. This can underlies a complete re-organization of the activities which compose the process and so, a modification of its way of running or of its objectives. Different existing approaches for improving these Business Processes help companies to understand how the different functions or processes are acting. Nevertheless most of them are focusing on the improvement of individual processes (Figure 1).

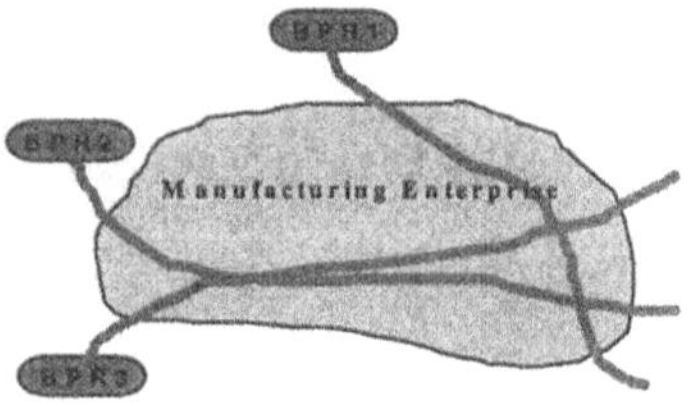

Figure 1 Business Process Re-engineering

Now, global performance is not a set of local business performances, and improving a local process does not underlie the fact that the performance of the whole system is improved. It is absolutely necessary to improve each business process according to the overall manufacturing enterprise performances, and not only according to the optimization of each process individually.

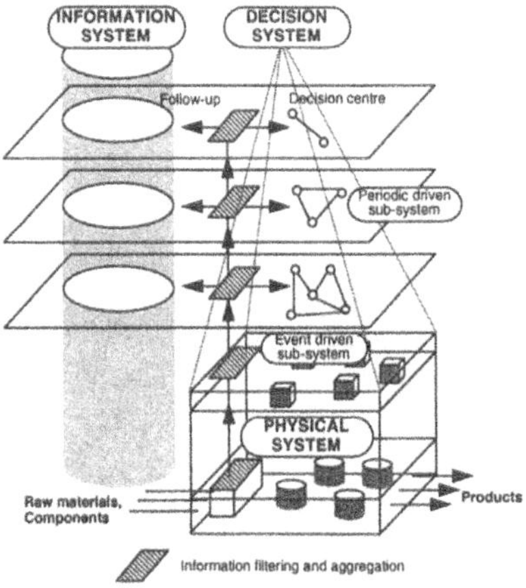

Figure 2 The GRAI conceptual model of the manufacturing system

This is the key idea of the GRAI approach which is based on the GRAI reference model. This model includes as well a global view of the system as a local view which have the same structure. The model identifies three systems in the manufacturing system : the physical system, the decision system and the information system (Figure 2).

The physical system has to transform raw materials, components, into the output Products through the resources. All along this transformation process, the physical system sends back the information, related to the process performance back, to the management system. This feedback information allows the management system to adjust its control.

The Decision System (DS), split up into decision making levels, according to several criteria, each level composed of one or several Decision Centers (DC).

The information system contains all information needed by the DS : it must be structured in an hierarchical way according to the DC structure.

The following part of the document describes a recent application of the GRAI approach for the re-engineering of a shop floor department in the most important French company in the domain of defense industry. Through this application, we will describe why GRAI approach is more powerful than traditional BPR approaches by ensuring in one hand Business Process integration through a multi-process improvement and in the other hand the coherence between physical part of the enterprise and its related decisional part

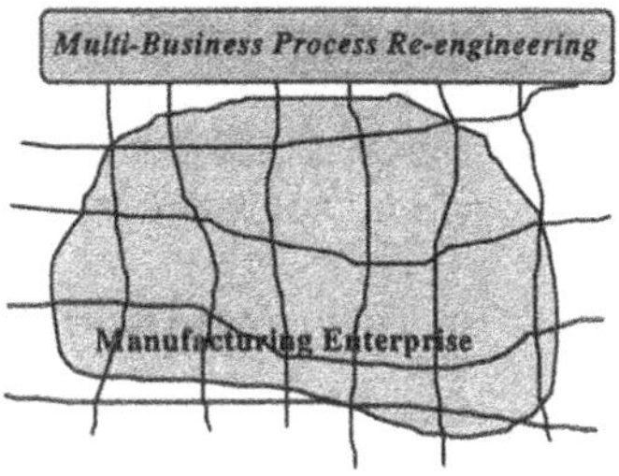

Figure 3 Multi-Business Process Re-engineering

Thus, instead of BPR we will use the concept of Business System Re-engineering (BSR) (Figure 3).

2 PRESENTATION OF A CASE STUDY

The studied shop floor department belongs to an important French Defense company. It works in the domain of mechanical industry and in the past, was acting as a subcontractor for the various divisions of the company. Due to the tremendous changes in this industrial sector, the re-structuring of the workshop and the re-orientation of the activities of this workshop was absolutely necessary in order to find new markets. Further more, important financial losses was questioning its future.

The context of the study was delicate because the whole enterprise was in total re-structuring, and the social context was difficult. After different applications of re-engineering studies which triggered few fruitless changes during two years, the majority of the employees was skeptical before the application of GRAI approach.

For this structural and psychological reasons, it has been decided to perform the Business System Re-engineering in combination with a training action.

The study performed with the GRAI approach was organized in three steps :

1. existing situation analysis : the modelling phase,
2. specification of the new shop floor department : the design phase,
3. implementation process and planning : the action planning.

The application of GRAI is also structured in terms of participants. It requires:

- a project board composed by the top management of the shop floor department (in fact, the manager of the shop floor department and the Director of the whole Manufacturing System). The goals of this group are to give the objectives of the study, and to orient the study according to the results of the main phases,
- a synthesis group composed of the main decision makers of the shop floor department : Purchasing, Engineering, Manufacturing, Supplying, Quality and Maintenance functions (8 persons) . This group has to give global information, to follow the development of the study, and to validate the results of the various stages. Their technical abilities and their suggestions will be used to guide the design of the new system. This group has in charge of analyzing and modeling the

overall organization of the workshop : physical system, main DC, relations between functions, hierarchical structure and coherence of the Production Management System (PMS).

- a specialist/analyst, whose job is, in particular, to collect all the data needed for the various phases, and to elaborate the proposals to the synthesis group. An analyst from the company was helping the specialist/analyst.
- the interviewees group which has to provide more details on specific parts of the shop floor department.

After the modelling phase, specific groups may be defined, according to the specific solutions to be developed.

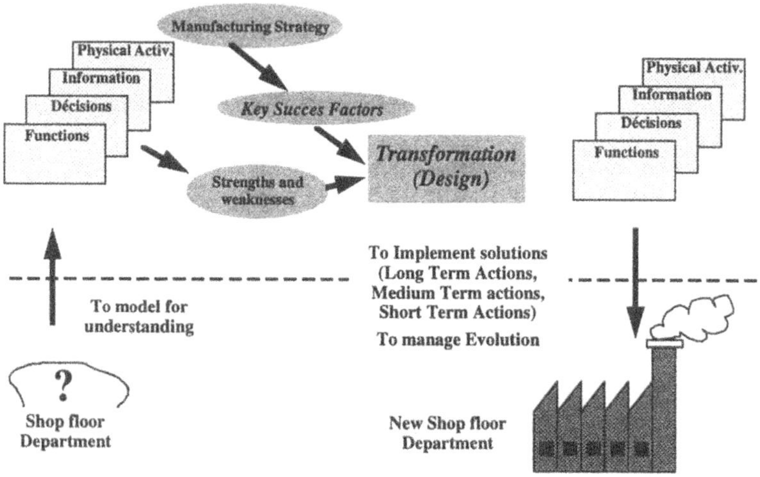

Figure 4 Overall re-engineering process in the GRAI approach

The main advantages of this approach are the following.

First, there is a set of formalisms, which allows the user to perform an integrated model of the shop floor department, according to the three sub-system identified in the conceptual model, plus the functional view. The GRAI grid gives an overview of the decisions made in the various functions. Through the links between these decisional activities, it is possible to study the effective integration of these various functions. Further more, for all Decision Centers, the activities to be performed are analyzed and designed according to the structure and the organization of the physical system (the controlled system). This ensures the coherence between the decisional system and the physical one. Finally, the information system is structured according to the requirements of all the Decision Centers.

Second, the structured approach provides an efficient guideline to manage the project (step by step approach) and it also ensures the validity of the model which are performed (top-down analysis followed by a bottom up one).

These two main advantages give to the GRAI approach a real power to perform BSR in an integrated way.

3 DESCRIPTION OF THE MAIN RESULTS

3.1 The modeling phase

The main objective of the modeling phase of GRAI approach is to determine the improvement tracks to be studied during the design phase. For that purpose, the study was organized in three steps :

1. Training to modeling techniques,
2. Application to the shop floor department,
3. Analysis of the performed models to evaluate the shop floor department situation.

The modeling phase provides a set of models and the functional view of the system :

- functional view of the workshop (Idefø formalisms),
- physical model of the workshop (Idefø formalisms),
- decisional model of the management structure (GRAI grid and GRAI nets formalisms),
- information system model (Entity/Relation formalisms).

The models of the shop floor department was elaborated using the knowledge of the Synthesis Group. In parallel, we have created a group of Foremen in order to perform the training at all levels and also to collect complementary information.

Two examples of model are presented figure 5 and figure 6 :

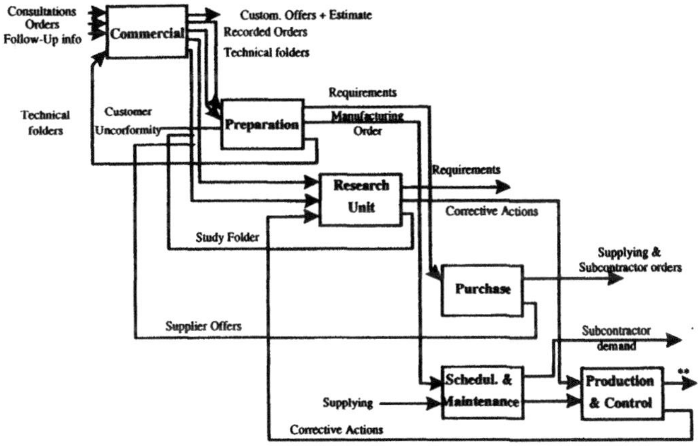

Figure 5 Example of the workshop functional model

These both models were elaborated from the knowledge of the Synthesis Group. Such a top-down approach allows to focus on the real problems, and it avoids to be swamped by a mass of irrelevant detail, which may cost time and money.

This top-down phase is completed by a bottom-up phase in order to collect information on the floor by interviews. But this complementary collect is done in the frame of the initial models performed through the top-down phase.

In this particular application, we have used the training session with the foremen to collect additional information.

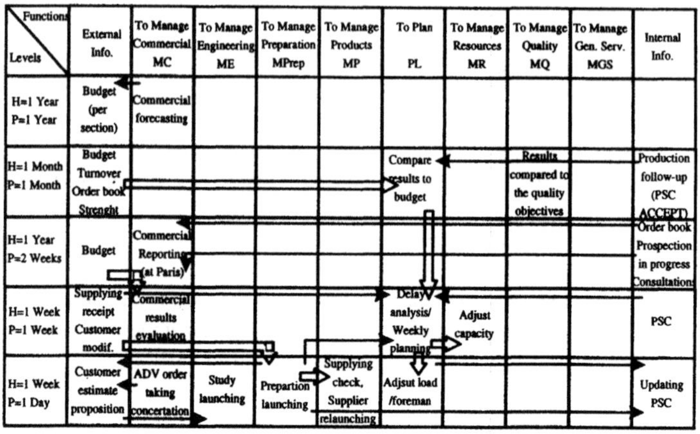

Figure 6 Example of the workshop decisional model

The GRAI oriented evaluation is performed on the following aspects :

- production management structure (decisional level and functional decomposition),
- coordination principles (decision frames, information flows),
- production flow analysis (physical system organization).

Based on this frame and on the resulting models, strengths and weaknesses have been identified. In particular the analysis of the decisional model has highlighted different problems for examples :

- there is no long term and medium term planning (it is detected on the GRAI grid),
- the yearly budget objective is not decomposed into operational objectives which leads to the fact that it is difficult to elaborate a Master Production Scheduling,
- the supplying level can not be efficient without a master production scheduling.

This evaluation has also been performed according to the business planning of the company. However, this business planning was not precise enough, and the information was not enough disseminated.

Finally, this step has allowed to identify the weaknesses of the studied workshop :

- lack of long term and medium term decision for anticipation,
- low flexibility and lack of anticipation due to a short term event driven management,
- insufficient cost evaluation and no meeting of lead time,
- concept of contract not enough used.

The strengths of the system have also been highlighted :

- real capacity for adaptation and reaction,
- good technical know-how,
- high quality product level,
- important human investment.

To improve the competitiveness of the workshop, seven improvement tracks have been proposed based on the evaluation of the weaknesses and after discussions with the Synthesis Group :

- strategic orientation and commercial actions,
- structuring of the physical system,

- organization of the "Scheduling department",
- implementation of a hierarchical planning system,
- product quality control definition,
- tools to support customer order engineering,
- training on Benchmarking procedures in order to improve some Business Processes.

3.2 The design phase

Before to start the design phase, the business planning of this workshop has been collected. This business planning was deduced from a market analysis and also from the evaluation of the strong points of the physical system (products and processes).

The design phase objectives was to eliminate the weaknesses, to look for practical solutions and to define the planning to implement the selected solutions.

The organization of this phase has been performed according to the 7 improvements tracks. For each track, a working group has been created, with participants from the both groups : the Synthesis Group, and the Foreman Group.

For each group, the following approach has been used :

- detailed analysis of the identified problems,
- analysis of existing studies and solution proposal,
- feasibility study,
- selected solution definition,
- proposal for an implementation planning.

Also, to ensure the overall coherence of the results, various meeting of the synthesis group were organized.

The results of the design phase is a global model describing the organization of the workshop, based on the conclusions of each working group. This model specifies for the three sub-systems (Physical, Decisional and Informational) the structure, the relations between the various functions of the shop floor department, and the responsibilities to be allocated.

Physical system

Based on the evaluation of the current situation, and based on the business planning, a product oriented organization has been selected. A detailed analysis has shown that the current available machine-tools were sufficient to ensure the re-organisation of the production cells (Figure 7).

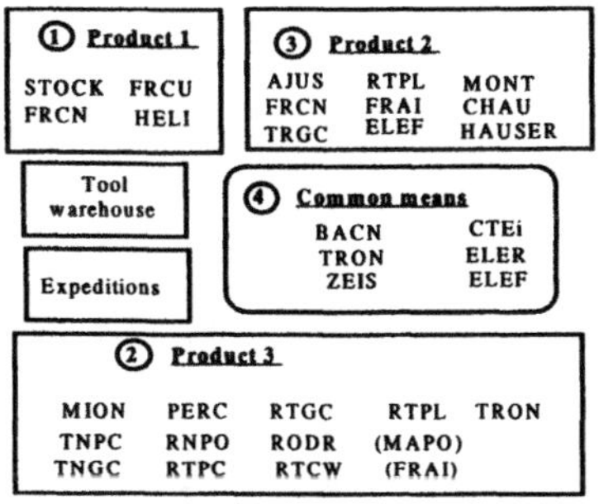

Figure 7 Production cell re-organisation

A layout has been proposed, taking into account the technical constraints related to specific machines. One constraint has led us to identify one common cell for the last activities of some manufacturing processes (especially quality checking activities).

Training requirements and the new team organization has also been defined.

Decisional system

The interest of the GRAI approach is to allow an integrated modeling. So, based on the structure of the physical system and based on the hierarchical planning principles, we have specified the decisional structure of the PMS. From this decisional structure, the various responsibilities in each function have been defined precisely (Figure 8).

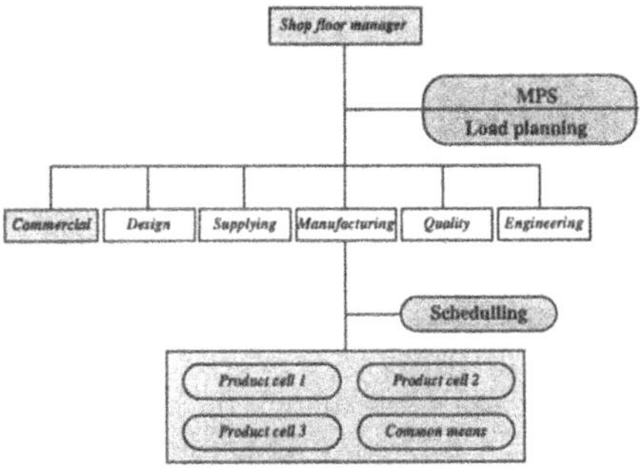

Figure 8 Function responsibilities regarding Production Management System

In this structure (Figure 9), the three basic planning levels and the relations between the various functions have been defined.

The detailed description of each planning level is based on the physical system, as it is seen by this level.

Functions / Levels	External Info.	To Manage Commercial MC	To Manage Engineering , ME	To Manage Products MP	To Plan PL	To Manage Resources MR	To Manage Quality MQ	Internal Info.
H=5 Years P=1 Year		Commercial strategy		Purchasing corporate management	Manufacturing strategy		Quality strategy	
H=1 Year P=1 Month	Orders (firm, forecast)	To negociate and to follow yeraly orders	Yeraly reporting analysis of forecast / performed	Supplying policy Suppliers selection	Master production scheduling	Training planning Industrial equipment [illegible]	Adaptation to requirements	Available capacity
H=4 Monthes P=1 Week	Firm orders Suppliers	To prospect new customers To follow quotations	To manage quotations	Short term supplying management	Capacity planning	To adjust capacity	To manage quality testing activities	Work in progress
H=3 Weeks P=1 Week	Firms orders	To manage tendering	To follow orders To make routine	To relaunch suppliers	Scheduling	To allocate resources	Testing implementation	

Figure 9 The GRAI grid : the decisional structure of the Production Management System

Information system

The Information System was deduced from the Physical and the Decisional Systems including some complementary collect of information. However, after the modeling phase, it appeared that it was necessary to implement a scheduling system, because of the new environment (customer order driven activity, important amount of work-orders, needs for simulation to evaluate lead-times), and also because the existing system was supposed to stop at the end of the following year.

So a specification book has been written. Three suppliers were contacted, and a demonstration of each selected tool has been organized for the end users.

3.3 Actions Plan

The overall result of this design phase is a set of organization principles and technical solutions. Such a result is not sufficient for a manufacturing manager, it is important to define how and when the solutions would be implemented. The elaboration of the Actions Plan aims at preparing the implementation of these solutions.

It consists in specifying a list of actions, with the related required investment, the responsible person, and the scheduling of the various steps.

These actions were defined according three horizons :

1. short term actions (1 month horizon)
2. medium term actions (2 to 18 months horizon)
3. long term actions (more than 18 month horizon).

4 CONCLUSION

During this study, a BSR has been performed on the whole manufacturing enterprise. Most of the individual business processes have been studied : customer order processing, workshop control, quality checking, customer order engineering processes, long term management process, product manufacturing processes, etc. All these processes have been designed in a coherent way, based on the GRAI approach, which allowed to relate all these processes.

Also, based on this integrated design, the manager of the workshop succeed in discussing about the CAPM to be implemented. At the beginning, the company top management was thinking about one software for all the workshops of the company ; a MRP based software. Finally, in this workshop, they have selected a simple scheduling package (specification book already mentioned) which is more relevant for their type of production (specific tooling in three main areas).

Three months after the end of the application, the workshop had increased its turn over (≈10%) and many improvements have been implemented (level of customer service rate, number of firm orders / Number of quotations).

This study which is exemplary, has supported the application of GEM (GRAI Evolution Methodology) which is a methodology developed in the frame of the EUREKA TIME GUIDE project (TIME : Tools and Methods for Integration and for Management of the Evolution of Industrial Enterprises - GUIDE : GUIDing the Evolution). GEM is an extension of GRAI approach and aims at providing the industrial companies a methodological support to steer and to guide their long term evolution process, mainly through modeling techniques, Self-Audit capability and Benchmarking techniques. The figure 10 summarizes the main steps of the structured approach of GEM.

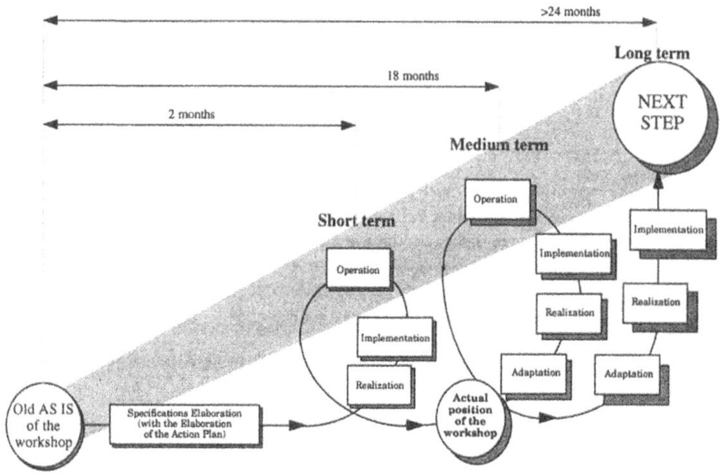

Figure 10 GEM structured approach

Industrials have already demonstrated their interest for GEM , particularly in the frame of the European ESPRIT program. The partners of the project REALMS II (Re-Engineering AppLication integrating Modelling and Simulation part II) which are ELVAL (Greece), BSW (Germany), NTUA (University of Athens - Greece), Siemens Nixdorf (Germany), IFAB (University of Karlsruhe - Germany) and the LAP/GRAI (France) have decided to use the methodology. The objective of the project is to finalize a methodology dedicated to the European SMI market through the application of GEM overall approach.

RÉFÉRENCES

CAM.I, Computer Aided Manufacturing International (1980), CAM-I -Architect's Manual: ICAM definition method, Computer Aided Manufacturing -International Inc. TEXAS, USA. 1980.

DOUMEINGTS G., MARCOTTE F., ROJAS H. : "GRAI Approach : a methodology for re-engineering the manufacturing enterprise" . *Re-engineering the enterprise* - edited by J. Browne and D. O'Sullivan - Chapman & Hall - 1995

LE MOIGNE J.L. (1974) "La théorie du système général. Théorie de la modélisation."

MARCOTTE F.(1995) "Contribution à la modélisation des systèmes de production : extension du modèle GRAI" - Thèse de doctorat - Université Bordeaux I - Octobre 1995.

MELESE J. "Approche systémique des organisations : Vers l'entreprise à complexité humaine". Suresnes, Editions Hommes et Techniques, 1979, 158 p..

ROBOAM M. (1988) "Modèles de référence et intégration des méthodes d'analyse pour la conception des systèmes de production" - Thèse de Doctorat, Automatique, Université Bordeaux I - Juillet 1988.

2

IT Support for the generic tactical decision making process of pricing in competitive consumer markets

Madan G. Singh and Nathalie Cassaigne
Computation Department, UMIST, Sackville St,
Manchester M60 1QD, UK
tel. 00 44 161 200 3347, 00 44 161 200 3358
fax: 00 44 161 200 3346
e-mail: M_Singh@mac.co.umist.ac.uk
N.Cassaigne@umist.ac.uk

Abstract

Sustainable industrial production, which is the main theme of this conference, is only possible when the firm manufactures goods which appeal to potential customers, at prices that they are prepared to pay. These prices and the resulting sales volumes should enable the firm to meet all its costs and make a profit whilst meeting its longer term strategic goals of capturing or retaining appropriate levels of market share.

In order to help firms in the consumer goods industries to survive and prosper in competitive and fast changing markets and to enable them to meet their strategic goals, it is necessary to provide them with tools that enable them to better use the tactical decision levers at their disposal. This paper focuses on the main short or medium term lever that such companies can use in order to cope with the competition and examines the information technology needs to support the use of this lever via the business process which it constitutes i.e. the Pricing Process.

In the paper we show how certain novel pricing decision support tools developed in Manchester could be used to enable the firm to take better (i.e. more profitable) pricing decisions as compared to the unaided decision making process. Examples have been provided from the petroleum and reference is given to work on other consumer markets.

1 INTRODUCTION

Modern consumer goods companies can manufacture their products to meet the changing needs of their market by using information technology to make their production processes extremely flexible. The main focus of this conference is to show how information technology can be or is being used for this purpose. Yet, the main driver of this need, i.e. the changing needs of the consumer and the increased competition is only addressed peripherally in this conference. In this paper, the ways of influencing the consumer in order to maintain profitable and sustainable production is the key element considered. Specifically, how to price products within their specific competitive market in order to meet the firm's strategic goals and maximize profits is the main theme of this paper.

The kinds of firms considered in this paper are firms who sell their goods directly to the consumer at prices that they chose. Thus firms who sell their goods via intermediaries (e.g. super markets) who set the final price are not considered here since in this latter case, there is often a price negotiation whereas we are primarily focusing on the situation where the vendor is one amongst a small number of competing vendors who sell similar products and each is able to set his own prices at which some consumers buy and others do not. Examples would be oil companies, banks, companies making over the counter pharmaceutical products etc. In the services sector, examples would include the pricing of telecommunication services, pricing of goods on the shelf in mass retailing, pricing in banking i.e., the setting of interest rates, pricing of insurance products etc.

Human beings have been pricing goods and services for thousands of years. However, virtually all pricing decision making is based on cost based or competition based approaches or some mix of the two. Computerization has merely provided spreadsheet like tools to enable costing and thus pricing to be carried out efficiently. Whilst such efficiency is desirable, it is surely more important to improve the **effectiveness** of such decision making. In this paper, we will focus on the **use of computers to improve the effectiveness of pricing decision making in competitive markets**. This will involve the processing of information and knowledge that is potentially available but not at present used for this purpose as well as the use of a learning loop which enables the decision support tools to adapt to changes taking place within the environment of the client company.

It should be emphasized that pricing in large consumer goods companies is not something which is done by one person but it is rather a process which involves the skills and judgments of a number of people coming from different departments within the company (e.g. finance for the accounting information, manufacturing to give an idea of existing stocks and state of the production cycle, marketing and sales to provide an input on the needs of the consumers etc.).The aim here is to examine the generic pricing process in consumer goods companies and provide IT tools to support it such that the supported process enables superior decisions to be made as compared to the unaided process.

Finally, it is worth noting that the pricing decision making that is addressed in this paper is not strategic pricing which is something that one might do in a planning mode in order to figure out whether it is worthwhile to develop a new product and what its functionality's should be and at what price it will be eventually sold if the product is in fact developed. Here we are dealing with **tactical pricing** of existing products that needs to be displayed in the market place in the immediate future.

In the rest of this paper, we start in section 2 with a description of the generic pricing process and the pricing decision making problem that is at the heart of this process. In section 3, we provide an outline of generic tools developed in the Decision Technologies Group at UMIST to support such decisions. In section 4, we show the specialization of this tool for the petrol pricing decision making process as well as the results from controlled trials of the use of this system to demonstrate the improvement in effectiveness of such decision making through the use of this tool. In section 5, examples from other industries are cited in order to demonstrate genericity of the concepts and tools. Some conclusions are provided in section 6.

2 THE GENERIC PRICING DECISION MAKING PROBLEM

The key elements for pricing decision making are[1,2]:

- Costs and how they vary with sales volume.
- Sales and how they vary with prices.
- Impact of competitor prices on the sales of the company's products.
- Interactions between the prices of certain products within the company's product
- Portfolio and sales of other products within the same portfolio (cannibalization effects).
- Strategic issues of: Price Positioning and Market Share goals

Information about **costs** comes from the accounting systems within the firm. However, typical accounting systems do not readily provide the expected costs for different sales levels. For this, a model is required which separates out the fixed and variable costs and then allows one to predict what the costs would be as the sales or other variables which affect the costs change.

.Information and knowledge about **the relationships between the sales and own prices** lies in principle with the sales managers and the marketeers of the company. The competitor analysis section within the marketing department typically possesses information and knowledge
about the likely impact of **changes in competitor prices on the sales of the company's products.**

Although most of pricing practice is based on the idea that the pricing of each product of the firm be considered independently of the other products of the firm, there is significant profit potential if we can model the inter-relationships between the products. For example, certain products within the portfolio of the company might be in very price-sensitive markets whilst others might be in less price-sensitive markets. This might allow for tradeoffs to be made between the products thus allowing for better pricing. Again the sales and marketing departments should possess the information and knowledge required to assess these relationships.

As far as the **strategic aspects** are concerned, two generic issues stand out i.e. **price positioning and market share goals.**

Price positioning is concerned with giving signals to consumers about the quality of the company's products and it typically involves ensuring that an appropriate relationship exists between the prices of the individual products of the company and those of competitors and also between the individual products of the company. Price-positioning is not changed very often since it is a part of the branding strategy of the company.

Market share goals represent the second key aspect of the company's strategy in that the overall market share of the company is a valuable commodity which is a significant component in the value of the company itself. It is a part of the "capital" of the company. Sales volume is an appropriate surrogate variable for market share.

It should be noted that there is always a tradeoff possible between profits and market share since more profits can be obtained at the expense of market share by increasing prices but in the longer term this is self defeating.

2.1 The idealized generic pricing process

In large or even medium sized firms, prices are not set by one person but are rather a group decision to which many actors contribute. It should be seen as one of the important business processes of the firm which typically is in dire need of re-engineering in order to improve its effectiveness.

The main reason for this need for re-engineering of the pricing process is that present day pricing processes were designed to solve the pricing decision making problem of the single decision maker for whom, as examined above, the pricing problem is far too complex to enable him or her to take all the above mentioned factors into account. The decision maker thus ignores most of the above factors and takes the pricing decision within the framework of satisfying rationality[3] by focusing only on costs or on the competition[4]. The advent of computers to support pricing decision making has merely re-confirmed the satisficing framework by providing spreadsheets for calculating cost based or competition based prices!

The pricing process is one of the key business processes in the firm since it is linked to the revenues of the firm: most other processes deal with the cost side of the firm. It's outputs are highly visible to the consumer and give significant "image" messages about the firm and signal the goals of the company to the competitors. Pricing, if done well, has the potential to significantly add value for the firm. Equally, if not done well, it can put into question the very survival of the firm.

In Figure 1 below, we shows what a generic idealized pricing process of the firm should look like in order to meet all the imperatives listed in section 2 above. It shows what information and knowledge is required for superior pricing decision making and from where within the company it comes from.

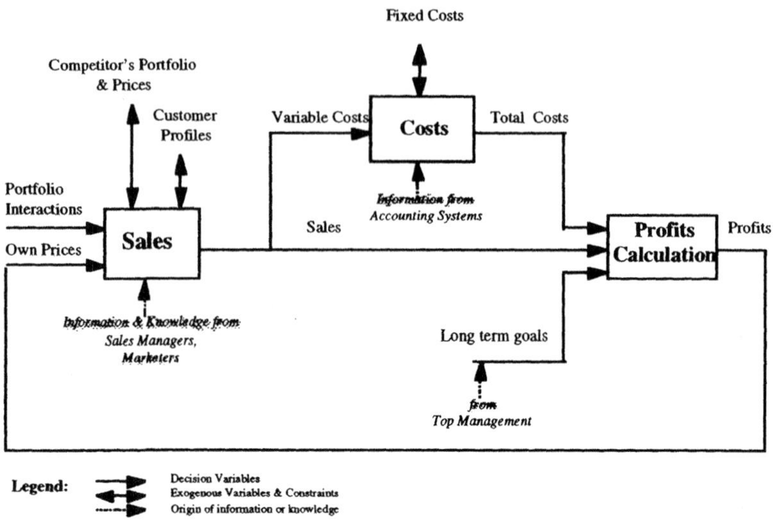

Figure 1 The generic pricing process.

Next we will examine the decision support systems developed within the decision technologies group at UMIST to support such an idealized generic pricing decision making process.

3 DECISION SUPPORT FOR THE GENERIC IDEALIZED PRICING PROCESS: THE PRICE-STRAT SYSTEM [5]

Price Strat is a piece of software that is sufficiently flexible to enable the user to tackle a variety of pricing issues in a coherent way. The flexible use of Price-Strat and the taking of it from one industry to another is what constitutes the Price-Strat methodology.

In order to build up a picture of demand, cost structure, company strategic objectives and competitors for any specific company in each of its markets, the Price-Strat software enables one to first construct a knowledge repository. This is built up relatively rapidly by managers defining their strategic aims, price positioning aims, their perceptions of competitors, etc. All this information is fed into the computer and used by the Price-Strat program to generate a series of 'what-if' pricing scenarios where the prices of the company's products vary as do those of different competitors. For each 'what-if' scenario, a group of the company's managers is asked to give their independent assessments of the impact on the company's sales and costs. In this way, Price-Strat is able to capture a significant element of the intuition and expert knowledge of experienced managers. The intuitive knowledge is combined with any other information available (e.g. past data and/or market research data) and this completes the knowledge repository.

The knowledge repository is then interrogated by pressing a button and within 1-2 minutes, the computer calculates the optimal prices which meet strategic constraints and maximise profits.

The user can then modify any strategic constraint and recalculate (within 1-2 minutes) the *opportunity cost* of shifting a strategic boundary. Equally, if a competitor makes a price move or the company's buying prices change, the new information can be fed in interactively and the resulting optimal prices for the company's portfolio of products can be calculated rapidly.

Experience with the application of Price-Strat in different industries shows that despite the differing views of the world of the various managers (head office, regional, marketing bias, finance bias, etc.), the optimal pricing strategies are close to each other if not identical, thus demonstrating the robustness of the approach.

3.1 Learning

The decisions recommended by Price-Strat can be implemented and, after a short time, the real sales data can be fed in. This provides a reality check. More importantly, Price-Strat uses this new data to enrich the knowledge repository and through it, keep track of the slowly changing factors.

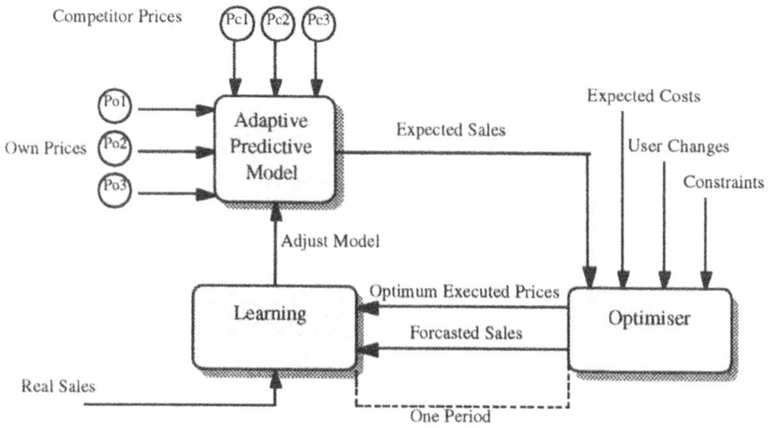

Figure 2 Price-Strat Architecture.

Figure 2 shows the generic structure of the Price-Strat Knowledge Support System for superior pricing.

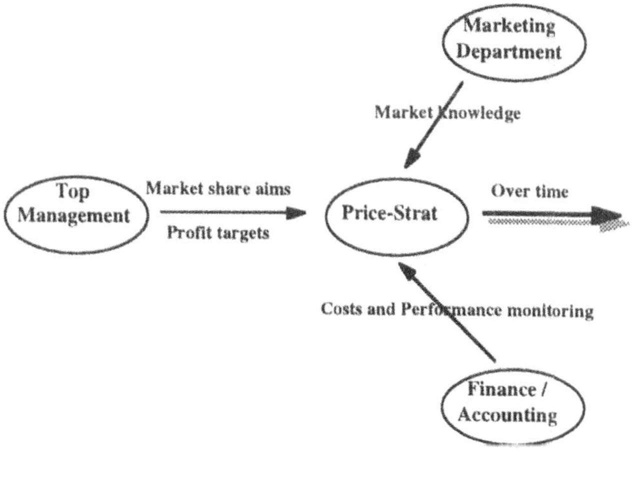

Figure 3 Price-Strat within the re-engineered pricing process.

Figure 3 shows how Price-Strat could lie at the heart of a re-engineered ideal generic pricing process. This figure depicts Price-Strat as providing intermediation between top management (as custodians of the company's strategy), the Marketing Department (providing knowledge and information about the market) and the Finance/accounting function providing costing information and knowledge as well as performance monitoring

Next, we show the application of this approach and system for the important area of petrol pricing.

4 KNOWLEDGE SUPPORT FOR PETROL PRICING [6]

To quote a senior oil company executive:

"One of the most pressing issues facing fuel retailers in Europe today is the continued pressure on prices, and thus profit margins, for which intensifying competition is a significant contributory cause"

One of the most noticeable aspects of this competition is the structural change to the market caused by the penetration into the fuel retailing market of food retailing Supermarkets.

Whereas fuel retailers place considerable emphasis on brand image, station quality, product improvements and added facilities, the key to profitable retailing is increasingly becoming pricing.

The Supermarkets have established their position as low cost providers of fuel, threatening other low cost retailers who do not have the brand image. The Majors are becoming leaner, offering keener prices than before. The Mini-majors are being squeezed. The density of Supermarket outlets is increasing, intensifying the competition between them, cutting their already tight operating margins.

No section of the fuel retail market is free from this pricing pressure.

Pricing is becoming the key tactical decision to protect the Retail business. Without careful individual price management Volumes & Profits will decline.

4.1 The Business needs

The most usual approach taken by fuel retailers is to keep prices in line with competitor prices. Although this does protect sales volumes, it does not enable the management to achieve profit aims. More specifically profit performance is harmed by price wars and the subsequent reaction of increasing prices (to recover profit margins) does result in the loss of sales volume.

To manage prices and to establish control over profit performance, fuel retailers aim to:-

Make available adequate up-to-date information.

- Use this information in a Pricing Process, that also uses all the skills and intuitive market knowledge available in the company, to promptly make effective price moves by grade and by outlet.
- Identify the tactics that will best achieve the strategic aims for the retail network in the competitive market place.

The Pricing Process not only needs to set effective prices, but also needs to:-

- Involve those staff with local market knowledge.
- Reassure the retail network management that the pricing image is being maintained and that the sales volume and profit performance goals will be achieved.
- Provide feedback to reassess the tactics of price, promotion, acquisition/disposal etc. that will best achieve the retail network's strategic aims.

The Price-Strat System has been specialized for this problem and this version is called PriceNet.

4.2 The Pricing Process using PriceNet

PriceNet is a Price Decision Support system specifically designed for fuel retailers. PriceNet provides the key benefits of:

1. capturing and enabling the explicit use of local market knowledge on the price/sales volume models that are then constantly updated by actual performance.
2. Enabling the network management to set achievable sales volume and profit targets within an explicitly defined price positioning strategy.
3. Providing prices for each grade and each outlet that will achieve the sales volume targets more profitably.
4. Focussing management attention on those outlets that need (or would benefit from) a price move.
5. Enabling alternative pricing options to be examined; quantifying the benefits and risks of implementing each option.
6. Providing sales and profit forecasts and price positioning information that enables the network management to review tactics.

The actual management of the pricing process involves the transmittal, on a daily basis, of the prices of competing stations in the neighbourhood of each of the company's stations, by telephone to the company's main frame computer which also has in it that day's costs of the products as traded on the Rotterdam market. All this information is downloaded into PriceNet via a flatfile. PriceNet is then run by a single analyst and his manager who between them are able to control and set prices at several hundred stations.

4.2.1 The PriceNet Price/Sales volume model

Whilst a large number of factors influence the sales of fuel, it is only price moves that cause rapid changes in sales volumes. PriceNet models the Price/Sales volume relationship in great detail.

The PriceNet model is built by estimating for each product the elasticities for:

- the influence of its own price on sales;
- the influence of each competitor's price on sales;
- the influence of each of the other product's price on sales.

PriceNet makes an initial estimate of these relationships using local knowledge of the market supplied by the sales management, combined with historical data of the actual sales, prices and competitor prices over the recent past (if it includes at least ten significant price changes).

At the end of each week the PriceNet models are updated with actual sales performance. This learning process adjusts the elasticities and accounts for any changes in the influence of the other factors on sales volumes.

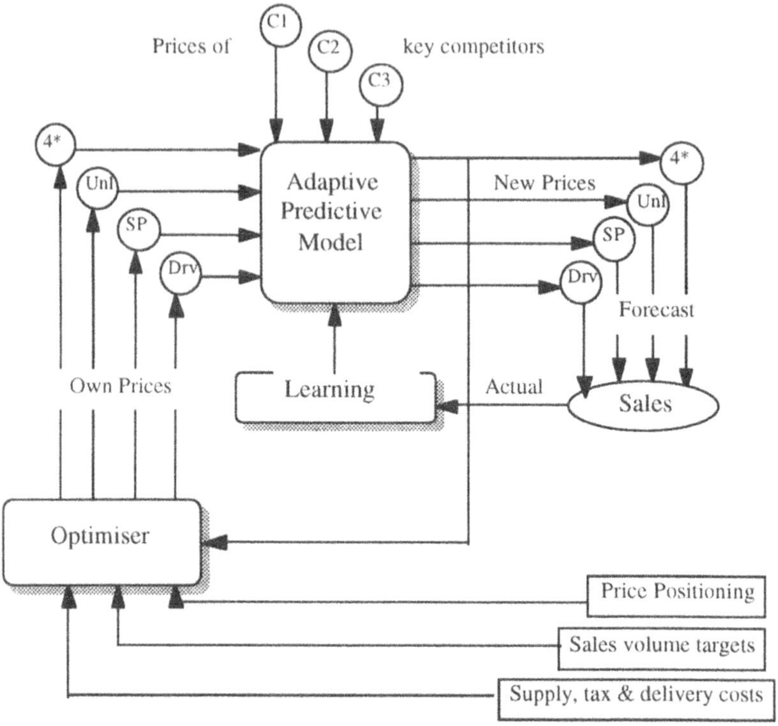

Figure 4 PriceNet site-specific price/sales model.

4.2.2 Representing the network strategy in PriceNet

PriceNet provides a forecast of this (coming) week's sales volume and profits, using the site-specific models, taking competitor price moves and changes in product costs into account.

PriceNet will offer 5 options, using prices for each grade and for each station that are within the network price positioning strategy; and which give the optimum profits for five levels of sales within the achievable range. The management can select the option that best achieves their sales volume and profit aims, or revise the price positioning strategy giving more latitude for price changes.

Period	Targets	Last 6 wk's actual	Last week actual	This wk's forecast Current Prices	This wk's forecast Proposed Prices
Sales	15438	14765	15081	14973	15132
Profits	244	220	280	295	292

The price positioning strategy is represented in PriceNet as:

- •min/max price limits for each grade
- •min/max price relationships with competitors for each grade
- •min/max price differentials between grades
- •maximum allowable price change
- •marketable prices

4.2.3 The Optimisation Process

Once the network tactics have been determined for this week, PriceNet will allocate the chosen network sales volume target across the retail outlets and optimize each site-specific model to determine the prices that will achieve the sales volume target within the price positioning strategy; and yield more profit.

PriceNet resets the price changes for any outlet that yields insufficient improvement over current prices; proposing the price moves for each outlet that best achieve the chosen network tactics.

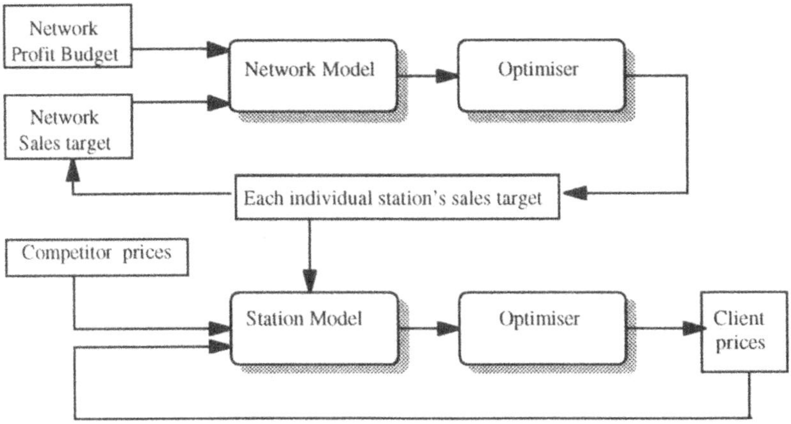

Figure 5 Group optimiser.

4.2.4 Focusing management attention

PriceNet provides the management with an exception report focusing attention on those outlets where a price move is proposed and flagging those outlets where competitors have made a significant price move or where the sales volume or profit performance has declined over the past six weeks.

Table 1: PriceNet exception report

Station	*Action*	*Improve profit margin*	*Protect sales volume*	*competitor or price moves*	*Sales Volume declining*	*Gross profit declining*	*Forecast note correct*
13567	opt	...	3.5	yes	yes	...	...
17893	...	...	...	yes	...	...	...
22678	opt	354	...	...	...	yes	...
33654	...	...	...	...	yes	...	yes
45632	opt	298	...	...	...	...	...
56903	opt	...	2.7	yes	...	...	...
etc.							

4.2.5 Reviewing pricing options

PriceNet provides for each outlet a 13 week history of own and competitor prices, sales volumes and profits performance. The display enables price move patterns and sales effects to be easily identified.

The sales and profit performance of alternative pricing options for each outlet can then be reviewed using the PriceNet model. The risks of competitor price reactions to the pricing option can be assessed.

PriceNet provides an output file of the price moves made by the Pricing Process.

Table 2: price history

	Grade	*Propose*	*Current*	*19 Nov*	*12-Nov*	*07-Nov*	*30-Oct*	*23-Oct*	*16-Oct*
Each key	4*								
competitors	Unld								
price	Super								
	Diesel								
Client Prices	4*								
	Unld								
	Super								
	Diesel								
Sales 000(s)	4*								
Litres	Unld								
	Super								
	Diesel								
Total Sales	4*								
Target Sales	Unld								
	Super								
	Diesel								
Gross Profits	4*								
	Unld								
	Super								
Total Profits	Diesel								

4.2.6 Summary network performance

PriceNet provides a sales volume and profit forecast for each outlet and for the retail network.

PriceNet provides a comparison of client and competitor prices. Enabling the confirmation, at a glance, that the price moves are in line with the desired network tactics.

Sales Volumes 000(s) Litres				***Profit Margin £000(s)***			
Objective	Last 6 week's actual	Last week actual	This week's f/cast	Objective	Last 6 week's actual	Last week's actual	This week's f/cast
15438	1476	1508	150	244	220	280	289
	5	1	76				

Client Prices				*Competitor Prices*		
6 week	Last week	This week	pence/litre	6 week	Last week	This week
55.9	56.1	56.2	4*	56.1	56.0	56.0
50.4	50.7	50.9	Unld	50.5	50.4	50.4
54.5	55.2	55.4	Super Plus	54.8	54.8	54.8
51.1	51.2	51.3	Diesel	50.4	50.3	50.2

4.2.7 Identifying effective tactics to achieve the strategic network aims

The PriceNet site-specific models clarify the pricing opportunities (or lack of them) for each outlet. This feedback helps the network

- For a client in the United Kingdom, where 90 day experiments were done on six sites in different parts of the country, average net margin increased by 33% whilst volume dropped by 1% within the clients strategic price positioning aims.

- At DTW areas and Jobber terminals, improvements in profitability over the period considered was 30% and 4%, respectively.

These experiments are described in detailed in [6]. They demonstrated the significant potential of Price-Strat to improve oil company profits through superior pricing.

Following on from these experiments, large scale experiments on petrol pricing were done on an oil major in the United Kingdom for a much larger number of sites over 1 year. As a result, Texaco (UK) Ltd. and Repsol UK have obtained a licence for this technology, and other oil companies are currently testing it in both Europe and North America.

5 APPLICATIONS OF THE GENERIC SYSTEM PRICE STRAT IN OTHER INDUSTRIES

As discussed previously, the generic PriceStrat system, which is in fact a toolkit, can be used to provide pricing decision support for the pricing process of any firm which sells directly to large number of consumers and is thus able to set its prices which consumers take or leave. in the previous section, we have seen that in the case of petrol pricing, in view of the fact that petrol prices change very often, it was worthwhile to specialise the PriceStrat toolkit into the PriceNet System for use within the Pricing Process of oil retailers. The same has been done for a number of other industries. For example, the Bank Price System [7] supports the pricing process of retail banks and helps them to set interest rates for savings and lending products.

The Tele-Price System has been designed to support the pricing process of telecommunication companies and enables such companies to set the call tariffs, line rental and hand sets.

The Retail price System has been designed to support the Category Management function in Mass Retailing.

All these systems are marketed by Knowledge Support Systems Ltd., a company set up to exploit the research of the Decision Technologies Group.

6 CONCLUSIONS

In this paper we have examined the key business process of pricing in consumer industries and seen that as organised at present in virtually any company, it is well worthwhile to re-engineer this process. The paper has focussed especially on the provision of IT support in general and software support in particular to enable one to improve not only the efficiency but also the effectiveness of this process. We have shown how the PriceStrat family of tools is well suited for providing this support and given examples of this from a number of consumer industries including petrol retailing.

7 REFERENCES

1. Nagle, T. (1987) The strategy and tactics of pricing. Prentice-Hall, Englewood California,

2. Singh, M.G. (1991) Knowledge support systems for smarter tactical decisions for pricing and resource allocation. IEEE Control Systems Magazine, pp. 3-7.

3. Simon, H.A. (1966) The New Sciences of Management Decisions (Harper & Row, New York, H.A.).

4. Kotler, P (1987). "Marketing management: analysis, planning and Control" Prentice-Hall, Englewood, California,.

5. Singh, M.G. and Bennavail, J.C. (1994) "PriceStrat: a knowledge support system for decision making during price wars", IDT Vol. 19, 4, pp. 277-296.

6. Singh, M.G. (1996). Knowledge support for profitable pricing in a competitive environment". Plenary paper, 4th Intl. Conference on the Cognitive Foundations of Economics and Management", Paris, September 1995. To appear in Information and Systems Engineering, Vol. 2,3.

7. Singh, M.G. (1994). Decision Technologies for supporting the interplay between Qualitative and Quantitative aspects of Managerial Decision Making. Proc. QUARDET 93, Barcelona, 16-18 June 1993. Also Mathematics, Computers and Simulation Journal, vol. 36, pp. 103-114.

8 BIOGRAPHIES

Madan G. Singh

Madan G. Singh B.Sc. (Exeter), M.Sc. (Manchester), Ph.D (Cambridge), Docteur es Sciences (Toulouse), C.Eng., FIEE, Fellow of the IEEE, FBCS, is the Chairman of the Computation Department and holds the chair of Information Engineering at UMIST, (the University of Manchester Institute of Science and Technology) in Manchester, England. He has been at UMIST since 1979 when he was elected to the chair of Control Engineering at the age of 33. He was the head of the Post Graduate Department "The Control Systems Centre", from 1981 to 1983 and from 1985 to 1987. Prior to that, he had been a Fellow of St. John's College Cambridge, an Associate Professor at the University of Toulouse and a researcher of the French CNRS.

Professor Singh was the first "non-American" President of the IEEE Systems, Man and Cybernetics Society (1994, 1995) and won *the 1993 Norbert Weiner Award "for truly outstanding contributions to research and scholarship in Information and Decision Technologies, Optimisation and Control and Systems Engineering"*. In February 1994, he was made a **"Chevalier dans l'ordre des Palmes Academiques" by a decree signed by the Prime Minister of France** and was graciously granted unrestricted permission by The Queen to wear the insignia of this honour. Hewas awarded the degree of ***Doctor of Engineering, honoris causa***, by the University of Waterloo, Canada in 1996.

Professor Singh edited the 10 volume "*Encyclopedia of Systems and Control*" for which he received an "*outstanding contribution award*" *from the IEEE SMC Society* . Indeed, the Reviewer in the prestigious journal Automatica said that "*...all in all, the systems and control encyclopedia constitutes a major milestone in the annals of systems and control and in the epistomology of technologically based decision making*". The project also led to his coordinating the publication of 8 Concise Encyclopedias, each on a different subject. In addition to the encyclopedia, Professor Singh has authored or co-authored **8 books and over 170 scientific articles and edited or co-edited a further 10 books**. He has also directed over 50 research or major consultancy projects yielding several million pounds of funding or fees. He is the co-editor-in-chief of the journal "*Information and Systems Engineering*" and is the editor of the quarterly *SMC Newsletter* (distributed to 5500 members worldwide). He is also a member of the editorial boards of 9 other journals and edits 3 book series. He was the co-editor-in-chief of the North Holland Journal: *Information and Decision Technologies* (1980-94).

He was a visiting Professor at INSEAD (France) (1980-94) and is or has been an invited or honorary Professor at: Ecole Normale Superieur de Cachan, Ecole Centrale de Lille, Theseus Institut (Sophia Antipolis), and the University of Aeronautics and Astronautics in Beijing. He is a member of the Scientific Councils of the IUSPIM (Marseille) and IUP MIAGe (Aix-en-Provence).

Professor Singh is also the Chairman of the UMIST Campus company Knowledge Support Systems Ltd. and of Control Sciences Ltd. These companies provide services related to the decision support systems and decision theoretic approaches developed by Professor Singh and his collaborators within the Decision Technologies Group at UMIST on Pricing, Advertising budget allocation, shelf space allocation in retailing etc. Through various licensing arrangements, these products are being used by corporations worldwide to enhance the quality of their decision making.

Professor Singh is a citizen of both the United Kingdom and France. He is married and has two children.

Nathalie P.L. Cassaigne

Degrees:

M.Sc., DEA "Management Sciences", DESS.

"Information Systems Engineering", University of Social Sciences and INP Toulouse

Business Experience:

Research Associate in the Computation Department at UMIST and UMIST

Consultant for various French local authorities since 1991

Research:

Information Systems and Tactical Decision Support System

3

The time based competition paradigm

B. Andries
L. Gelders
Centre for Industrial Management
Celestijnenlaan 300A
3001 Leuven - Heverlee
Tel.: 32 (16) 32.25.67
Fax.: 32 (16) 32.29.86
bertrand.andries@cib.kuleuven.ac.be
ludo.gelders@cib.kuleuven.ac.be

Abstract

In this paper we discuss two important aspects of Time Based Competition: lead time reduction in manufacturing and new product development. In the first part - lead time reduction in manufacturing - we describe an organizational structure that is based on the order penetration point and time, and a technique - Value Added Analysis - that is aimed at the elimination of waste. In the second part we explain how some of the concepts of JIT can be applied to reduce cycle times in new product development.

Keywords

Time based competition, new product development, manufacturing management

1. INTRODUCTION

The three C's, customer, competition and change, have created a new world for business. It is becoming increasingly apparent that organizations designed to operate in one environment cannot be expected to work well in another (Hammer and Champy [1993]).

The nature of competition itself has changed : while price, the driving force of the sixties and seventies, is still important, it is often not crucial. Even (product) quality, the paradigm of the eighties, cannot be seen as the major competitive factor anymore. In fact price and quality can be seen as enablers - you simply need them to participate on the market - but leadership will be achieved through other competitive aspects that are less related to the product itself but more to how and when the product is offered. Gelders et al. [1995] performed a study within Belgian companies and concluded that the competitive factors, "the winners", could be classified into two broad categories :

- "Quality of logistics". This includes:
    - Lead time and delivery reliability. Both are important and interrelated : surveys have demonstrated that (industrial) customers accept longer lead times if (and only if) delivery reliability is assured.
    - Service quality. "How is the product brought to us and what can we expect from the supplier during its life cycle?". After sales service, customer complaints, order tracking, quality and customization of packaging, quality of paperwork (invoicing, reports, etc.) are all becoming increasingly important
    - Customization. Most cars today are produced on order, Gateway assembles all its PC's on order, ... Benjamin et al. [1995] foresee a future where we will interactively place orders through "information superhighways" directly to the producer. This opens the door to full customization, both in product and service.
- Product diversity. *"The ability to rapidly, effectively and continuously introduce quality products that meet customer needs will be the mark of the most successful companies now and in the future"* (Choperena [1996]).

When we look closely at these competitive factors we see that one of the common denominators is time. Companies that are able to reshape their processes to fasten the throughput time not only have a time-wise competitive advantage but also obtain a superior performance along other competitive dimensions (price, quality, flexibility, etc.). Many companies have shortened their lead times in the early nineties under pressure of early time based competitors or to be able to meet the constraints of JIT-deliveries. The reengineering wave has forced and helped many companies to rethink and streamline their administrative processes. But these are only pieces of the chain linking the supplier to the customer. New battles are now being waged in product development, distribution and service (Blackburn [1991]) on the one hand and into the integration of these time based ideas in the supply chain as a whole on the other hand

Within this concept every aspect of the business chain has to be reevaluated, not only from supplier to customer but also from idea to market, from strategy to organizational structure, etc. And all this with only one thing in mind, the customer.

Not all areas are always as important however. Stalk [1993] warned us for the problems related to an exaggerated focus on time in his HBR-publication "Japan's dark side of time". Distribution for example is crucial for a chain of department stores, new product development (and time to market) is crucial in electronics, while manufacturing lead time is essential in production to order companies or JIT-suppliers. But in the automotive sector e.g. all three aspects are of prime importance.

In this paper we analyze the impact of the Time Based Competition (TBC)-paradigm on two parts of the chain : manufacturing and new product development.

2. TIME BASED COMPETITION IN MANUFACTURING

A lot has changed in manufacturing within the time span of a single career. In order to understand the difficulties of managers to adapt in the difficult world of permanent (r)evolution it is useful to remember briefly the history of the last 45 years. In the early

1960s, because of rising wages and technological innovation, the industrial nations moved into the era of scale-based strategies, using specialized machinery to produce large batches at low costs. Thanks to advancements in computing power MRP systems made their entry pushing materials down the line without too many considerations for capacity except for the fact that time should not be lost setting up the machines. This resulted in huge inventories and lead times, complex materials management but relatively low unit costs. By the mid-seventies manufacturers (mostly in the United Kingdom) adopted a concept developed in the USSR: the focused factory (Skinner [1974]). By focusing on specific products and key elements of production competence, complexity was reduced, thereby enabling these focused manufacturers to attain higher productivity and lower costs than their broad line competitors. Concepts as group technology, cellular manufacturing made their entry. However as product diversity increased and product mix became highly volatile the inherent weaknesses of this production system became apparent: it lacked the volume flexibility to adapt to a changing market. The recognition of this limitation was the driving force of yet another revolution in manufacturing: the introduction of the flexible factory. By successfully introducing flexibility the Japanese were able to abolish the trade-off between scale and variety. This was the beginning of the well documented JIT-era. Finally, according to Stalk [1990] and Blackburn [1991] the ability to produce many different products within short time spans created the need to introduce even more products at faster rates and to reducé lead times accordingly in the other areas of the supply chain (distribution, administration, ...). This means TBC in manufacturing inevitably relies heavily on the concepts of the flexible factory and JIT.

2.1. Time Based Manufacturing Logistics

Variety and availability are important factors in today's markets. Customers want a vast spectrum of products and they want them now! But how do producers have to adapt ? Both factors are contradictory from a producer's point of view : variety means that the number of different products is high while availability means (traditionally) that you have to keep these products in stock. When the number of products increases, the number of products in inventory increases and hence the inventory value (and cost). The reliability of the forecasts will decrease with the number of products to be forecasted. As a consequence the amount of unsold and obsolete inventory will increase together with the stock-outs. Moreover, increased diversity means production in small batches which increases the proportional part of setups and lowers overall productivity.

This spiral of increasing logistic costs and decreasing efficiency can be avoided. The answer lies in redesigning the logistic chain towards responsitivity and lean production: a good localization of the Order Penetration Point (OPP), a Time Based Strategy (TBS) downstream the OPP and the use of concepts like commonality, load and throughput oriented order release are key factors in this process (Andries & Gelders [1995]).

The OPP is the stage in the production line from where production is on order. Production in front of the OPP is for general purpose and based on forecast while every part produced or assembled behind the OPP is dedicated to a customer.

When the OPP is located close to the end of the line customer demand is fulfilled from inventory. Inventory replenishment and production planning are based on forecast of demand during lead time. The safety stocks are high because they are proportional to uncertainty about customer demand and production lead times. When the OPP is moved

to the beginning of the line production is on order. Inventory of finished goods is low (or non-existent) but customer service is critical. Manufacturers who assemble to order typically have their OPP situated somewhere in the middle of the supply chain.
When deciding about the location of the OPP one has to take two opposite forces into consideration : the delivery times imposed by the market and the inventories.
When a company's order cycle time (i.e. the lead time *from OPP to customer*) is longer than the market delivery time, this company will tend to react by moving the OPP (= its inventories) downstream (i.e. closer to the market).
Raw material inventories are cheaper than end item inventories. From an economical point of view it is interesting to move the inventories upstream. But how far upstream ? A obvious solution is to locate the OPP at the convergence point of the Bill Of Materials. This convergence point is located at that place in the supply chain where the amount of parts, components, sub-assemblies or assemblies is the lowest. The exact location of this point depends of the type of industry. There are three basic types (Figure 1 and Table 1) :

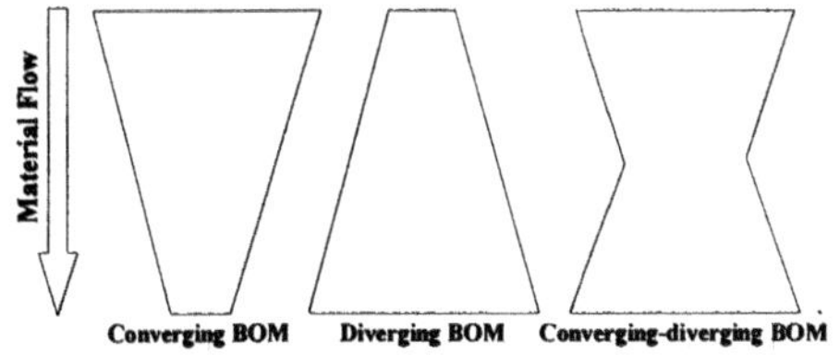

Figure 1

Type of BOM	Number of raw materials	Number of end items	Explanation and example
Converging	High	Low	Audio-visual industry : different parts assembled into a (relatively) small number of end-items. The OPP is located at the end item inventory, production is to stock.
Diverging	Low	High	Process-industry (e.g. Petro-chemicals). Small amounts of raw-materials are combined into large amount of end items. Production is (or should be) to order
Converging-diverging	High	High	A large amount of parts is combined into a small amount of assemblies that can be combined into lots of different end-items. The automobile industry is a typical example, but also most of the industries with a transition from process to discrete manufacturing.

Table 1

Based on the concept of the OPP a strategy is developed. The idea behind this strategy is that not all the parts of the supply chain have to be totally focused on time but only

these parts that are visible for the customer. In the other parts lead time reduction is to be seen as a means of cost reduction and not as an objective in itself.
Figure 2 illustrates the concept of a strategy based on the OPP and time. The supply chain can be divided into three distinctive parts, each with specific characteristics and objectives. First we shortly describe the different parts. Control methodologies will be dealt with in the sequel.

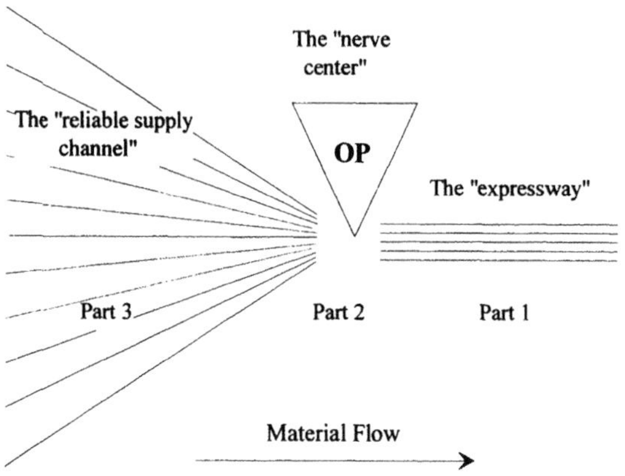

Figure 2

The "expressway". Production is on order. Delivery reliability, responsitivity, throughput and speed are *the* key issues. Lead time reduction is automatically translated into an equal reduction of delivery time. The control methodology is specifically designed for these characteristics.
The "nerve center". Within a time based strategy, all control and production management must start from this point : the performance of the downstream part (the expressway) depends on the material availability at this point
The "reliable supply channel". Delivery reliability is the key issue. Production is based on forecasts. Both lead times (because of cycle stock and forecast accuracy) and lead time variability (safety stocks) are important.
We can now discuss the production control strategies in the different parts of the chain.

A. The Production downstream the OPP

The control and planning of this part of the line depends on the OPP. The objectives of the planning is to maximize the service level while keeping the lead time beneath the promised lead time (imposed by the market). Four concepts are of prime importance here :

a) Setup and lotsize reduction
b) A flow oriented manufacturing layout
c) A high VAT (Value Added Time) / lead time value.
d) A low variability in the material flow
e) A dynamic priority system based on due dates

We discuss topic a, b, c and d and we refer to the scheduling literature for topic e.

a) Setup and lotsize reduction

Setup times are an important driver of lead times and flexibility. Basically the cost of variety is the cost of setups. Eliminate the setups and the cost of variety disappears.
Machine changeovers are not standardized activities in most factories. According to the SMED-principles (Single Minute Exchange of Die), a setup should be a well considered task, like any other process in production. SMED is a method to reduce setup times. The standard procedure for SMED analysis is very easy to understand and doesn't require long training to perform. The implementation steps are very similar with Deming's famous "wheel for quality improvement". The procedure can be summarized into four steps: planning, observation, analysis and action.
In the planning stage, a planning team is formed and goals are specified. Both top (production)management and setup- and machine-operators should be part of the team from the very beginning of the SMED-project. Management participation is important because SMED will involve machine downtime and loss of production and eventually investment. Workers participation is important because they will have to perform the setups. Early involvement is recommended to make them understand the reasons and motivations of the SMED program.
In the next phase, the actual setup is observed. This observation is usually based on a video recording of the setup. A detailed description of the different setup operations is added to the visual record. The analysis phase starts with a time and method study of the operations. Every activity should be broken down into basic motions before the SMED-analysis can start. Shigeo Shingo [1985] distinguished 5 basic principles :

1. Classify internal versus external activities. Internal activities are activities performed while machines are down. External activities may be performed while the machine is operating.
2. Complete external activities prior to setup (while the machine is running). Moving all external activities from in-line to off-line can result in substantial reduction of setup time (Shingo reports savings from 30 to 50 %).
3. Convert internal to external activities by modifying the setup organization. One technique is to double pieces of equipment. Perform the setup on one piece while the other is running.
4. Install parallel or simultaneous operations. This usually involves adding manpower.
5. Smooth and simplify the setup. In this phase training and teamwork and technical improvements are essential.

The goal of principle 1 to 3 is to reduce the tasks that must be carried out while the machine is down. Principles 4 and 5 minimize the time while the machine is down. One should notice the extreme simplicity of the principles. Most technical and organizational improvements suggested by Shingo require very little research or investments ("Simplify first, then automate."). Setup time reduction is the first step to continuous improvement. Once batch sizes and hence WIP is reduced other problems will show up (according to the "water and the rocks" analogy). These problems (the rocks) will prevent further lead-time reduction. The rocks on the factory floor are layout, quality and planning.

b) Flow oriented manufacturing layout

Large batches cause high work in process. Work stations tend to look like islands surrounded by oceans of inventory. The consequence is a slow and jerky flow of materials through the factory. Machines don't have to be connected nor synchronized because of large buffers which make control superfluous. The distance between work stations becomes very large and communication is reduced to a minimum. Many classical factories are organized in process-centers. Products are brought from process to process along complicated routings. Space, time and inventory have become insurmountable obstacles for local communication between work stations and divisions. Performance is measured by local performance measurement systems which strengthen the sub optimality of the systems. Centralized computer control becomes unavoidable.
Small batch sizes and frequent changeovers rectify these drawbacks. Oceans of inventory disappear, space is created and work stations can be moved closer. With connected instead of isolated work stations, a synchronized material flow becomes possible. Parts are manufactured and moved to the next station when they are needed. Downstream demand pulls the parts and the components through the manufacturing system. If demand is not too variable and a limited number product families can be identified (based on low changeover times within the family and/or similar routings) the factory can be divided into parallel flow lines which simplify the task of planning and (potentially) reduce lead times even further thanks to elimination or reduction of the setups within the product class. A good example of this strategy was given by an electronic component manufacturer in Belgium. Historically the layout of the factory had been organized as a job shop. However faced with an increasing number of products (> 300), long changeover times and an explosion of packaging possibilities (> 50) planning became difficult if not impossible to manage. Careful examination of the product mix and routing permitted to identify 5 product families for production and 3 families for packaging. Reorganizing of the job shop into a flow shop with 5 parallel lines permitted to reduce lead times (and WIP) by 75%. Only one additional machine had to be bought and 3 lines had to share to 2 bottleneck machines. Three flexible packaging machines were also bought. The space freed permitted to move the raw material inventories from the outside warehouse into the plant facilitating material handling and control.

c) A low VAT/lead time

The Value Added Time is the sum of the processing times of the value added steps. We will further discuss a technique aimed at increasing the ratio VAT/lead time. In this paragraph we describe an order release strategy that leads to better lead time control.
The ratio VAT/lead time is maximized by (among other things) reducing the queueing effect to its minimum. But one should be cautious when increasing the fraction. Limited queues are necessary to keep machine utilization at acceptable levels. A realistic target level for VAT/lead time lies between 0.25 and 0.4. Queue time can be limited by keeping WIP levels under a certain (low) level. The amount of WIP on the other hand is strongly dependent on the company's release strategy (Figure 3) :

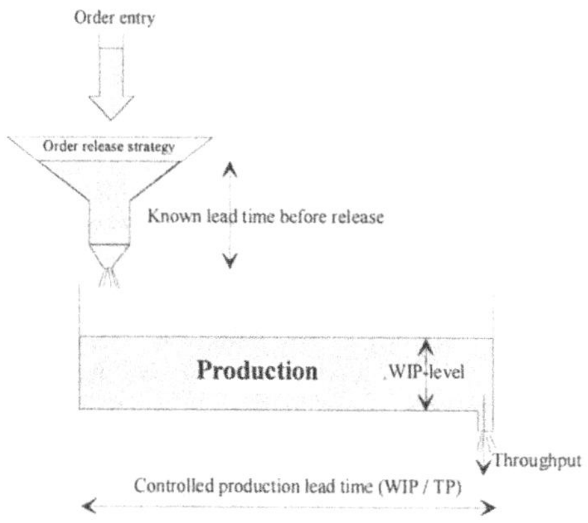

Figure 3

The principles of this methodology are very easy to understand (Wiendahl et al. [1992]). The production lead time is directly proportional to the amount of WIP (Little's law : lead time = WIP / throughput). If WIP is kept low and constant then lead times will be short and constant (and vice versa). Moreover with low WIP, control of material flow is facilitated, priority rules are easier to apply, costs of non quality are reduced,...i.e. variation is reduced and predictability is increased.

The consequence of reduced WIP on the production floor is an increased backlog of orders in the planning phase. But this has advantages too : customers order changes have less influence on production, priorities can be changed before order release, etc.

The planning lead time can be calculated too (backlog / throughput). This means that reliable approximations of total lead times can be calculated and delivery dates agreed upon with the customer. Order tracking is simplified by transparency both in planning and production lead time.

d) Material flow variability

Material flow variability has a strong negative impact on production lead times. It can be shown that interarrival time variability and process time variability have a dramatic impact on lead times.

But processing and arrival variability are not the only type of variation in production. We also have :

Machine - breakdowns influence the material flow in different ways :

- it reduces global capacity, directly in case of a bottleneck machine, indirectly if it causes starvation of a bottleneck machine
- it creates a lumpy material flow that results in WIP concentration at some places and starvation at other places
- it increases the amount of WIP : companies with unreliable machines tend to release more orders to avoid starvation ("raise the waterlevel to hide the rocks").

Synchronization problems are caused by missing parts at assembly stations. Many manufacturers "solve" this problem by increasing buffer capacity.

Customer order changes interrupt the processing. They have a highly disturbing effect when they occur during the production phase. It also originates a negative spiral :

planning will release more orders (i.e. more WIP) to avoid idleness when order changes are frequent. This increases lead times which increases the probability of changes occurring in the production phase....This spiral can be reverted into a positive one. If WIP is kept low, most changes will occur during the planning lead time.
Material availability is one of the key issues of the planning at the OPP : " Do not release an order unless all parts are available". Some companies disassemble less urgent orders (cannibalization) to obtain the necessary components. The consequences of this policy on the material flow should be clear....A better solution is to add a "throughput module" to the planning software. This module selects orders from the order list as a function of the material availability and the due date. A printed circuit board manufacturer in Belgium dramatically improved its lead time performance after installing a "throughput optimizer" that checked component availability before release for its 2000 different PCB types.
Any lead time reduction program that fails to recognize the importance of these factors is bound to have only marginal results.

B. OPP inventory management and production upstream the OPP

Key issues here are forecasts, inventory policies and cost reductions. The first aspect is important for the production in the supply channel while the second factor determines the number of different components and the quantity of each component to be kept in inventory at the OPP. Both must be focused towards maximum material availability at the OPP at reasonable cost (third factor). In the remainder of this paper we suggest solutions to those problems.

a) Forecasts

The accuracy of forecasts depends mainly on three factors :

1. The forecast technique. The number of available techniques being almost unlimited, we refer to specialized literature.
2. The number of products to be forecasted.
3. The forecast horizon. Forecast accuracy decreases when the horizon increases. The forecast horizon corresponds to the delivery lead time of the supply channel.

We will come back to point two later. Point three is important because little attention has been paid to lead time effects on forecast accuracy until now. Most production and inventory models (or lotsizing models) are based on the "best" trade-off between inventory and setups (e.g. economic order quantity models). Lotsizes increase when setup costs are important also causing WIP and lead times to increase. Long lead times increase the forecast horizon hence reducing forecast accuracy. Reduced forecast accuracy is automatically translated into an inventory problem at the OPP. This increases costs and reduces material availability at the OPP and hence overall reliability of the supply chain. Lead time related costs should be included in the lotsize calculations. This can be achieved by including lead time effects into the lotsizing calculations and adding a forecast inaccuracy cost that depends on the lead time.

b) Inventory policy at the OPP

The key aspect here is material availability. Three questions remain to be answered :

1. Which items/components/parts should be kept in stock ?

2. Is it possible to reduce the number of different items ?
3. How many of each item should we have in stock ?

The most important issue in choosing which items to keep in stock is the material availability *as a function of the end product.* We have mentioned yet that component availability was a conditio sine qua non for order release. This means that the only way to realize a short delivery time is to keep the components in stock (or to work on a JIT-base with the suppliers). Question one can thus be stated differently : for which end items do we want short delivery times ? Though this depends on the policy of the company, we think that every company should consider how far it is prepared to go to fulfill customer requests. Is it rewarding to keep components in inventory for every "exotic" product type ? Has the company any idea of the cost of short delivery times for those products ? Techniques like Activity Based Costing help to provide answers to these problems. These techniques attribute logistic costs to the products that drove those costs, providing arguments to remove components of slow moving, high cost products from stock. Another valuable tool is the PARETO-analysis that highlights the important segments of products from the great majority of less significant items. One could say, as a general inventory policy, that no components should be kept in stock for the last 5 or 10 % of demand (unless the component or end item is needed for strategic reasons). This means that the end item demand and the logistic costs determine whether a component will be held in stock or not.

As a consequence of this policy there will be two OPP in the line : one located before the assembly stations and responsible for (e.g.) 90 % of turnover and one situated at the beginning of the supply chain (raw materials) dedicated to end items with low demand (and long delivery times).

Another helpful technique to reduce OPP-inventory is standardization and commonality at that point. Commonality is achieved by using common components for different end items. Though commonality is important in all parts of the supply chain, it is at the OPP that the biggest returns can be achieved.

The last point of interest in this part of the chain is production control. Production to forecast, focuses on material availability and the (usually) large number of parts makes this part of the chain especially suited for a MRP(II) type of control. On the other hand, pull systems can be utilized for components with stable demand (especially those with a high degree of commonality). We think MRP(II) production control combined with JIT-aspects for components with stable demand is the best type of control for this part of the chain.

2.2. Value Added Analysis (VAA)

Value added analysis is a straightforward method aimed at reducing waste(d time) in the supply chain. This is done by focusing on the value adding activities in the supply chain and by trying to eliminate all activities that do not add value to the product or service. But, ... before a company can start with the elimination of waste, a thorough definition of value and value added (time) should be provided.

The most common definitions assume that value is added to a product whenever the product undergoes a transformation (that is valuable for the customer) or when a product is brought closer to the customer. Every operation that cannot be included in these two classes is wasteful and should be eliminated or at least reduced. But one should be very careful with the classification of value-added activities. Each step must be critically evaluated: is this step really value adding ? Can't we avoid or simplify it ? Table 2 provides a few examples of traps to avoid.

Step	Value Added because:	NOT 100 % -Value Added because:
Product quality inspection	Product of unknown quality is *transformed* into a product of known quality	Inspections are superfluous if process reliability is high ("zero-defect")
Two serial production steps	The steps are needed to manufacture the product	Could both steps be combined ?
One production step	The step is needed to manufacture the product	Could the step be avoided through e.g. Design for Manufacturing ?
One production step	The step is needed to manufacture the product	Is the step performed at nominal capacity ?
Material handling	Products must be stored or transported to the next workstation	Why does the product need to be stored ? Is it possible to place both workstations next to each other ?

Table 2

Once value has been defined within the company the actual value analysis can start: as Shapiro put it in his 1992 paper: **Staple yourself to an order.**

Time and efforts are wasted in the supply chain. Typical values of VAT-ratio's (= the sum of the value added processing times divided by the total lead time) range from 1 to 5 %. This percent is even lower if products are considered instead of orders or lots and if end item inventory turnover is included in the total lead time. This means that an order spends at least 95 % of its lead time waiting for value to be added. If one wants to reduce lead time one must work on these 95 %. Lead time reduction by increasing production speeds only affects 5 % of the total value. The problem is we can work twice as fast but we can't wait twice as fast!

The best way to identify these 95 % is to "staple" yourself to an order and follow it through the whole supply chain from order entry to delivery (in most cases it will even be useful to backtrack the order up to the MPS cycle or the raw materials forecast).

This approach is necessary for two important reasons :

1. The contradiction between horizontal product flows and vertical organizational structures:
   - The products flow horizontally from one department to the other while the information flows through the hierarchical decision structure causing delays and incompatibilities.
   - Vertical knowledge gaps. The only persons who can oversee the whole supply chain are high in the hierarchy, meaning that they have a strongly simplified view on the process while operational people only have a limited view on the chain.

2. Prioritization. It is only by following the order that one can understand the importance of priorities on lead times and especially lead time variation.

We believe that traditional process mapping techniques (Data Flow Diagrams, flow charts, ...) will often fail to reveal these (very important) lead time influencing factors because they rely too much on people's view on the process which is fundamentally different from the process seen from a product's perspective. DFD's do not include a time dimension while flow charts only include a sequence of events and not the timing of these events.

When one looks at the process from a product's perspective one will invariably notice that the flow of the order consists of an endless repetition of the same building block : **a tandem queue**. This block is represented in Figure 4:

WBP	PT	WBT	T

WBP : Wait Before Process
PT : Processing Time
WBT : Wait Before Transport
T : Transport

Figure 4

- **Wait before process** (WBP) consists of queuing before a "workstation" (machine, manual process, administrative process,...). It is NON value added. Usually the sum of all WBP's is responsible for more than 75 % of total lead time.
- **The processing time** (PT) is the time during which the product is "transformed", and is therefore value added. If a value added process is eliminated (e.g. by better design), all the other components of the tandem queue disappear, resulting in substantial lead time reduction.
- **Wait before transport** (WBT) is the time a product has to wait before it is moved to the next "workstation". It is NON value added.
- **Transport** is the action of taking the product to the next workstation. Transport can be achieved by material handling systems but also electronically when the workstations are administrative systems (in this case the product is information). If the next "workstation" is the customer, this step is value added. In most other cases it is not adding value.

The whole actual value analysis can then be divided into 4 remaining steps : order-stapling, identification, improvement and iteration.

A. Order-stapling

As has been said, the order must be followed from its first appearance in the company until it has left the system (or until the order is paid if financial cycles are investigated too). Shapiro [1992] distinguishes 10 different steps in the order management cycle:

1. Order planning
   - sales forecasting
   - capacity planning
2. Order generation
3. Cost estimation and pricing
4. Order receipt and entry
5. Order selection and prioritization
6. Scheduling
7. Fulfillment
   - procurement
   - manufacturing
   - assembling
   - testing
   - shipping
   - installation or delivery
8. Billing
9. Returns and claims
10. Post-sales services

Table 3

This presentation is interesting because the classical "order cycle" only starts from step 4. In the remainder we will pay no further attention to steps 8 to 10.
Steps 1 to 3 are important, but difficult to trace because they happen before the actual order is even placed.
When performing a value analysis it is recommended to follow the order "on the floor" *personally*. But one must be careful. It does not mean following the "physical" order, wherever the product is but "following the logical order on its critical path". Figure 5 illustrates the difference between the "physical" flow of the order and the logical flow :

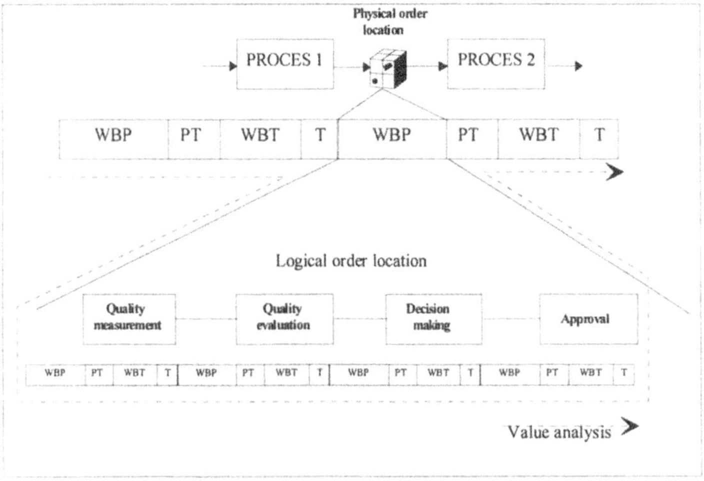

Figure 5

While the order is waiting on the production floor a sample is sent to the quality department for inspection. The logical location is then at the quality department and the value analysis must follow that flow. When two processes are in parallel the critical path should be followed.

Once all paths are identified, follow-up documents can be designed to perform further data gathering. Four control points are needed for every tandem queue that must be measured (one at the end or the start of every component of the queue). If the processing times are known and processing time variability is negligible, three measurement points are sufficient.

Figure 6 shows an example of an analysis performed at a chemical plant. This part of the study involved 12 operations performed in 9 different activity centers (called departments). 36 measurement points were required in the first phase (identification of opportunities) of the lead time reduction project. These 36 measurements were reduced to 25 to be registered using 3 different follow up documents. After the identification phase, 9 control points were chosen to monitor improvements. These points were necessary to be able to recognize individual department responsibilities in the improvements.

Three product families were identified based on routing characteristics. Two orders of each family were personally (and concurrently) followed to identify critical paths and design proper follow-up documents. 36 orders were then traced with the documents resulting in the data shown in figure 5. From the lead time and VAT / LT figures planned lead times were chosen as objectives for the lead time reduction.

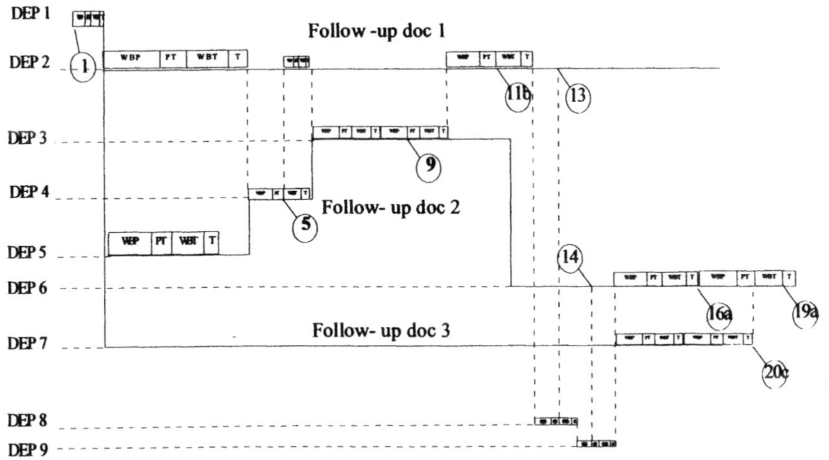

From → To	1 → 5	5 → 9	9 → 11b	11b → 13	13 → 14	14 → 16a	16a → 20c	20c → 19a
Actual LT (days)	6.4	1.81	0.34	2.33	0.45	18.30	1.76	1.93
VAT / LT	17%	0.6 %	5%	7%	42%	93%	0.01%	26%
Planned LT (days)	1.5	0.5	0.3	0.5	0.45	18	0.5	1

Figure 6

B. Identification

Now, the best opportunities for lead time reduction can be identified. This identification can be made by means of two proactive diagnostic performance measures :

- Lead time
- Ratio VAT/lead time

Both measures can be used for single processes (1 tandem queue), groups of processes or even complete departments or plants.
The first measure states where the greatest absolute lead time reduction can be achieved while the greatest relative opportunity is given by the second measure.

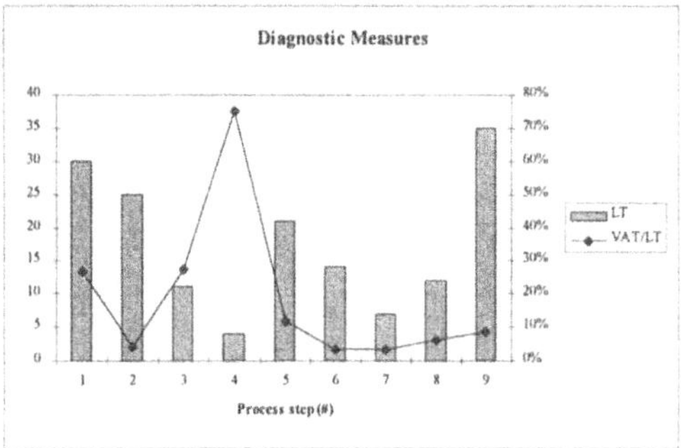

Figure 7

From Figure 7 we see :

- Processes 1, 2 and 9 give the best opportunities for absolute lead time reduction
- Processes 2, 6, 7, 8 and 9 give the best opportunities for relative lead time reduction
- No significant lead time reduction can be expected from processes 3 and 4
- Conclusion : processes 2 and 9 are the ones to start with

It is important to remark that the ratio VAT/LT is meant as a diagnostic performance measure (i.e. to determine where improvements can be made) and *not* a control measure (i.e. to control the improvement process) because improvement could impact both the nominator and the denominator. Lead time (or VAT) can be used for both purposes, diagnostic and control.

The next step is the identification and the removal of the causes of waste. Although this step is probably one of the most important ones it is also highly case dependent and difficult to discuss in general terms. We therefore refer to the case study literature.

3. TIME BASED COMPETITION IN NEW PRODUCT DEVELOPMENT

Traditionally Western management has always assumed that there is plenty of time for new product development : all options were taken into consideration, all alternatives were generated and evaluated, different scenarios were played off and tons of reports had to go up and down the hierarchical structure of the company. But meanwhile product life cycles kept shrinking and Japanese competitors were introducing cheap and high quality products much faster than before. This led to a situation where the product development cycle exceeds the life cycle of the previous generation and where Western companies lost market share to the Japanese. Figure 8 illustrates the evolution of the PLC in the computer industry

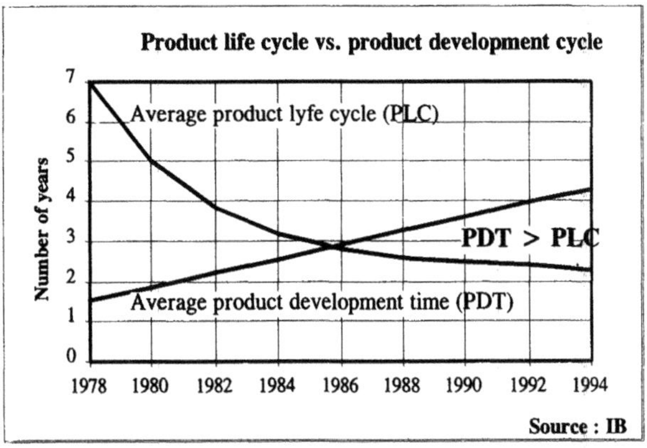

Figure 8

A time based focus has to be introduced in the whole supply chain including new product development

One of the key characteristics of Time Based Competitors is their ability to get new products to the market before their competitors can. They create products which often provide value exceeding customer demand and they make a profit doing so.

A comparison of Japanese and Western companies (especially in engineering intensive organizations) reveals that their new product introduction rate is a few times higher than their Western competitors. Clark [1987] stated that the Japanese can develop a new vehicle about 18 months faster than Americans. The product life cycle of an automobile has an average duration of about 6 years. It is easy to understand the competitive advantage of Honda for example who brings a new car to market every 4 years on average. And indeed, late market introduction can cause severe loss of returns. Mc. Kinsey and Co. [1985] reported that a 6 month late introduction can cause a 33 % decrease in net profits. Another known and painful example are FORD's billions of dollars of lost revenue due to the late introduction of the Mondeo as a replacement of the aging Sierra.

Research on fast product development is still in an embryonic stage. The different principles which lead to time compressed manufacturing processes are well known, some of them were described in the previous paragraphs. The same guide-lines for new product development seem to be non existent. However a few concepts seem to crystallize :

1. The necessity of an environment focused on change and innovation
2. The use of information technology which providing the most performing design tools.

And also, according to Christopher Meyer [1993] : *"The only way to increase profit and reduce costs, while concurrently improving product development speed, is to fundamentally change the product development process itself."*
The Japanese quickly noticed that the very same JIT-ideas that had been successfully used in manufacturing could be transposed to NPD. Indeed the JIT-model is very attractive for new product development. First of all, the basic goals of JIT - the elimination of waste, simplicity, total quality and speed - are key attributes in all parts of the company. Second, the experience gained in production can be transposed to new product development. Third, JIT incarnates the KAIZEN principle of incremental and continuous improvement. The model for the management of the NPD presented in the following is a translation of some the JIT-principles to the development phase.

3.1. Concurrent Engineering

In most Western companies new product development consists of consecutive or serial steps. In the very first stages design-engineers are responsible for the conceptual design and the drawings. The blueprints and/or the prototype are delivered to the production engineers to prepare large scale production in a next stage. After this step the people of planning and logistics are asked to join the project to prepare the bill of materials and resolve inventory and capacity problems. In the last phase marketing and sales perform the market introduction. This phased approach of product design, development, manufacturing and marketing has important drawbacks. One problem is the very long product development cycle. Another problem lies in the fact that 90 % of the budget costs are sunk before production engineers get involved in the process. Lack of market feasibility due to late involvement of marketing is the third drawback.
The major drawbacks of "consecutive engineering" can be resolved by concurrent engineering. Broadly defined, concurrent engineering is the simultaneous design of products and their engineering, manufacturing and marketing processes (Figure 9). In contrast to the serial approach, with concurrent approach to engineering, multi-functional teams attack all aspects of product development simultaneously. Information is released in incremental units, the batch size for information transfer from one design stage to the next is reduced to the smallest unit possible. As such this step is very similar to the batch size reduction in production lead time reduction.

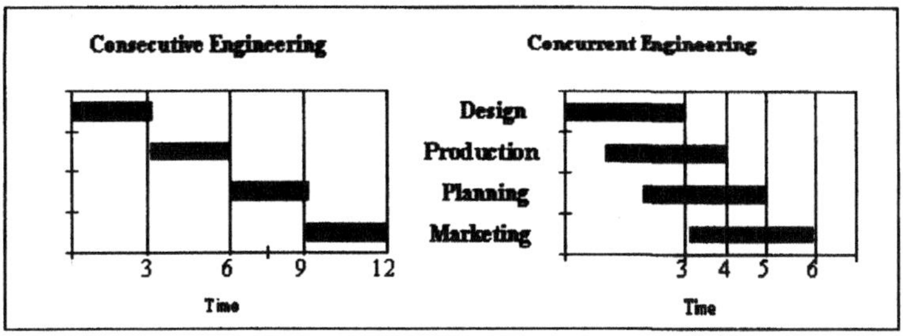

Figure 9

The major benefits of concurrent engineering are:

- Reduced cost of change
  Catching mistakes early in new product development is very important, the later a change or revision is made the more expensive it is. During the initial phase the designer can make changes easily, often with little more than a few computer keystrokes. But later in the process, when the tooling for production has been made changes become very expensive. Business week has reported that the cost of change for a major electrical product increases with one order of magnitude with each phase of its development
- Reduced development lead time

However small batch information processing and concurrent engineering are not easy to implement and are accompanied with serious risks. Releasing partial information increases the likelihood that erroneous information is sent downstream. Flexibility of the team workers is very important because they will have to work with incomplete information. People from different departments will have to overcome their reluctance of working together and to release incomplete information to critical eyes. Modern information technology and severe procedures must be introduced to prevent people to work with outdated information,etc.
Experts say it may take two full product development cycles to realize most of the benefits associated with concurrent engineering.

3.2. Setup time reduction

A new product introduction can be seen as a changeover from an old product to a new one, and the goal is to make the transition as quickly and efficiently as possible. There are four basic activities that apply to changeover time reduction in new product development :

1. Internal versus external activities
   SMED tries to reduce changeover times by performing as many activities as possible before or after the actual setup. Within new product development this can be achieved by performing development, manufacturing or marketing activities while the design phase is still being fine-tuned. Much of the work on manufacturing equipment or layout and on equipment procurement can be done during the design phase and hence be converted into off-line activities. These principles have been covered yet in the paragraph about concurrent engineering.

2. Eliminate and simplify tasks

   Every non value added activity must be identified and eliminated. Value added analysis has proven its usefulness in improving manufacturing processes. Early Time Based Competitors discovered that many of the delays in design, development and testing can be reduced by eliminating activities such as redesigning existing parts and prototyping.

   Design engineers are prone to reinvent the wheel. However one of the easiest ways to reduce costs and save time is to use off-the-shelve components or common parts in different designs (i.e. commonality). Moreover an additional benefit of commonality will arise later in production.

   When analyzing total new product development lead-time it is not unusual to see prototyping being responsible for 30 % or more of total lead-time. As a consequence firms today are questioning the role of prototypes and are searching for alternatives, and finding them.

   Modern technologies like solid modelling or simulation offer effective alternatives. Solid modelling for example allows designers to check fit, relative size, clearances and to get an idea of functionality. High resolution monitors provide very realistic images of designs which permit to evaluate aesthetics and make former clay models obsolete. Moreover solid modeling encourages (early) changes in the design.

   Simulation and analysis software provide the ability to evaluate concepts while they are still bits in the computer. Using the geometry created by CAD, such software can analyze product performance and suggest areas for improvement. Finite elements analysis is generally considered the most valuable tool for this purpose. It is used for whole ranges of problems, from the usual structural analysis to thermal, dynamic, magnetic and fluid analysis.

   When prototyping is unavoidable, NC-programming and rapid prototyping use a variety of technologies (e.g. stereolithography) to more directly turn CAD geometry into a physical part.

3.3. Teamwork

One of the greatest barriers to overcome in developing a concurrent engineering environment is the ability of people to get along and work with each other. Indeed, the small batch information processing approach places higher premium on teamwork than the conventional and sequential approach. As a firm moves toward overlapping activities information flows within and between departments become more complex and cross-functional teamwork becomes increasingly important. The choice and training of the product development team has taken a critical dimension. One of the first steps to developing a cross-functional team is to break down the barriers that exist between specialties (Hammer and Champy[1993]). These teams are composed of people from marketing, sales engineering and manufacturing, who have no idea what the people from other groups do and how they perform. It is therefore necessary to train team members on how to operate.

3.4. Information flow and layout

In most companies, new product development is organized as a job-shop. Design is done in one area, drawings are then transmitted in batch to the next department in the pipeline (engineering) and so on. With concurrent engineering and cross-functional teams, people from different departments must be put together and work together simultaneously. This means that the layout must be reevaluated for every project in a process of ongoing improvement. People must get used to be relocated after each project. Managers must accept to dedicate their designers to small numbers (ideally one) of projects. Some companies even remove people from their home offices to put them at work together in isolation for the duration of the project.

3.5. Automation

Automation should be viewed as the culminating, not the initiating, activity in efforts to reduce development time. In accordance to the principle "Simplify first, then automate", automation is the last step in lead-time reduction programs. When automation is introduced without making the changes in the organizational structure or the design team, it may result in adding expensive technology that need to be restructured (islands of technology cause sub-optimality without improving global efficiency). In many companies CAD/CAM procedures have merely automated the paper system without improving overall performance. One must not forget that when a drawing is idle, time is wasted no matter whether the drawing is idle in electronic form or in paper form. Figure 10 describes the different phases of design and their key technologies.

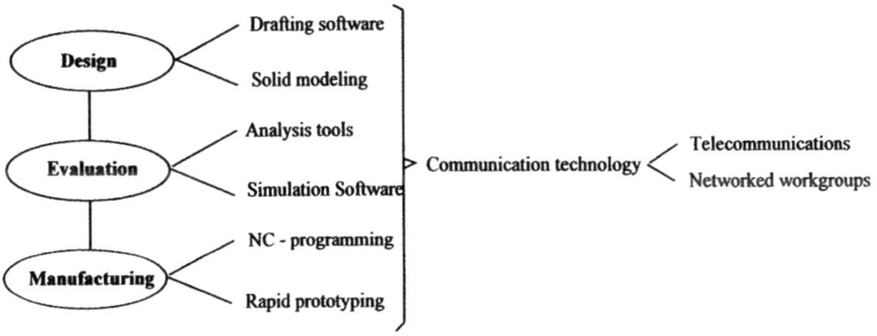

Figure 10

3.6. Quality and continuous improvement

Introducing quality in manufacturing means " Do it right the first time". For new product development this principle can be translated into " Do the early part right ". i.e. do the up.front work thoroughly even at the price of lengthening that phase, because correcting an error in a later phase can be very expensive. Many accounting people and budget cutters will likely scream at an approach that calls for thorough initial planning. They may say that the best time to save money is early in the project because that leaves a lot of time to revise things later ("We can always produce a few more " in large batch manufacturing). That is completely incorrect however. It is much more cost-efficient to spend the money up-front to do the job right.

Another guarantee of quality in JIT or time based manufacturing is provided by the small batch sizes and empowerment of workers. Thanks to small process and transfer batches products arrive quickly at the next workstation where faulty parts are identified.

The operator has then the authority to stop the line and identify the cause of the defect. The same concepts apply to concurrent engineering where information and blueprints are released early to the next phase where mistakes, functional problems, etc. are detected. Within teams meetings are organized frequently to deal with these problems.
The third dimension of quality is continuous improvement (KAIZEN, Imai [1986]) or, "learn from your mistakes". Senge [1990] recently emphasized the importance of organizational learning. Learning experience from each project should be gathered and passed on to the next project and so on. When development cycles are short, the learning cycle is accelerated creating a faster learning curve. Moreover as more emphasis is put on communication and contact between departments the information is spread faster and more efficiently.
And last but not least the (ten) concepts of quality and **total design process control** (Sleeckx [1993]) are summarized in Table 4.

Concept	Technique(s)
Listen to the customer	• Quality function development • Quality = what the customer wants ≠ perfection !!!
Work with product teams	• Multi-disciplinarity • Dedication and co-location • Autonomous • Concurrent engineering
Integrate process design	• Design for assembly • Design for disassembly • Design for manufacturing
Optimize co-makership	• Strong long term business relationships • Concurrent engineering "over the walls"
Work with digital product models	• STEP (Standard for Exchange of Product data • EDI (Electronic Data Interchange) • Lotus notes or similar software
Integrate CAD/CAM/CAE	• Integration ≠ interfacing • CNC - rapid prototyping
Use simulation wherever possible (correct mistakes before you *really* make them)	• Aesthetic modelling in real world environment (remember Jurassic park) • Finite elements • Interference control • Kinematic analysis
Use inherent quality control	• FMEA (Failure Mode and Effect Analysis) • Taguchi methods
Accelerate decision making	• Lean management • Autonomous teams
Improve continuously	• KAIZEN • Learning cycles • Deming wheel

Table 4

4. CONCLUSION

Just in time, fast cycle manufacturing, quick response, business process reengineering, concurrent engineering, agility, lead time management, lean manufacturing, time based competition, etc. whatever the name is, in essence it all boils down to the same: time. If you are able to bring competitive products faster to the customers than its competitors, you will get the deal. In the previous paragraphs we discussed some of the techniques available for lead time reduction and control within manufacturing and new product development. We choose these techniques among several others for two reasons: first because they are easy to implement in most environments without high investments and second because we successfully tested and implemented these techniques in projects conducted by the Center for Industrial Management in the process, the discrete and the service industry.

5. BIBLIOGRAPHY

Andries, B. and Gelders,L. (1995), Time Based Manufacturing Logistics, *Logistics Information Management*, Vol. 8, No. 3, pp. 30 - 36

Benjamin, R., Wigand, R. (1995), "Electronic Markets and Virtual Value Chains on the Information Superhighway", *Sloan Management Review*, pp. 62-72

Blackburn, J.D. (1991), *Time-Based Competition - The Next Battle Ground in American Manufacturing*, Business One Irvin, Homewood, Illinois

Choperena, A.M. (1996), Fast Cycle Time - Driver of Innovation and Quality, *Research Technology Management*, Vol. 39, No.3, pp. 36 - 40

Clark, K.B., Chew, B. and Fujimoto, T. (1987),"Product Development in the World Auto Industry", *Brookings Papers on Economic Activity*, Vol. 3,pp. 729-777

Gelders, L., Mannaerts, P. and Maes, J. (1994), Manufacturing Strategy, Performance Indicators and Improvement Programmes, *Int. J. Prod. Res.*, Vol. 32, No. 4, pp. 797 - 805

Hammer, M. and Champy,J., (1993), *Reengineering the Corporation*, Harper Business, New York

Imai, M. (1986), *KAIZEN: The Key to Japans Succes*, Mc. Graw Hill, New York

McKinsey & CO (1985), *Triad Power : The Coming Shape of Global Competition*, Flammarion, Paris

Meyer, C. (1993), *Fast Cycle Time: How to Align Purpose, Strategy and Structure fir Speed*, Free Press, Macmillan Inc., New York

Senge, P. (1990), *The Fifth Discipline*, Double Day, Doubleday Dell Publishing Group, New York, New York

Shapiro, B.P., Rangan, K.V. and Sviokla,J.J. (1992), Staple Yourself to an Order, *Harvard Business Review*, pp. 113 - 122

Shingo, S. (1985), *A Revolution in Manufacturing: The SMED System*, Mass. Productivity, Cambridge

Skinner, W. (1974), The Focused Factory, *Harvard Business Review*, May-June 1974, pp. 113 - 121

Stalk, G. (1990), *Competing against time :how time based competition is reshaping global markets*, Free Press, New York,

Stalk, G.JR. and Webber, A.M. (1993), Japan's Dark Side of Time, *Harvard Business Review*, July -August 1993, pp. 93 - 103

Wiendahl, H.P.and Glassner, J. (1992), Petermann,D., Application of Load-Oriented Manufacturing Control in Industry, *Production Planning Control*, Vol. 3, No. 2, pp. 118 - 129

4
Virtual reality in manufacturing

Dr. A. K. Bejczy
Jet Propulsion Laboratory, MS198-219
California Institute of Technology
Pasadena, CA 91109-8099
Phone: (818) 354-4568
Fax: (818) 393-5007
e-mail: bejczy@telerobotics.jpl.nasa.gov

Abstract

Computer graphics based virtual reality techniques offer powerful and very valuable task visualization aids for planning system set-up and system processes, for previewing system and device actions, and even for seeing some invisible things in a complex action or production environment. Virtual reality techniques connected to CAD data banks can also offer 3-D simulation based programming and control tools for computer-controlled devices and can provide real-time interactive operator interface to those devices. Virtual reality techniques can also be applied to quick product prototyping by high-fidelity product modeling

Keywords

Computer graphics, visualization, modeling, operator interface, computer programming and control.

1 INTRODUCTION

Visualization is a key problem in many areas of science and engineering. The emerging 'virtual reality' technology, which is a major technical idea and undertaking of the 1990s, can help resolve or lessen many visualization problems.

The term 'virtual reality' (VR) essentially denotes the creation of realistic looking world scenarios by the use of computer graphics simulation techniques which in turn can be connected to Computer Aided Design (CAD) data banks. A distinctive feature of VR products is that the synthetic world scenarios are not static but dynamic: the user can interact with them in various ways in real time.

VR technology already found its way in many application areas like entertainment, advertising, arts, medicine, robotics, and lately also, in manufacturing. This article is focused at applications or application potential of VR in manufacturing. In the first part of the article actual robotic/telerobotic application examples are quoted to illustrate the application potential of VR in manufacturing, since robotics to a large extent is connected to manufacturing. The second part of the article specifically addresses manufacturing topics.

2 VIRTUAL REALITY IN ROBOTICS

The role of VR in robotics includes (i) *task visualization* for planning system set-up and actions, (ii) *action preview* prior to actual motion execution, (iii) *predicting motions in real time* in telerobotics when there is a considerable communication time delay between operator control station and remote robot, (iv) *operator training* prior to exercising hardware for learning system operation, (v) *enabling visual perception of non-visible conditions and events* like multi-dimensional workspace constraints, hidden proximity motion states and actual contacts, (vi) *constructing virtual sensors* for training and for safety under actual operations, and (vii) to serve as a *flexible operator interface modality* replacing complex switchboard and analog display hardware.

Dual-arm workcell

A VR system was developed for a dual-arm workcell of the Advanced TeleOperation Project (ATOP) at the Jet Propulsion Laboratory (JPL) a few years ago. The workcell was surrounded by a gantry robot frame providing two-d.o.f. mobility for a stereo and two mono TV cameras, each on a pan-tilt base, in three orthogonal planes. Camera focus, zoom, and iris can also be controlled remotely. The workcell housed two eight-d.o.f. robot arms, each equipped with a 'smart' robot claw-hand which can sense both the grasp force at the base of the claws and the three orthogonal forces/moments at the base of the hand. The task environment in the workcell simulated the Solar Maximum Satellite Repair (SMSR) which was actually performed by two astronauts in space suits in the Space Shuttle cargo bay in 1983. (Note that this satellite was not designed for repair!). The goal of the experiments in the ATOP workcell was to show how to do the same repair in a teleoperation mode. To achieve this goal, the use of computer graphics offered an indispensable help in many ways.

Altogether, twelve major sequential subtasks are implied in the Main Electrical Box (MEB) change-out in the SMSR mission, and each subtask requires the use of some tool (screwdriver, scissors, etc.) This corresponds to a large set of task and control visualization off-line analysis by VR techniques, also taking account of workspace, arm kinematics, use of tools, and viewing constraints. It is noted that the motion of the robot arms' graphics images during task and control simulation are controlled by the same inverse/forward kinematics control software that controls the actual hardware system.

Figure 1 illustrates a graphics operator interface to the control of the dual-arm telerobot system for SMSR experiments, as a result of the off-line VR task and control simulation analysis. The subtasks are listed in the lower right insert, four preview control options are listed in the upper left insert, and the upper right insert lists some current status messages. It is noted that the graphics image of the two arms is updated in real time at a rate of at least 15 Hz so that the operator can always view the full system configuration even when all TV cameras are focused and zoomed on some particular subtask details. More on the SMSR task and control graphics visualization and the resulting graphics operator interface work can be found in Bejczy et al. (1990), Fiorini et al. (1993), Kim et al. (1991), Kim (1993).

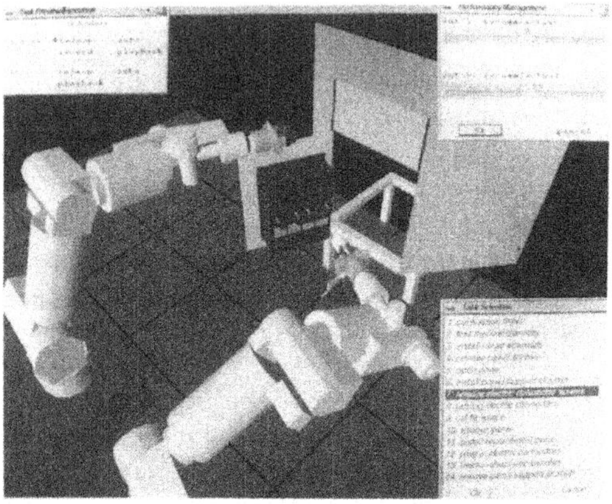

Figure 1. Graphics Operator Interface for Preview and On-Line Visualization, with Task Script Titles and Redundancy Mgmt. Status

Two major conclusions resulted from the SMSR VR task and control visualization off-line analysis: (i) The full task can only be performed with the given dual-arm setting if the base of either the dual-arm system or the task board (preferably the task board) is movable in two directions in a plane by about ±45 cm in each direction and rotatable by about ±20 degrees, (ii) a small TV camera mounted to the base of the right arm's end effector would considerably contribute to the effectiveness and safety of the performance of several subtasks.

VR calibration

The actual utility of VR techniques in telerobotics to a high degree depends on the fidelity by which the graphic models represent the telerobot system, the task and task environment. A high fidelity VR calibration methodology has been developed by the ATOP project at JPL. This development has four major ingredients. First, creation of high-fidelity 3-D graphics models of remotely operated robot arms and objects of interest for robot arm tasks. Second, high-fidelity calibration of the 3-D graphics models relative to given TV camera 2-D image frames which cover the sight of both the robot arm and the objects of interest. Third, high-fidelity overlaying of the calibrated graphics models over the actual robot arm and object images in a given TV camera image frame on a monitor screen as seen by the operator. Fourth, high-fidelity motion control of robot arm graphics image by using the same control software that drives the real robot.

The second ingredient is based on a TV camera calibration technique, using the Newtonian ideal pinhole model for image formation by the camera, which is a linear model. The camera calibration is performed by using the manipulator itself as a calibration fixture. The operator enters the correspondence information between 3-D graphics model points and 2-D camera image points of the manipulator to the computer. This is performed by repeatedly clicking with a mouse a graphics model point and its corresponding TV image point for each corresponding pair on a monitor screen which shows both the graphics model and the actual TV camera images. To improve calibration accuracy, several poses of the manipulator within the same TV camera view can be used to enter corresponding model and TV image points to the computer. Then the computer computes the camera calibration parameters. Because of the ideal pinhole model assumption, the computed output is a single linear 4x3 calibration matrix for a linear perspective projection.

Object localization is performed after camera calibration, entering corresponding object model and TV image points to the computer for different desired TV camera views. Again, the output is a single linear 4x3 calibration matrix for a linear perspective projection.

The actual camera calibration and object localization computations are carried out by a combination of linear and nonlinear least-squares algorithms, and the computational procedure depends upon whether an approximate initial solution is known.

After completing camera calibration and object localization, the graphics models of both the robot arm and the object can be overlaid with high fidelity on the corresponding actual images in a given TV camera view. The overlays can be in a wire-frame or solid-shaded polygonal rendering with varying levels of transparency, providing different visual effects to the operator for different task details. In a wire-frame format, the hidden lines can be removed or retained by the operator, dependent on the information needs in a given task.

These high fidelity *fused* virtual and actual reality displays became very useful tools for predicting and previewing robotic actions under communication time delay, without commanding and moving the actual robot hardware. The operator can generate visual effects of robotic motion by commanding the motion of the graphics image of the robot superimposed over TV pictures of the live scene. Thus, the operator can see the consequences of motion commands in real time, before sending the commands to the remotely located robot. This calibrated virtual reality display system can also provide high-fidelity *synthetic* or artificial TV camera views to the operator. These synthetic views make critical robot motion events visible that otherwise are hidden from the operator in a given TV camera view or for which no TV camera view is available.

VR performance demonstration

The performance capabilities of the high-fidelity graphics overlay preview/predictive display technique described above was demonstrated on a large laboratory scale in May, 1993. A simulated life-size satellite servicing task was set up at the Goddard Space Flight Center (GSFC) in Maryland, and controlled 4000 kilometers away from the JPL ATOP control station. Three fixed TV camera settings were used at the GSFC worksite, and TV images were sent to the JPL control station over the NASA Select Satellite TV channel at video rate. Command and control data from JPL to GSFC and status and sensor data from GSFC to JPL were sent through the Internet computer communication network. The round-trip command/information time delay varied between four to eight seconds between the GSFC worksite and the JPL control station.

The task involved the exchange of a satellite module. This required inserting a 45 cm long power screwdriver, attached to a robot arm, through a 45 cm long hole to reach the module's latching mechanism at the module's backplane, unlatching the module from the satellite, connecting the module rigidly to the robot arm, and removing the module from the satellite. The placement of a new module back to the satellite's frame followed the reverse sequence of actions.

The experiments were performed successfully. The calibration and object localization errors at the critical tool insertion task amounted to 0.2 cm each, well within the allowed error tolerance. More on the VR calibration methodology and on the above-quoted experiments can be found in Kim et al. (1993) and in Bejczy et al. (1995). Figures 2 and 3 illustrate some of the overlay scenarios and a synthetic view.

Visualization of the non-visible

Recently a technique was developed for visualizing the highly constrained and complex geometric motion capabilities of a dual-arm system. The two arms work together on an object in a closed kinematic chain configuration. The motion space analysis has to take account of the base placement of the robot arms, object dimensions, object holding patterns and constraints. It turns out that this constrained motion space can be visualized as a complex 3-D object with hidden unreachable holes or cavities of varying shapes. The developed technique is an *inverse* computer vision procedure in the sense that it creates rather than recognizes visual forms. More on this technique can be found in Tarn et al. (1995).

Figure 4 shows a 3-D transparent view solution of a particular case problem, from different view angles. The upper left 3-D semi-transparent display is a perspective view such that the x-axis is to the left, the y-axis is to the right, and the z-axis is to the top. Two thick, black patches, seen to the left and right across the center are the unreachable spaces hidden in the outer light gray reachable space. The bigger transparent light gray object is the reachable space, and the lighter gray empty space parallel to the y-axis passing through the center corresponds to the forced unreachable space that avoids collision with robot bases.

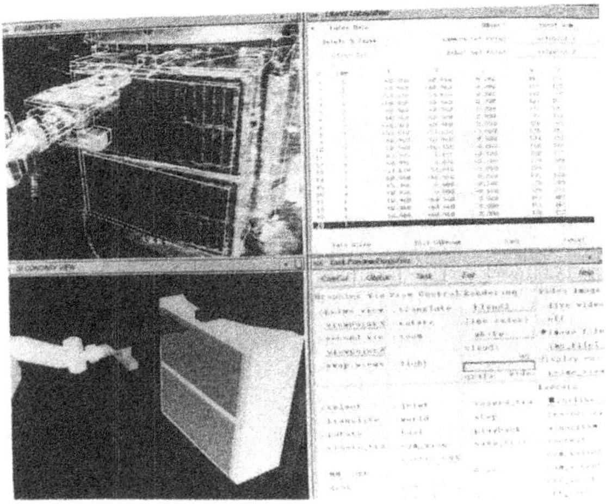

Figure 2. A Calibrated Graphics Overlay, with a Related Synthetic View.

Figure 3a. Predictive/Preview Display of End Point Motion.

Figure 3b. Status of Predicted End Point after Motion Execution, for the Same Motion Shown in Figure 3a, but from Different Camera View.

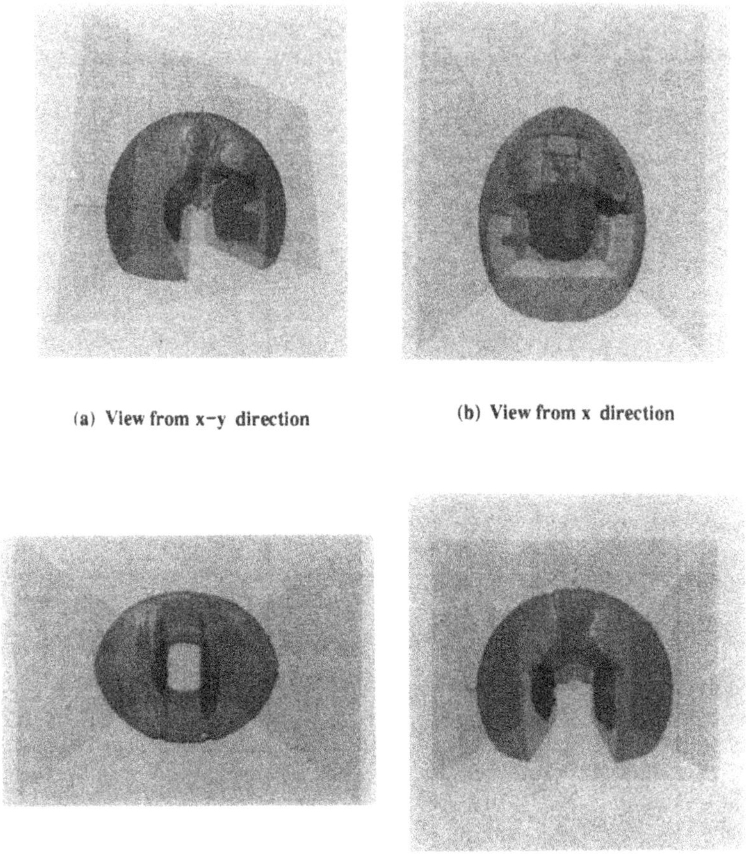

(a) View from x−y direction

(b) View from x direction

(c) View from z direction

(d) View from y direction

Figure 4. Four views of the 3D transparent motion space for a rod of length 0.5 m.

Virtual sensors

The notion of 'virtual sensors' is referred to the simulation of sensors in a computer graphics environment that relate the simulated robot's interaction with the simulated objects. Two examples are quoted here: proximity sensing and contact force sensing.

Computer graphics simulation of *proximity sensor* signals is a relatively simple task since it only implies the computation of distance from a fixed point of a moving robot hand in a given (computed) direction to the nearest environment or object surface in the graphics 'world model'. For this simulation, the TELEGRIP™ software package of Deneb Robotics, Inc., has been adopted. The package provides an excellent interactive 3-D graphics simulation environment with CAD-model building, workcell layout, path designation, and motion simulation. It also provides various functions related to object distance computations and collision detection. TELEGRIP CLI (ascii-text Command Line Interpreter) commands include INQUIRE COLLISIONS (reporting collision and near miss status) and INQUIRE MINIMUM DISTANCE (reporting minimum distance between parts or devices). TELEGRIP GSL (Graphics Simulation Language) also supports ray cast ( ) function that computes the intersection distance from a point in a specified (ray cast) direction. Using these object distance computation functions, various proximity sensors can easily be simulated. More on this can be found in Bejczy et al. (1995).

Force/torque sensor can also be simulated by computing virtual contact forces and torques for given simulated geometric contact models. In general, an accurate simulation of virtual contact forces and torques can be very computation-intensive, but an approximate simulation, for example, a simplified peg-in-hold task, can be accomplished without difficulty, described in Bejczy et al. (1995). In this graphics simulation, the hole and its support structure are assumed to be rigid with infinite stiffness, while the robot hand holding the peg is compliant for all three Cartesian transitional axes and also for all Cartesian rotational axes. It is further assumed that the compliance center is located at a distance L from the tip of the peg with three lateral springs k_x, k_y, and k_z and three angular springs k_{mx}, k_{my}, and k_{mz}. Both the operator-commanded and the actual positions of the peg are described by the position of the compliance center. For a given operator-commanded peg position, the actual peg position after compliant accommodation can be different, depending upon the current state of the peg: whether the peg is currently in the hole or not. For the peg-not-in-hole state, two conditions are considered: no-touch or peg-on-wall. For the peg-in-hole state, four conditions are considered: no-touch, peg-side one-point contact, peg-tip one-point contact, or two-point contact.

Figure 5 shows a force-reflecting virtual reality training display for a peg-in-hole task. Contact forces and torques are computed and reflected back to a force-reflecting hand controller in real time. They are also displayed on the upper left corner of the display screen at 20 Hz rate.

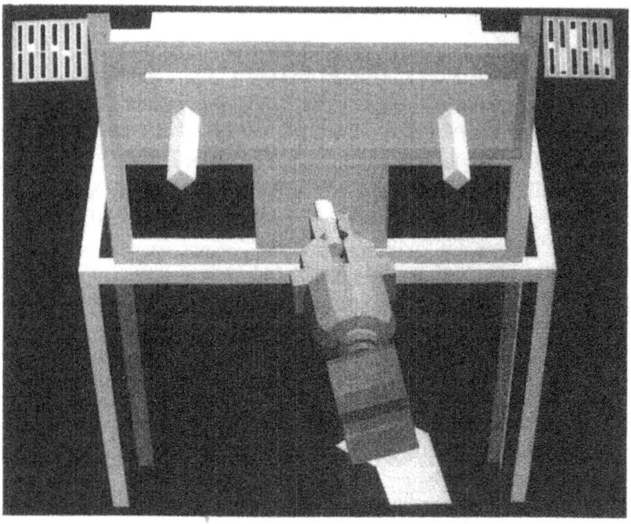

Figure 5. Force-Reflecting VR Training Display; Simulated Contact Force/Moment Values Shown in the Upper Left Bar Display.

3 VIRTUAL REALITY AND MANUFACTURING

From an engineering viewpoint, manufacturing as a technological activity comprises four classes of engineering task: product design, system set-up design for production process, actual set-up of production system, and operation of the actual production system. VR techniques, connected to CAD data banks, can substantially contribute to the success and cost-effectiveness of all four classes of manufacturing engineering tasks in a very efficient way. The efficiency of VR techniques applied to manufacturing is rooted to two basic capabilities of VR methods: high-fidelity visualization providing model and situation representation in the 3-D space and the power of interactive handling of large data sets in real time or near real time.

Product design

VR techniques offer the opportunity to expand the product design capabilities of CAD techniques by new technical dimensions to satisfy ergonomic, aesthetic, assembly, customization and quick prototyping considerations and requirements. In fact, VR techniques connected to CAD data banks can elevate product simulation to or near to a proof-of-concept level, saving considerable time and cost.

Automobile manufacturers already capitalized on these capabilities very successfully during the past five to six years. To keep the volume of this article within limits, only one advanced example is quoted here. Kalawski (1993) reports the efforts of researchers at the British Aerospace Brough Laboratory. They created a Rover 400 model car interior by using CAD-like objects and a graphical programming language. The system included a high-resolution HRX EyePhone, a DataGlove, a Convolotron for 3-D sound, and a set of SGI computers for graphics rendering. To reinforce realism of VR simulation, the researchers used a real car seat. Therefore, the designers were able to study the ergonomics of the car interior accurately and change the position and orientation of virtual parts when needed. It was then possible to redesign the interior of the car while immersed in the graphics simulation environment. For example, the designers could grab and move the steering wheel assembly from the "right-hand" drive position of the UK, Japan and Portugal to the "left-hand" drive position of the US and the rest of the European countries.

Production system design

A production system consists of several passive and active cooperating devices, parts to be treated, moved and assembled, a logical sequence of actions, a nominal time line of actions, and the capability of recovering from a variety of possible errors. Moreover, it is often required to design a production system with some flexibility in mind relative to changes in the product to be produced by the system. A flexible manufacturing system typically also employs robots. VR techniques, connected to CAD data banks, offer very valuable opportunities and capabilities not only for the design but also for the real-time simulation, control and monitoring of flexible robotic manufacturing systems. Good examples of VR simulation of flexible manufacturing cells are described in Apostoli et al. (1993), Apostoli et al. (1995) and Moran et al. (1995). These quoted examples also cover some programming, control and monitoring aspects of a manufacturing cell using VR simulation which gives 3-D representation of the global geometry and of the changing geometric relations in a working cell.

An other good example is reported by Strommer et al. (1993), using a transputer network for development of control of robots in a manufacturing environment. In the reported effort, the computer control programming is done on a *virtual* robot and its *virtual* environment. The environment can be a manufacturing cell, and utilizing all available feedback elements of the human-machine interface. The interface is divided into a human-computer module and a connection to the real world which contains a robot of the same type as the virtual robot. The information loop between computer and real world is then closed by sensor data coming from the real world. After control code development and debug, the code is stored and downloaded to the real robot which then executes in on the real environment.

Production system operation

First of all, the operation of computer controllable production systems poses a complex programming problem. There are some commercially available software products that can help develop programming of production systems using VR methods. Examples are the IGRIP™ and TELEGRIP™ software packages of DENEB Robotics, Inc. (P.O. Box 214687, Auburn Hills, MI 48321-4687, USA; Internet address: support@deneb.com). These products use open architecture, industry standards and network protocols, and offer application-specific options.

They also demonstrated meeting acceptance criteria imposed by automotive companies and functioning in real time.

An interesting and quite active research theme today is the programming of assembly tasks by recognizing human teaching operations. Examples can be found in Takahashi et al (1992), Tso et al. (1995) and Tung et al (1995). This type of programming approach also can be elevated to higher levels and generalized as outlined in a system description by Ogata et al. (1996). That system operates in two phases: the teaching phase, and the execution phase. In the teaching phase, the system learns a task in a specific environment from a human operator using a VR teaching interface, then translates it to a general task description in the task interpreter, and stores it. In the execution phase, the system retrieves the general task description, observes the environment where the task will be executed, modifies the learned motion in the task planner, and executes the task.

4 CONCLUSION

Manufacturing appears to be a major beneficiary of VR technology and of its evolving capabilities in many ways. It can considerably improve design and production processes and introduce new approaches to the programming style of computer controlled manufacturing systems, rendering the programming effort more efficient and more transparent.

The application of VR technology in manufacturing is not yet a fully evolved capability and accepted practice, and it requires some investment in proper education and teaching, and in appropriate commercially available hardware/software tools. The cost of appropriate commercial tools is on a decreasing trend as we know it from the price history of the last few years. The real progress and success will depend on human attitude and education which in turn calls for an effective teaming of VR and manufacturing technologists.

5 REFERENCES

Apostoli, R.S., Moran, D.D., Trombotto, G.D., Abate, F.J. (1993), A wholly designed and built flexible manufacturing cell (FMC) for training and research purposes. *12th Brazilian Congress of Mechanical Engineers;* Brazilia, Brazil; pp. 1603-1606.

Apostoli, R.S., Luciano, C.J. (1995), Manufacturing Simulation in a flexible system. *4th Symposium on Low Cost Automation;* (IFAC-AADECA); Bs. As., Argentina; pp. 446-451.

Bejczy, A.K., Kim, W.S., Venema, S. (1990), The phantom robot: predictive displays for teleoperation with time delay; *Proc. IEEE Int'l. Conf. on Robotics and Automation,* Cincinnati, OH.

Bejczy, A.K., Kim, W.S. (1995), Sensor fusion in telerobotic task controls; *Proc. IEEE Int'l Conf. on Intelligent Robots and Systems*, Pittsburgh, PA.

Fiorini, P., Bejczy, A.K., Schenker, P.S. (1993), Integrated interface for advanced teleoperation; *IEEE Control Systems Magazine*, Vol. 13, No. 5.

Kalawski (1993), *The Science of Virtual Reality and Virtual Environments;* Addison-Wesley Publ. Co., U.K., 405 pp.

Kim, W.S., Bejczy, A.K. (1991), Graphics displays for operator aid in telemanipulation; *Proc. IEEE Int'l. Conf. on Systems, Man, an Cybernetics*, Charlottesville, VA.

Kim, W.S. (1993), Graphical operator interface for space telerobotics; *IEEE Int'l. Conf. on Robotics and Automation*, Atlanta, GA.

Kim, W.S., Bejczy, A.K. (1993), Demonstration of a High Fidelity Predictive/Preview Display Technique for Telerobotic Servicing in Space; *IEEE Transactions on Robotics and Automation, Special issue on space robotics* (October).

Moran, O.D., Rosales, A.G., Luciano, C.J. (1995), Real-time control of a tutorial flexible manufacturing cell; *4th Symposium of Low Cost Automation;* (IFAC-AADECA); Bs. As. Argentina; pp. 311-315.

Ogata, H., Tsuda, M., Mizukawa, M. (1996), Teaching 6-dot tasks for robot operating assembly in the same category; to appear in *Proc. 8th Int'l. Conf. on Advanced Robotics*, Monterey, CA.

Strommer, W., Neugebauer, J., Flaig, T. (1993), Transputer-based workstation as implemented for the example of industrial robot control; *Proc. of Interface to Real and Virtual Words Conference*, Montpellier, France; pp. 137-146.

Takahashi, T., Ogata, H. (1992), Robotic assembly operation based on task-level teaching in virtual reality; *Proc. IEEE Int'l. Conf. on Robotics and Automation;* pp. 1083-1088.

Tso, S.K., Liu, K.P. (1995), Automatic generation of robot programming codes from perception of human demonstration; *Proc. IEEE Conf. on Intelligent Robots and Systems;* pp. 23-28.

Tung, C.P., Kak, A.C. (1995), Automatic learning of assembly tasks using a DataGlove system; *Proc. IEEE Conf. on Intelligent Robots and Systems;* pp. 1-8.

Acknowledgment: This work was carried out at the Jet Propulsion Laboratory, California Institute of Technology, under contract by the National Aeronautics and Space Administration (NASA).

Brief Author Biography

Bejczy, Antal K., is a Senior Research Scientist at the Jet Propulsion Laboratory, Pasadena, CA, and an Affiliate Professor, Department of Systems Science and Mathematics, Washington University, St. Louis, MO. He received B.S. in EE at the Technical University of Budapest, Hungary; M.S. in Natural Sciences, University of Oslo, Norway; Ph.D. in Applied Physics, University of Oslo, Norway in 1963. He was a Research Fellow and Senior Resident Research Associate at the California Institute of Technology in 1966-1969, at the Dept. of Applied Sciences. He joined JPL in 1969. First, he worked on advanced space mission studies, then turned his interest on robotics and teleoperation in 1972. He published about 150 journal and conference papers mainly on topics in dynamic modeling, sensing, control, and human-machine interaction in telerobotics, including telepresence and virtual reality; 12 chapters in various edited volumes in robotics, and co-edited one book on "Parallel Computation Systems for Robotics." He was president of the IEEE Council on Robotics and Automation in 1987, and serves as an ADCOM member in the IEEE Robotics and Automation Society since 1990. He organized numerous special sessions on robotics and teleoperation at IEEE conferences. He became an IEEE Fellow in 1987. He received the NASA Exceptional Service Medal in 1991. He received over 40 NASA Technical Innovation awards, and holds 6 U.S. patents. He is General Chairperson of the "8th International Conference on Advanced Robotics" to be held in July, 1997.

PART TWO

Virtual Enterprises I

5

Cross Border Enterprises: Virtual and Extended!

Ingrid Hunt, Dip. B.Sc. M.Sc.
CIMRU (Computer Integrated Manufacturing Research Unit)
University College Galway, Ireland.
E-mail: Ingrid.Hunt@UCG.IE

Alexandra Augusta Pereira Klen, Dra. Eng.
GRUCON (Research and Training Group in the Areas of Numerical Control and Manufacturing Automation)
Federal University of Santa Catarina , Brazil.
E-mail: aklen@mbox1.ufsc.br

Jiangang Zhang, B.E., M.Eng.Sc.
CIMRU, (Computer Integrated Manufacturing Research Unit)
University College Galway, Ireland.
E-mail: Zhang.Jiangang@UCG.IE

Abstract

Many industries today are facing the challenge of the concept of the "survival of the fittest". In order to achieve success and market share enterprises must quickly adopt new ideas and concepts. Presently, cross border enterprises is the emerging topic for business and manufacturing research and application. Cross border enterprise can be seen as the broad name for two new manufacturing paradigms, the Extended Enterprise and the Virtual Enterprise.

Keywords

Cross Border Enterprise, Extended Enterprise, Virtual Enterprise, EDI.

1 INTRODUCTION

In today's changing world, enterprises need to survive in an ever volatile competitive market environment. Their success will depend on the strategies they practice and adopt. Every year, new ideas and concepts are emerging in order for companies to become successful enterprises. Cross Border Enterprises is the new 'hot' topic arising in the business process world at present. Many terms have been coined together and are being driven in the popular business press to describe this new strategy of conducting business, i.e. Extended Enterprise (Browne et al., 1995; O'Neill and Sacket, 1994; Busby and Fan, 1993; Caskey, 1995), Virtual Enterprise (Goldmann and Preiss, 1991; Parunak, 1994; Goranson, 1995; Doumeingts et al., 1995), Seamless Enterprise (Harrington, 1995), Inter-Enterprise Networking (Browne et al., 1993), Dynamic Enterprise (Weston, 1996) and so on. Many people have argued that they mean the same thing, just using different words. Others feel they are different. But how different are they? In this paper the authors will present some basic lines required from this new strategy for conducting and coordinating distributed business processes (DBP), as well as trying to clarify the particularities of two of the widest spread terms related to it: Virtual and Extended Enterprise.

2 CLUSTERS OF PRESSURES

The business world currently faces an increased trend towards globalisation, environmentally benign production and customisation of products and processes, forcing individual enterprises to work together across the value chain in order to cope with market influences.

International competition is rapidly increasing, therefore replacing national competition due to open markets, with increased size, accessibility and homogeneity. Now with the reduction in trade barriers, the harmonisation of standards and laws and with easy access to many markets, competitors can be in every quarter of the global market. So, the means of production are increasingly accessible throughout the world, regardless of national boundary: production technology, financial capital, marketing information, engineering capabilities, etc. Today products are being designed in Ireland, engineered for production in Portugal and produced in Brazil. Reflections of many organisation representatives show that fuelled by advances in information technology and improving communication infrastructures, this trend will continue at an accelerated rate. They ask: in order to support globally distributed manufacturing, what capabilities do the managers, the employees and the manufacturing systems need to have?

Society and government regulations are placing greater pressure on manufacturers to create processes and products that are environmentally friendly. Enterprises who develop processes and programmes that focus on waste management and its reduction, recycling and reconditioning etc. will gain a competitive edge over its competitors. Whereas ten years ago a producer saw his responsibility ending after sales service, more and more producers today are managing the entire life cycle of their product, from raw material back to raw material. Once again it is asked: to support this new environmental awareness, what capabilities do the managers, the employees and the manufacturing systems need to have?

The world is experiencing an explosion of product variety available to consumers today. The Japanese are now introducing new model cars in three years, from concept to showroom (Womack et al., 1990). Think of the computer business. There are thousands of software applications, each one turns your portable PC into a different, customised product. Because the production methods and technologies are more flexible, companies are better able to produce what customers want, when and where they want it. Yet again they ask: to support this new customer orientation, what capabilities do the managers, the employees and the manufacturing systems need to have?

These global trends call for complete re-examination of the whole approach to production management methods, strategies and structures rather than continue focusing on perfecting methods or systems that implicitly assume that the manufacturing environment consists of one company, one location producing disposable products for stable mass markets. Increased globalisation, sensitivity to the environment and attention to customers demand a more holistic view of business, seeing them as part of a broad network of activities. The new reality in production management is that products can be designed, developed and produced virtually anywhere.

On the one hand, from an interorganizational view, this leads to the configuration of extended/virtual enterprises (Nagel and Dove, 1991; Goldman, Nagel and Preiss, 1995) whose success is based on flexibility, speed and the emerging paradigm of Agile Manufacturing to align and integrate processes with business partners in order to provide customers with exceptional value. On the other hand, from an intra-organisational view, it can be said, analogously, that this leads to the configuration of virtual production areas (Pereira Klen, 1996). From here now some important aspects related to the interorganizational view will be the subject of study.

3 WHAT IS A CROSS BORDER ENTERPRISE?

Theorists have spent a short time of their studies, considering the perspective of creating, operating and dissolving a network of enterprises; the accompanying quotations suggest that there is a considerable agreement in research being carried out in this area. Actually, the question is more related to the term used to define the networking of enterprises than to anything else. Should it be called 'extended' or 'virtual'?

Scholars have proposed a variety of definitions of cross border enterprises. Here is a small sample:

- "... where core product functionalities are provided separately by different companies who come together for the purpose of providing a customer-defined product/service." (Browne et al., 1995).
- "... it means integration on a grand scale - a scale that transcends traditional external and internal corporate boundaries." (Harrington, 1995).
- "... when a company uses electronic networking to form close ties with suppliers, distributors and customers." (Landon, 1994).
- "A rapidly configured multi-disciplinary network of small, process-specific firms configured to meet a window of opportunity to design & produce a specific product." (Parunak, 1994).
- "...when several enterprises at dispersed geographical location, engaged in the manufacture of various products at different scales of operations and possessing diverse recourses bind together. ... Problem solving and decision making are conducted by flexible teams cutting across the individual enterprises and distributed over time and space." (Janowski and Acebedo, 1996).

To answer the proposed question, perhaps it is interesting to begin studying the 'phenomena' and afterwards worrying about how to call it. The next section touches on some of the main topics related to the Extended/Virtual Enterprise.

4 NETWORKING ENTERPRISES

Browne et al. (1995) have identified five issues that represent for them the future shape of manufacturing systems. So, enterprises should be aware of the following points when moving in the direction of coordinating and conducting distributed process business:

- Reduced product life cycles,
- Time-based competition,
- Total product life cycle,
- High quality organisations and people,
- Manufacturing strategy.

Product Life Cycles

Customer awareness of technological development on a global scale, electronic communications, close coupling of the customer to the manufacturing source have all led to reduced product life cycles. When moving in the direction of coordinating and conducting distributed process business, enterprises will have to be aware that the reduction in product life cycles will have to raise the flexibility in their enterprises to new levels. The flexibility required by enterprises will be through enterprise agility and accessing external resources through the cross border enterprise.

Time based-competition

The quicker a product can be placed on the market the higher the price the product can demand. The generation, communication and application of ideas and their application in products is the core of fast time to market. When moving in the direction of coordinating and conducting distributed process business, enterprises must be open to the communication input increasing reliance on the cross border enterprise, including input from the customer.

Total Product Life Cycle

Manufacturing enterprises are being forced to take responsibility for the total life cycle of the product, including the environmental effects and the costs of disassembly, disposal or refurbishment. Networking enterprises can be seen to be responsible for and involved in the integration of the whole product life cycle, from material procurement, to production, to customer service and finally to end-of-life product handling.

Creating effective organisations

The cross border enterprise focuses on the horizontal communication chain in order to form effective organisations. Teams of people will cooperate across the value chain. Therefore each member (link) in the chain understands how its activities add value to the customer. Browne et al. (1993) state that the hierarchical authority systems prevalent in vertically integrated industrial firms are slow to identify rapidly changing customer requirements and to react quickly with creative product offerings.

Now, the once upon a time vertical businesses and organisations, are quickly becoming a thing of the past. Businesses are now becoming flatter i.e. horizontal as opposed to vertical. The cross border enterprise welcomes the concept of a more closely knit environment for the integration of the enterprises.

Adopting an appropriate manufacturing strategy

The determination and implementation of good manufacturing strategic decisions in any enterprise has and will be recognised as the one of the sources of advantage in today's competitive market.

Summarising, the greater the cooperation across the value chain, the greater the satisfaction of the customer needs. The challenge now in industry is to aspire to this new trend. But Browne et al. (1995) shows that many manufacturing industries have already been carrying out their business to allow for this new concept. For example, Just in Time (JIT), Manufacturing Resource Planning (MRP), World Class Manufacturing (WCM), Benchmarking, etc. have all helped in the understanding of this concept. Florida et al. (1993) tells us that in order to conduct and coordinate distributed business process the enterprise should be a knowledge-based organisation that uses the distributed intellectual strengths of its members, suppliers and customers. As well as its *technical* and *communication* resources the enterprise depends on *knowledge* and *trust* for its success.

4.1 Extended/Virtual Enterprise: Frameworks

A framework can be used to give a broad view of complicated systems and concepts. The design of a framework for cross border enterprise concepts will aid in the understanding and application of Extended/Virtual Enterprise. It's beyond the scope of this paper to detail the description of frameworks for the Extended/Virtual Enterprise or to try to produce some new ones. It is hoped that in the near future an Extended/Virtual enterprise framework will be developed through the work carried out by the SCM+[1] consortium. Two of the existing frameworks which briefly describes the Extended/Virtual Enterprise concepts are presented here. They are shown in the following two figures. Figure 1 is a model of an Extended Enterprise framework which shows manufacturers, component suppliers, service providers, and customers integrated to provide quality products and services. The Extended Enterprise has a long term corporation among its business partners. Figure 2 is a model of a Virtual Enterprise framework which shows different organisations coming together to exploit market opportunities. These business partners are capable of prospering within the Virtual Enterprise environment.

[1] SCM+: INCO-DC Project: 950880 Beyond Supply Chain Management in Food Industry.

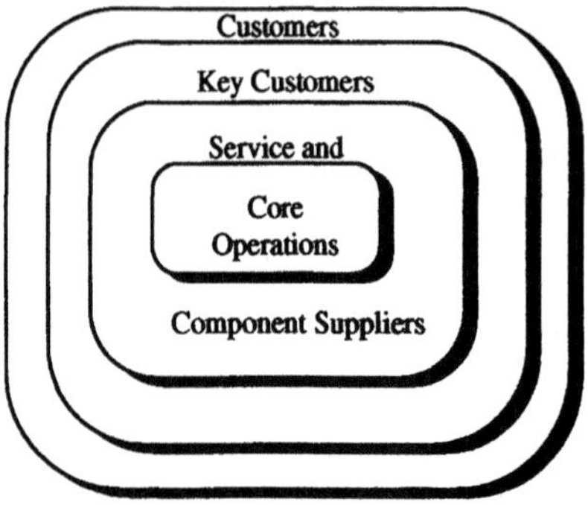

Figure 1 Extended Enterprise (Browne et al., 1995)

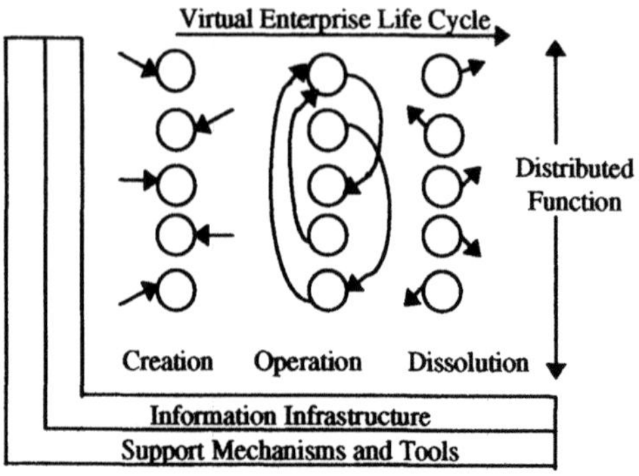

Figure 2 Virtual Enterprise (Parunak, 1994)

4.2 Helping to Keep Up the Good Work: the Technological Resources

The use of information in each enterprise transaction poses unique challenges. Each enterprise will have its own information residing in existing enterprise applications also called legacy systems and therefore it needs to be accessed by other partners, within the enterprise network, in a common format. The business process world welcomes the development and emergence of Electronic Data Interchange (EDI) which supports the networking of suppliers of raw materials, producers, processors, distributors and customers. EDI allows an efficient and inexpensive two-way flow of information between the 'nodes' (partners of the enterprises network). Tighter control over suppliers, distributors, manufacturers etc. in both local and global markets can be maintained.

So, the network of enterprises is enabled by state-of-the-art information and communication technology which are being used in today's business environment. According to the Forrester Research report (1994), by the year 2002, companies will spend 45% of their Information Systems and Information Technology (IS/IT) budget on applications spanning company boundaries, to include a complex web of relationships between a company, its partners, customers, suppliers and its markets. This reiterates what has been already said where the use of technology has to be encouraged for sharing and exchanging information right across individuals and organisations. Therefore, the new strategy of conducting business drives the requirements of communication technology to higher levels. Cassidy (1993), indicates that the these requirements can be met, for instance, by the information super highways now being established.

It is perhaps clear that the concept of networking enterprises success should be built on a foundation of shared up to date information which is frequent and accurate. One would feel that without this information the concept would not be truly successful. But this is not enough. Even more, enterprises will have to take advantage of tacit knowledge and the art of trusting people whom you do not see.

4.3 Sustainable Competitive Advantage: the Human Being

O'Neill et al. (1994) quite rightly agree that the scope of managing the network of enterprises is global, which looks for the integration of the skills and contributions of those companies and individuals belonging to the value network. They continue to tell us that management of these networks explore the synergy necessary to satisfy the diversity demanded by customers and to innovate, not just products, but also management practices. In this way, this is a concept that many top managers have been trying to practice for many years, establishing closer links with customers, suppliers and trading partners. Therefore one would feel that this is not a really technical challenge but a good challenge for the management of these enterprises.

So, people are important in the success of this new concept of networking enterprises. In the extended and virtual manufacturing environment, customers are welcomed into the manufacturing process. The philosophy of this concept encourages people throughout the supply chain to participate in decision making for the product life or process. It has already been mentioned that the management hierarchy should be horizontal to succeed. The management hierarchy of the enterprises network aims for a flat decision making organisations.

But, do these networks have business and manufacturing strategies? Busby et al. (1993) describes the first as an incremental process, with planning, implementation, evaluation and revision representing small steps, done almost simultaneously. Goldman et al. (1991) defines the second as an approach called cooperative or proservice manufacturing.

So, cross border enterprises offers customer support with the ***manufacturing seen as a specialised form of service***, where the integration of competencies of all the people involved in the manufacturing processes achieves economies of scale. Consequently, from these strategies the enterprises network is able to provide tailored products to satisfy the specific needs of their customers.

4.4 Extended/Virtual Enterprise examples

Although it is a trend today that enterprises are moving to become Extended and Virtual, sometimes it's difficult to clearly define the difference between conventional enterprises and Extended/Virtual enterprises. For many entrepreneurs the moving process is obvious and natural. Today enterprises have, both the traditional features and new emerging concepts. There are many enterprise partnerships today who operate in the Extended/Virtual enterprise environment. The followings are two examples of how organisations are moving towards the Extended/Virtual enterprise concept,

Supply Chain Management (SCM) in food industries

A recent SCM project is attempting to create an Extended Enterprise environment for food industries. The aim of the project is to integrate the supply chain of a food industries across their organisation's boundaries. The integration will include small food suppliers, a central distribution centre, warehouses, and supermarkets. This will lead to seamless information and product flow among all business partners in this organisation who are striving to achieve an Extended Enterprise environment.

A manufacturers' services company

This company is a customer-oriented, world wide provider of electronics manufacturing services and support for emerging and multinational original equipment manufacturers (OEMs). They see their enterprise as a global service provider with a small-company philosophy. This enterprise wants to create a network of links with their customers through technology, which will help all of them to integrate and come together when necessary in order to provide products and services based on shared goals and shared rewards. Therefore, they are working towards a Virtual Enterprise concept which enables them to explore common market opportunities.

5 CONCLUSIONS

Now, enterprises can no longer be seen in total isolation. Individual enterprises must work together across the value chain in order to meet customer needs. The challenge for the future of any enterprises is to consider the new concept of conducting business and to facilitate inter-enterprise networking across the value chain. In doing so, smaller firms will be able to gain the economies of scale of larger companies.

Caskey (1995) describes the concept of the Extended Enterprise as no longer treating our suppliers and customers as 'them' instead they are treated as 'us'. When these Extended Enterprises become dynamic i.e. adapting their processes to the customer's current demands or particular product development then the enterprise can be called a Virtual Enterprise.

A different way of describing the terms 'virtual' and 'extended' is to focus on the word 'temporary'. The Virtual Enterprise can be seen as a temporary 'getting together' of enterprises who come together to put a product quickly on to the market while the Extended Enterprise gives us a more permanent 'togetherness' of enterprises. The Extended Enterprises depend on organisation stability and the relationships across the value chain. The Virtual Enterprise on the other hand depends on high technology and advanced information systems in order to successfully integrate the business partners.

The authors have come to the conclusion that the concepts discussed in this paper (virtual enterprise and extended enterprise) are complementary ones. Therefore, let us conclude that a successful Extended Enterprise should lead the way for an organisation to take the role of a successful Virtual Enterprise.

6 ACKNOWLEDGEMENTS

The authors wish to thank all the members of the SCM+ project, INCO-DC 950880 for giving them the opportunity to write this paper.

7 REFERENCES

Browne, J., Sackett, P.J. and Wortmann, J.C. (1995) Future Manufacturing systems-Towards the Extended Enterprise, *Computers in Industry*, 25, 235-254.

Browne, J., Sackett, P.J. and Wortmann, J.C. (1993) The System of Manufacturing: A prospective study, November 25.

Busby, J.S. and Fan, I.S. (1993) The Extended Manufacturing Enterprise: Its Nature and its Needs, *International Journal of Technology Management*, Special Issue on Manufacturing Technology: Diffusion, Implementation and Management, Vol.8 Nos.3/4/5, 294-308.

Caskey, K.R. (1995) Cooperative Distributed Simulation and Optimisation in Extended Enterprise, *IFIP WG5.7 Conference Proceedings.*

Cassidy, J. (1993) Bell Rings Multimedia Revolution, *The Sunday Times*, 17th Oct., 2-3.

Doumeingts, G., Ducq, Y, Clave, F. and Malhene, N. (1995) From CIM to Global Manufacturing, *Proceedings of CAPE'95*, 379 - 388.

Florida, R. and Kenny, M. (1993) The New Age of Capitalism: Innovation mediated-Production, *Futures*, Vol.25 No.6, July/August, 637-651.

Forrester Research: Computing Strategy Report (1994) The New Customer Connection, September.

Goldman, R. and Preiss, K. (1991) 21st Century Manufacturing Enterprise Strategy: An Industry-Led View, Iacocca Institute, Leigh University, OH, 7-13.

Goldman, S., Nagel, R. and Preiss, K. (1995) Agile Competitors and Virtual Organisations, Strategies for Enriching the Customer, Van Nostrand Reinhold.

Goranson, H.T. (1995) Agile Virtual Enterprise - Best Agile Practice Reference Base, Working Draft of January 2.

Harrington, L. (1995) Taking Integration to the Next Level, *Transportation and Distribution*, Vol. 36, Iss.8 Aug., 26-28.

Janowski, T. and Acebedo, C.M. (1996) Virtual Enterprise: On Refinement Towards an ODP Architecture, *Pre-Proceedings of Theoretical Problems on Manufacturing Systems Design and Control*, Lisbon, June.

Landon, N. (1994) Electronic Bonding, *International Business*, March, 26.

Nagel, R. and Dove, R. (1991) 21st Century Manufacturing Enterprise Strategy, Iacocca Institute Report, Vol.1.

O'Neill, H. and Sackett, P.J. (1994) The Extended Manufacturing Enterprise Paradigm, *Management Decision*, Vol.32, No.8, 42-49.

Parunak, V., (1994) Technologies for Virtual Enterprises: A Proposal for a NIST ATP, NCMS - National Center for Manufacturing Sciences.

Pereira Klen, A.A. (1996) An Approach to Conceptualise Learning Enterprises in the Manufacturing Sector, *Doctoral Thesis*, UFSC.

Weston, R. (1996) Model Driven Configuration of Manufacturing Systems in support on the Dynamic, Virtual Enterprise, Plenary Session - *Third Conference on Concurrent Engineering, Research and Applications (CE96/ISPE)*, Toronto, Canada, August.

Womack, J.P., Jones, D.T. and Roos, D. (1990) The Machine that Changed the World, Simon & Schuster Inc..

8 BIOGRAPHY

Ingrid Hunt graduated from Sligo Regional Technical College in 1994 with a B.Sc. in Quality Assurance and graduated from the University of Ulster in 1995 with an M.Sc. in Manufacturing Management. After graduating she worked as a Quality Manager in a manufacturing organisation. She joined CIMRU in May 1996 and is currently Project Manager for Esprit and INCO-DC projects in the area of Logistics and Supply Chain Management. She is currently working on her Ph.D. at CIMRU, in the area of the Virtual Enterprise and Extended Enterprise.

Alexandra Augusta Pereira Klen is a research scientist at GRUCON in the University of Santa Catarina in Brazil. She obtained a B.Sc. degree in Mechanical Engineering in 1988 from this university. She was an invited scientist at BIBA in Germany from April 1991 to September 1994. She obtained a Ph.D. degree for her work done in the area of 'An Approach to Conceptualise Learning Enterprises in the Manufacturing Sector' in 1996 from the University of Santa Catarina. She currently works on ESPRIT and INCO-DC projects.

Jiangang Zhang received his B.E. degree in Automation Control from Tsinghua University of P.R. China in 1989 and then worked at National CIM Research Centre of China (CIMS-ERC) for four years. He joined CIMRU in 1993 and got his M.Eng.Sc. degree in Computer Integrated Manufacturing (CIM) in 1995. He is currently a research assistant and is working on his Ph.D. degree at CIMRU. His research interests include CIM Methodology, Extended/Virtual Enterprise, Electronic Commerce, Internet Business, etc..

6
Towards the Virtual Enterprise in Food Industry

*L. M. Camarinha-Matos *, R. Carelli**, J. Pellicer**, M. Martín***

**New University of Lisbon and Uninova*
Faculty of Sciences and Technology
Quinta da Torre - 2825 Monte Caparica, Portugal
E-mail: cam@uninova.pt

****Universidad Nacional de San Juan*
Instituto de Automática
Avd. San Martin - Oeste 1109, 5400 San Juan, Argentina
E-mail: RCARELLI@inaut.unsj.edu.ar

Abstract

The case for the development of the Virtual Enterprise concept in the Agribusiness sector is presented. Current tendencies and the strategy of an European - Latin American project (SCM+) are discussed. A preliminary Petri net model of the interactions in a food supply chain is illustrated as part of the approach to assess current situation and characterize the information infrastructure needs. Finally, a summary of the major functional requirements for a Virtual Enterprise reference architecture is presented.

Keywords

Virtual enterprise, supply chain, Petri nets, Agribusiness, Electronic Commerce.

1 INTRODUCTION

The paradigm of virtual enterprise is becoming more and more important in the manufacturing industries nowadays.The manufacturing process is not any more carried on by a single enterprise, but each enterprise is just a node that adds some value (a step in the manufacturing chain). In virtual enterprises, manufacturers do not produce complete products in isolated facilities. They operate as nodes in a network of suppliers, customers, engineers, and other specialized service functions. Virtual enterprises materialize by selecting skills and assets from different firms and synthesizing them into a single business entity.

This tendency creates new scenarios and technologic challenges, specially to Small and Medium Enterprises (SMEs). Under classical scenarios these SMEs would have big difficulties -- due to their limited human and material resources -- to have access to or to use state of the art technologies but, at the same time, the new electronic-based inter-relationship imposes the use of standards and new quality requirements.

The effort being put on the implantation of high speed networks (digital high ways), supporting multimedia information, open new opportunities for team work in multi-enterprise - multi-site networks. This new scenario brings new requirements in terms of exchanging, sharing, managing and controlling information related to the products being produced. One key area is control; for example, access rights to product information, scheduling of information access, control of interactions. These requirements impact the design, availability and cost of appropriate distributed information logistics infrastructures.

As for the agriculture and food industry, this move towards virtual enterprises is not so developed yet. Nevertheless to the SMEs involved in food processing industry, it is becoming clear that a real competitive advantage can only be achieved and sustained through the creation of relationships and strong information links and co-working among the enterprises involved in the various steps of the value chain. In our view the creation of **virtual enterprises** is inevitable in order to efficiently utilize all of the relevant human, organizational, and business resources and to facilitate the necessary interdependencies between the suppliers (farmers and others), manufacturers, distributors, sellers and ultimately customers.

It shall be noticed that, besides the simplified supply chain (Fig. 1), producing a consumer product in modern food industry involves producing a sequence of intermediate products / services. Each of these intermediate components is used as an ingredient or raw material to make the next product in sequence (value chain).

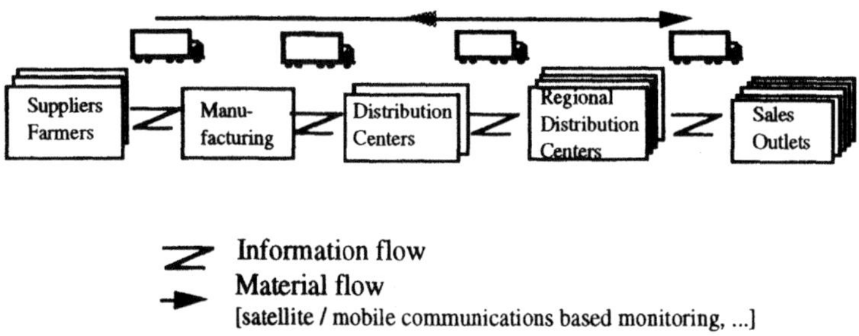

Figure 1 Simplified supply chain.

For instance, in the production of canned tomato paste, the enterprises which might work together to constitute a virtual enterprise include wholesalers and retailers, can producers and their supply chain, food processors and packers, farmers and fruit growers, transportation and delivery enterprises. That is the realization of the **agribusiness concept** defined as the set of all operations involved in the processing and distribution of the agropecuary products, the production operations within the farm and the storing, processing and distribution of all agropecuary products and their derivatives. This increases the need for greater communication and coordination among specialists in value-added partnerships. Emerging Information Technology capabilities such as EDI-Electronic Data Interchange, federated multi-agent architectures, STEP-Standard for Technical data Exchange on Products, etc., support the emergence of the virtual enterprise. Inter-enterprise and intra-enterprise networking across the value chain can be supported by today's advances in computing and telecommunication technologies and their emerging integration into the Information Superhighway.

This paper presents the approach and some preliminary results of the work being developed in the context of the SCM+ (Beyond Supply Chain Management in Food Industry) project (SCM+ 95). This is a cooperative project funded by the INCO-DC program of the European Commission and involves research groups and companies from the European Union and Latin America. Due to the complexity of the addressed problem, in our opinion, the design and development of a Virtual Enterprise concept in food industry should be addressed through a two-phase approach. The first phase is devoted to study and analyze food value added chains in

Latin America and carry out a feasibility study on the potential for an Extended Enterprise approach. The consortium is planning to address the second phase of the project, namely the possible implementation of a pilot Extended Enterprise involving the various actors in the chain, viz. farmers, processors, distributors, etc., as a follow up project to SCM+. Clearly, the results of the feasibility study stage are absolutely necessary input to a possible second phase and indeed follow on project.

The SCM+ project involves several diverse application domains - tomato-paste, olives, coffee, milk, marmalade, etc. - within the food industry, including SMEs involved as associated partners in the project. As a first step, this study concentrates on the strong existing links between the producers (farmers) and the food processing enterprises in Latin America. The project is studying the current situation and analyzing the requirements set by the technological, organizational, social, business process reengineering (BPR), etc. aspects of this industry and the needs to be considered in applying the supply chain management and Virtual Enterprise approach. This stage will be followed by the evaluation of the impacts and benefits that are likely to result for the SMEs in food production through the creation of an integrated network and associated information network within the supply chain. In parallel to the problem analysis and the study and design of a virtual enterprise for this industry, some efforts are being spent on the evaluation of other complementary functionalities that can be supported by an information network, such as on the effect of the management of seed distribution, integrated pest management, and multi-media based training on agricultural techniques and advanced technologies. Due to the limited resources available, the effort devoted to these complementar aspects is, however, minimal.

Today the Latin American economy offers a challenging and fluctuating market. This market is dynamic in terms of the demand and the financial and social states of its consumers, competitive and hard to forecast in terms of the capital and investments in the short term and long term products and rapidly improving in terms of the technology development, the quality and expertise of its technical and supervising manpower, the equipment it employs, and its participation in the global market. Global economic associations such as the MERCOSUL / MERCOSUR are expected to have major impacts in the region.

In addition to the timely, competitive, and cost effective production, an important aspect of the Virtual Enterprise 'model' is the potential it offers to help those involved in the very early stages of the food chain, namely the fruit producers and farmers. Developing and successfuly commercializing differentiated products that appeal to consumers is a surviving condition for some peripheric regions but this requires the organization of increasingly complex partnerships (Holt 95). A Virtual Enterprise approach would involve the food processors working very closely with their supply chain, namely the farmers and fruit producers and supporting them with invaluable advice on the choice of seed and plant, disease control, protection of plants against insects, and the selection, supply and use of appropriate fertilizers etc. Such an approach clearly is necessary to upgrade the productivity (and therefore the economic and social conditions) of these farmers and to ensure regular supply of high quality raw materials to the food processors.

2 GENERAL OVERVIEW

2.1 Current developments

The application / development of the concept of virtual enterprise in the Agribusiness / Food Industry is still very limited. One of the few related works is the CYBERFARM initiative (University of Illinois at Urbana-Champaign), which is still a simple experiment on the use of Internet to support farmers (www.ag.uiuc.edu/aim/demo/ccnet.html).

On other economic sectors, however, a growing interest can be noticed (Walton 96).
The most important initiative is the NIIIP - National Industrial Information Infrastructure Protocol (NIIIP 96), which is a very large American project, involving 18 partners (enterprises, universities and government organisations), led by IBM and with a budget of 60 million US $. NIIIP goals are: to adopt and develop, demonstrate and transfer into wide-spread use the technology to enable industrial virtual enterprises. This project, now on its second year, has produced already a huge amount of documentation, publicly available (www.niiip.org), including a reference architecture, characterisation of various scenarios for virtual enterprises and detailed proposals regarding the various components of the architecture.

At the European level the situation has evolved from an initial "fragmented" approach to more integrated projects. Two of these projects are the recently started Esprit PRODNET II and LogSME. The SCM+ consortium intends to keep contact with these projects. Other related projects are the Esprit X-CITTIC and PLENT.

At national level, various recent initiatives can be found from which major results are expectable in the near future.

Finally, it shall be mentioned that there is a large number of activities in other areas that can provide supporting technologies, like Workflow and EDI.

2.2 Modeling challenges

As the Virtual Enterprise paradigm is new, there aren't yet specific tools, methodologies or an established modeling tradition in this area. A pragmatic approach is to "borrow" methods and tools used in the software engineering and manufacturing engineering areas. But it is our belief that adaptations are necessary. Establishing a platform to a Virtual Enterprise is not creating something from the scratch. Enterprises already exist as well as their cooperation practices and some support tools. Therefore, the process has to start with a Situation Analysis, envolving modeling and evaluation of current systems and processes. Then a Requirements Specification can be made, mainly focused on the desired improvements and having as guideline a Reference Architecture. Some promising tools from other areas are:

Modeling Tools

i. Classical tools.

Generic CASE tools - Classical Systems Analysis methodologies and modern CASE (Computer Aided Software Engineering) tools offer graphical representations for functional and data models.

Manufacturing Engineering tools - A number of tools have been used in the practice of manufacturing systems engineers and researchers. Some of the most popular are:

Functional modeling: SADT / IDEF0, CIM-OSA BP/EA, DFD - Data Flow Diagrams, etc.

Data modeling: EE/R (Extended Entity Relationship model), IDEF1, EXPRESS - the modeling language used in STEP, Object Oriented representations, etc.

Dynamic modeling: Petri Nets, in various "dialects"/extensions, STD (State Transition Diagrams), IDEF2, IDEF3, Rule based systems.

Telecommunications Engineering tools - A number of tools developed in the area of telecommunications and distributed systems to model protocols and the behavior of dynamic distributed systems can be useful for modeling some aspects of the Virtual Enterprise platform.

-SDL - Specification and Description Language . Although it is standardized by ITU (International Telecommunication Union), SDL is not restricted to describing telecommunications services, but it is a general purpose description language for communication systems.

-LOTOS - Language of Temporal Ordering Specification. A formal specification technique for specifying concurrent and distributed systems, consisting of a language for specifiying processes and an algebraic specification language called ACT ONE. G-LOTOS - Graphical representation of LOTOS concepts

-Estelle.

ii. Formal approaches

The use of formal methods with a strong algebraic foundation is calling the attention of complex systems developers. Some of the most popular methods used nowadays are: Z, VDL (Vienna Development Method) and RAISE (Rigorous Approach to Industrial Software Engineering). These formal approaches are starting to appear in the VE area [Janowski 96].

Reference architectures

A Virtual Enterprise Reference Architecture will be a general model describing an abstract Virtual Enterprise in various stages of its development, with each stage described from various points of view (functions, information, resources, behavior, etc.). Once a Reference Architecture is established, the model of a particular VE could be derived or instantiated from it using an appropriate methodology. Again, some approaches and concepts can be borrowed from other areas:

i. Model-based architectures

CIM architectures - Several attempts to establish Reference Architectures for industrial enterprises have been made. These architectures, being focused on a single enterprise, do not fit with the needs of virtual enterprises but can be useful in two ways:

1. Giving a general overview of the global structure, in its multiple views, of a generic industrial enterprise (a node in the virtual enterprise network).
2. Suggesting an approach for the definition of a VE Reference Architecture.

Some of the most well known Reference Architectures are:

- CIM-OSA - Open Systems Architecture for CIM
- PERA - Purdue Enterprise Reference Architecture
- GRAI - GRAI Integrated Methodology
- GERAM - IFIP/IFAC Generic Enterprise Reference Architecture and Methodology

GERAM is not another proposal for an enterprise reference architecture, but it is meant to organise existing enterprise integration knowledge, unifying existing architectures rather than intending to replace them.

Multi Agent Systems

One approach to VE modeling is to consider it as a multiagent system (each node in the network as an intelligent agent). In this way it is worthwhile looking at results coming from the MAS community, namely in terms of:

-Organizational approaches for MAS

-Interaction languages for MAS. For instance the KQML language.

ii. Technology-based architectures

In the specific area of Virtual Enterprises some attempts are already being done in order to establish Reference Architectures. The most significant (in size) is NIIIP (National Industrial Information Infrastructure Protocols). Another example can be found in the Esprit PRODNET project.

To create a support infrastructure for VE, it is therefore necessary to put a considerable effort in modelling and defining a reference architecture.

The general approach chosen in this project starts at the analysis of current procedures followed by food industries and agro-producers - farmers/suppliers, existing production management and control software, and the evaluation of its extension towards the cooperation network philosophy taking into account the reality of SMEs and the context of developing countries. Existing results, namely in terms of standards for EDI (Electronic Data Interchange) and STEP (STandard for Exchange of Product model data), are being evaluated for their suitability for adoption as standards applied to the food industry. Other results from Integration in Manufacturing, (the Extended Enterprise), Business Process Reengineering (Design of Supply

Chains), the management of renewable resources in food production, sustainable improvement in agriculture and agro - industrial production, and crop disease prevention will also be integrated.

To illustrate the applicability of one "classic" modeling formalism to this application domain, next chapter introduces some preliminary results of using Petri nets to capture and assess the dynamic process behavior of an example supply chain.

3 APPROACH TO MODELING THE SUPPLY CHAIN USING PETRI NETS

The interrelationship between the enterprises participating in the value chain, is mainly governed by events which represent the forwarding or reception of relevant information -data, orders, receipt acknowledgements, etc.-, as well as the physical interchange of products. Therefore, the main focus in searching adequate dynamic modelling methodologies and tools should be set on the representation of discrete-event systems. Additionally, continuous system representation could complement the models, when trying to refine the model in order to realistically represent the time lapses involved in some events (Bernhardt, 1994 ; Carelli, 1995). The model should be able to capture the dynamic nature of the interactions as well as an adequate representation of the dynamics of each component of the value chain.

This section presents an approach for dynamically modelling the supply chain in the context of the extended enterprise concept. The model is a simplified version of a real food company described with some detail in (Carelli, 1996a). Different elements and relationships of the chain have been considered, including particular cases as the model for the supplier of raw material, the supplier of supplies and commodities for production, the client as well as the producer-factory relationship considering different situations, the supplier-factory and the factory-client relationships (Carelli, 1996b). In the following, a very simple producer-factory relationship model is presented.

In recent years, many techniques for modelling discrete-event systems have been developed. Among them, the following ones can be mentioned: Markov processes and their imbedded Markov chains, Petri Nets, queuing networks, automata and finite-state machines, finitely recursive processes, min-max algebra models, and discrete event simulation and generalised semi-Markov processes (Cao, 1990), (Ho, 1989).

Petri nets represent a tool for modelling systems with interacting and concurrent components. It represents an analytic and graphical approach useful for modelling, analysing and simulating discrete event dynamic systems. This section makes use of Petri nets for dynamically modelling the value chain applied to the food industry, considering as a first approach two elements of the chain: the producers and the factory.

3.1 Approach to Modelling The Supply Chain: Producer-factory relationship model.

Characteristics: Producers of tomatoes from Northern Argentina (production: 25,000 Tn) offer their production to the factory in San Juan-Argentina. The factory places an order for 100 trailers. San Juan producers (21,000 Tn) offer their production to the factory. This latter approves a shipment of 840 trailers. Mendoza producers (26,500 Tn) offer their production. The factory places an order for 1060 trailers. With all this raw material, the factory plans to elaborate 50,000 Tn tomato sauce. The Petri net model is presented in figure 2.

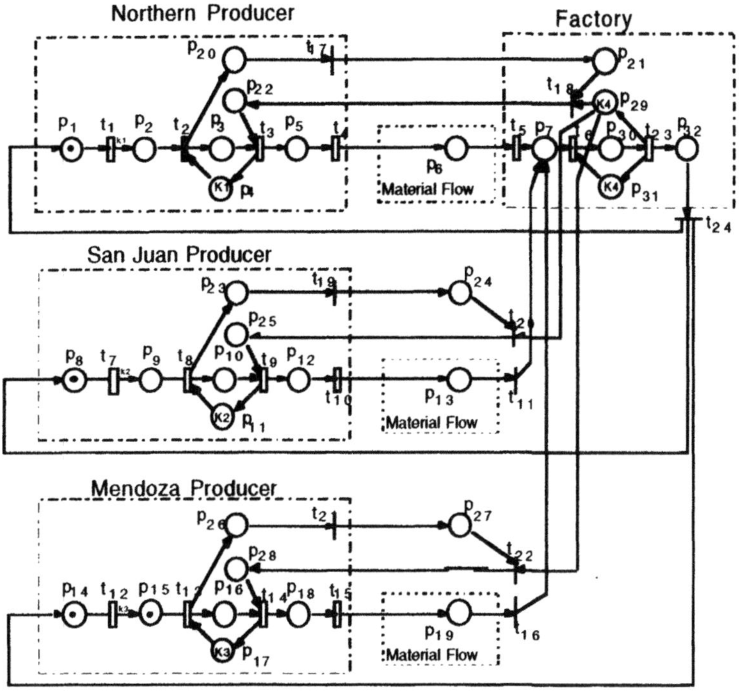

Figure 2 Producer-Factory chain PN model.

PN elements description: Following, the interpretation of places and transitions of the Petri net model is given.

Places

p1 : Initial condition for Northern producers.

p2 : Production enabling.

p3: Tomatoes ready to be cropped (Northern producer).

p4: Total tomato volume to produce (Northern producer).

p5: Tomatoes cropped to be shipped (Northern producer).

p6: A 25 Tn tomato trailer is sent from the Northern producer to the factory.

p7 : Tomato trailers in queue at the factory.

p8 : Initial condition for San Juan producers.

p9 : Production enabling.

p10: Tomatoes ready to be cropped (San Juan).

p11: Total quantity of tomatoes to produce (San Juan producers).

p12: Tomatoes cropped to be shipped (San Juan producers).

p13: A 25 Tn tomato trailer is sent from the San Juan producer to the factory.

p14 : Initial condition for Mendoza producers.

p15 : Production enabling.

p16: Tomatoes ready to be cropped (Mendoza).

p17: Total quantity of tomatoes to produce (Mendoza producer).

p18: Tomatoes cropped (Mendoza producer).

p19: A 25 Tn tomato trailer from the Mendoza producer is sent to the factory.

p20: Offer from the Northern producer.

p21: Offer from the Northern producer is received at the factory.

p22 : The order for one 25 Tn tomato trailer delivered to the Northern producer

p23: Offer from the San Juan producer.

p24: Offer from the San Juan producer is received at the factory.

p25 :A 25Tn tomato trailer order is delivered to the San Juan producer

p26: Offer from the Mendoza producer.

p27: Offer from the Mendoza producer is received at the factory.

p28 : A 25 Tn tomato trailer order is delivered to the Mendoza producer

p29: Monitor of tomato orders.
p30: Tomatoes in stock at the producer's, ready for shipping to the factory.
p31: Quantity of tomatoes to be processed.
p32: The factory produces 50,000 Tn tomatoes.

Transitions :
t1: Enabling for starting production in the North.
t2: A Northern farmer produces 25 Tn of tomatoes.
t3: 25 Tn of tomatoes are cropped.
t4: A 25 Tn tomato trailer is sent to the factory.
t5: Trailer arrives at the factory.
t6: Trailer is unloaded in the factory's warehouse.
t7: Enabling for starting production in San Juan.
t8: A San Juan farmer produces 25 Tn of tomatoes.
t9: 25 Tn of tomatoes are cropped in San Juan.
t10: A 25 Tn tomato trailer is sent by the San Juan producer to the factory.
t11: Trailer arrives at the factory.
t12: Enabling for starting production in Mendoza.
t13: A Mendoza farmer produces 25 Tn tomatoes.
t14: 25 Tn of tomatoes are cropped in Mendoza.
t15: A 25 Tn tomato trailer is sent by the Mendoza producer to the factory.
t16: Trailer arrives at the factory.
t17: A Northern farmer tells the factory that he has available 25 Tn of tomatoes.
t18: The factory approves and places an order for this 25 Tn tomatoes offer.
t19: A San Juan farmer tells the factory that he has available 25 Tn of tomatoes.
t20: The factory approves and places an order for this 25 Tn tomato offer.
t21: A Mendoza farmer tells the factory that he has available 25 Tn tomato.
t22: The factory approves and an order is placed for this 25 Tn tomato offer.
t23: 25 Tn tomato are processed.

PN initial marking :
k1: 100 trailers unit.
k2: 840 trailers unit.
k3: 1060 trailers unit.
k4: 2000 trailers unit.

Conflict Resolution. A typical conflict arises when the factory receives offers from a larger number of producers than needed for the planned industrialised production. Therefore, the factory has to decide which producers are to fill the needs. In the Petri net model, this conflict is evidenced by a conflict in transitions t1 to tn respecting place p1 (see Fig. 3). A supervisory model becomes imperative, and this is inserted at a higher Petri net layer.

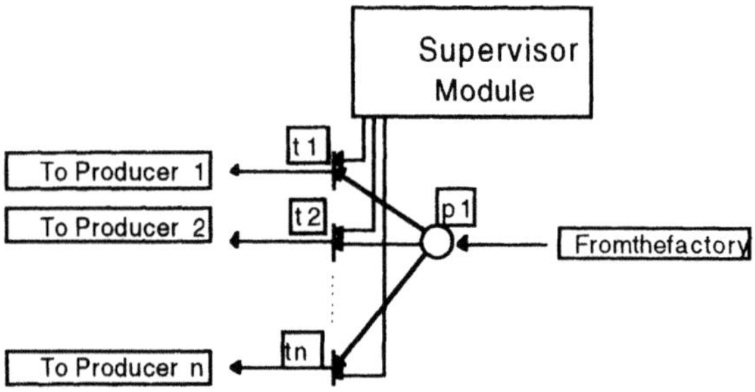

Figure 3 Supervisor Module for solving conflicts.

In the model of extended enterprise concept, certain decision-making mechanisms are necessary in order to modify or fit the model's configuration to any given situation. A proposal for solving this, though still under elaboration, consists in supervisor module for modifying the Petri net models for each producer and their interrelationships with the factory.

3.2 Simulation Example

To show the feasibility of modelling the linking between actors in the value chain using Petri nets, the PASCELL software (Colombo, 1994) is used to simulate the above described model. This software allows the edition, modelling, simulation and analysis of discrete event systems through Petri nets.

Values for a real factory has been taken for the simulation, considering a scaling in time and product volumes, in order to simplify this example. A 50-day cropping season was chosen. The Northern region produces in the first five days the 5% of the factory´s needs. At day 3, San Juan begins production of 42% of the factory's needs. At day 15 Mendoza begins production of 53% of the factory's needs.

A total production of 500Tn is taken with 5Tn trailer capacity, thus making 5 trailers from Northern Argentina, 42 trailers from San Juan and 53 trailers from Mendoza

Fig. 4 shows the Gantt diagram for the first 55 hours of simulation. The bars corresponding to T1, T6 y T10 are referred to the initial conditions for the beginning of production in the three areas considered. The bar corresponding to T2 refers to the Northern production. Lines T15 y T16 represent the communication links between the producers and the factory. It is noted there is one line for each 5Tn trailer the factory places an order for. The bar T3 is referred to the forwarding of trailers from the Northern producers. Bar T5 refers to the unloading of trailer into the factory's warehouse .

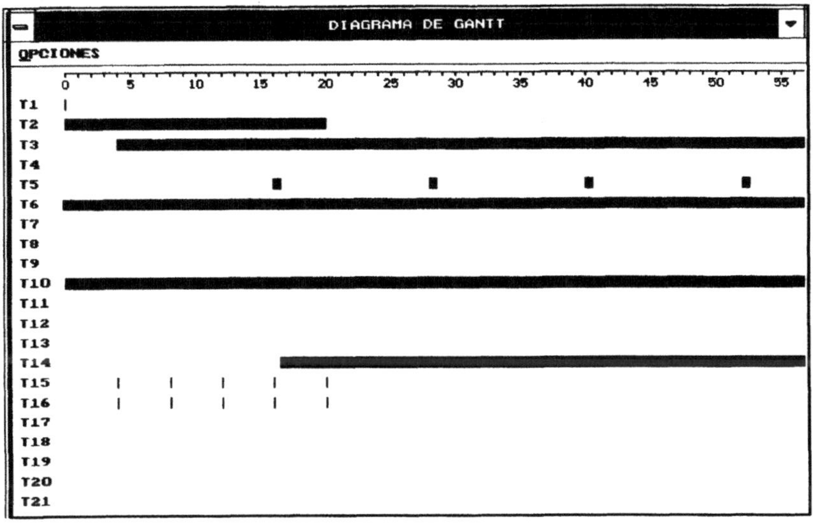

Figure 4 Gantt diagram.

Fig. 5 shows the Gantt diagram continued from Fig 4. At hour 64, the last Northern trailer is sent to the factory. (bar T3). At hour 68, San Juan production begins (bar T7). Bar T8 shows the beginning of San Juan trailer shipments. Bar T14 has no interruptions due to the continuity of the industrial production at the factory's plant.

In this chapter, Petri nets have been proposed as a promising tool for modelling and analysing the value chain producer-factory within the virtual enterprise concept. Petri nets provide useful models which capture the precedence relationships and structural interactions of stochastic, concurrent and asynchronous events. Conflicts can be modelled and deadlock situations in the system can be detected. Petri nets are a mathematical and graphical tool with a well developed mathematical and practical foundation. Timed Petri nets allow for the quantitative analysis of the value chain models. A first approach for modelling the food value chain has been presented and illustrated through a simulation example.

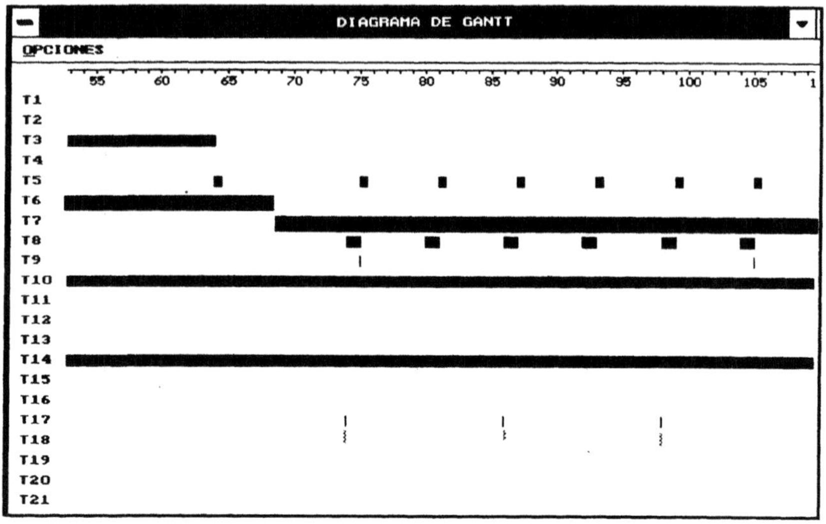

Figure 5 Gantt diagram (continuation).

4 TOWARDS AN ARCHITECTURE

The establishement of a support platform for VE can be analysed from various perspectives:

• **Virtual Enterprise creation**

Considering a VE as a temporary and dynamic (i.e., with variable "geometry") organization that appears in reaction to a business opportunity and is disbanded on the completion of this business process, a number of supporting functionalities can be identified. For instance:

-Selection of potential partners and negotiation of agreements
Various policies can be considered, like an open market approach (public call for tenders), based on privileged lists, etc..
The negotiation process, although strongly interactive as the ultimate decisions are human made, could be supported by the contract-net approach of the Distributed Artificial Intelligence community.

-Provisions for joining / leaving a VE, including the appropriate management of all consequences in terms of fulfilment of agreed duties and information access rights.

-Definition of the network topology and coordination approach.
Different levels of integrated coordination can be envisaged. For instance, a star-like structure with a dominante node, or a more "democratic" value-chain without dominant enterprise.

-Auxiliary functionalities to monitor and parametrize the network behavior. One of the important issues here is the identification of who will perform this function in the case of a "democratic" value-chain.

- **Virtual Enterprise operation**

Once a VE is established, the following points are some important goals to be achieved by such web interconnection from farmers to point of sales and crossing all intermediate nodes - manufacturers and distribution centers.

a. Efficiency of orders flow - with the extensive use of EDI standard, instead of fax, to support the orders follow up and with the possible adoption of STEP for the exchange of product data, as a support standard to the flow of information between nodes.

b. Follow up of orders evolution - fax or other non-electronic media lead to a non-effective monitoring of the orders evolution. Delays in orders processing, temporary incapacity of a supplier, the need to re-adjust delivery times, etc., are some factors that point to the need for a more flexible and reliable interchange of information. A client node might want to know the status of processing of its orders at the supplier's site in order to prevent any difficulties for itself.

c. Management and exchange of orders information among SMEs - a distributed information management system can guarantee the cooperation between the nodes, in the form of :
- an open computational infrastructure;
- sharing and information exchanging between all nodes, but preserving their local autonomy;
- privacy and safety for some parts of nodes' information.

d. Distributed and dynamic scheduling - scheduling is one of the most important activities in any manufacturing/production system. In the context of virtual enterprise in food industry, there is a scheduling function at each node and a scheduling at the network level. Breakdown of production originated by some disease or other unexpected event, changes in priorities, etc., may require the transfer of orders from one node (producer) to another. Thus, there is a need for an infrastructure which supports negotiation between nodes for workload distribution / contract assignment. Therefore, besides a distributed scheduling system, it is also expected a decision support system that helps a node to make decisions based on an updated knowledge of the current status of the network and historic information regarding the performance, quality, and reliability of each node (supplier) (Rabelo 96).

Of course this functionality is calling for a high level of coordinated control (with the consequent reduction of the local autonomy) for which some companies might not be ready yet.

e. Quality information
Sharing and keeping track of quality information allong the value chain.

- **Complementary benefits**

As mentioned before, once a networked infrastructure is established, a number of complimentary services could be supported as well. Examples:

-Distance training
-Dissemination of information on IPM (Integrated Pest Management)
-Advice on seasonal procedures
-Etc.

In this particular area, and due to the isolated conditions of some farms, it is foreseeable that a large number of services can emerge on the network.

Current research activities in SCM+ are addressing these issues in an attempt to define a reference architecture for a VE particularly suited for this business domain.

5 CONCLUSIONS

The virtual enterprise paradigm seems to be a promising, even mandatory, approach to help the agribusiness sector facing current market challenges.

Although there is a great deal of interest for virtual enterprises in many manufacturing areas, the situation is still in its early stage when it comes to the food industry.

The SCM+ project is an international cooperation initiative aiming at characterising the needs, defining a reference architecture and implantation mechanism, and evaluate the foreseeable impacts and organizational changes.

The project is still in its early phase, but preliminary results show that it is possible to adapt various modelling techniques and methodologies developed or used in other areas, like CIM.

6 REFERENCES

Bernhardt R., Schreck G., Colombo W., Carelli R. "Integrated Discrete Event and Motion Oriented Simulation for Flexible Manufacturing Systems". *Studies in Informatics and Control.* Vol. 3, Nos. 2-3, pp. 209-217, September 1994.

Cao X.-R. and Ho Y-Ch. Models of discrete event dynamic systems. *IEEE Control Systems Magazine,* Vol.10, No.4, June 1990.

Carelli R., Colombo W., Bernhardt R., Schreck G. "Discrete Event and Motion - Oriented Simulation for FMS". BASYS '95. *Balanced Automation Systems. Architectures and Design Methods.* Ed. Chapman and Hall. pp. 363-372. Brasil. 1995.

Carelli R., Martín M., Pellicer J., Masiello A., Oviedo W. "Frutos de Cuyo : Characterization of the relationships with suppliers and customers", 1st. Technical Report, SCM+ Project, May 1996.

Carelli R., Pellicer J. "Feasibility study of Petri Nets Modelling of the Supply-Production-Client Chain", Technical Report, SCM+ Project, May 1996.

Colombo W., Pellicer J., Martin M. and Kuchen B. "Simulator for FMS using Timed Petri Nets" (In Spanish : Simulador de Sistemas Flexibles de Manufactura usando Redes de Petri Temporizadas). *Proceedings on X Congreso Brasilero de Automática - VI Congreso Latinoamericano de Control Automático,* Vol. 1, pp. 100-102. Río de Janeiro, Brasil. September, 1994.

Ho Y.C., Ed. Special Issue on Discrete Event Dynamic Systems. Proc. IEEE, Jan. 1989.

Holt,D.A. ; Sonka, S.T. - Virtual Agriculture: Developing and transferring agricultural technology in the 21st century. University of Illinois, www.ag.uiuc.edu/virtag1.html.

Janowski, T.; Acebedo, C.M. - Virtual Enterprise: On refinement towards an ODP architecture, Workshop on Theoretical Problems on Manufacturing Design and Control, Costa da Caparica, Portugal, Jun 1996.

NIIIP - The NIIIP Reference Architecture (Revision 6), Jan 1996, www.niiip.org.

Rabelo, R.; Camarinha-Matos, L.M. - Towards agile scheduling in extended enterprise, in *Balanced Automation Systems II: Implementation challenges for anthropocentric manufacturing,* Chapman & Hall, Jun 1996.

SCM+ - SCM+: Beyond Supply Chain Management - Technical Annex, INCO-DC Programme, Sept 1995.

Walton, J.; Whicker, L. - Virtual Enterprise: Myth & Reality, J. Control, Oct 96.

PART THREE

Virtual Enterprises II

7

Information flows and processes in a SME network

F. Bonfatti, P. D. Monari, R. Montanari
Department of Engineering Sciences - University of Modena
Via Campi 213/B - 41100 Modena (Italy)
tel +39 59 378514, fax +39 59 378515
e-mail bonfatti@unimo.it

Abstract

The paper presents and discusses the main informational activities that take place in a network of small-medium manufacturing enterprises. The network management process is split into elementary steps, grouped according to the entity in charge of them: the network coordinating unit or the single node. The identified steps are those that ensure high reactivity of the network as a whole to customer requests, and the required autonomy and equal rights to the network nodes. The states of the network result strongly related to those of two basic informational elements: the customer order and the node task. The analysis has been carried out to provide requirements for the software tools that should support the network activities.

Keywords

Virtual enterprise, SME, information flow, information management, Esprit project

1 INTRODUCTION

In recent years the concepts of holonic system (Koestler, 1989), virtual enterprise (Rolstadas, 1994), distributed manufacturing environment (Hirsch, 1995) and similar are gaining increasing importance and diffusion. A promising type of virtual enterprise is that obtained by organizing small-medium sized enterprises (SMEs) into a coordinated network (Bonfatti, 1995). The benefits that come to the single SME from joining a network include access to new business opportunites, knowledge exchange, costs and risks sharing, improvement of specialization.

Enterprise network management is the subject studied by the EP project 20723 - PLENT (Planning Small-Medium Enterprise Networks) of the European ESPRIT programme. The project is aimed at designing and developing a set of software tools (PLENT, 1997) to support the activities of a SME network in terms of:

- workload distribution among nodes to satisfy customer orders,

- reactive process planning at the nodes to meet the network manufacturing objectives,
- evaluation of node performances to ascertain their reliability.

According to the PLENT approach, the network organization relies on three basic principles (PLENT, 1996).

- The network is a stable structure. It means that only *a regimen* behaviours are studied, while transient events such as network constitution or modification are not considered.
- The nodes have ensured equal rights. All the nodes are potentially enabled to perform the same functions, although they differ for the tasks they can execute.
- Each node maintains its complete autonomy. Modification of the current organization, including its information system and procedures, is not required.

As a matter of fact, SMEs present positive qualities to behave like network nodes thanks to their lean structure, adaptability to market evolution, active involvement of versatile human resources, habit to establish subcontracting relations, good technological level of their products. But a major obstacle to coordination is the traditional individualistic attitude of SMEs, the same attitude that, in other times, gave them the motivation for starting and growing their business. This attitude leaves each enterprise completely alone in facing marketing, purchasing, design, engineering and technological innovation problems. In addition, other problems arise from historical distrusts between enterprises traditionally in competition with each other.

This situation cannot be overcome by a centralized planning system, acting on behalf of the single nodes. It would result in overlapping the local systems and subtracting node responsibilities, thus introducing unbearable interferences with the single node policy. In addition, this solution would require a centralized detailed knowledge of production plan progress of all the nodes and, consequently, a burdensome information exchange.

For opposite reasons, a system relying on spontaneous self-regulation mechanisms, such as those proposed for large enterprises that decentralize their business, would keep open the distrust problems cited above. This kind of network is based on horizontal and vertical communications between nodes that do not ensure a global system vision and do not remove the current interdependencies between enterprises (where the largest or most powerful enterprise, or that supplying a critical component, determines the supply policy).

It is the aim of the PLENT project to propose a SME network organization that ensures fast response to costumer request while keeping the required independence degree of the node enterprises in due consideration. This approach is based on the neutral application of a number of clear, objective rules, agreed upon by all the network nodes, concerning task distribution and node reliability evaluation. In other terms, the key aspects are two: coordinated planning and node behaviour measurement. The choice of the most suitable decision-making and optimization algorithms comes later and is, in some sense, less important.

The management of this type of network considers the roles of three fundamental entities, customer, coordinating unit and nodes, and the interactions between them (Bonfatti, 1996).

Customer (C)

Customers represent the external entities which activate the network processes. All the communications between the customer and the network are managed by the coordinating unit, hence the customer does not see nodes, with the only exception of that delivering the final product.

Coordinating unit (CU)

The CU main function is to collect external orders and decide how to involve the network nodes in the manufacturing process. To this aim, it must keep an updated view of the current nodes state and reliability, and it is directly informed by nodes whenever problems arise. Finally, the CU has to manage all the information which is necessary to evaluate network nodes performances.

Nodes (N_i)

Their duty is to carry out the tasks which are assigned to them by the CU, so as to deliver to the next node (or to the customer, if the node is the last one) the right quantity of product within the due date. Moreover, every node has to inform the CU about problems causing delays or wastes of material. Short and medium term planning of manufacturing activities are basic steps of node behaviour.

The modalities of interaction of these entities, the information they exchange, the processes they execute in normal and emergency conditions, constitute basic requirements for software design. Hence, the first phase of the PLENT project has been devoted to a detailed study of these aspects. The paper reports the results of this study. Section 2 examines the states assumed by the two main informational elements: customer order and node task. Then, Sections 3 and 4 present, by means of data flow diagrams, the processes that take place within the single node and the coordinating unit, respectively, as well as the information they exchange.

2 ORDERS AND TASKS

The network external demand is represented by customer orders addressed to the CU, requiring a specific type product. Each order is defined by a triplet $<p, q, dd>$ (product type, ordered quantity, due date). The states that the order can take during its processing are reported in the state diagram of Figure 1.

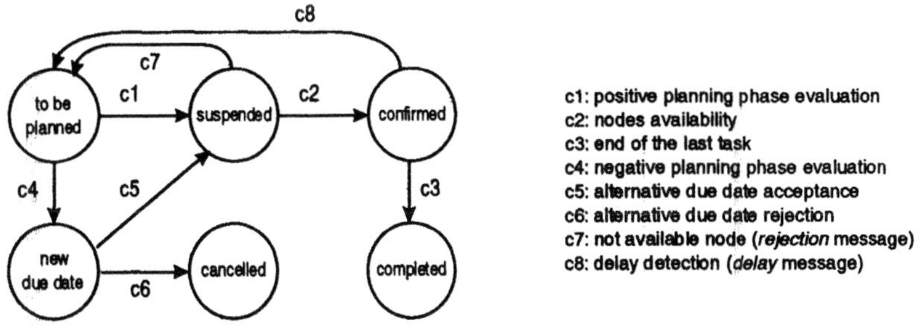

Figure 1 State diagram of the customer order.

Each time a customer sends a request to the CU, a new order instance is created whose initial state is *to be planned.* Then, the assignment process decomposing the order into *n* tasks is executed, and the order assumes the *suspended* state. In fact, only when all the nodes involved in tasks distribution confirm their availability, the order state becomes *confirmed.* Finally, after all the manufacturing steps have been accomplished, the order changes its state into *completed.*

If the CU is not able to determine an executable task distribution, the order passes from the *to be planned* state to the *new due date* state. Starting from this state, and according to the customer decision, the order can assume the *to be planned* state value again or be *cancelled.*

Transitions c7 and c8 occur whenever a new planning phase is required. This case occurs when a node rejects the assigned task or introduces delays. Replanning activities can determine new task distributions (c1) or the suggestion of alternative due dates (c4).

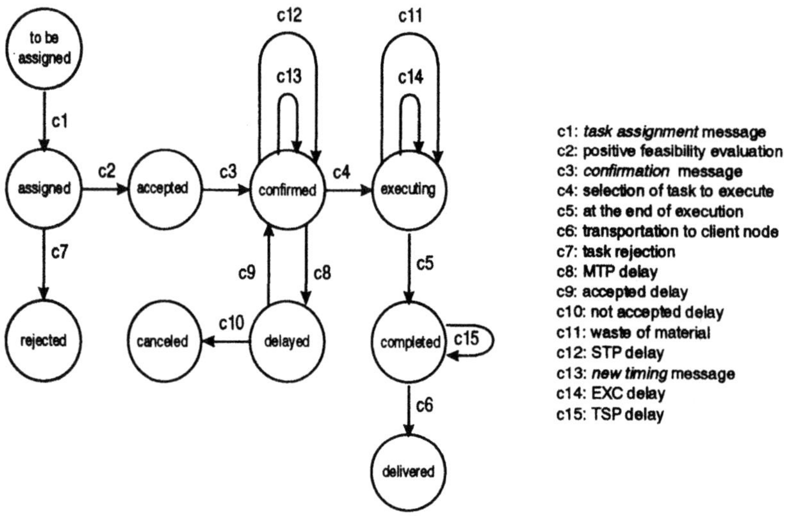

Figure 2 State diagram of the single task.

In its turn, every tasks assigned to a node is represented by the triplet <*p, q, dd*> (phase type, quantity, due date). Its evolution occurs according to the rules represented in Figure 2. At the end of the assignment process, the task (whose value is *assigned*) is communicated to the involved node which verifies (manually or automatically) its feasibility. As soon as this, and all the other tasks resulting from a customer order, is *accepted* by the relative node, the order changes its state from *suspended* to *confirmed,* and the tasks itself is *confirmed* by the coordinating unit back to node of interest.

Starting from this situation, the node will inform the CU on task evolution whose correct sequence of state values includes *executing, completed* and *delivered.* Instead, if the tentatively assigned task is rejected, its state changes from *assigned* to *rejected* (and, therefore, the order to which it belongs is replanned).

A confirmed task is planned for execution by the involved node. If a delay is detected (c8) the task state becomes *delayed* until a decision is taken by the coordinating unit (either the delay is accepted or the task assignment is cancelled). Another task replanning may occur if the delay is detected at the scheduling time. The consequence of replanning is that sometimes the CU must communicate to nodes new timings of already confirmed tasks.

3 NODE PROCESSES

With the PLENT approach nodes are expected to maintain a high autonomy of local planning, on the medium and short term horizons, and of task execution. They are required to interact with the CU in order to make this aware of any significant state change of the assigned tasks and, more important, of any possible delay or material waste that can affect the order progress. CU is in charge of managing all synchronization problems the single node cannot face autonomously. This satisfies a specific requirement coming from real cases, where the burden of frequent and informal communications between nodes is considered one of the major drawbacks of distributed manufacturing with respect to centralized organization.

Network behaviour is described by a list of conditions which activate certain processes, those which take place inside the nodes or the coordinating unit. The triggering condition associated with a process is constituted by:

- a message coming from an external entity (e. g. a communication about a task assignment between the coordinating unit and a certain node or an order deletion directly decided by a customer),
- an event that occurs inside the examined entity (e. g. a delay during manufacturing inside a node or the expiring of the scheduled time between two medium term plannings).

As result, each process can involve in its turn external entities sending them new messages. The main functions performed by a node are represented in the data flow diagram of Figure 3.

ACC: acceptance/rejection of an assigned task

This process is executed whenever a task assignment message comes from the CU. Since CU carefully considers the node declared availability and previous assignments, task acceptance is the expected answer (the task is recorded as accepted but not yet confirmed). Task rejection will make node position in the network worse (the task is recorded in the task log file).

The decision about acceptance/rejection of the task, on behalf of the node, can be automatically reached (e. g. by means of a medium-term planning simulation) or according to heuristic criteria: what really results critical is the node rapidity to send the answer to the CU. This, after having calculated tasks distribution among n nodes to carry out a certain order, waits for the confirmation of all the nodes involved. Only after the n acceptance messages have been received by the CU, this sends a confirmation signal.

Usually, node answer will be positive, hence the ACC will result in updating the data structure storing the related tasks. At the end of the process, the node waits for the CU confirmation signal.

CFM: accepted task confirmation

The triggering condition of the CFM process is the confirmation/rejection signal simultaneously sent to all the nodes involved in the tasks distribution. If all the assigned tasks have been accepted, then the CU sends to them a confirmation signal. Instead, if a node has rejected its duty, it is necessary that the CU performs a new planning.

If we suppose that *m* of the *n* tasks have been rejected (a quite improbable event) then the CU can manage this situation in two ways: (i) by assigning the *m* tasks to other nodes, that should confirm in their turn without modifying the distribution involving the *n* - *m* tasks already assigned; (ii) by calculating a new task distribution involving also some of the *n* - *m* tasks already assigned; (iii) by failing in searching new assignments which are able to satisfy customer requests, in which case it is necessary to negotiate a new due date with customer (moreover, all the tasks are deleted and afterwards are assigned again according to a new distribution). In the last two cases, the CU can deny a task assignment even if it was already accepted by a node. Therefore, the examined task is removed from the set of those accepted.

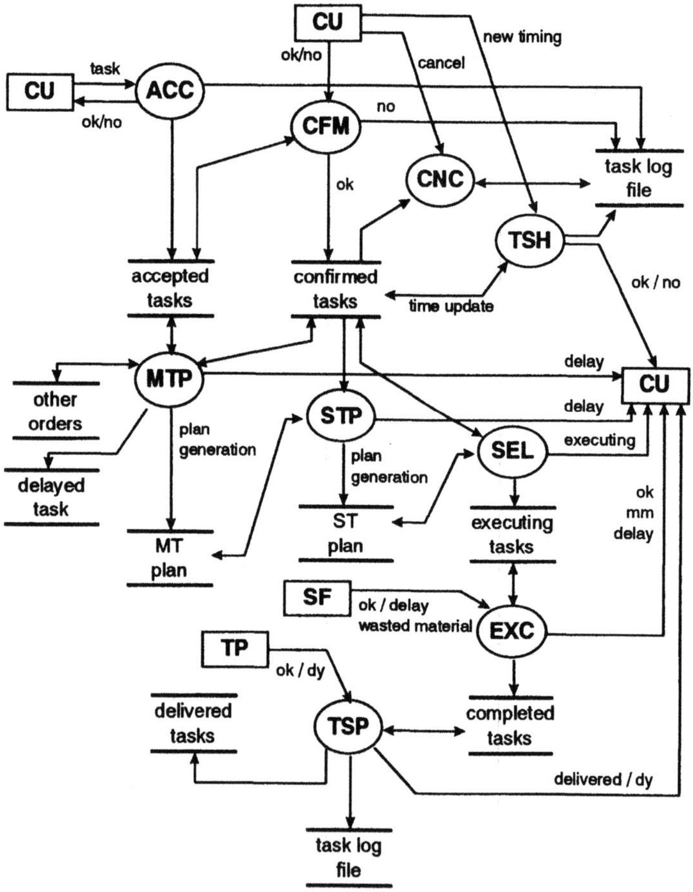

Figure 3 Diagram of the node processes.

MTP: medium term planning

Unlike the two activities seen above, whose activation depends on signals coming from the external enviroment, the MTP process is cyclically performed (e. g. two times a week) in order to provide forecasts regarding tasks execution and information about purchase orders.

The process is applied to all the orders received by a node, from the CU or from other customers, directly. This process takes into account either the confirmed tasks or the accepted ones which have not yet been confirmed. The process results in a medium term planning where each order or task refers to the temporal axis which determines the beginning of the working. The accuracy of the start and end times strongly depends on the characteristics of the MTP package, which can differ according to the node examined.

At the end of the planning process some tasks can change their state according to their finishing dates as they are determined by MTP. In particular, if the finishing date preceedes the due date, no state change occurs, otherwise some perturbations involving network behaviour can raise. Whenever a node is not able to carry out a certain task within the period fixed by the coordinating unit, the involved task is marked as delayed and a message is sent to the CU. Then, the CU will try to distribute these phases among the nodes following the examined one, by means of partial replanning.

STP: short term planning

Similarly to the MTP process, also this process is executed cyclically (usually, once a day) in order to schedule the production orders that refer to the next hours. The STP process only takes into account tasks whose state assumes the confirmed value and whose start date (that fixed by the MTP process) falls within the temporal horizon considered by short-term planning.

The result of the STP process is a plan where the examined tasks are placed on the temporal axis showing their execution sequence. Besides, the level of detail reached during STP is higher than the one corresponding to MTP since the start and end dates considered by the latter are only foreseen but not definitively fixed.

The STP process allows the node to know whether its tasks can be carried out within the date fixed by the CU. Shall this condition not occur, the corresponding signal has to be sent by the examined node to the coordinating unit. The CU, in its turn, has to warn the nodes following the examined one in the network, against the perurbations determined by the delay that affects the materials delivery.

SEL: selection of the task to be executed

It is the aim of this process to select the tasks to be executed: hence, their state passes from the confirmed value to the executing one. The coordinating unit is informed on tasks state change.

EXC: control of tasks to execute

Whenever a task assigned to a node is completely accomplished, without determinimg perturbations, a positive acknowledgment is sent from the shop-floor. This message triggers another one, sent by the node to the coordinating unit, showing that the new value assumed by the task state is the executed one.

Nevertheless it could happen that a task is not carried out within the time that has been previously fixed due to materials acquisition or working problems. In this case, the node has to inform the CU on the problems occurred (by means of messages like delay or missing materials). The CU will try to overcome the drawbacks by assigning some tasks to other nodes or negotiating new requests with the customer.

TSP: transportation to the next node (or to the customer)

Every time a task is completely executed, its product is delivered to the entity responsible for the transportation to the next node (or to the customer). After this activity execution, the task state becomes *delivered.*

CNC: task cancellation

Whenever the coordinating unit deletes a task that have already been assigned (and accepted) to a node, because of some problems such as those seen above, the CNC process of the examined node has, in its turn, to delete the task from the accepted ones.

TSH: task timing shifted

Every time a node informs the CU on a possible delay, the coordinating unit tries to calculate a new task distribution involving the tasks assigned to the node considered and causing the delay. Hence, it is necessary a partial re-planning that takes into account only the nodes that have not yet executed their tasks so as to be able to move up the date that refers to the production order release. Therefore, the TSH process takes place in these nodes. Other nodes that are not able to update the production order release are only informed by the CU on possible perturbations.

4 COORDINATING UNIT PROCESSES

The main functions carried out by the coordinating unit arerepresented in the data flow diagram of Figure 4. Processes ASS, RRP, MDLY, SDLY are all related to workload distribution to the nodes, based on the following principles:

- One order at a time. Splitting customer orders into node tasks is performed order by order. Since the external request should be answered in the shortest time, the CU cannot cumulate it with other orders for optimization purposes, but must process it immediately.
- Task assignment at the latest. Assignment of order tasks to nodes is carried out according to a backward approach, taking into account the tasks already distributed to nodes. Thus, the network nodes receive the communication of their involvement with the widest time margins and can plan the respective activities in the best conditions.
- Balanced node involvement. The network nodes have equal rights, hence the assigned workloads are proportional to the respective available capacities, considering also the already assigned workloads and the node reliability. The daily available capacity of a node with respect to a given phase is that declared by the node itself (possibly corrected, as we shall see), decreased to take into account previous assignments.
- Node internal concurrency. The resources of a node can be used to perform operational phases in alternative; it means that assignment of a workload for a given phase subtracts resources to the execution of the other phases. In order to take into account these situations, whenever a task is charged onto a node the capacity of the relative phase is

decreased, together with a proportional decrement of the capacities of all the concurrent phases.

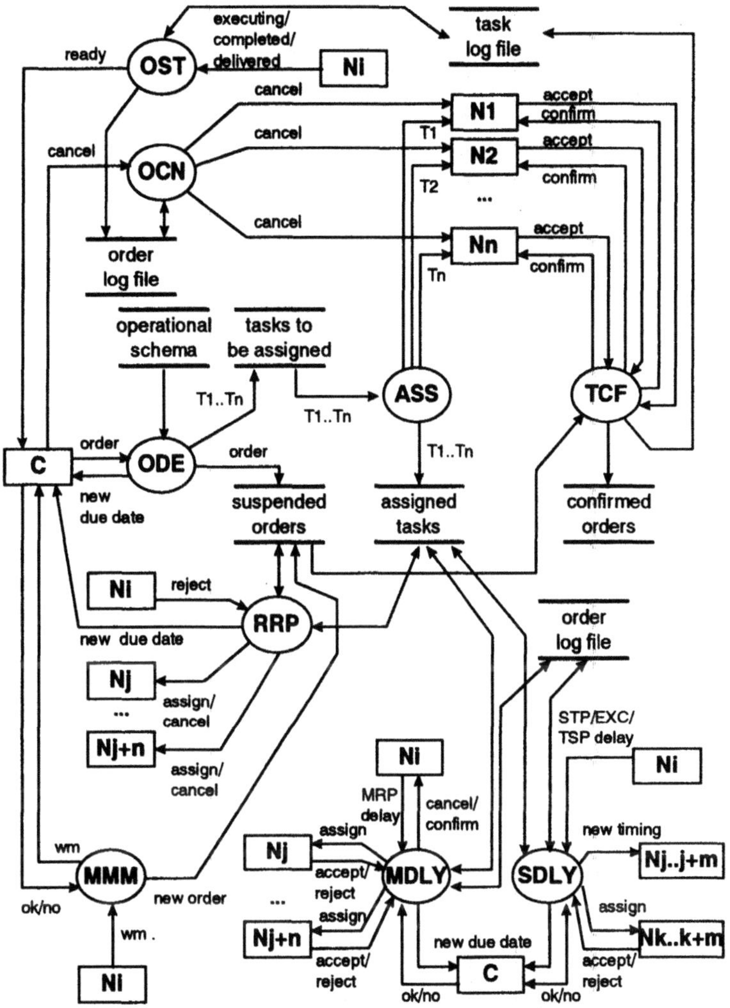

Figure 4 Diagram of the coordinating unit processes.

Some particular events could require new partial or global planning in charge of the coordinating unit. In particular, new replanning is necessary every time that the network receives a new order, or a task is rejected by a node, or the MTP process of a node points out a delay, or the EXC process of a node points out wasted materials.

Instead, if the delay is due to STP, EXC or TSP, the coordinating unit only informs the corresponding nodes on the time shifting that affects the material due date.

ODE: customer order decomposition into tasks

The ODE process is executed whenever the coordinating unit receives an order from a customer. It is the aim of this process to apply a proper algorithm that calculates a tasks distribution according to the information related to the network operational schema. The new order assumes the *suspended* state until the corresponding tasks are accepted by the nodes.

ASS: tasks assignment to network nodes

This process starts after the proper workload distribution has been calculated by the above algorithm. During this phase the coordinating unit informs all the nodes, involved by the workload distribution, on the tasks assigned to them. Obviously, more tasks can be assigned to the same node.

TCF: tasks confirmation to the nodes and the customer

After task distribution, the coordinating unit waits for the answers from the nodes involved in the distribution itself. It is clear that the answers have to reach the CU within a certain time, usually fixed by the network management. If all the nodes send to the CU a message whose meaning is task accepted, the CU, in its turn, can start the confirmation process that definitely assigns the tasks to the nodes, as in a two phase commit communication protocol.

The order corresponding to the examined tasks changes its state from the suspended to the confirmed values. Afterwards, a confirmation signal is sent to the customer responsible for the order.

RRP: replanning due to task rejection

If one (or more) of the above nodes rejects the tasks assigned by the CU, it is necessary a new assignment process in order to replace this node (these nodes). The new planning phase could determine a new tasks assignment involving also the tasks already assigned, or another negotiation with the customer.

Obviously, at the end of the re-planning phase it is necessary to perform the TCF process again. Moreover, it could be impossible to find out a new distribution tasks after rejection: in this case, the coordinating unit has to negotiate with the customer another due date.

OST: order and tasks state monitoring

This process allows the coordinating unit to update the task and the order state using the information associated to the messages coming from the nodes. As soon as the enterprise in charge of the operational schema root communicates that its duty has been accomplished, the CU informs the customer on product delivery.

OCN: order cancellation

Whenever some problems are detected, the coordinating unit has to inform the customer on the need of fixing a new due date.

If the customer decides to delete the order, the OCN process provides the coordinating unit to inform the nodes involved in the tasks distribution on the deletion of their tasks; moreover, the order and these tasks assume the cancelled state.

MDLY: replanning due to MTP delay

This process is executed by the coordinating unit every time problems occur during the medium-term planning performed by a single node. In this case the node communicates to the coordinating unit the delay its MTP has foreseen and the CU starts a new planning phase that is managed by the MDLY process.

During this replanning phase, redistribution of the tasks that follow the delayed ones, can involve all the nodes following the one that has detected the problem. The result of the process is a new distribution involving all the tasks that follow the one that has determined the delay. Moreover it is possible that the CU assigns this task to another node. After having obtained the new distribution, all the nodes considered by the new distribution have to send to the CU, within a fixed time interval, a confirmation message: this message is received by means of the TCF process.

The tasks that have been previously assigned to the nodes following the one responsible for the delay cannot be removed, since this delay do not depend on these nodes. Therefore the CU can only require an effort to them in order to anticipate the due date that refers to their products: this kind of effort will be payed by assigning a positive score to the examined nodes.

If the coordinating unit is not able to find out a new proper distribution it is necessary a new negotiation phase with the customer.

SDLY: management of delays due to STP, EXC, TSP

This process (similar to that above) is performed by the coordinating unit as reaction to the messages coming from the nodes that have detected problems during short-term planning, task execution or transfer phase.

Since the delay can be detected when the task is almost accomplished, the CU cannot remove the phase from the node duty or move up all the tasks that have to be executed by the following nodes. Instead, the coordinating unit has to identify the two following sets of nodes: the set of node whose tasks are still managed by the medium-term planning; the set of nodes whose tasks are already taken into account by the short-term planning or are running.

As result of a re-planning phase all the nodes belonging to the first set will be required to anticipate their task ends (this is performed by means of assign, accept, confirm messages) so as to balance the delay due to the nodes of the second set. As to the nodes belonging to the second set, they are informed (by means of a new timing message) on their tasks moving on the time axis.

If some nodes cannot accept the results of the re-planning phase it is necessary (just as in the above process) to inform the customer on the delay.

MMM: material wastes management

This process is executed whenever a material waste message is sent to the CU by a node. It is the aim of this process to decrease the quantity of the ordered product provided that the customer is satisfied. Otherwise, it is necessary to create a new order that is able to balance the wasted material.

5 CONCLUSIONS

The control architecture adopted for the activities carried out in a network of small-medium manufacturing enterprises has been presented in terms of information flows and processes involving the network basic entities, namely customer, coordinating unit and nodes. This solution has been identified to meet two fundamental aims which take into account the peculiar nature of SMEs:

- Autonomy. Every network node is an enterprise used to operate alone on a market populated of many competitors, often located in the same industrial district. This results in a strong spirit of independence that might interfere with the participation in the network. Only an organizational paradigm which leaves the single node free to decide its role in the network, ensuring at the same time the correct application of the agreed management rules by a neutral coordinating unit, can overcome distrust and possible difficulties.
- Sustainability. The new organizational paradigm should not require significant modifications of the node organization. The node must remain free to participate in different networks and also maintain its direct customers. The benefits ensured by the virtual enterprise (wider market, incentive to improve productivity and quality) shall not be payed with a loss of flexibility and initiative. As soon as further organizational models will be identified, the single enterprise must still be in condition to try them.

A validation activity of the proposed model has been carried out on three european networks (in Italy, Spain and Hungary) in order to test the model applicability, verify the planning paradigm principles and receive indication from end users about model enhancement and correction. The study cases range from small batch production of oleodynamic components to one-of-a-kind production of transfer machine tools for the automative industry.

At present, the PLENT project is completing the software design phase, and the development phase is going to start.

6 REFERENCES

Bonfatti F., Monari P.D. (1995) Planning small-medium enterprise networks. International Conference on Improving Manufacturing Performance in a Distributed Enterprise (IMPDE '95), Edinburgh.

Bonfatti F., Monari P.D., Paganelli P. (1996) Coordination function in SME networks. International Conference on Architectures and Design Methods for Balanced Automation Systems (BASYS '96), Lisbon.

Koestler A.(1989) *The Ghost in The Machine*. Arkana Books.

Rolstadas A. (1994) Beyond year 2000 - production management in the virtual company. IFIP Transactions B-19 on *Production Management Methods*, North Holland.

Hirsch B. E. et al. (1995) Decentralized and collaborative production management in distributed manufacturing environments, in *Sharing CIME Solutions* (eds. J. H. K. Knudsen et al.), Ios Press.

PLENT Consortium (1996) *Network Policy Modelling*. Deliverable D01.

PLENT Consortium (1997) *Software Design*. Deliverable D02.

8

Developing IT-enabled virtual enterprises in construction

A.Grilo
PhD Candidate, University of Salford
Bridgewater Building, Salford, M5 4WT, UK, e-mail:
a.grilo@mail.telepac.pt

L.T. Almeida
Professor of Instituto Superior Técnico
Av. Rovisco Pais, 1000 Lisboa, Portugal

M.P. Betts
Professor of University of Salford
Bridgewater Building, Salford, M5 4WT, UK

Abstract

The achievement of IT-enabled virtual enterprises in construction is a fundamental step in order to raise productivity within construction projects, and therefore reduce costs, reduce lead times, and raise quality. Based on the patterns of successful IT-enabled virtual enterprises on manufacturing and service industries, this paper analyses what constraints and facilitates the initiation and development of electronic links within construction virtual enterprises.

Keywords

Construction, virtual enterprises, electronic links, networks

1 INTRODUCTION

The emerging of IT-enabled virtual enterprises in manufacturing and service industries has been a response to the global and increasing competition, and higher complexity of products/services. The complex management of these networks of firms is supported by Information Technology, which facilitate and stimulate inter-organisational communications. Competitive advantage of firms is increasingly determined by the effective management of the interdependencies within the virtual enterprise, and therefore the strategic role of IT is shifting from internal operations to inter-organisational functions.

Construction virtual enterprises existed well before emerging in manufacturing/service industries. Regardless of the ineffective management of the interdependencies and of poor information flows, electronic links are absent of construction virtual enterprises. Main contractors do not seem to embrace the potential of IT in order to exchange information with designers, manufacturers, material suppliers, subcontractors, etc.

This paper aims at contributing to a better understanding of the lack of electronic links within construction virtual enterprise. A model is presented which allows a comparison between successful IT-enabled virtual enterprises in manufacturing/service industries and construction cases.

2 IT-ENABLED VIRTUAL ENTERPRISES

One of the most relevant and smooth innovations in manufacturing firms has been the emergence of the concept of virtual enterprises. Current goods manufacturing see over 60% of the value of the product added not by the manufacturer who sells the final product, but by other manufacturers, usually designated as suppliers. The outsourcing of value adding activities to specialist firms, which manufacture at better quality and lower costs components and/or sub-assemblies, has shift responsibility from manufacturing departments and production planning departments to the purchasing departments and to suppliers. Purchasing has become supply management. Examples can be found in automotive industry (Lamming, 1993), in clothing, book publishing, electronics, etc. (Miles and Snow, 1992).

There is no unique form or definition for virtual enterprises, but in a general way it is considered as a network of organisations that exists for the development and manufacture of one or few products/services. These are usually characterised by having intermediate forms between markets (competitive arm's-length supply) and vertical integration. Virtual enterprises may be organised around a core firm, usually a large firm, which co-ordinate the network of firms, usually SMEs, or may be self-organising between firms, with no enduring leader (Harrison, 1994).

Hence, in order to obtain sustainable competitive advantage, companies have to look beyond internal boundaries, as competitive advantage lie partly within a given organisation and partly in the larger network of firms around it (Venkatraman, 1991). Their greatest challenge is to know how to manage effectively the interdependencies and relationships of the network, and efficiently the internal business processes. Firms must regard logistics processes (physical and information flows) as important as production processes. A major facilitator to the co-ordinating function is Information Technology (IT), by linking electronically through EDI, interactive on-line databases, Internet, video-conferencing, E-mail, etc. the organisations belonging to the virtual enterprise. IT can facilitate and stimulate inter-organisation communication, by providing more adequate, timely, and accurate information. The benefits from the use of IT for the coordination of the virtual enterprise vary accordingly to the technology but in a generic way, they can reduce costs, shorten lead times, and raises quality (Venkatraman, 1991).

3 THE CONSTRUCTION PROJECT VIRTUAL ENTERPRISE

Although new to the manufacturing firms, the virtual enterprise concept has long been used in the construction industry. Indeed, temporary networks of "co-operating" organisations have existed for a long time in the construction industry, usually denominated by temporary multiple organisations (Cherns and Bryant, 1983). In a traditional and simplistic way, construction projects are split into the design and construction phases. The architect is the project leader, developing a design and design management responsibility and devolving a role for specialist design and management functions to engineers, surveyors and the like. On the construction phase a main contractor is responsible for the completion of the entire project and procures the services of specialised firms. These specialised firms are subcontractors who carry out work on-site, and suppliers who manufacture products off-site and deliver them as needed. In a generic way, for each project, a virtual enterprise is assembled in order to build the product.

Construction managers and researchers stress that the construction organisation is the main responsible factor for the problems of low productivity and innovation of the industry (Hillebrant and Cannon, 1990; Atkin and Pothecary, 1994). Current procurement and organisational developments in the sector are increasingly focusing on more effective management of the virtual enterprise through developments such as design and build, multi-disciplinary design and project management, etc. (Latham, 1994). But although some progress has been achieved, there has not been significant improvements.

Looking from the example of manufacturing and service virtual enterprises, we must conclude that the problem is not grounded on the construction organisational structure but on the way construction virtual enterprises are managed, which is supported by inefficient information flows. Indeed, a good project management must pay as much attention to the production as to the overall information flow. Increases of 15% to 20% in the performance of projects (cost, quality, speed, customer satisfaction) may be accomplish if the information flow within the network of relationships in the virtual enterprise are carefully addressed and improved (Latham, 1994). The geographical distance between headquarters of the companies and the construction sites, which have a limited existence in time and space, makes it difficult to have on-line information on planning of activities and resources and their control. Moreover, little attention has been given to the establishment of effective communication and logistics processes involving designers, contractors, subcontractors and material suppliers.

A contrasting difference between manufacturing/service virtual enterprises and construction virtual enterprises is that the former are enabled and supported by IT. Indeed, a benchmarking study to the largest contractors in the UK revealed that contractors did not have any type of electronic links with their subcontractors, material suppliers, manufacturers, designers, etc. (Betts et al, 1995). Another study has shown that although there are electronic links between building merchants and their suppliers (manufacturers and material suppliers) there was no electronic links whatsoever with contractors (Baldwin, 1995). In the UK, despite some recent national initiatives to develop electronic trading between contractors and their suppliers, the results have been widely disappointing. Again, common explanations advocate that the industrial structure and organisation of construction does not encourage electronic links: its industry fragmentation; fragmented clients; unique nature of each project; and lack of IT standards are often blamed (see e.g. Construct IT - Bridging the Gap, HMSO, 1995; or Baldwin et al, 1995).

4 CONSTRAINTS AND FACILITATORS OF IT-ENABLED VIRTUAL ENTERPRISES

In our perspective, more coherent and robust theoretical explanation is required in order to better understand the lack of electronic links on construction virtual enterprises, with focus on the large main contractors-suppliers interactions, but also to pave the way forward. Thus, it is important to analyse *what constraints and facilitates the initiation and development of IT-enabled virtual enterprises, and how electronic links influence the organisation of virtual enterprises*. The approach used is based on the development of a generic model and the construction case is made by comparison with successful IT-enabled virtual enterprises.

The complexity of the inter-relationship and interdependencies of firms belonging to a virtual enterprise and the definition of its boundaries is a complex research issue. The network of firms which support a virtual enterprise may be compound by hundreds of firms, and even thousands if we consider the next level of suppliers (the suppliers of main suppliers). This complexity means that the boundaries of research will need to be defined in such a way as to allow to tackle the complexity. In this work we tackled the problem by defining a series of dyadic interactions (one-to-one relations) between the firms, essentially between a hub firm (e.g. a buying/core firm like a retailer, automotive assembler, main contractor, etc.) and the suppliers. These dyadic interactions are embedded within a web of relationships which may shape their patterns. Finally, this web of relationships is embedded on the environmental context (e.g. industry culture, technological trends, etc.). The model presented here relates only to the dyadic level, and is based on literature research. Empirical validation is currently being held.

The model defines that there are four main conditioning factors that constraint and enable the initiation and development of electronic links on dyadic interactions on virtual enterprises, as depicted in Figure 1.

Governance structure

This factor addresses the issues related with the co-ordination mechanism of the dyadic interaction. It is related with the power dependence of firms, i.e. if one firm is dominant over the other or if the power is balanced. It is also about the relationships. These can be seen as lying in a continuum. At one extreme, relationships are competitive, free-market, mostly distant or adversarial, with low levels of trust, and usually short-term. On the other extreme, relationships are co-operative, collaborative, there is mutual trust, and a commitment to the long-term interaction.

Information exchange

Refers to the information that is transferred between the hub firm and supplier and vice-versa. It is related with the type of information that is shared, which may be operational information such as enquiries, delivery times, project planning, cash management, etc.; tactical information such as joint process improvement, joint targeting of objectives, sales forecasts, etc.; and strategic information e.g. new businesses new products, strategic partnering, etc.. Another important issue is the information quality i.e. related with the information adequacy, timeliness, accuracy, and transparency.

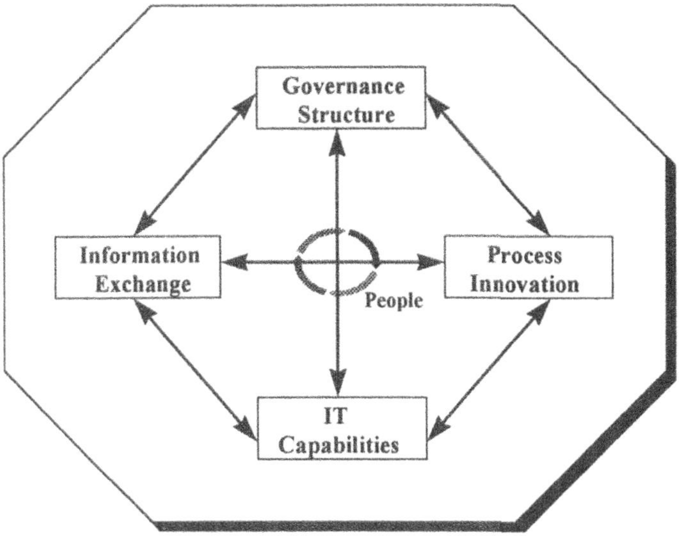

Figure 1 Conditioning factors for the development of an IT-enabled dyadic interaction

Process innovation
It refers to the changes which occur on internal business processes and related with the deployment of electronic links. The scope of the changes in the processes is an important issue and these can vary from incremental changes, simple automation or elimination of tasks, to radical re-engineering of logistics and purchasing functions, etc. Also, the benefits that firms obtain from process changes is an important issue, and these can vary from operational benefits like cost reduction, reduced life cycle, to strategic benefits such as more flexibility and responsiveness, more and better trade, etc.

Technology
It refers to the technology utilised for the electronic exchange of information/data, i.e. EDI, E-mail, interactive on-line databases, groupware, Internet, etc. It also addresses the internal IT capabilities of firms.

In Table 1 a comparison is made between the patterns of the conditioning factors of successful IT-enabled dyadic interactions and of construction dyadic interactions.

Table 1 Patterns of the conditioning factors

Successful IT-enabled dyadic interactions	*Construction dyadic interactions*
Governance	**Structure**
• Hub firm has buying power over supplier • More co-operative, collaborative and mutually trustful relationships. Commitment to long-term relations based on formal agreements.	• Main contractor has little buying power over suppliers. • There is little co-operation and collaboration and relations are usually adversarial. Firms are usually committed only to the project duration.
Information	**Exchange**
• Hub and suppliers tend to share more than operational information, often tactical and sometimes strategic. • Information is more accurate, adequate and timely. There is also more openness/transparency.	• Contractor and suppliers share the minimal information possible. Restricted to the project operational information • Project information is usually not available when is needed, it is often not very accurate, and it is definitely not transparent.
Process	**Innovation**
• There is usually incremental changes in internal logistics and purchasing processes. Some radical changes on processes happen. • Hub firm seeks essentially operational benefits, which can be high even with incremental changes. Suppliers achieve essentially strategic benefits.	• Contractors do not see purchasing and logistics as core processes. There is a lack of believe that improvements on these processes may bring benefits. • General believe that benefits may only be achieved if radical changes are made to the construction process
Technology	
• Open technology and simple, widely available and mature technology like EDI, E-mail and interactive on-line databases are the most common. • Hub firms have usually considerable internal IT capabilities. Suppliers have some IT capabilities and are often helped by the hub firm.	• Contractor's IT capabilities are usually good but lack any expertise on electronic communications. Suppliers have usually very low IT capabilities. • There is not many construction specific IT standards, but technology from other industries may be easily adapted.

The patterns of successful IT-enabled dyadic interactions are rarely in place before the full deployment of electronic links, and it is obscure which were the initial patterns, the causes and the effects. However, a sharp difference can be found between what is happening in construction dyadic interactions and dyadic interactions of successful IT-enabled virtual enterprises. We believe that IT-enabled virtual enterprises emerge more easily where the necessary effect upon the patterns of each conditioning factor will be lower in order to achieve the "successful pattern". In a generic way, dyadic interactions on construction virtual enterprises require large "investments" in order to achieve the pattern of successful IT-enabled dyadic. It is this gap between the required pattern and the real pattern of conditioning factors that is hindering the development of construction IT-enabled virtual enterprises (Grilo et al, 1996).

Moreover, conditioning factors interplay with each other in a complex way, creating self-reinforcing systems that can either facilitate or hinder the development of IT-enabled dyadic in order to exemplify.

- Collaboration, mutual trust, and long term commitment usually lead firms to share more tactical and strategic information, and also on a more open way. If firms share information about internal processes they can more easily envisage better ways of working together and achieve radical and dramatic changes with electronic links. A good example is given by the automotive assemblers that have been working with some major suppliers in order obtain lean production and supply (Lamming, 1993). In construction virtual enterprises, adversarial and short-term relationships lead to basic project operational information sharing, which is usually of low quality and non-transparent, and not allowing firms to envisage efficient ways to deploy electronic links in order to improve business processes.
- The deployment of complex and/or close IT systems for electronic links, such as proprietary groupware, graphical images, and advanced or proprietary EDI systems usually implies a high level of collaboration between firms, which will be difficult to achieve if companies have adversarial attitudes. Moreover, companies will not be willing to invest capital, time and resources if they have not some sort of guarantee that they will be trading on a long-term basis. On the other hand, open and widely available systems like those from the Internet do not usually require that firms have long-term, collaborative relationships.
- Often, supplier firms make process changes and deploy IT systems because they are coerced by hub firms, which have usually buying power over them. They often are also coerced to share more information than they were initially willing to. In construction virtual enterprises, main contractors do not have enduring buying power over suppliers, as it last only to the project duration. Thus, they can not coerce their suppliers in order to deploy IT systems or improve business processes.

Finally, it is important to remark that although the pattern of the conditioning factors of dyadic interactions may be greatly influenced by the web of relationships involved on the virtual enterprise, which are usually influenced by environmental characteristics (like industry culture, technological trends, etc.), these patterns are basically a matter of individual choice of firms. Against industries' culture and way of working, pioneering firms in automotive, retailing, clothing, etc. industries had to develop IT-enabled dyadic interactions by investing in collaborative relationships, by sharing tactical and strategic information with their suppliers, by acknowledging that improvements in logistic and purchasing functions could contribute to the overall business performance, and developing their own IT systems with (by then) immature technology.

Unarguably that current characteristics of construction virtual enterprises do not facilitate the emerging of enabling patterns on conditioning factors. However, it is basically a matter of choice of construction firms. Some positive evidence has been emerging recently. The partnering concept has been developing in the US, Australia and UK, although on a very restrict way (Bennett and Jayes, 1995). In Sweden, the large contractor Skanska on a re-engineering effort has identified the logistic and purchasing processes as the ones with greatest potential for improvement (Betts, 1996). They have established long-term formal agreements with major suppliers and are developing with them electronic links.

5 CONCLUSIONS

This paper addressed the problem of the lack of electronic links within construction virtual enterprises. It analysed what constraints and facilitates the initiation and development of IT-enabled enterprises, and why construction firms have not electronic links. In a generic way, it is concluded that the electronic linking of construction virtual enterprises requires that main contractors and suppliers develop more collaborative, trustful and long-term committed relationships. Construction firms will need to share more than basic project operational information, with more quality and transparency. Another crucial factor is the acknowledgement by construction firms of the importance and potential benefits of logistic and purchasing processes to the overall construction process. Finally, construction firms must develop IT capabilities as far as electronic communications are concerned. Following examples on other industries, collaborative and co-operative efforts between main contractors and suppliers on technology development are crucial.

6 REFERENCES

Atkin, B. and Pothecary, E. (1994), Building Futures; University of Reading, Dept. of Construction Management and Engineering

Baldwin, A., Thorpe, A, and Carter, C. (1995), Data Exchange in Construction: EDI, MFE or IE? in *Developments in Computational Techniques For Civil Engineering*, pp 11-15, B.H.V Topping, Civil-Comp Press, Edinburgh, UK

Bennett, J. and Jayes, S. (1995), Trusting the Team: Best Practice Guide to Partnering in Construction, CSSC, University of Reading

Betts, M., Atkin, B. and Clark, A. (1995), Best Practice Report: Supplier Management; Construct IT - Centre of Excellence, HMSO

Cherns, A. and Bryant, D. (1983), Studying the Client's Role in Construction Management; *Construction Management and Economics*, **2** , pp117

Grilo, A., Betts, M. and Mateus, M. (1996), Electronic Interaction in Construction: Why Is It Not a Reality?; *in CIB Proceedings of the Construction on The Information Highway*, Bled, Slovenia

Harrison, B. (1994), Lean and Mean: The Changing Landscape of Corporate Power in the Age of Flexibility; BasicBooks, New York

Hillebrant, P.M and Cannon, J. (1990), The Management of Construction Firms; Macmillian

Lamming, R. (1993), Beyond Partnership: Strategies for Innovation and Lean Supply; Prentice Hall

Latham, Sir M. (1994); Constructing the Team; HMSO

Miles, R.E. and Snow, C. (1992), Causes of Failure in Network Organizations; *California Management Review*, **Summer**, pp53-72

Venkatraman, N. (1991); IT-induced Business Reconfiguration, in *The Corporation of the 1990s: Information Technology and Organizational Transformation*, Oxford University Press

9

Using an Informal Ontology in the Development of a Planning and Control System - the Case of the Virtual Enterprise

António Lucas Soares, Jorge Pinho de Sousa, Américo Lopes Azevedo, João Augusto Bastos
INESC / Manufacturing Systems Engineering Unit, Rua José Falcão, 110 - 4300 Porto, Portugal
email: {asoares, jsousa, aazevedo, jbastos}@inescn.pt

Abstract

This paper describes a preliminary experience in trying to improve the communication and to reach language agreements in the context of a large R&D trans-national project, comprising partners with different academic and industrial cultures. This kind of projects has difficulties in that, besides cultural and language problems, the project teams undertake work largely by themselves, with reduced interaction with each other. The use of ontologies has a great potential in reducing those problems. Within this research work our goal is to extend the communication role of an ontology towards the mediation of the development actors "world views". In doing so, we intend to improve the engineering of an intrinsically complex software system, particularly in the requirements identification, system specification and system design phases, overcoming some of the referred difficulties. It is presented an extract of the Virtual Enterprise ontology currently under construction, focusing in some core definitions relevant for the purpose of developing a distributed planning and control software system. An example concerning the development of such a system is described in order to demonstrate the usefulness of the ontology based systems development.

Keywords

Information systems development, ontologies, planning and control, virtual enterprise

1. INTRODUCTION

This paper aims at reporting on some preliminary experience in trying to improve the communication and to reach language agreements in the context of a three year RTD trans-national project (E.20544, X-CITTIC, 1996/1999 - A Planning and Control System for Semiconductor Virtual Enterprises). This project aims at designing a model for planning and controlling an industrial Virtual Enterprise (VE) that should lead to the design of an architecture and to the development or selection of a set of software tools to be tested on two of the partners which are large complex organisations on the semiconductor industry.

The project involves three industrial companies on the semiconductor sector (GEC Plessey Semiconductors, TEMIC, ALCATEL-MIETEC), one software company (Nimble), and three academic institutions (Imperial College, INESC, Fraunhofer Gesellschaft IPA). The semiconductor industry clearly has features of the so-called Virtual Enterprise, in particular concerning distributed sites and different ownership situations of the plants. The manufacturing processes involve, in general, a set of sites (that may be located in different regions and countries), subcontracting and outsourcing, and create therefore quite complex logistic problems that have to be explicitly taken into account in planning processes. Moreover, due to the increasing demanding requests of customers, in a highly competitive environment, it becomes very important to have flexible planning systems and reliable control mechanisms. Quick response to the customers needs and to unpredictable changes in production conditions is also a need.

An Ontology for the VE is being developed as a first step towards constructing a platform for agreement on concepts and on the terminology to be used inside the project, as well as a way to structure the concepts to be used in the design. This is an attempt to acquire, represent and explore knowledge, by providing a core of basic concepts and language constructs, with precise purposes, i.e. to facilitate the project progress and to improve its efficiency and effectiveness.

2. A PLANNING AND CONTROL SYSTEM FOR THE VIRTUAL ENTERPRISE

Many manufacturing enterprises are becoming global businesses covering multiple manufacturing sites consisting of shop floors, subcontractors and suppliers. For that reason, planning and control activities are very complex, and have to take place both within the enterprise and across the whole supply network in order to achieve high levels of performance. However, this planning and control task is not covered well by state-of-the-art solutions. Moreover, the logistics associated to manufacturing products in several different plants and subcontracted companies become an important practical issue.

2.1 Project Objectives

The main deliverable of the current project will be a planning, scheduling and control system aiming at improving manufacturing performance in a virtual enterprise by integrating the heterogeneous manufacturing systems and by bridging the gap between the higher planning levels and the lower shop floor control levels. The project addresses the problem of planning and control of a distributed enterprise and aims at providing an integrated toolbox for distributed planning and scheduling that can be used in the operational control of manufacturing processes.

This toolbox will provide manufacturing solutions and services leading to shorter customer order lead times, improved delivery precision, reduced inventory and stocks, improved resource utilisation, quick response to customer enquiries, cost reduction in the logistic processes.

These targets will be realised by implementing:

- means for improved customer due date calculation;
- a multi-site planning tool to co-ordinate local activities across the virtual enterprise network;
- an integrated toolset for local planning and control as well as the management of shop floor deviations and contingencies based on the selection of existing tools;
- the complementary development of missing links and an integration methodology;
- the consolidation of the distributed schedules through methodologies and tools for the negotiation between different modules;
- an integrating infrastructure for planning and scheduling, supporting co-operative software applications;
- novel interfacing capabilities with existing systems;
- a real-time information link bridging the gap between planning and shop floor execution processes.

Major advances will be required over state-of-the-art and will lead to new concepts and tools for distributed planning and scheduling. In addition, system openness, interfacing to other enterprise IT systems and hardware vendor independence will all be given the utmost priority in order to facilitate future flexibility to change.

2.2 Project context

The different partners have various cultures, with normal problems in "natural" language communication. Concerning the industrial partners, the different companies have organisations and languages that can be substantially different. Even if the general objectives and goals of the system under development were set in a precise manner, this was not the case of its scope. The very nature of the project, involving the idea of Virtual Enterprise, leads to difficulties in defining the entities covered by the concept and the roles to be played by those entities.

From the point of view of the project management there are obvious difficulties. The project teams work largely by themselves, with reduced interaction with each other. Periodical technical meetings are held with the purpose of aligning objectives and synchronising the various tasks. These co-ordination aspects are in practice an additional problem in terms of an efficient and smooth course of the project.

3. ON THE CONCEPT OF VIRTUAL ENTERPRISE

In the past, manufacturing was essentially organised on a local level. With the fast development of transportation and communication means, an expansion process took place across countries and continents, naturally leading to the creation of global markets and global manufacturing systems. Nowadays, more and more companies are being organised as manufacturing networks of different units, namely, plants, logistic centres and storage facilities. In order to produce cost effective products within the time scales requested by the customers, most of the companies are using subcontractors spread all over the world.

The concept of Virtual Enterprise was born inside this environment, but it is obviously ambiguous and hard to define. We should however try to set a precise definition, for operational and communication purposes. In the literature, we can find several approaches to these issues, but they are not completely satisfactory in solving the ambiguity and imprecision associated to the terms used in the domain. First of all, it is interesting to review some terms related to the VE. The APICS dictionary defines Virtual Corporation and Virtual Factory (APICS, 1995):

"Virtual Corporation - (...), the capabilities and systems of the firm are merged with those of the suppliers, resulting in a new type of corporation where the boundaries between the suppliers systems and those of the firm seem to disappear.(...).

Virtual Factory - (...). It is a transformation process that involves merging the capabilities and capacities of the firm with those of its suppliers.(...)."

Based on the GERAM (Generic Enterprise and Reference Architecture and Methodology) entities, Ferreira (1995) presents a framework that allows the consideration of the *extended enterprise*. This is achieved by labelling products, the manufacturing enterprise, the engineering design enterprise, and the strategic enterprise, as new entities of the reference architecture. In this context, the concept of *extended enterprise* encompasses not only the integration of processes from different manufacturing enterprises, but also their integration with the engineering enterprise and the strategic management enterprise. As one would expect, it may happen that the previously mentioned entities are in fact different enterprises, thus reinforcing the need for integration within the extended enterprise.

In Browne et al. (1995) the concept of *extended enterprise* is also explored. The *extended enterprise* is viewed as "an expression of the market driven requirements to embrace external resources in the enterprise without owing them". The emphasis is put in the integration of manufacturing and distribution planning and control systems. A similar concept is that of *supply chain*. For EIL (1995) the *supply chain* of a manufacturing enterprise is a world-wide network of suppliers, factories, warehouses, distribution centres through which raw materials are acquired, transformed and delivered to customers.

Another related interesting concept is that of *network organisation* in which *separate firms*, each retaining its own authority in major budgeting and pricing matters, function as integral parts of a greater organisation co-ordinated by a *core firm*. In this context, the different firms, spread along a value-added chain (e.g., suppliers, manufacturers, and distributors), tend to function in a way that stresses complementary, ongoing relationships, and reciprocity (Ching et al., 1993).

In the literature, much more related definitions and concepts can be found, and in the context of this project we came to the following first operational definition: "The Virtual Enterprise is a subset of units and processes within the supply chain which behave like a single company through strong co-operation towards mutual goals".

First, it is clear that we are talking of a physically distributed network of units (several sites) and possibly (likely) owned by different companies. Secondly, it should also be clear that we focus on the manufacturing aspects of the company, or on aspects that are strongly related to manufacturing. However these ideas are probably not crucial, as they do not capture the key features for planning / control purposes (which are the concerns of the project). As counter-examples, one can consider several distinct production units (supposed to be managed separately) standing on the same site, or the ability to largely control subcontractors not really owned. Therefore, as a first, non-formalised approach, we propose a slightly different view on the concept, based on the following ideas. We start by considering a set of *units*: these are separately managed entities (plants, subcontractors, transport companies, warehouses, "customers", ...) that have some kind of relationship to our business. As opposed to the concept of "node" (in a network) that has a "mathematical" connotation, the term "unit" better reflects the manufacturing terminology. For each of these units, we define the *level-of-*

control (*loc*), in some rough, partially subjective way, say in 0-1 scale. This means that our *core* company has a *loc* of 1, a given plant may have a *loc* of 0.7, a subcontractor a *loc* of 0.3, a "customer" might have a *loc* of 0.1 or close to 0. Then we have, for the different units, some kind of "measure" of "ownership" towards the main (core) management entity (of course, this is not directly related to real ownership). This setting should allow us to define the VE in a kind of "fuzzy" way, not imposing a strict definition of the borders (type of "you are inside the VE, or you are out in the world...).

Nevertheless, for management purposes, it is important to set a "threshold", i.e. a minimum level of "ownership", below which we would say the unit is outside of the VE even if we can not forget it in our management processes (this should be the case of "normal" customers, but not the case of very closely related special customers, or some with whom partnerships have been established). In this context, a subcontractor could be of different types (related to his *loc*), allowing different levels of information share and control.

It should be noted that the VE Ontology, currently under construction, does not yet formalise these ideas, but we feel they can be usefully incorporated and used in practice.

4. THE NEED FOR MEDIATION

In section 2 we saw how the specific case of a planning and control system development for a Virtual Enterprise run into difficulties given the profusion of concepts and terms and the confrontation of different perspectives. Our experience showed what others had already reported, in a different context, that terminological confusion breeds conceptual confusion and vice-versa (Bradshaw et al.). This is mostly true in phases of conceptual brainstorming and perspectives confrontation which characterise the early phases of the system development process. One way to improve this process is to agree *a priori* with a conceptual and terminological core, setting the ground for the subsequent discussion. Such a core would assume a mediation role between the development actors. This led us to the use of ontologies in implementing the referred mediation scheme. An example of mediation in a broader techno-organisational context is reported in (Soares and Mendonça, 1996).

Ontologies are an active research area in the field of Artificial Intelligence supporting a modelling view of knowledge engineering as opposed to a transfer view (Guarino, 1995). In this perspective knowledge engineering is viewed not as the capture and storage of something extracted from an expert's mind but as the result of a constructive modelling process of an objective reality. Different definitions of ontology can be found in the literature. A concise definition that is becoming widely accepted is the one proposed by Gruber (1993): "an explicit specification of a conceptualisation". A good synthesis of what an ontology is appears in (Uschold and Gruninger, 1996), quoting a source in the SRKB mailing list: "Ontologies are agreements about shared conceptualisations. Shared conceptualisations include conceptual frameworks for modelling domain knowledge, content specific protocols for communication among inter-operating agents and agreements about the representation of particular domain theories. In the knowledge sharing context, ontologies are specified in the form of definitions of representational vocabulary" (Gruber, 1994).

4.1 Purpose of the Ontology

Ontologies have been applied so far in a range of works with several purposes. Knowledge sharing and reuse are the base line directly or indirectly referred in each of them. Sharing arises when a common conceptualisation of a given domain is essential to the undertaking and co-ordination of activities within that domain. Sharing implies some sort of communication between different people, people and implemented computational systems, different implemented computational systems (Uschold, 1995). Reuse can be viewed as the step forward towards a generalised sharing, through formalisation mechanisms. Our purpose in building an ontology for this project is to improve the communication between partners concerning the requirements identification, specification and design phases of the planning and control system. With this we expect to achieve a faster agreement on the system's conceptual model and a more consistent use of terms and concepts throughout the software development. Eventually, when a more mature state of this ontology is reached, we intend to codify it and make it available for reuse, in so contributing for the clarification of the Virtual Enterprise concept. As a side effect, we also expect to contribute to the research of a less explored role of an ontology: the one of communication medium between people.

4.2 Issues in the Ontology Development

In the construction of the Virtual Enterprise Ontology we followed whenever possible the methodology for developing ontologies outlined by Uschold and King (1995). This methodology includes the following steps: identify purpose, build the ontology (capture, code, integrate existing ontologies), evaluation and documentation. According to our goals, we concentrated in the purpose identification and building steps, particularly the capture phase. The later is probably a crucial step in the process, and consists in identifying the key concepts and relationships in the domain, producing precise text definitions, identifying terms, and reaching an agreement on these issues. Another important aspect in the capture phase is the inclusion/integration of other ontologies. In our case, the Enterprise Ontology (Uschold et al., 1996) was particularly helpful and was thoroughly used in the VE Ontology construction. The Plan Ontology (Tate, 1995), although in a draft state, was also used.

5. AN EXTRACT OF THE VIRTUAL ENTERPRISE ONTOLOGY

We present now an extract of the VE Ontology, focusing in some core definitions relevant for the purpose of developing a planning and control system. It should be noted that this ontology is still under construction, and dynamically evolving. At this stage, a certain degree of non-formalism is kept.

As a first step towards defining this ontology, we have started by an attempt to list all possible entities involved in what might be the VE. As a basis, we have taken concepts from a related, more precise idea, that of a Supply Chain. Then we have tried to match those concepts and terms with those in already existing more general ontologies, such as the Enterprise Ontology (EO). The VE Ontology is committed to the Meta-Ontology defined in the EO.

It should be noted these are still preliminary steps, intended to illustrate the general approach followed in this work. We adopted the following notation: words in upper case are terms defined in the VE Ontology; words in upper case and italics were included from other ontologies (their formal definition is not presented, as their usual interpretation is enough for the purposes of this explanation).

Networked/Extendend Organisations

VIRTUAL ENTERPRISE: is a set of inter-related VE UNITS that are totally or partially committed to some common *PURPOSE*.
Notes:
1. The relation between VE UNITS can be one of *OWNERSHIP* or a looser one of belonging to a common supplier chain.

VE UNIT: An *ORGANISATIONAL UNIT* belonging to the set that composes a VIRTUAL ENTERPRISE. A VE UNIT may be a manufacturing/processing plant, a subcontractor, a transport provider, a supplier or a warehouse.
Notes:
1. The term *OU* is here used in its broader sense i.e., as an UNIT providing the a complete set of products or services (e.g. a plant, a manufacturing unit, a warehouse, etc.)

Plans management

***PLAN:** an ACTIVITY SPECIFICATION with an INTENDED PURPOSE.*
Notes:
1. See notes under ACTIVITY SPECIFICATION.

***PLANNING:** an ACTIVITY whose INTENDED PURPOSE is to produce a PLAN.*

PRODUCTION PLAN: is a *PLAN* whose *INTENDED PURPOSE* is to specify when and where PRODUCTION ACTIVITIES will be performed. A PRODUCTION PLAN produces an assignment of *ACTIVITIES* to VE UNITS and *TIME INTERVALS*.

ROUGH PRODUCTION PLAN: is a PRODUCTION PLAN based on aggregate information and setting *TIME INTERVALS* for production in the VE UNITS considered to be feasible candidates.*

FINE PRODUCTION PLAN: is a PRODUCTION PLAN based on more detailed information, and setting *CALENDAR DATE's* for production in specific VE UNITS.*

GLOBAL PLANNING: an *ACTIVITY* whose *INTENDED PURPOSE* is to produce PRODUCTION PLAN's

LOCAL PLANNING: an *ACTIVITY* whose *INTENDED PURPOSE* is to produce PRODUCTION *PLAN's* for a single VE UNIT.

CAPACITY: is an *ATTRIBUTE* of a *RESOURCE* representing what is available for allocation to *ACTIVITIES* over time. The CAPACITY of a *RESOURCE* constrains the number of *ACTIVITIES* that it can simultaneously support.

RESOURCE: is an *ENTITY* with some amount of CAPACITY that is allocable to *ACTIVITIES* over time

Orders management

ORDER: it is a CUSTOMER request for PRODUCT, issued to a SUPPLIER.. An ORDER contains one or more ORDER ITEMS.
Notes:
1. An ORDER contains general information such as the CUSTOMER identifier, number of items, shipping address, etc.

ORDER ITEM: an item within an ORDER. This includes reference to the PRODUCT SPECIFICATION, the quantity of PRODUCT to produce, and the CUSTOMER's requested DELIVERY DATE.

CUSTOMER ORDER: is an ORDER in which the CUSTOMER is an EXTERNAL CUSTOMER.
Notes:
1. A CUSTOMER ORDER can be in one of the following states: under negotiation, confirmed, planned, being executed, completed.

PRODUCTION ORDER: is an ORDER in which the CUSTOMER is an INTERNAL CUSTOMER.
Notes:
1. A PRODUCTION ORDER can be in one of the following states: under planning, planned, being executed, on schedule, late, stopped, cancelled and completed.

PRODUCT REQUEST: is a statement defining a CUSTOMER's needs, in terms of PRODUCT types, quantities and DUE DATES.

CUSTOMER: the requirer of a PRODUCT. He expresses is requirements in terms of ORDERS. A CUSTOMER can be an INTERNAL CUSTOMER or an EXTERNAL CUSTOMER depending on its commitment to the VIRTUAL ENTERPRISE.

SUPPLIER: a provider of goods or services.

PRODUCT: is the result of a manufacturing *ACTIVITIES* chain. A manufacturing enterprise *PURPOSE* is to make PRODUCTS.

ENQUIRY: is an *ACTIVITY* that an EXTERNAL CUSTOMER *EXECUTES* with the *PURPOSE* of obtaining information about PRODUCTS, prices and DELIVERY DATES. This process eventually leads to placing a COSTUMER ORDER.

QUOTATION: is a statement of price, description of PRODUCTS, DUE DATES, and other terms of sale, offered by a supplier to a prospective CUSTOMER.

DUE DATE: is the *CALENDAR DATE* when PRODUCTS are available for delivery.

DELIVERY DATE: is the *CALENDAR DATE* when PRODUCTS are delivered to the *CUSTOMER*.

ORDER PROMISING: an *ACTIVITY* whose *PURPOSE* is to produce a QUOTATION in response to an EXTERNAL CUSTOMER REQUEST.

AVAILABLE-TO-PROMISE: is a quantity of PRODUCT available to allocate in a given *TIME POINT* to satisfy a CUSTOMER REQUEST. This quantity is calculated on the basis of the company's inventory and actual and planned production.

6. THE VE ONTOLOGY AS A MEDIATOR IN THE SYSTEM DEVELOPMENT LIFE-CYCLE

As referred above, one less explored role of an ontology is the one of communication medium between different people working together for a given purpose. This is the case of a (software) system development process where teams composed by end users and system developers have to collaborate throughout the development life cycle. In this research work our goal is to extend the communication role towards the mediation of the development actors "world views". In doing so, we intend to improve the engineering of an intrinsically complex software system, particularly in the requirements identification, system specification and system design phases, overcoming some of the difficulties described before.

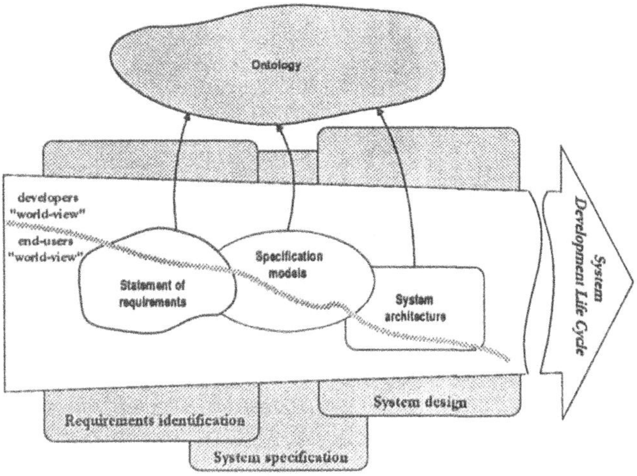

Figure 1 - Use of the VE Ontology in the Planning and Control system development

Figure 1 shows the general approach to the use of ontologies in the system development process. During the development life cycle end-users and developers world views constraint the purpose, scope and goals of the development object, as well as the discourse to describe and reason about the domain/system. Though the purpose, scope and goals of the system must be minimally agreed early in the project, conceptual differences and language misunderstandings and redundancies are important obstacles in a smooth evolving of the process. The degree of influence that each world view have in the development life cycle phases is qualitatively represented in the figure by the two areas separated by the thick grey line. End-users world-view has more influence in the requirements identification and specification phases whilst developers world view is more influential in the system design phase. System development phases and milestones are shown overlapping each other to symbolise both the fuzziness of the borders between phases and a desirable system development approach based on evolutionary prototyping. The role of the ontology as mediator is materialised in the concepts and terminology used in the system development milestones - statement of requirements, specification models and system architecture, which are expressed according to the terminology and structure defined by that ontology.

6.1 An Example of Ontology Based Systems Development

The example we are about to present is intended to demonstrate the usefulness of the ontology based systems development approach in improving the analysis, specification and design phases of the project. It is focused on a set of activities that can be designated by *Global Planning and Control*. We begin by transcribing the statement of requirements (at a very broad level) of for this set of activities using the terminology defined in the VE Ontology:

"...*Global Planning and Control* must include a set of activities designated by *order promise* that should respond to a CUSTOMER ENQUIRY by commitment to a DUE DATE, PRODUCT QUANTITY, PRODUCT QUALITY and PRODUCT PRICE i.e., a CUSTOMER QUOTATION. These promises are used to build up a rough PRODUCTION PLAN. A ROUGH PRODUCTION PLAN is a PLAN containing suggested TIME

INTERVALS for production in the UNITS which are considered to be feasible candidates at an aggregate level. The ROUGH PRODUCTION PLAN is revised every time a CUSTOMER ENQUIRY is received or when an exception occurs requiring to repair the ROUGH PRODUCTION PLAN (see details on exceptions below).

The ROUGH PRODUCTION PLAN is used by a *global* PLANNING set of activities in the generation of a FINE PRODUCTION PLAN for each UNIT in the VIRTUAL ENTERPRISE. A FINE PRODUCTION PLAN is a PRODUCTION PLAN with a high degree of feasibility with respect to throughput, slack, conformance to the original PLAN and cost. The PURPOSE of the *global* PLANNING set of activities is to generate FINE PRODUCTION PLANS.

In order to adjust the ROUGH PRODUCTION PLANS and CUSTOMER QUOTATIONS in face of unexpected constraints a set of reactive ACTIVITIES are needed. This set of activities is designated by *exceptions handling*. Examples of exceptions may be: unrealised forecasts and unexpected UNIT CAPACITY change ..."

The statement of requirements is the starting point for a semi-formal specification of the planning and control system. To illustrate this, the well known OMT (Object Modelling Technology) methodology (Rumbaugh et al., 1991) was followed and its functional model used to translate the statement of requirements into the systems specification functional part. An extract of the model is shown in Figure 2. The use of the VE Ontology assures a conceptual and terminological coherence and consistence of concepts and terminology between the two milestones.

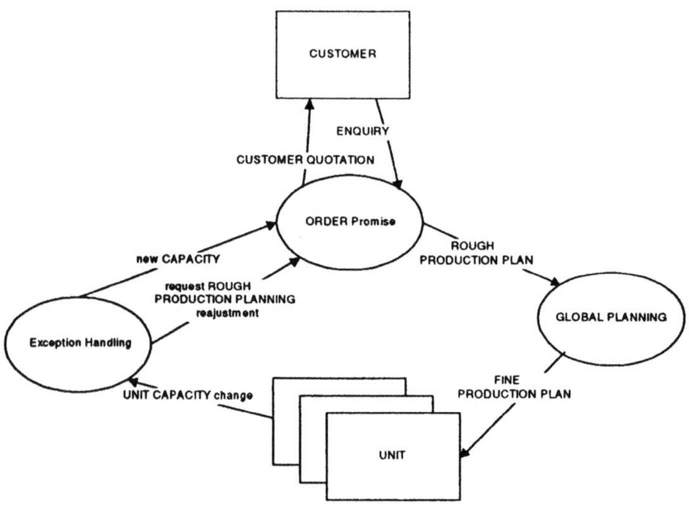

Figure 2 - Extract of the Global Planning and Control functional specification model (OMT)

7. CONCLUSIONS AND FURTHER DEVELOPMENTS

The preliminary experience reported in this paper shows that the use of an informal specifically created ontology has a great potential for mediating the development of large, complex software systems, by providing a platform for agreement on the language to be used inside the project, and also as a way to structure the concepts to be used in the design.

This approach was applied to a large R&D trans-national project, involving partners with different academic and industrial cultures, aiming at designing a planning, scheduling and control system for the Virtual Enterprise (VE). Significant improvements are being obtained by using this approach, particularly in the requirements identification, system specification and system design phases.

As a first step towards constructing this ontology, the concept of Virtual Enterprise was fully discussed, leading to a clarification of the entities and terms to be used in the project.

The development of an ontology is a complex and time consuming task, that is dynamically evolving over time. The work done is far from having reached a stable and consolidated outcome. Further work is needed throughout the project, as well as the continuous improvement of the approach described in this paper.

8. REFERENCES

APICS, 1995, APICS Dictionary, 8th Edition. American Production and Inventory Control Society, Inc.

Bradshaw, J.M., Boose, J.H., Shema, D.B., 1992, Steps Toward Sharable Ontologies for Design Rationale. AAAI-92 Design Rationale Capture and Use Workshop, San Jose, CA, July.

Browne, J., Sackett, P., Wortmann, H., 1994, Industry Requirements and Associated Research Issues in the Extended Enterprise. European Workshop on Integrated Manufacturing Systems Engineering, IMSE'94, 12-14, Grenoble.

Ching, C., Holsapple, C.W., Whinston, A.B., 1993, Modeling Network Organizations: A Basis for Exploring Computer Support Coordination Possibilities. Journal of Organizational Computing, 3(3), 279-300.

EIL (Enterprise Integration Laboratory), 1995, The Integrated Supply Chain Management Project. http://www.ie.toronto.ca/EIL/eil.html.

Ferreira, J.P., Mendonça, J.M, 1995, Modelling and Simulation of Manufacturing and Information Aspects of the Extended Enterprise. Proc. of the Int. Conf. on Improving Manufacturing Performance in a Distributed Enterprise: Advanced Systems and Tools, Edinburgh, U.K.

Gruber, T.R., 1993, A Translation Approach to Portable Ontology Specifications. Knowledge Acquisition, 2:199-220.

Guarino, Nicola, Formal Ontology, Conceptual Analysis and Knowledge Representation. To appear in International Journal of Human and Computer Studies, edited by N. Guarino and R. Poli.

Rumbaugh et al., 1991, Object-Oriented Modeling and Design, Prentice Hall.

Tate, A., 1995, Towards a Plan Ontology, Technical Report AIAI - The University of Edinburgh.

Uschold, M., King, M., 1995, Towards a Methodology for Building Ontologies. Technical Report AIAI-TR-183, The University of Edinburgh.

Uschold, M., King, M., Moralee, S., Zorgios, Y. The Enterprise Ontology. Technical Report AIAI-TR-195, The University of Edinburgh, 1996. [http://www.aiai.ed.ac.uk /~entprise/enterprise/]

Uschold, M., Gruninger M., 1996b, Ontologies: Principles, Methods and Applications. Knowledge Engineering Review, 11 (2).

Soares, A.L., Mendonça, J.M., 1996, Multiple-Perspective Knowledge Representation in Techno-Organisational Development Processes, in Koubek and Karwowski (Eds.), Manufacturing Agility and Hybrid Automation-I, IEA Press, Louisville, Kentucky, USA.

9. BIOGRAPHY

António Lucas Soares is an Electrical Engineer teaching at the Faculty of Engineering of the University of Porto and doing research work at INESC-UESP. He holds a MSc. in Industrial Automation and is currently undertaking its PhD work in the area of collaboration models for techno-organisational development. The main research interests are the interdisciplinary collaboration in the development of integrated manufacturing systems and socio-technical systems.

Jorge Pinho de Sousa is an Assistant Professor at the Faculty of Engineering of the University of Porto and researcher at INESC-UESP. He holds a Ph.D. in Operations Research (Université Catholique de Louvain, Belgium), 1989. His main interests are on Operations Research, Combinatorial Optimization, Project Management, Industrial Organization, and in Production and Operations Management.

Américo Lopes de Azevedo obtained the degree of Electrical and Computers Engineer at the Faculty of Engineering of the University of Porto and is doing research work at INESC-UESP. He has been teaching in the Electronics and Computers Department where is assistant since 1988. Currently, he is preparing his PhD with a work in the area of manufacturing systems. His main interests lie in the Integrated Manufacturing area.

João Bastos obtained the degree of Mechanical Engineer at the Faculty of Engineering of the University of Porto and is doing research work at INESC-UESP. He holds a MSc. in Industrial Informatics and is currently undertaking his PhD work in the area of intelligent manufacturing systems.

PART FOUR

Learning and Genetic Algorithms

10

Federated Knowledge Integration and Machine Learning in Water Distribution Networks

H. Afsarmanesh(1), L.M. Camarinha-Matos(2), F.J. Martinelli(2)
(1) University of Amsterdam
Kruislaan 403, 1098 SJ Amsterdam, The Netherlands.
email: hamideh@wins.uva.nl
(2) New University of Lisbon and Uninova,
Faculty of Sciences and Technology
Quinta da Torre, 2825 Monte Caparica, Portugal
email: cam@uninova.pt; fjm@uninova.pt

Abstract

Water distribution networks are geographically distributed systems, with greater heterogeneity in terms of control structures, management strategies, and varying geometry with continuous expansion and changes in demand along their life. Due to these characteristics, water distribution companies face the problem of data and knowledge integration related with control and optimal exploitation. A few European and International RTD projects are currently focused on the design and development of a next generation system to support the control, optimal operation and decision support of drinking water distribution networks. In the context of a two year RTD project - WATERNET*, an evolutionary knowledge capture for advanced supervision of water distribution network is being developed. The WATERNET system, assists the distributed control of water management network to minimize the costs of exploitation, guarantee the continuous supply of water with a better quality monitoring, save energy consumption and minimize natural resources waste. This system comprises of several subsystems: a distributed information management subsystem, a machine learning subsystem, an optimization subsystem, a water quality monitoring subsystem, a simulation subsystem, and a supervision system that integrates these subsystems in order to assist the decision making and optimal operation of the network. This paper first provides a high level global description of this on going research on water management system. Then, it concentrates on the current stage of development on two specific subsystems in the project; the distributed information management and the machine learning subsystems.

* The work described here has been partially supported by the ESPRIT project 22.186 - WATERNET, whose partners are: ESTEC, UNINOVA, SEBETIA, WBE, Univ. of Amsterdam, ADASA Sistemas, Univ. Politecnica de Catalunya, ALFAMICRO, and Univ. of Naples.

Keywords
Federated cooperative databases, integrated information sharing, knowledge integration, machine learning systems, water distribution networks, intelligent supervision

1 INTRODUCTION

Most water supply industries nowadays lack a global "overview" of the status of the production and the distribution system. It is difficult, costly, and takes a long time to recognize failures in the system, to identify the non-optimized operations and resource wastes, and to take the proper recovery and maintenance actions. Control of the system is often carried out locally, based on the operators experience. Typically, there is none or little coordinated control in order to assure a continuous supply, meet the quality standards, save energy, optimize pipeline sizes and reduce wastes. Existing systems for control and/or monitoring of the water supply and distribution are heterogeneous and of different levels of automation and reliability. Furthermore, these systems cannot be easily linked together in order to extract the needed integrated information.

However, today the consumers demand higher performance for the water supply. Recent research (Stroomloos 1994) shows that the society tends to become more and more dependent on public supplies, e.g. the water and power, while it becomes less capable to deal with interruptions (The vulnerability paradox). The same research points to the need of water companies to improve continuity and quality of water supply. The water industry will gain extensive economical benefits by: minimization of energy-consumption, the rational and effective failure diagnosis and maintenance, minimization of mixing zones of water of different qualities, minimization of detention time in the distribution system, and the rational expansion / modification of the network structures.

A network for water distribution presents a wide set of specific problems and requirements. Some of the important problems identified and addressed in the WATERNET project includes the following:

- The network is distributed through a wide geographical area; e.g., an average normal system may comprise of close to 900 Km of main distribution pipes.
- Because of the geographical structure of the covered area and the available sources and consumption requirements, each system is different, and as a consequence it employs specific mechanisms for control and management.
- In order to guarantee the continuous supply of water, several pumping stations and reservoirs must be controlled and managed in an integrated way.
- The quality of the water must be guaranteed from the capture points to the consumers.
- The network is in permanent expansion because of the increase in the number of big consumers: factories, industrial sites, etc.
- Other sources of uncertainty that must be considered are due to the disruption (total or partial) of pipes, or even pirate derivations.
- Criteria for regular and repair equipment maintenance must be met in order to optimize the exploitation of the service.

Based on the above requirements, WATERNET is developing a modular Supervision System featuring a Distributed Information Management Subsystem, an Optimization Control Subsystem, a Learning Subsystem, a Simulation Subsystem, and a Water Quality Monitoring Subsystem distributed control architecture.

An improvement in any of these problematic areas through the use of an effective drinking water distribution system produces a huge economic benefit for both the water industry and the consumers. As an illustrative example, with the mere application of a distribution optimization software, specially developed for the drinking water network of Barcelona city in Spain (Quevedo 1988), a 15% reduction of the previous distribution costs was achieved. This resulted a 2 K ecu per day cost reduction for that industry.

This paper first provides a high level global approach to the knowledge capture for advanced supervision of water distribution networks in WATERNET project. It further focuses on two specific subsystems of distributed information management and machine learning. The paper also provides some examples from the environment describing the applicability of these subsystems to distributed water control problems.

The remaining of this paper is organized as follows. Section 2 provides an integrated control approach to production and distribution of water. In this section a high level description of the work planned in the WATERNET project is presented. Section 3 describes a distributed/federated information management framework for the water control network. A summary of the PEER federated integration architecture is also provided in this section. Section 4 describes the learning system for the water management application. And finally, Section 5 concludes the paper.

2 PROJECT SUMMARY

2.1 Project Objectives

The WATERNET project approaches the problems addressed in Section 1, and introduces an architecture and necessary mechanisms to support these requirements. In specific the design and development of WATERNET environment is based on the support for the following key technical issues:

a) Distributed multi-agent control system

Due to the geographical distribution and open/dynamic structure of the water distribution systems, emerging results from the area of Distributed Information Management Systems (Federated Cooperative databases, Integrated Information Sharing, etc.) are applied in the proposed system.

b) Functionality control and management system

Optimization, forecasting, diagnosis, prognosis, and recovery are obvious components for an advanced supervision system. Real-time situation assessment of the dynamic, hard to measure systems is being explored here.

c) Multi-paradigm learning system

In water distribution networks, domain models are not enough to ensure all supervision requirements. As in other areas, the codification of knowledge can be limited. For certain sub-problems, a database of solved cases can be the sole source of knowledge. In practice, the paradigm of Programming by Human Demonstration, namely a set of programmed predefined complex systems that show the particular examples of desired human behavior, have proven to help in certain complicated problems (Camarinha-Matos 1996). If traditional expert systems are often criticized for being limited in their abilities to surpass the level of existing experts, learning systems have the potential to discover new relationships among concepts, therefore exceeding the performance of experts. This does not mean that expert knowledge is being ignored. Rather, the technical approach in the WATERNET project consists of both traditional programming and programming by demonstration.

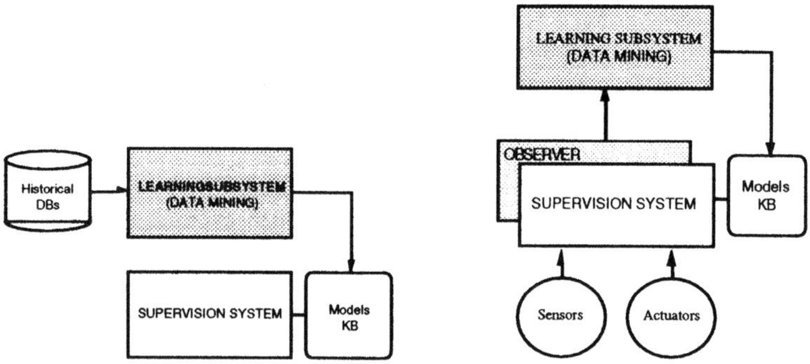

Figure 1 a) Learning from historical data b) Continuous learning

2.2 Current situation

A typical modern control and management system for water supply industry is a distributed network of stations computationally linked to a central control station. In the SMAS Sintra, a Portuguese water industry, remote stations are all points in this network that serve both as the local controller and the observation unit. According to their characteristics, a local computer system is configured and equipped with the necessary interfaces in order to control and manage every node locally. This local system applies a proper control algorithm and monitors all the sensorial data relevant for its task. A log of commands, alarms, sensorial values e.g. pressures, levels, flows, etc. is gathered locally, and transmitted periodically to the control unit database. Sensorial data is collected at short intervals - usually once every few minutes. Because of the functions offered by the communications network, any anomaly detected by this system is automatically reported to the control unit, and a proper recovering procedure is executed by the human supervisor if necessary and if available.

A control unit is localized in a control room of the water distribution services and permits the supervision of an area by the operators. It consists of a set of workstations connected by a local area network, the communications front-end, and the data storage equipments. The operator uses a graphical interface to dialogue with the system. A diagram of the network and related stations is displayed on the workstation screen. Using the mouse, the operator can enter direct commands to a remote station in order to visualize its status of operation and perform any necessary action. The main job of the operator at a control unit, is to monitor the network and to observe the incoming alarm messages. According to the messages received, he can start the necessary action in order to recover, and if necessary he can be helped with a mobile brigade.

At each remote station, the latest data produced locally is maintained in a local database. This database can later be used by several users related to maintenance, planning, etc. Each user scans the data in order to find the information relevant for its needs. At intervals specified by the operator, the control unit polls all the remote stations for collecting their data gathered locally and stores it in its database. This scenario corresponds to a state of the art installation. However, in some cases, the situation can be less automated. At present, some water distribution systems do not even include an integrated control system, and only rely on the local human operators and spoken communications with a control unit. In the case of SMAS Sintra, at present there is only one central control unit that supports the supervision of the entire network. On the other end of the spectrum, there are water companies, e.g. the WBE, a Dutch water company, that require a more complicated system.

2.3 Need for advanced control and management systems

a) The problem of optimal operation

This problem consists of deriving strategies for the control elements in a water system (valves, pumps, turbines) so as to achieve the service objectives (meeting consumer demands with appropriate pressure levels) with the minimum possible cost. These strategies must be found ahead of time based on demand predictions. The mathematical formulation of this problem requires an optimization of a cost function which is generally non-linear and includes several constraints. Thus, the mathematical formulation is difficult and needs further research and even the evaluation of alternative techniques, e.g. the machine learning.

b) Controlling large water distribution systems

While the hardware system for telecontrol from a central dispatch is reasonably efficient at this stage, the problems of management of large amounts of information, scheduling the control of water transfer from the sources to the consumers with appropriate pressure levels ahead of time, estimating the state of the network from the readings of a limited number of sensors and dealing with breakdown alarms are still largely unsolved. Currently, the main emphasis of research in this area is on solving the optimal scheduling problem. Even for relatively small networks, this is an extremely complex problem, due to the nonlinearities involved and the computationally intensive process.

c) Information management

In addition to the problems of scheduling and control, the problems of information management have become more apparent with the accumulation of experience in telecommunication and telecontrol in the 90's. In particular, the availability of large amounts of both numeric and symbolic information, the complex structured knowledge that needs to be shared to support the distributed control and management of water supply units, and the heterogeneity of the nodes in the network and their distinct and various tasks, has made it necessary to use more advanced data management technology than have previously been employed.

The problems of management and control of water networks continue to be a challenging one for most European cities. From one side, no generally applicable techniques have been yet derived for the problems tackled in the past. On the other side, the significant progress in the computing power and data management techniques can now be applied to the water industry. It is possible to envisage a solution for the whole new range of management and control problems, which had not yet been dealt with in previous telecontrol systems. The application of new technology, in turn improves the ability of water authorities to provide a good service to the consumers taking maximum advantage of the limited natural resources available.

Extracting knowledge from available or being collected data, by application of machine learning / data mining techniques, seems to be a promising approach.

3 DISTRIBUTED INFORMATION MANAGEMENT SUBSYSTEM (DIMS)

Modern Water supply and water distribution management heavily depends on the support of a strong information management system. A distributed/federated information management framework needs to be designed and developed to support the cooperation and information exchange among different sites and their activities involved in intelligent water distribution control network. This network of sites must be open and dynamic to support its expansion, as the number of sites in the water distribution network increases. Consequently, it must support the management of the dynamic and incremental evolution of the information within the integrated network. The sites in the network can be autonomous since they may pre-exist to the development of the water distribution system, serving a specific purpose, and must be able to retain their local control and autonomy. For instance, there are sites that provide certain information and/or services, e.g. maintenance, complaint handling, demographic information, maps and geographic information, etc.

The sites are also heterogeneous due to the variety of their functionalities ranging from the production, distribution, and maintenance to data gathering, control, simulation, diagnosis, prognosis, learning, optimization, and planning. In the current design of the DIMS subsystem, it is assumed that the information of each site in the network is represented using a common information-representation model and can be accessed using a common access-language. Existing standards, such as common interface with the IDL (of ODMG) and common exchange architecture with CORBA (of OMG), may need to be used to support possible database model heterogeneity, which is outside the scope of the current research and this paper.

Each site in the network is being extended with a layer that supports the federated information sharing and exchange among different sites. In the development of the federated architecture, an existing software prototype, the PEER system, developed at the University of Amsterdam (UvA) is applied. PEER is a federated object-oriented information management system (Afsarmanesh 1993), (Tuijnman 1993), (Afsarmanesh 1994) that is designed and prototyped at UvA; its development is based on the AIM (Agent Information Management) software developed by UvA in ESPRIT - ARCHON project (Wittig 1992).

3.1 Management of information in water control network

In water supply and distribution network, typically the information is gathered, processed, and stored in geographically distributed sites that are called units in this section. Every unit serves a specific function in the system and thus units are of different kinds. In order to perform its tasks, every unit relies on accurate and proper integration of remotely collected data and the availability of up-to-date shared information, at the time when it is needed.

Three main specific kinds of units can be identified. Every production site comprises a "remote unit" that collects the local data and executes some local level of control. A "control unit" performs the control supervision and control of the water supply and distribution system. for a number of remote units. There is a third kind of unit, the "auxiliary unit" that assists the tuning and optimizing tasks of different functionalities of the system. Examples of auxiliary units include the machine learning unit, the simulation and optimization unit, the water quality monitoring unit, etc..

The units mentioned above are interconnected by a communication intranet network. Some of these units are tightly and some loosely coupled. For instance, the remote units are tightly-coupled with the control units. All the information gathered and stored at any remote unit is also transfered and stored in a control unit. In some existing water management networks, the remote units are only connected to the control units, while in some others they are also connected with each other and with the auxiliary units in a network. On the other hand, some units are only loosely-coupled and they only share and exchange a part of their information with each other as required. In order to properly perform their tasks, for instance, the water quality monitoring units need to access a part of the information stored in the control units.

3.2 Need for information sharing and exchange among nodes

The proper functionality of all units involved in the water control systems depends on the sharing and exchange of data with other units. The autonomy of some units need to be preserved, for instance, the supervision unit is an autonomous unit, while the remote unit has only partial control over its functionality and takes orders from the control units. Similarly, the heterogeneity of information representation and classification needs to be supported. In general, the same piece of information is viewed differently by two units, and different level of details are associated with it. For instance, to the water quality monitoring unit, all the details of every water quality sample are very important piece of information, while to the alarm handling and maintenance units, only a part of certain water quality samples are of interest and relevance if they are the cause of alarm situations.

3.2.1 Example cooperation scenario

Typically auxiliary units require to share and access information from the control units, and the remote units (either directly or through the control unit) in the network. Consider the learning subsystem defined in Section 4 of this paper that performs several types of tasks. For instance it supports the production planning, alarm handling, preventive maintenance, etc. For every one of these tasks, learning algorithms need to perform a number of subtasks, for which they require access to different data from certain other units in the network.

In more details, the task of providing preventive maintenance information, involves the two major subtasks of: (1) Characterization of devices life cycle, and (2) identification of behavioral changes in devices. Now, subtask (1) requires the access to maintenance history of devices, that is stored in the control unit, while subtask (2) requires the access to up-to-date current device status, flow, etc. that is coming from the remote unit. Once the data of every unit is classified and handled through its federated PEER layer, their interoperability and information exchange are properly supported by the system and the logical and physical distribution of the information over the units is completely transparent to any user in the network.

3.3 PEER federated integration architecture

The PEER federated information management system supports the management, and sharing and exchange of information in a network of loosely/tightly coupled nodes. Each node in the federation network can autonomously decide about the information that it locally manages, how it structures and represents this information, and which part of its local information it wishes to export and share with other nodes. Each node can import information that is exported by other nodes and then transform, derive and integrate (a part of) the imported information to fit its interest and corresponds to the local interpretation. PEER is a pure federated system; namely there is no single global schema defined on the information to be shared by different nodes in the network, and there is no global control among nodes. Clearly, this definition can be made more restrictive based on the configuration of an application environment. The PEER information integration infrastructure helps the human users in a cooperative team, by supporting the information integration at different levels of granularity, e.g. for the global task, or among different activities and sub-activities. PEER supports both the loose coupling and the tight coupling between nodes.

A PEER layer is defined and augmented to every node that needs to share and exchange information with any other node in the network. In the PEER layer of a node, the information is then structured and defined by several kinds of `schemas', as shown in Figure 2. The user needs to identify (1) the information that is available locally in the node (LOC schema), (2) the information that the node needs to access remotely and so the node imports it (IMP schemas), (3) the information that is available locally and other nodes need to access, so the node exports it (EXP schemas), and (4) the integration of the LOC information with the IMP information to create a coherent pool of information (INT schema) that is needed to be accessed by this node. At the development stage, the design of these schemas, makes both the information managed by a node and the interfaces to other nodes explicit. However, once the PEER layer is developed, the integration facility of PEER, its distributed schema management, and its distributed query processing (Afsarmanesh 1993), (Tuijnman 1993), (Afsarmanesh 1994) make the distribution of information and the heterogeneous information representations among different nodes totally transparent to the user.

Information integration in PEER is supported by a declarative specification using the PEER Schema Definition and Derivation Language SDDL (Afsarmanesh 1993). The SDDL language supports the integration, derivation, interrelation, and transformation of types, attributes and relationships from their sources.

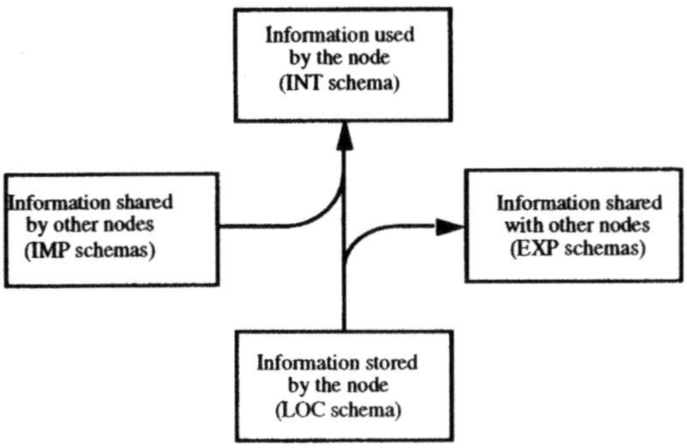

Figure 2 The architecture of the information managed by a node

A prototype implementation of the PEER system is developed in the C language and includes two user interface tools (Afsarmanesh 1994), a Schema Manipulation Tool (SMT) and a Database Browsing Tool (DBT).

3.4 Information analysis for building a PEER layer

One difficulty in building a DIMS for water management systems is due to the fact that the typical existing subsystems lack a recorded semantic description for the data that they work with. Namely, there are no schemas (meta-data) or textual description for the stored data. The gathered information is usually dumped in files and the knowledge about their semantics is only carried by the expert operators who need to interpret the stored data and issue commands, both at the local site and at the control unit.

In order to build a PEER layer for any unit in the network, first step is to identify all the information that is handled by the unit locally. The next step is to pin point the data that this unit needs to access (import) from the other units in the network, and then to identify the operations applied to the data (local or imported) that in turn supports the functionality of the unit.

So far, a preliminary analysis of some information management requirements at the SMAS-Sintra water industry is achieved. The goal of this analysis is to focus on the design of the PEER layer for the remote units. Once this process is completed, the information will be classified/reclassified by the object-oriented database model of PEER and will be represented through corresponding schemas (LOC, IMP, EXP, and INT) at the PEER layer of such units. The analysis results identifies the static and dynamic characteristics of information handled in a remote unit in SMAS industry. Namely, a complete set of all the gathered and handled data; data structures that are currently defined on the data and a textual description for the data items; operations that are currently applied to the data; and other system functionalities is being identified. Following are preliminary the results of this analysis.

a) Variety of data gathered:
Data sensored on the level, flow, pressure, motor power, energy and valve reading, the alarm data etc. represent some of the various kinds of information collected at the unit.

b) Frequency and the amount of gathered data:
The status of all the sensors is read and gathered once every few minutes during the normal operation, generating huge amount of data. In irregular and exceptional situations, e.g. the maintenance situation, requires even more data collection than the periodic reading. This data constitutes the historic database.

c) Simple Data structures currently available:
All information is stored in files, so the record formats are the only structures used to classify and store the information.

d) Operations applied on the data at present:
The alarm handling, historic data search, and data transfer requests are example operations at the unit.

e) Some logged System functionalities:
Logs of actions taken, and control commands are kept at present in the remote unit.

All data gathered at this stage belongs in the LOC schema of the remote units. At present, in the case of SMAS-Sintra there is no cooperation and information sharing among the remote units themselves. Also, there is no communication line between the remote units and other auxiliary units. Therefore, the remote units do not import any information from outside and their main purpose is to collect relevant information and to export them to other units. Thus, the EXP schema of a remote unit is the same as its LOC schema.

4 LEARNING SUBSYSTEM

In most advanced water distribution management systems, the main implemented aspects deal with the problem of control and daily exploitation in which the main objective is the guarantee of a continuous supply. The experience shows that, even with such systems, there are frequent requests to change the control algorithms and control strategies of the remote stations in order to refine the entire system's exploitation. Mainly these requests are based on the analysis of the collected data, for example when the distribution services find evidences such as: if they change the current pumping strategy they can pump the same amount of water in a less expensive way.

In some countries there are even various levels for cost of consumed energy according to the phases of the day. Of course, these conclusions can only be reached a posteriori; namely after the system is running for some time. Originally, and without necessary collected information on a running system, it is not possible to achieve such optimizations. In some other situations a priori decisions can be made to support the change in the water consumption, e.g. to support the seasonal behaviors. These aspects support the idea of developing learning tools for exploitation of the collected data. With such tools the water distribution control technicians can concentrate on the optimization and planning aspects, and to explore the gathered data in order to build a realistic model of the system. Some of the relevant aspects to be learned are:

- Models describing the behavior of specific nodes, mainly those closer to the high concentration of consumers.
- Models to relate the alarms as symptoms of future breaks (prognosis): in pipes, pumps, etc.
- Models for daily forecast of water demands, permitting optimal management of production.
- Consumer trends models, helping future planning of expansion of the network.
- Diagnosis models that help to identify and locate problem areas in the network.

4.1 Examples of learning tasks

There is therefore a large number of potential opportunities for application of learning techniques in the water distribution system. Some examples of learning tasks are:

4.1.1 Support for Production Planning

In this area a number of problems can be considered, all of them related to the planning of water production and distribution or the optimization of this process:

a) Identification and characterization of pattern repetitions along the time (seasonal patterns)
This task is intended to detect temporal repetitions of patterns in collected data:

- Patterns at specific time periods along the day.
- Patterns along the week.
- Seasonal patterns along the year.

Example:

During weekdays, between time H1 and H2, the average consumption reaches the level C.

Or, in a rule form:

IF weekday(D) and T >=H1 and T<=H2 THEN expected_consumption = C.

A characterization of such patterns may be an important input for the production planning activity. This knowledge can help in:

- Avoiding periods of lack of supply.
- Optimizing the production costs, by anticipation of customer needs.

b). Identification of tendencies (changes) in consumption
The objective here is to detect any monotonic changes in the average values of consumption for equivalent time periods. This information gives an idea on how the network is evolving (in terms of demand) and may be a good tool for the network expansion / restructuring activities. Superimposing (comparing) the seasonal patterns with average values for different (but equivalent) time periods can give an idea on how the network is evolving.

Example:

Average flow C in subnet N is steadily increasing X% per year.

c) Identification of working period of devices
Objective: Identification and evaluation of the working periods of devices (groups/pumps) in order to reduce the operation costs. This is specially useful in regions with different energy prices along the day.
Example:
Detection of all devices operating in periods of high price energy.

4.1.2 Identification of the behavior of human operators

Learning the operators' behavior can be useful for:

- Training of new operators.
- Giving advise in critical situations to less experienced operators.
- Contributing to the optimization of the algorithms of local controllers.

The existence of a reference model of good behavior may be necessary in order to validate the examples. Or at least some a-priori qualitative model should be used.
Examples of such learning tasks are:

a) Actions as a function of physical system parameters
To learn which actions are typically performed by operators as a function of read values of system parameters. For this purpose, the values of system variables in a time window before realization of each action have to be analyzed. Similarities of values of some variables in all time windows before a certain kind of action may induce the conclusion that such values were the reason for the decision of applying that action. Nevertheless, such conclusion needs an assessment by a human expert or a validation against an a-priori qualitative model of the system.
Example:
IF level of reservoir R reaches the value L AND group P is on
THEN switch pump P off.

b) Actions as a function of time
Detect / identify actions that are performed in specific times.
Example:
Every weekday group B should be switched on at 18:00.
Or:
IF weekday(D) AND time(D) = 18:00 THEN switch_on(B).

4.1.3 Monitoring and alarm handling

In this area, a set of tasks related to the acquisition of supervision knowledge are considered. For instance:

a) Alarm detection
a.1) Alarm detection based on readings of system variables
Detect primary causes of alarms, i.e., identify explanations for alarms based on the observation of variables evolution before the alarm situation.
Example:
Relationship between an alarm and low pressure level, high flows or pipe break.

a.2) Alarm detection function of time
To detect alarms that occur with some timely repetitive pattern. For instance, alarms due to insufficiencies of the distribution network or due to a bad production plan.

Example:

Every Saturday, near time H, the minimum level alarm of reservoir R goes on.

a.3) Consequences of alarms not processed
Identify what can happen to the system if some alarms are not handled by the operators. The operators may ignore some alarms by assuming these alarms have no consequences. One approach is to verify if any action was done in a time interval after the occurrence of the alarm. If not, we can consider that the alarm was discarded and then a careful analysis of the system behavior, before and after the alarm, is necessary.

Example:

If some maximum level alarms are not handled, other level alarms may be generated and an overflow of reservoirs may occur.

a.4) Inter-relationships between alarms
To detect inter-relationships between alarms that occur in a short time interval. Such alarms may represent several failures or a single failure that provoked various alarms.

Example:

IF a reservoir R is too full THEN the maximum level alarm goes on
AND pressure on pipe D alarm ALSO goes on.

b) Identification of actions related to alarms
b.1) Actions that cause alarms
To identify actions that may cause the system to enter (even temporarily) in an abnormal state. This might be useful in order to create protection or advise mechanisms.

Example:

A few minutes after the start of pomp P, the minimum level of reservoir R goes on.

b.2) Actions after alarms
To identify which actions, i.e., which recovery plan, should be performed for each alarm situation. For this purpose it is necessary to analyze all actions performed after an alarm. On the other hand, it may happen that different operators have different strategies for problem solving after an alarm occurs.

Example:

After a maximum level alarm for reservoir R, pumps P1 and P2 must be switched off.

4.1.4 Preventive maintenance

For instance:

a) Characterization of devices life cycle
Identify the influence of the operating conditions on the life cycle of devices, namely in terms of malfunctioning situations, maintenance, etc. This knowledge might help identifying persistent

problems in a specific region (which need particular attention), problems with a given equipment supplier, the average life time of each class of equipment, etc.

Example:

Valve V, which normally operates at pressure P1, can have a useful life X times larger if operating at pressure P2.

b) Identification of behavioral changes in devices
To identify the evolution of devices behavior with age / time of use.

Example:

Relationship between the time of operation of a group and the volume of pumped water.

4.1.5 Improvement of user satisfaction

To characterize special anomalies, like frequent ruptures in pipes, from the analysis of users' complains and information. Example: Frequent ruptures in the same pipe / zone may suggest its replacement.

4.1.6 Other possibilities

Some additional learning tasks could be:

a) Identify the range of values for system variables in normal operation
To learn which is the range of values, in normal operation, for each system variable. This information contributes to the characterization of what is a normal state and is useful for the supervision activity. These values may also be used for simulation and optimization purposes.

Example:

The pressure in pipe P should be between P1 and P2.

b) Detection of abnormal variations of some variable
To detect and explain (if possible) unexpected variations of some variable, by comparison to the expected pattern.

Example:

A pressure was almost steady for a while and it had a sadden change in spite of no actions have been done.

The above list is a starting point, which needs to be assessed by the end users in terms of usefulness, priorities, availability of data and a-priori knowledge, etc. On the other hand, other learning tasks may be missing in this list, but the approach was to use this list as a first step in an iterative process.

4.2 Preliminary experiments

This section describes some preliminary experiments on machine learning applied to historical data from one of the SMAS-Sintra stations (Pedra Furada). The objective was mainly of a preparatory nature rather than an attempt to generate useful knowledge out of the data. Therefore, the motivation was to:

- Understand the structure and contents of the historic data repositories;

- Identify basic data transformations needed before these data can be applied to standard learning algorithms;
- Start testing the behavior of some algorithms with a selected subset of data.

Training Data:

```
**ATTRIBUTE AND EXAMPLE FILE**

LN-11: (FLOAT)  -> Level
LC-6-1: (FLOAT) -> Flow
LC-6-2: (FLOAT) -> Flow
LP-1: (FLOAT)   -> Pressure
Action: "AG_18(Off)" "AG_18(On)" "None";
@

1.5  4.2  0.4  2.8  "None";
1.3  ?    ?    ?    "None";
?    4.2  0.1  2.7  "AG_18(On)";
?    4.5  0    2.8  "None";
1.3  13.5 0.7  2.8  "None";
1.5  13.1 1.1  2.8  "None";
1.5  12.4 1.1  2.9  "None";
1.5  12.6 1.1  2.9  "None";
1.8  13.1 1.1  2.7  "None";
1.8  13.1 1.1  3.6  "None";
2    ?    ?    ?    "None";
?    13   0.9  3.2  "None";
?    1.9  0.1  3.4  "AG_18(Off)";
2    4.7  0    3.8  "None";
2    4.4  0.4  2.9  "None";
2    4.4  0.3  2.8  "None";
2    4.4  0.4  2.8  "None";
2    4.4  0.4  2.9  "None";
1.8  4.2  0.4  2.8  "None";
1.8  4.4  0.4  2.8  "None";
1.5  4.5  0.4  2.8  "None";
1.5  3.7  0.4  2.8  "None";
1.5  4.5  0.4  3    "None";
1.3  ?    ?    ?    "None";
?    4.4  0.3  2.9  "AG_18(On)";
?    4    0    2.9  "None";
1.3  13.8 0.1  3    "None";
1.3  13.1 1.1  3    "None";
1.3  13.6 1.1  2.8  "None";
1.3  13.1 1.1  2.8  "None";
1.5  13   1.1  2.7  "None";
1.5  13.3 1.1  2.8  "None";
1.5  13.3 1.1  2.8  "None";
1.5  13   1.1  3    "None";
1.5  13   1.1  3    "None";
1.5  13.1 1.1  2.9  "None";
2    ?    ?    ?    "None";
?    12.8 1.1  2.9  "None";
1.8  13.6 0    2.9  "None";
?    1.4  0.1  2.9  "AG_18(Off)";
1.8  4.2  0.3  2.9  "None";
1.5  4.4  0.4  3    "None";
...
...
```

Figure 3a

Generated rules:

```
**RULE FILE**
@
Time: [ Mon Sep 16 19:47:07 1996 ]
Examples: g18p_cn2.txt
Algorithm: UNORDERED
Error_Estimate: LAPLACIAN
Threshold: 0.00
Star: 5
@

*UNORDERED-RULE-LIST*

IF      LC-6-1 < 3.40
  AND 0.05 < LC-6-2 < 0.15
  AND LP-1 > 1.75
THEN    Action = "AG_18(Off)" [7 0 1.25]

IF      LN-11 > 1.90
  AND   LC-6-1 < 2.70
THEN    Action = "AG_18(Off)" [4 0 3.75]

IF      LN-11 > 1.90
  AND   2.30 < LP-1 < 2.75
THEN    Action = "AG_18(On)" [0 1 1.75]

IF      4.20 < LC-6-1 < 4.60
  AND   0.05 < LC-6-2 < 0.25
  AND   LP-1 > 3.45
THEN    Action = "AG_18(On)" [0 2 0.62 ]

IF      LN-11 < 1.90
  AND LC-6-1 > 2.70
  AND 0.05 < LC-6-2 < 0.15
  AND LP-1 < 1.55
THEN    Action = "AG_18(On)" [0 1.50 0.62]

IF      LN-11 > 1.90
  AND   LC-6-1 < 8.60
  AND   LP-1 < 1.55
THEN    Action = "AG_18(On)" [0 0.50 2.25]

...
...
```

Figure 3b

In this project the intention is not to create new learning algorithms but rather to adopt, assess and, if necessary, adapt standard learning systems to this particular application domain. It is our belief that more than new algorithms, the main challenges are to:

- Identify techniques for pre-processing of raw data in order to extract high level features;
- Integrate multiple learning tasks;
- Integrate the learning system with other functionalities (Supervision Systems Optimization System,etc.).

Before a commitment to any particular learning technique or a decision about any necessary changes to classical algorithms is made it is necessary to make a first (empirical) trial evaluation. In this way, a set of inductive algorithms, generating decision trees or rule sets, were selected and applied to a subset of data. This work is still in a very early phase but it helped in getting a more clear feeling about the problem. It has to be pursued and repeated with other classes of algorithms. Lets consider an example:

The learning task considered in the following example is the characterization of situations that lead an operator to perform some action. Figure 3a shows a partial example set selected for the operation of one particular pumping group. Figure 3b shows the set of rules learned by a CN2-style (Clark 1991) algorithm.

Training Data:

```
**ATTRIBUTE AND EXAMPLE FILE**

LN-11: (FLOAT)  -> Level
G_18_status: "On" "Off"
Action: "AG_18(Off)" "AG_18(On)" "None";
@

1.5  "Off"  "None";
1.3  "Off"  "None";
?    "Off"  "AG_18(On)";
?    "On"   "None";
1.3  "On"   "None";
1.5  "On"   "None";
1.5  "On"   "None";
1.5  "On"   "None";
1.8  "On"   "None";
1.8  "On"   "None";
2    "On"   "None";
?    "On"   "None";
?    "On"   "AG_18(Off)";
2    "Off"  "None";
2    "Off"  "None";
2    "Off"  "None";
2    "Off"  "None";
2    "Off"  "None";
1.8  "Off"  "None";
1.8  "Off"  "None";
1.5  "Off"  "None";
1.5  "Off"  "None";
1.5  "Off"  "None";
1.3  "Off"  "None";
?    "Off"  "AG_18(On)";
?    "On"   "None";
1.3  "On"   "None";
1.3  "On"   "None";
1.3 13.6 1.1  2.8  "None";

...
...
```

Figure 4a

Generated rules

```
**RULE FILE**
@
Time: [ Mon Sep 16 20:04:12 1996 ]
Examples: g18a_cn2.txt
Algorithm: UNORDERED
Error_Estimate: LAPLACIAN
Threshold: 0.00
Star: 5
@

*UNORDERED-RULE-LIST*

IF      LN-11 > 1.90
  AND   G_18_status = "On"
THEN    Action = "AG_18(Off)" [4.50 0 16.50]

IF      LN-11 < 1.40
  AND   G_18_status = "Off"
THEN    Action = "AG_18(On)" [0 4.50 10.50]

IF      1.40 < LN-11 < 1.90
  AND   G_18_status = "Off"
THEN    Action = "None"  [0 2.25 173.75]

IF      LN-11 < 1.90
  AND   G_18_status = "On"
THEN    Action = "None" [4.50 0 102.50 ]

...
...
```

Figure 4b

The training data was obtained from the real system and submitted to the learning algorithm after some pre-processing. This example illustrates the viability of using conventional learning algorithms to extract operation rules. However, these rules must be assessed by the domain experts.

Considering that pressures and flows are consequence of the pumping process and not variables that directly influence the decision making, a refinement can be made. Figure 4 shows this situation. For this test the parameters used are the reservoir level and status information on the group 18 (On or Off).

In previous tests unknown values were fed directly to the learning algorithm. Considering that during the sample interval (5 min., in this case) the changes in the water level are small, a new test was performed replacing each unknown ("?") by the previous known value. In this case, for the same training set, the following rules were generated:

IF LN-11 > 1.9 AND G_18_status = "On" THEN Action = "AG_18(Off)"

IF 1.65 < LN-11 < 1.90 AND G_18_status = "On" THEN Action = "AG_18(Off)"

IF LN-11 < 1.40 AND G_18_status = "Off" THEN Action = "AG_18(On)"

IF LN-11 > 1.40 AND G_18_status = "Off" THEN Action = "None"

IF LN-11 < 1.65 AND G_18_status = "On" THEN Action = "None"

(DEFAULT) Action = "None"

As mentioned before, other tests were performed with other algorithms not shown here due to lack of space, but further work is necessary.

4.3 Model driven learning

It is widely recognized that applying standard inductive learning algorithms to real world problems is quite an art. Such algorithms are unable to take into account background knowledge and, therefore, it is up to the engineer to perform a set of preparatory actions / transformations on raw data in order to get useful rules or decision trees. Generated knowledge has to be assessed by a domain expert. This process is usually interactive and iterative, very time consuming and might require addition / removal of training examples, modification of the example description language or even modification of parameters of the learning algorithm. This situation has been reported by several authors.

It is also recognized that a-priori or background knowledge could be used to guide the induction process and then contribute to a more effective way of generating accurate knowledge. In our approach we plan to follow a strategy similar to that of Clark & Matwin (Clark 1993) which use Qualitative Models to guide inductive learning. A Qualitative Model (QM) (Dague 1995) tries to represent in a simplified way the "macroscopic" behavior of a physical system.A QM is a graph whose nodes represent parameters of the system and the arcs represent their relationships. Arcs have labels indicating a "qualitative" relationship between two parameters. One example of such labels is:

+ => monotonically increases with
- => " decreases "

A classical illustration:

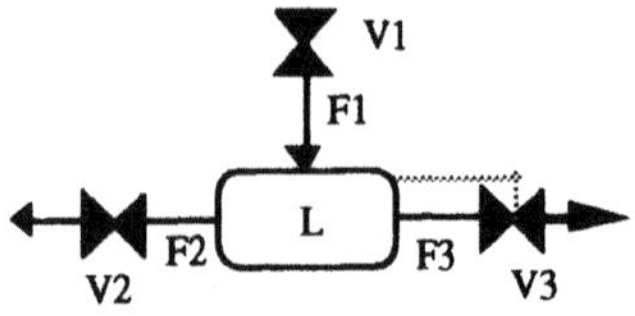

L- level of reservoir
Fi - flow in pipe i
Vi - opening of valve i

Figure 5a Hydraulic example

Causal relationships between variables:

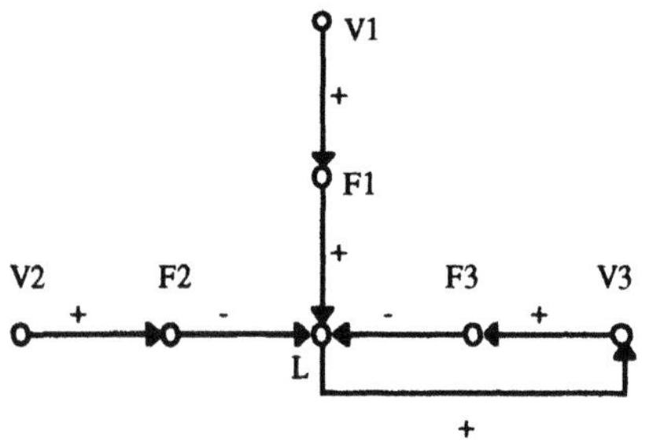

Flow F1 increases with the opening of valve V1
Level L increases with the flow F1
The opening of V3 depends on the level L
...

Figure 5b Causal graph

Other authors use other labels:

$X \xrightarrow{Q+} Y$ Y monotonically increases with X

$X \xrightarrow{I+} Y$ the rate of change of Y (dY/dt) monotonically increases with X

Q- and I- are the inverse monotonic relationships.

The purpose of establishing a QM is not to support simulation or prediction. Our main goal is to guide induction (restricting the search space, for instance). In this way, it is not mandatory that a complete qualitative model be defined. We can even think of relationships that we cannot classify in terms of Q or I labels. This means, we know (or guess) there is a relationship between the two parameters but we don't know its nature. Nevertheless this knowledge is useful, as it gives information about the parameters that must be considered in the training set.

In our current work the following examples of qualitative relationships are being used:

$$A \xrightarrow{+} C$$

$$A \xrightarrow{-} C$$

$$(A1 \mid A2) \; \& \; B \xrightarrow{+} B$$

These examples show relations between physical system variables. These relations represent influences (positive or negative) between variables; they do not represent the exact function that specifies the inter-relationship between the variables.

In this kind of relations we can use discrete or continuous variables and use logical relations to represent influences from various variables over a studied variable.

5 CONCLUSIONS

In this paper a general approach to the development of the next generation water distribution management systems was summarized. Two key components of such systems, the distributed information management and learning subsystems, were discussed and their initial functional requirements established.

A federated information management framework was described to properly support the cooperation and information exchange among different involved sites and their activities in intelligent water distribution control network. The autonomy, heterogeneity, need for loose and tight coupling, and the cooperative information sharing were identified as the characteristics of the network nodes. Three kinds of units, namely the "control unit", the "remote unit", and the "auxiliary unit" were described. Further, the results obtained in the analysis steps for building the interoperation layer among the units were presented.

Preliminary experiments with existing historic data show that control knowledge and optimization guidelines can be extracted from these data by the application of conventional machine learning techniques. However, it became also evident that transformations of raw data (extraction of high level features) are a key point in this process, although no general methodology is available. This will require extensive experimental work. The identification of the adequate data transformations required by the learning subsystem will be an important input to the specification of functionalities of the information management system. Complementarily, the use of incomplete qualitative models is foreseen as a promising approach to guide the learning process and to reduce the iterative assessment effort regarding the "quality" of the acquired knowledge.

ACKNOWLEDGEMENT

This work is funded in part by the European Commission, via the Esprit Waternet project. The authors also thank ESTEC and SMAS-Sintra for the supply of historic data and fruitful discussions.

Fernando José Martinelli also thanks CNPq (Brazilian Council of Research and Development) for his scholarship.

6 REFERENCES

Afsarmanesh, H. et al. (1993), Distributed Schema Management in a Cooperation Network of Autonomous Agents. In Proceedings of the 4th IEEE Int. Conf. on "Database and Expert Systems Applications DEXA'93", Lecture Notes in Computer Science (LNCS) 720, pages 565-576, Springer-Verlag, September 1993.

Afsarmanesh, H., Wiedijk, M., and Hertzberger, L.O.(1994) Flexible and Dynamic Integration of Multiple Information Bases, in *Proceedings of the 5th IEEE International Conference on "Database and Expert Systems Applications DEXA'94"*, Athens, Greece, Lecture Notes in Computer Science (LNCS) 856, pages 744-753. Springer Verlag, September 1994.

Camarinha-Matos, L.M., Seabra Lopes, L., Barata, J. (1996) Integration and Learning, in Supervision of Flexible Assembly Systems, *IEEE Transactions on Robotics and Automation,* Vol. 12, N. 2, April 1996, pages 202-219.

Clark, P. and Boswell, R. (1991) Rule induction with CN2: Some recent improvements, in *Machine Learning - EWSL-91* (ed. Y. Kodratoff), pages 151-163, Berlin, 1991. Springer-Verlag.

Clark, P. and Matwin, S. (1993) Using Qualitative Models to Guide Inductive Learning, in: *Proceedings of the 10th International Machine Learning Conference (ML93).* (ed. P. Utgoff), San Mateo(CA-USA): Morgan Kaufmann Publishers inc., 1993. p.49-56.

Dague, P. (1995). Qualitative Reasoning: A Survey of Techniques and Applications. *AI Communications*, v.8, nrs.3/4, p.119-192. Sept./Dec. 1995.

Quevedo J. et al. (1988) A contribution to the interactive dispatching of water distribution system, in *International Symposium on AI, Expert Systems and Languages in Modelling and Simulation*, pp41-46, Barcelona, 1987.

Stroomloos (1994) *Stroomloos*. Rathenau Institute, The Netherlands, 1994.

Tuijnman, F. and Afsarmanesh, H.(1993) Management of shared data in federated cooperative PEER environment. *International Journal of Intelligent and Cooperative Information Systems (IJICIS)*, 2(4):451-473, December 1993.

Wittig, T. (1992) *ARCHON: An architecture for Multi-Agent Systems*, Ellis Horwood, 1992.

11
Exploration of an unknown environment by a mobile robot

Rui Araújo, and A. T. de Almeida
Institute for Systems and Robotics (ISR), and
Electrical Engineering Department, University of Coimbra
Pólo II, Pinhal de Marrocos, 3030 Coimbra, Portugal
Telephone: +351 (39) 7006276/00. Fax: +351 (39) 35672.
email: rui@isr.uc.pt

Abstract

Mobile robots can make important contributions to the integration of industrial operations. The improvement of mobile robot autonomy will lead to many benefits which include a decrease in programming effort, simplifications on shop floor design, and the lowering of costs and time to market. Additionally this will free human resources which may be applied to the areas of product development. All those factors will ultimately contribute for enabling more agile and sustainable industrial production. In this article we demonstrate the validity of the parti-game self-learning approach for navigating a mobile robot, by finding a path to a goal region of an unknown environment. Initially, the robot has no map of the world, and has only the abilities of sensor-based obstacle detection and straight-line movement. Simulation results concerning the application of the learning approach to a mobile robot are presented.

Keywords

Learning control, path finding, mobile robot

1 INTRODUCTION

An important attribute for autonomous robots, is the ability to navigate their environments, which are usually unknown the first time robots face them. In this article we treat the problem of controlling an autonomous mobile robot, so that it reaches a goal location in an unknown 2D environment. This is called the *robot path finding problem* where a path that avoids collisions with obstacles must be generated between an initial and a goal robot configurations.

Many control methods are based on mathematically modelling the system/plant to be controlled. In the case of mobile robotics for example, it is usual to assume world knowledge, generally appearing in the form of a *global map* of the world over which *path planning* algorithms operate. However it is difficult to provide the robot with a global map model of its world. The difficulties may arise from various reasons. The *map building* operation is itself a difficult separate problem. It may become even more complicated if the robot environment changes. Also, if the robot control system requires the introduction of the global map, this may become a tedious and time consuming programming task.

On the other hand, *reactive systems* (e.g. (Brooks 1986)) don't require the existence of a global map. Those systems are organised as a layered set of task-achieving modules. Each module implements one specific control strategy or *behaviour* like "avoid hitting anything" or "keep following the wall". Each behaviour implements a close mapping between sensory information and actuation to the system. However, basic reactive systems suffer from two shortcomings. First, they are difficult to program. Second and most important, pure reactive controllers may generate inneficient trajectories since they choose the next action as a function of the current sensor readings, and the robot's perception range is limited. To address this second problem, some control architectures combine *a priori* knowledge and planning, with reaction (e.g. (Arkin 1990)).

Robot programming and control architectures must be equipped to face unstructured environments, which may be partially or totally unknown at programming time. An interesting possibility, is *learning* to act over the world by self-improvement of controller performance. This may be viewed as a concept of *self programming* in which control of a complex system in principle does not need extensive analysis and modelling by human experts. Instead the system makes use of sensory information, and learns to interact with the world by constructing it's own abilities.

In this paper we demonstrate the effectiveness of the Parti-game self-learning approach (Moore & Atkeson 1995) to the specific case of learning a mobile robot path from its initial position, to a known goal region in the world. Also the algorithm does not have any initial internal representation, map, or model, of the world. In particular the system has no initial information regarding the location or the shape of obstacles. We demonstrate that the mobile robot can learn to navigate to the goal having only the predefined abilities of doing straight-line motion, and obstacle detection (not avoidance) using its own distance sensors.

The organisation of the paper is as follows. Section 2 presents the learning controller architecture. In section 3 we present the experimental environment around which we performed some experiments concerning the application of the exploration algorithm. Section 4 presents results of simulations using the controller for learning to navigate the robot to a goal region on an unknown world. Finally in section 5 we make some concluding remarks.

2 LEARNING CONTROLLER ARCHITECTURE

The problem we wish to solve in this work may be stated as follows: The mobile robot is initially on some position on the environment, or world, and then it must learn a path to a known goal region in the world. Also the algorithm does not have any initial internal representation, map, or model, of the world. In applying the algorithm, we assume that the initial abilities of the mobile robot are only two. First, it is able to perform *obstacle detection* operations, i.e. to detect obstacles that may obstruct its normal path. Second, the mobile robot is able to perform *straight line movements* between its current position and some other specified position in the world. Two comments to this second ability are in order. First, this ability implies the knowledge of the robot current position, and second the movements may fail because of the presence of an obstacle that is detected.

The mobile robot controller architecture used in this work, is based on the application of the parti-game learning controller algorithm (Moore & Atkeson 1995) to the specific

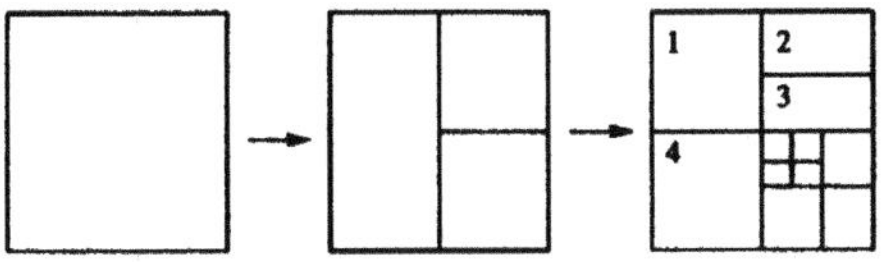

Figure 1 Subdividing the state-space.

case of learning a mobile robot path. In this paper we demonstrate the effectiveness of this approach to the above stated problem.

The parti-game algorithm has application to difficult learning control problems in which: (1) We have continuous and multidimensional state and action spaces; (2) "Greedy" techniques (e.g. trying to act in opposition to an error) and hill-climbing techniques (e.g. optimisation-based training of a neural networks) can become stuck, never reaching the goal; (3) random exploration can be intractably time-consuming; (4) We have unknown, and possibly discontinuous, system dynamics and control laws.

Additionally the algorithm has some restrictions that however, do not prevent its application to our problem: (1) The dynamics are deterministic. (2) The task is specified by a known goal region on a multidimensional state-space. (3) A feasible solution is found, not necessarily an optimal path according to a particular criterion. (4) A local greedy controller is available, which we can ask to move greedily towards a desired state. However, there is no guarantee that a request to the greedy controller will succeed. On our case for example, it is possible for the greedy controller (the "straight-line mover") to become stuck because of an obstacle found by the robot.

2.1 The Algorithm

The Parti-game algorithm is based on partitioning the state-space. It begins with a large partition. Then it increases the resolution by subdividing the state-space (see figure 1) where the learner predicts that a higher resolution is needed. As usual a *partitioning* of the state-space is a finite set of disjoint regions, the union of which covers the entire state-space. Those regions will be called cells, and will be labelled with integers $1, 2, \ldots, N$. In this paper we will assume that the cells are all axis aligned hyperrectangles (rectangles in our 2D case). Although this assumption is not strictly necessary, it simplifies the computational implementation of the algorithm. A ***real-valued state***, s, is a vector of real numbers in a multidimensional space (2D space in our case). Real-valued states and cells are distinct entities. For example the right partitioning on figure 1 is composed of eleven cells, which are labelled with numbers $1 \ldots 11$. Each real-valued state is in one cell and each cell contains a continuous set of real-valued states. Let us define **NEIGHS**(i) as the set neighbours, or cells which are adjacent to i. In figure 1 **NEIGHS**$(1) = \{2, 3, 4\}$. When we are at a cell i, applying an ***action*** consists on actuating the local greedy controller "aiming at cell j". A cell i has an associated set of possible actions that is defined as **NEIGHS**(i). Each action can thus be labelled by a neighbouring cell.

The algorithm uses an environmental model, which can be any model (for example, dynamic or geometric) that we can use to tell us for any real-valued state, control action, and time interval, what will be the subsequent real-valued state. In our case the "model" is implemented by the mobile robot (which can be real or simulated), and takes the current position, and position command, to generate the next robot position.

Let us define the **NEXT-PARTITION**(s, j) function that tells us in which cell we end up, if we start at a given real-valued state, s, and using the local greedy controller, keep moving toward the centre of a given cell, j, until either we exit the initial cell or get stuck. Let i be the cell containing the real-valued state s. If we apply the local greedy controller "aim at cell j" until either cell i is exited or we become permanently stuck in i, then

$$\textbf{NEXT-PARTITION}(s,j) = \begin{cases} i \text{ if we became stuck} \\ \text{the cell containing the exit state otherwise} \end{cases}$$

In our work, the test for sticking performs an obstacle detection with the distance sensors of the mobile robot (see section 3). In other systems, sticking could be tested by seeing if the system has not exited the cell after a predefined time interval.

Let **CENTER**(i) be the real valued state at the centre of the hyperrectangle corresponding to cell i. Define **NEXT**(i, k) as the cell where we arrive if, starting at **CENTER**(i), we apply the local greedy controller "aiming at cell k":

$$\textbf{NEXT}(i,k) = \textbf{NEXT-PARTITION}(\textbf{CENTER}(i), k)$$

In general **NEXT**(i, k) $\neq k$ because the local greedy controller is not guaranteed to succeed.

Define the *cell-length* of a, possibly not continuous, path on the state space, as the number of cell transitions that take place as we go through the path. When the system is on the real-valued state s of a cell i, one of the key decisions that the algorithm has to take, is to choose the cell at which the system should aim using the local greedy controller. The method for making this decision is called a *policy*. As a first attempt to solve the problem we could choose always aiming at a next cell such that the shortest path (in terms of cell transitions) to the goal cell is traversed. The shortest path, $J_{SP}(i)$, from each cell i, to the goal cell, could be defined as:

$$J_{SP}(i) = 1 + \min_{k \in \textbf{NEIGHS}(i)} J_{SP}(\textbf{NEXT}(i,k))$$

except,

$$J_{SP}(i) = 0\,, \quad \text{if } i = \textbf{GOAL}$$

The $J_{SP}(i)$ values can be obtained by a shortest path method such as Dijkstra's algorithm (Horowitz & Sahni 1978) or by Dynamic Programming (Bellman 1957). The next cell to aim is the neighbour, i, with the lowest $J_{SP}(i)$. It is clear that, with this method, once we arrive at a cell i, we should placed at its **CENTER**(i) before we restart applying the local greedy controller aiming at the next cell. But this is not a realistic assumption because the cumulative effect of the system model, and the local controller may not allow us to reach the centre of the cell from the entry point. For example, in our case of a mobile robot, there could be an obstacle. Thus, once we enter cell i we must to assume that we may have to start aiming at the next cell, from a real-valued state $s \in i$ that is different from **CENTER**(i). Besides the impossibility of reaching **CENTER**(i), it may be also impossible to reach the next cell in the shortest path (**NEXT**(i, k)), if we don't start from **CENTER**(i). Therefore **NEXT**(i, k) is not the complete picture of what may happen if we aim at cell k, starting from a real-valued state $s \in i$. What we see is that, the shortest path may be infeasible, but this method would not consider other, possibly longer, but successful paths that could exist.

Since there is no guarantee that a request to the local greedy controller will succeed, each action has a set of possible *outcomes.* The particular outcome of an action depends on the real-valued state s, from which we start "aiming". The outcomes set, of an action j in cell i, is defined as the set of possible next cells:

$$\textbf{OUTCOMES}(i,j) = \left\{ k \;\middle|\; \begin{array}{l} \text{exists a real-valued state } s \text{ in cell } i \text{ for which} \\ \textbf{NEXT-PARTITION } (s,j) = k \end{array} \right\}$$

We can now, using a worst case assumption, define the shortest path from cell i to the goal, $J_{WC}(i)$, as the minimum number of cell transitions to reach the goal assuming that, when we are in a certain cell i and the our intended next cell is j, an adversary is allowed to place us in the worst position within cell i prior to the local controller being activated. The $J_{WC}(i)$ shortest-path is defined as:

$$J_{WC}(i) = 1 + \min_{k \in \textbf{NEIGHS}(i)} \; \max_{j \in \textbf{OUTCOMES}(i,k)} J_{WC}(j) \tag{1}$$

except,

$$J_{WC}(i) = 0\,, \quad \text{if } i = \textbf{GOAL} \tag{2}$$

The $J_{WC}(i)$ values can be obtained by the minimax algorithm (Horowitz & Sahni 1978) or by Dynamic Programming. The value of $J_{WC}(i)$ can be $+\infty$ if, when we are at cell i, our adversary can permanently prevent us from reaching the goal. By definition such a cell is called a *losing cell.* With this method, the next cell to aim is the neighbour, i, with the lowest $J_{WC}(i)$. Using this approach we are sure that, if $J_{WC}(i) = n$, then we will get n or fewer transitions to get to the goal starting from cell i. However, the method is too much pessimistic, because that, regions of a cell that will never be actually visited, are available for the adversary to place us. But those may be precisely the regions that lead to an eventual the failure of the process. So although this method guarantees success if it finds a solution, it may often fail on solvable problems.

Next we will describe **Algorithm 1**, that reduces the severity of this problem by considering only all empirically observed outcomes, instead of all possible outcomes for a given cell. Another argument contributing to this solution, is that as a learning algorithm, it is more important to learn the outcomes set, only from real experience on the behaviour of the system. Besides that, it could be difficult or impossible, to compute all possible outcomes of an action. Whenever an **OUTCOMES**(i,j) set is altered due to a new experience obtained, equations (1) and (2), are again solved in order to find the path to the goal. Before an action is experienced, we can not leave the **OUTCOMES**(i,j) set empty. In these situations we use, the default optimistic assumption that we can reach the neighbour that is aimed. **Algorithm 1** (see figure 2) keeps applying the local greedy controller, aiming at the next cell, on the "minimax shortest path" to the goal, until either we are caught on a losing cell ($J_{WC} = \infty$), or reach the goal cell. Whenever a new outcome is experienced, the system updates the corresponding **OUTCOMES**(i,j), and equations (1) and (2) are solved, to obtain the, possibly new, "minimax shortest path". Step 6.1 computes the next neighbouring cell on the "minimax shortest path" to the goal. Algorithm 1 has three inputs: (1) The current (on entry) real-valued state s; (2) A partitioning of the state-space, P; (3) A database, the set D, of all previously different cell transitions observed in the lifetime of the partitioning P. This is a set of triplets of the following form: (start-cell, aimed-cell, actually-attained-cell). At the end Algorithm 1 returns three outputs: (1) The

ALGORITHM 1
REPEAT FOREVER

1. **FOR** each cell i and each neighbour $j \in$ **NEIGHS**(i), compute the **OUTCOMES**(i, j) set in the following way:
   - **1.1 IF** there exists some k' for which $(i, j, k') \in D$ **THEN**:
     OUTCOMES$(i, j) = \{k \mid (i, j, k) \in D\}$
   - **1.2 ELSE**, use the optimistic assumption in the absence of experience:
     OUTCOMES$(i, j) = \{j\}$
2. Compute $J_{WC}(i')$ for each cell using minimax.
3. Let i := the cell containing the current real-valued state s.
4. **IF** i = **GOAL THEN** exit, signalling SUCCESS.
5. **IF** $J_{WC}(i) = \infty$ **THEN** exit, signalling FAILURE.
6. **ELSE**
   - **6.1** Let $j := \underset{j' \in \mathbf{NEIGHS}(i)}{\mathbf{argmin}} \; \underset{k \in \mathbf{OUTCOMES}(i,j')}{\mathbf{max}} J_{WC}(k)$
   - **6.2 WHILE** (not stuck and s is still in cell i)
     - **6.2.1** Actuate local greedy controller aiming at j.
     - **6.2.2** s := new real-valued state.
   - **6.3** Let i_{new} := the identifier of the cell containing s.
   - **6.4** $D := D \cup \{(i, j, i_{new})\}$

LOOP

Figure 2 Algorithm 1.

updated database of observed outcomes, D, (2) the final, real-valued system-state s, and (3) a boolean variable indicating SUCCESS or FAILURE.

We see that Algorithm 1 gives up when it discovers it is in a losing cell. One of the hypothesis of the algorithm is that all paths through the state space are continuous. Assuming that a path to the goal actually exists through the state-space, i.e. the problem is solvable, then there must be an *escaping-hole* allowing the transition to a non-losing cell and eventually opening the way to reach the goal. This hole has been missed by Algorithm 1 by the lack of resolution of the partition. A hole for making the required transition to a non-losing cell, can certainly be found on the cells at the borders between losing and non-losing cells. Taking this comments into account, the top level **Algorithm 2** (see figure 3), divides in two the cells in the borders, in order to increase the partition resolution, and to allow the search for the mentioned escaping-hole. This partition subdivision takes place between, each successive calls to Algorithm 1 that keep taking place while the system does not reach the goal region. Algorithm 2 has three inputs: (1) The current (on entry) real-valued state s; (2) A partitioning of the state-space, P; (3) A database, the set D, of all previously different cell transitions, observed in the lifetime of the partitioning P. This is a set of triplets of the form: (starting-cell, aimed-cell, actually-attained-cell). At the end, Algorithm 2 returns two outputs: (1) The new partitioning of the state-space, and (2) a new database of outcomes D.

ALGORITHM 2

WHILE (s is not in the goal cell)

1. Run Algorithm 1 on s and P. Algorithm 1 returns the updated database D, the new real-valued state s, and the success/failure signal.

2. IF FAILURE was signalled **THEN**

- **2.1** Let Q := All losing cells in P ($J_{WC} = \infty$).
- **2.2** Let Q' := The members of Q who have any non-losing neighbours.
- **2.3** Let Q'' := Q' and all non-losing members of Q'.
- **2.4** Split each cell of Q'' in half along its longest axis producing a new set R, of twice the cardinality.
- **2.5** $D := D + R - Q''$
- **2.6** Recompute all new neighbour relations, and delete from the database D, those triplets that contain a member of Q'' as a start point, an aim-for, or an actual outcome.

LOOP

Figure 3 Algorithm 2.

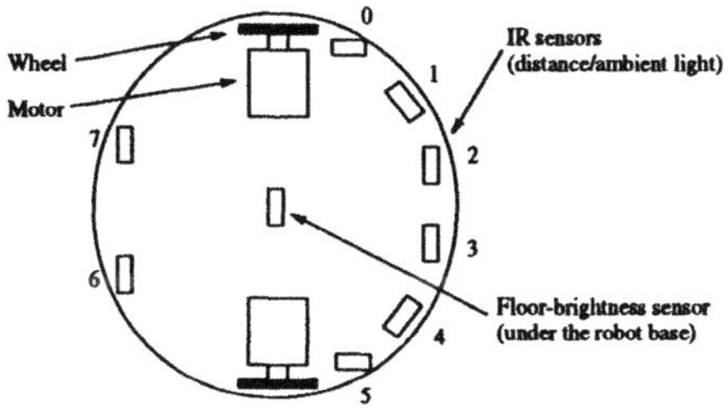

Figure 4 The Khepera miniature mobile robot.

3 EXPERIMENTAL ENVIRONMENT

In this section we give details of the experimental environment that we used to prove, the effectiveness of the algorithm presented in section 2 to solve the robot path finding problem that was formulated on the same section 2.

This work is based on the *Khepera* miniature mobile robot (Mondada, Franzi & Ienne 1993). The circular shaped Khepera mobile robot (see figure 4) has two wheels, each controlled by a DC motor that has an incremental encoder and can rotate in both directions. Each motor can take a speed ranging from −10 to +10. The unit is the (encoder pulse)/10ms that corresponds to 8 millimetres per second. Additionally, for physical supporting purposes the robot has two small balls placed under its platform. The mobile robot includes eight infrared proximity sensors placed around its body, two pointing to the front, four pointing to the left and right sides, and two pointing to the back side of the robot – see figure 4. Each proximity sensor is based on the emission and reception of infrared light. Each receptor is able to measure both the ambient infrared light (with the emitter turned off) and the reflected infrared light that was emitted by the robot itself. The sensor measures do not have a linear characteristics and depend on external factors such as objects surface properties, and illumination conditions. Each sensor reading returns a value between 0 and 1023. A value of zero means that no object is perceived, while a value of 1023 means that an object is very close and almost touching the sensor. Intermediate sensor values may give an approximate idea of the distance between the sensor and the external object.

The results reported in this paper were achieved using the "Khepera Simulator Version 2.0" (Michel 1996). This simulator allows a Khepera robot to evolve on a square world of 1000 mm of side width, where an environment for the robot may be created by disposing bricks, corks and lamps. Robot dynamics are not considered. The simulator uses a geometric model which is clearly adequate for the study of the learning approach.

To calculate its output distance value, a simulated robot sensor explores a set of 15 points in a triangle in front of it. An output value is computed as a function of the presence, or absence, of obstacles at these points. A random noise corresponding to $\pm10\%$ of its amplitude is added to the output light value. The light value output of a sensor is computed as a function of both the distance and the angle between the sensor and the light source. A $\pm5\%$ random noise is added to the resulting value. In our simulations we do not use corks and lamps, but only bricks. Also we do not use the light value reading from the sensors but only the distance value.

For the sticking condition test primitive, that is required in the algorithm described in section 2, we use $d(0), \ldots, d(5)$, which are the distance values of robot sensors 0 through 5 respectively (front - see figure 4). After performing tests with the mobile robot we concluded that it is appropriate to consider that the robot was stuck if at least one of the distances $d(0), \ldots, d(5)$ increases above 700, 400, 900, 900, 400, 700 respectively.

With respect to the motors, the simulated robot simply moves accordingly to the speed set by the algorithm of the user program. A random noise of $\pm10\%$ is added to the amplitude of the motor speed, and a $\pm5\%$ random noise is added to the change of direction (angle) resulting from the difference of speeds of the motors.

4 SIMULATION RESULTS AND DISCUSSION

In this section we present simulation results concerning the application of the learning approach described in section 2. The approach is applied to the problem of navigating a mobile robot to a goal region on an unknown environment. We present two examples.

In example 1 (see figures 5(a), 5(b), and 5(c)). There is a wall between the starting position of the mobile robot and the goal region. As can be seen by the trajectory on figure 5(a), the mobile robot does lots of exploration in the first trial. This is because the system does not have any initial knowledge about the world. In spite of this, the robot is already able to reach the goal in the first trial. As the robot performs more trials it, from experience, acquires knowledge about the "map" of the world. It happens that in this example, at the beginning of the fourth trial, the knowledge is already in its final form, and no further knowledge acquisition is performed. The resulting trajectory of the robot on the fourth trial can be seen in figure 5(b). The final path even though not being optimal (e.g. in terms of shortest geometric travelling path) can subjectively be said to be "weakly optimal" and certainly "not strongly suboptimal". The corresponding final partition can be observed in figure 5(c). We can see that this partitioning, is a representation of the world suitable to the learning algorithm. We also see that the robot does not explore unnecessary areas.

The same comments may apply to example 2 where the world is more complicated

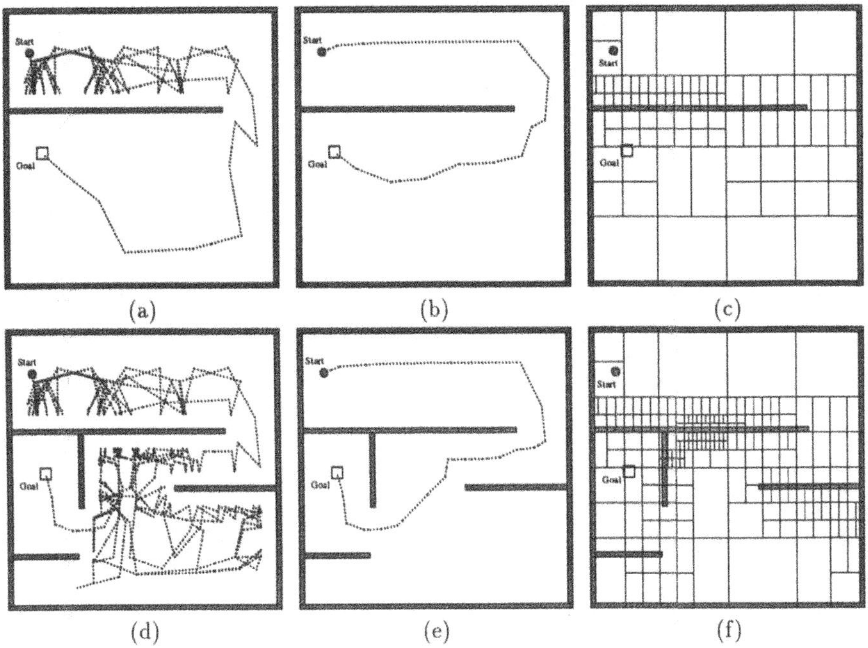

Figure 5 Simulation results concerning example 1: robot trajectory on (a) trial 1, (b) trial 4, and (c) final partition of the world. Simulation results concerning example 2: robot trajectory on (d) trial 1, (e) trial 4, and (f) final partition of the world.

(see figures 5(d), 5(e), and 5(f)). This example requires more initial exploration of the algorithm (figure 5(d)). However, similarly to example 1, on the fourth trial (figure 5(e)) there is also no further exploration and the followed trajectory is in its final form. From the final partitioning of the state-space (figure 5(f)) we see that, again, the system does not explore unnecessary areas.

In our experiments, the learning controller was implemented in the C programming language on a separate process taking approximately 5000 lines of source code, and communicating with the simulator process via the pipe interprocess communication mechanism. This translated to a reasonably small requirement of program memory. Data memory clearly depends on the size and complexity of the robot environment. However the multi-resolution partitioning of the state-space, and the database required by the algorithm clearly leads to an efficient use of data memory. The computational effort (time and other aspects) is not strongly restriction, and will be discussed and evaluated on a separate paper, where the effectiveness of some algorithm improvements will be shown.

We will now discuss the response of the method to a changing environment. Such a change can be seen as a union of one or more changes, each belonging to one out of two possible classes. On class 1, a new obstacle is created on a previous free-space location. Changes of class 2 correspond to the opposite, i.e. an obstacle is removed creating a free area on the state-space. The method is clearly able to overcome a change of class 1. In fact, suppose that an obstacle is created on a location that is currently being used by the robot path to the goal. In response, the method just restarts and continues the exploration process. This will lead to an alternate path to the goal, provided that it exists. We state that, although in some situations the method may be able to overcome situations of class 2, this does not hold in general. The reasoning for this is beyond the scope of this paper.

5 CONCLUSION

In this article we have demonstrated the validity of a learning approach for navigating a mobile robot, by finding a path to a goal region of an unknown environment. Simulation examples have shown that the method is able to build a kind of "map" of the environment without exploring details unnecessary for the execution of the given task. The examples have also shown that in a static environment, the utilisation of the already accumulated information, enables simultaneous learning of a path to a goal, with a decrease on the number of necessary steps in the consecutive trials. This is in contrast with other approaches where some prior topological (and other) knowledge of the environment is used (e.g. (Koenig & Simmons 1996)). Still other approaches don't require a prior map. For example (Beom & Cho 1995) propose a reactive approach that is based on a fuzzy system that learns to coordinate different behaviours. With this method there is no resulting map of the world. However due to the nature of this method, it probably has a higher ability to handle dynamic environments when compared with the approach we used.

REFERENCES

Arkin, R. C. (1990), Integrating Behavioral, Perceptual, and World Knowledge in Reactive Navigation, *in* P. Maes, ed., 'Designing Autonomous Agents: Theory and Practice from Biology to Engineering and Back', MIT Press, Cambridge, MA, pp. 105–122.

Bellman, R. (1957), *Dynamic Programming*, Princeton University Press, Princeton, New Jersey.

Beom, H. R. & Cho, H. S. (1995), 'A Sensor-Based Navigation for a Mobile Robot Using Fuzzy Logic and Reinforcement Learning', *IEEE Trans. Syst. Man Cybern.* **25**(3), 464–477.

Brooks, R. A. (1986), 'A Robust Layered Control System For a Mobile Robot', *IEEE Journal of Robotics and Automation* **2**(1), 14–23.

Horowitz, E. & Sahni, S. (1978), *Fundamentals of Computer Algorithms*, Computer Science Press, Inc, Potomac, Maryland.

Koenig, S. & Simmons, R. G. (1996), Unsupervised Learning of Probabilistic Models for Robot Navigation, *in* 'Proc. IEEE Int. Conf. on Robotics and Automation', pp. 2301–2308.

Michel, O. (1996), *Khepera Simulator Version 2.0 User Manual.* Freeware Mobile Robot Simulator Written at the University of Nice - Sophia Antipolis by Olivier Michel. Downloadable from the World Wide Web at `http://wwwi3s.unice.fr/~om/khep-sim.html`.

Mondada, F., Franzi, E. & Ienne, P. (1993), Mobile Robot Miniaturisation: A Tool for Investigation in Control Algorithms, *in* 'Experimental Robotics III, Proc. 3rd Int. Symp. on Experimental Robotics, Kyoto, Japan', Springer Verlag, London, 1994, pp. 501–513.

Moore, A. W. & Atkeson, C. G. (1995), 'The Parti-game Algorithm for Variable Resolution Reinforcement Learning in Multidimensional State-spaces', *Machine Learning* **21**(3), 199–233.

12

Genetic Algorithms - Constraint Logic Programming. Hybrid Method for Job Shop Scheduling

K. MESGHOUNI[1], P. PESIN[2], S. HAMMADI[1], C. TAHON[2], P. BORNE[1]

[1] *Ecole Centrale de Lille*
LAIL - URA CNRS D 1440
BP 48
59651 Villeneuve d'Ascq Cedex - France

[2] *Université de Valenciennes et du Hainaut Cambresis*
LAMIH - URA CNRS n° 1775
Le Mont Houy
BP 311
59304 Valenciennes Cedex - France

e-mail : mesghouni@ec-lille.fr, pesin@univ-valenciennes.fr, hammadi@ec-lille.fr, tahon@univ-valenciennes.fr, p.borne@ec-lille.fr

Abstract

The job-shop scheduling problem is one of hardest problem (NP-complete problem). In lots of cases, the combination of goals and resources has exponentially increasing search space, the generation of consistently good scheduling is particularly difficult because we have very large combinatorial search space and precedence constraints between operations. So this paper shows the cooperation of two methods for the solving of job shop scheduling problems; Genetic Algorithms (GAs) and Constraint Logic Programming (CLP). CLP is a concept based on Operational Research (OR) and Artificial Intelligence (AI). It tends to rid itself of their drawbacks and to regroup their advantages. The GAs are searching algorithms based on the mechanics of natural selection, they employ a probabilistic search for locating the globally optimal solution. That starts with a population of randomly generated chromosomes, but the difficulty resides in the creation of initial population. This paper explains how to use the CLP to generate a first population and we apply the (GAs) to provide a job shop scheduling minimizing a makespan (Cmax) of the jobs.

Keywords

Genetic algorithms, Initial population, Assignment and scheduling problems, Constraint Logic Programming

1 INTODUCTION

Production planning and job-shop scheduling involve the organization of production process, the coordination and control of the interaction of production factors and their transformation into products over time. Planning is mostly concerned by the description of the product routings for the real manufacturing process. Job-shop scheduling, which is one of the most important problem for companies in the domain of production management, focuses mostly on time-related aspects.

Moreover the scheduling problem is very hard to solve. It is then quite difficult to obtain an optimal solution satisfying the real time constraints using exact methods such as CLP because they require considerable computation time and/or complex mathematical formulation (Carlier, 1988). So GAs are used to solve job-shop scheduling problem and gives "good" solution near the optimal one. This paper presents the cooperation of two methods GAs and CLP for the solving of job shop scheduling problems.

The GAs employ a probabilistic search for locating the globally optimal solution. They have lots of advantages. They are robust in the sense that they provide a set of solutions near of the optimal one on a wide range of problems, they can easily be modified with respect to the objective function and constraints. Unfortunately, it is difficult to apply the traditionally GAs for the scheduling problem :

- There is no practical way to encode a job-shop scheduling as a binary string that does not have ordering dependencies. We fell that the chromosome representation used should be much closer to the scheduling problem. In a such cases a significant effort must be done in designing problem specific 'genetic' operators; however, this effort would pay off in increased speed and improved performance of the system. Therefore in this paper we have incorporated the schedule specific knowledge in operators and in chromosome representation. For this reason the parallel representation of the chromosome was created (Mesghouni 1996).
- How to find an efficient way to create an initial population satisfying all the constraints ?

CLP is a concept based on Operational Research (OR) and Artificial Intelligence (AI). It tends to rid itself of their drawbacks and to regroup their advantages. CLP provides solving facilities for Constraint Satisfaction Problems (CSPs) ; a CSP consists of a set of variables and a set of constraints on those variables . CLP allows to find all the combinations of values of each variable which satisfy all the constraints.

The features of CLP are :

- extension of PROLOG by adding solves on new domains such as finite domain restricted terms, boolean terms, linear rational terms,
- concept of constraint resolution which replace the concept of unification of PROLOG,

- active treatment on constraints thanks to the "test and generate" principle which induce a reduction of solving times,
- CLP don't give one (unique and optimal) solution but is able to give a set of solutions which satisfy a set of constraints.

This paper is organized as follows : The second section contains a detailed description and formulation of our flexible job-shop scheduling problem. The GAs and their operators are described in the third section. The fourth section contain a description of the CLP. The description of our hybrid process is presented in the fifth paragraph. Finally the last section contains an illustration of our method through some examples.

2 DESCRIPTION OF THE PROBLEM

The data and the constraints of this problem are (Hammadi, 1991) :

- There are n jobs, indexed by j, and these jobs are independent of each other,
- each job j has an operating sequence which is called g_j ,
- each operating sequence g_j is an ordered series of x_j operations $O_{i,j}$ for i = 1... x_j,
- the realization of each operation $O_{i,j}$ requires a resource or machine selected from a set of machines, $M_{i,j} = \{M_{i,j,k}$, k = 1,2, ...,M / M = the total number of machines existing in the shop$\}$ this implying the existing of the assignment problem.
- the processing time $P_{i,j,k}$ of an operation depends on the chosen machine,
- a start operation runs to completion (non preemption condition),
- each machine can perform operations one after the other (resource constraints).

3 GENETIC ALGORITHMS

3.1 Concepts

Genetic algorithm is a relatively new approach of optimum searching. GAs inherits its ideas from evolution of the nature. They are a parallel and global search technique because they simultaneously evaluates many points in the search space, they are more likely to converge towards a global solution. These algorithms are developed in a way to simulate a biological evolutionary process and genetic operators acting on the chromosome. They begin with a population of randomly generated strings and evolves towards a better solution by applying genetic operators; reproduction, crossover and mutation (Goldberg, 1989). They are so efficient that they can find a nearly optimum solution even for a large scale problem. For solving the traditional NP-completed problem some modification are indispensable (Michalewiez, 1992). In this paper genetic algorithms are mandatory by the following components :

- a specific genetic representation (or encoding) depending on the problem for determining the feasible solutions of the job-shop scheduling optimization problem (Syswerda, 1990),
- an efficient way to create an initial population of potential solutions,
- original genetic operators that alter the composition of children during reproduction.

3.2 Representation of the chromosome

Traditionally, chromosomes are simple binary vectors. This simple representation is not convenient for the complex job-shop scheduling problem when we must respect a certain order. A better solution is to change the chromosome syntax to fit the problem. In the case of our problem, the chromosome is represented by a set of parallel machines and each machine is a vector which contains the assignment operations for this machine. The operations are represented by three terms. The first one is the order number of the operation in its operating sequence, the second one is the number of the job, to which this operation belongs and the third one is the starting time of the operation if its assignment on this machine is definitive (Mesghouni, 1996). This starting time is calculated taking into account the resources constraints.

3.3 Design of crossover operator

In nature, crossover occurs when two parents exchange parts of their corresponding chromosomes. In a genetic algorithm, crossover recombines the genetic material in two parents chromosomes to make two children in order to generate a better solution.
The child one is given by the following algorithm (Mesghouni, 1997) :

```
step 1 :
        Parent 1, Parent 2 and the machine Mk are randomly selected.
step 2 :
        {Oi,j}k of Child 1 = {Oi,j}k of Parent 1
        i = 1
step 3 :
        /* M is the total number of the machines */
        While (i < M ) do
        Begin
                If ( i ≠ k ) then
                Begin
                        Copy the non existing operations of Mi of parent 2 into child 1
                        i = i + 1
                End
        End
step 4
        i = k
        If ( any operation of child 1 is missing ) then
                Scan Mk of parent 2 and copy the missing operation
        End if
```

To obtain child 2 go to step 2 and invert the role of parents 1 and 2.

3.4 Mutation

Mutation plays an important role in genetic algorithms, it introduces some extra variability into the population, it typically works with a single chromosome, it always creates another chromosome (Mesghouni, 1997).

4 Constraint Logic Programming

4.1 Concepts

CLP (Fron, 1994) is a concept based on Operational Research (OR) and on Artificial Intelligence (AI). It tends to rid itself of their drawbacks (lack of flexibility of the cost function, one unique and optimal solution if it exists, if it doesn't exist there is no response, impossibility to introduce qualitative criteria for the OR and unsuitability for numerical problems, high solving times for AI) and to regroup their advantages (ability to process numerical constraints for RO and power of data representation, independence data/data processing, solving of problems unsolved by classical algorithms for AI). We can sum up the different features of CLP :

- CLP is an extension of PROLOG by adding solvers on new domains such as finite domains restricted terms, boolean terms, linear rational terms, The concept of constraint resolution replace the concept of unification in PROLOG (Llia, 1991),
- The treatment on constraints is active thanks to the "test and generate" principle which induce a decrease of solving times (Esquirol, 1995),
- CLP doesn't give one (unique and optimal) solution but is able to provide a set of solutions which satisfy a set of constraints (Fron, 1994).

4.2 CLP and CSP

CLP provides solving facilities for CSP (Constraint Satisfaction Problems). A CSP (Boronad, 1993) consists of a set of variables, a set of definition domains for those variables and a set of constraints of those variables. As a consequence, you can solve a problem by this process :

- definition of variables and constraints,
- a priori propagation of the constraints,
- production of solutions according to definition domains and constraints.

So a scheduling problem can be regarded as a CSP, where variables are beginning time of each task, domains are definition domains of each variable and constraints are various constraints.

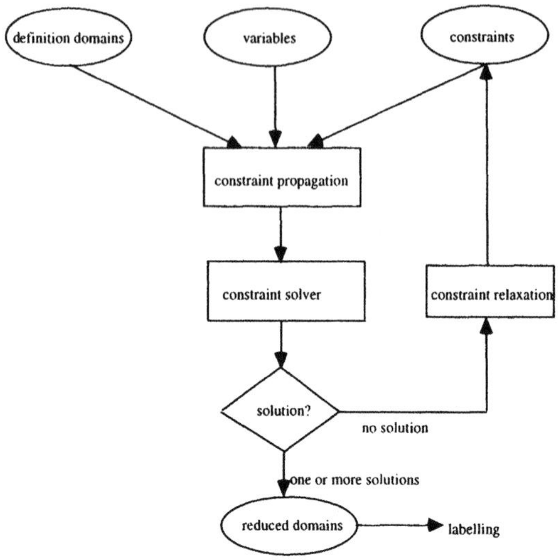

Figure 1 : Describe the structure of CLP using in scheduling

4.3 CLP and scheduling

In the field of scheduling, we can distinguish five main types of constraints (where $S_{i,j}$ = starting time of operation i of job j, $P_{i,j}$= processing time of operation i of job j after the choice of the resource, $I_{i1,j1-i2,j2}$=interval period between operation i1 of job j1 and operation i2 of job j2, $q_{i,jRt}$=amount of resource R selected for task i of job j at t, Q_R = the total amount of resource R available)

- relative location of an operation : to place an operation in relation with an other or with the others ($S_{i2,j2} - S_{i1,j1} >= I_{i1,j1-i2,j2}$ or $S_{i2,j2} - S_{i1,j1} >= P_{i1,j1}$ in the particular case),
- absolute location of a task : to place a task absolutely during the time ($S_{i,j} >= T$ where T is a given moment),
- disjunctive constraint : to force tasks not to be executed at the same time (if $S_{i1,j1} + P_{i1,j1} > S_{i2,j2}$ then $S_{i2,j2} + P_{12,j2} <= S_{i1,j1}$; if $S_{i2,j2} + P_{i2,j2} > S_{i1,j1}$ then $S_{i1,j1} + P_{i1,j1} <= S_{i2,j2}$),
- cumulative constraint : to force not to use the maximum amount of a resource available, for the set of all tasks using this resource($\sum q_{i,jRt} <= Q_R$),
- affectation constraint : to express the fact that one task must be processed by one resource and only one ($\sum A_{i,j,k} = 1$ during all the processing time of task i of job j, where $A_{i,j,k}$ is the affectation variable, k is the number of the resource, i,j means the task i of job j, $A_{i,j,k} = 0$ when task i of job j is not processed by k, $A_{i,j,k} = 1$ when task i of job j is processed by k).

CLP provides facilities to express all these constraints. All the constraints above can easily be expressed by predicates. We give the four main ones in the particular context of the constraint programming language CHIP (Constraint Handling in Prolog). The # means we are working on CHIP finite domains :

- succession constraints : succession ($S_{i2,j2}$,$S_{i1,j1}$, $P_{i1,j1}$):$S_{i2,j2}$ #>= $S_{i1,j1} + P_{i1,j1}$

- disjunctive constraints : disjunctive ($S_{i1,j1}$, $P_{i1,j1}$,$S_{i2,j2}$, $P_{i1,j1}$) : $S_{i1,j1}$ #>= $S_{i2,j2}$ + $P_{i2,j2}$
  disjunctive ($S_{i1,j1}$, $P_{i1,j1}$,$S_{i2,j2}$, $P_{i2,j2}$) : $S_{i2,j2}$ #>= $S_{i1,j1}$ + $P_{i1,j1}$.
- cumulative constraints (for a given resource) cumulative (Ld,Ldur, Lres, Lf,Lsur, Hau, Fin, Valinterm), with Ld:list of the starting times of the tasks processed by the given resource, Ldur:list of the duration of tasks, Lres: amount of resource given to each task, Lf:list of the ending times of the tasks processed by a given resource, other parameters are not very important in our context,
- affectation constraints (for a given task i of job j and for kn machines k1,k2,...,kn where i of job j may be processed) : affectation ($A_{i,j,k1}$,$A_{i,j,k2}$,...,$A_{i,j,kn}$) :-Ai,j,k1 + Ai,j,k2 + Ai,j,k3 +...............+ Ai,j,kn #= 1.

4.4 Hybrid process

After selecting an appropriate genetic representation of solutions to the problem, we should now create an appropriate initial population of solutions. This can be done in simple GAs in random way. However, some care should be taken in our problem because a set of constraints must be satisfied. Therefore a population of potential solutions found by CLP method is accepted as a starting point of our evolution GAs

In this paper, we present a cooperation of the CLP and the GAs to solve the flexible job-shop scheduling problem; we proceed in two steps.

Step 1 : We use CLP in order to provide a set of beginning solutions(set of scheduling) according to figure 1

Step 2 : we apply the (GAs) to solve a job shop scheduling minimizing a makespan (Cmax) of the jobs using a set of solutions found by the CLP like an initial population.

The first step is consists in two substeps :

- Finding a set of assignment (one for each task) with unlimited capacity which minimize the ending time of the scheduling. In this step succession constraints (for the tasks of each job) and assignment constraints (for each task of the scheduling problem) are applied. Ending time of scheduling is fixed to the minimum and all the possible sets of assignments are enumerate. An other solution consists in finding a set of assignment randomly for bigger size problem.
- Finding a set of scheduling with each assignment found above with limited capacity which minimize the ending time of the scheduling. In this step, cumulative constraints (for each resource) are applied. In this step succession constraints (for the task of each job) and assignment constraints (for each task of the scheduling problem) are applied

In the second step, we take this set of solutions(scheduling), we introduced it in GAs like an initial population, and we apply the genetic operators to improve the solution toward a optimal one.

5 NUMERICAL EXAMPLE

This section report on experiments made to show the performance of the proposed hybrid method. For the CLP we use the constraint programming language CHIP (Constraint Handling in Prolog) to generate a set of feasible schedule in a reasonable computing time. This set of solutions represent a good initial population for the GAs which is implemented in a C-language program to provide quickly a near optimal schedule.

5.1 Little size problem

In this section we compare the results given by still only GAs, and the solution given by our hybrid method.
We study the following example, 5 machines and 3 jobs.
The operating sequences of these jobs are

Job 1 : O1,1, O2,1, O3,1.
Job 2 : O1,2, O2,2, O3,2.
Job 3 : O1,3, O2,3.

and the processing time of these operations are shown in table 2

	M1	M2	M3	M4	M5
O 1,1	1	9	3	7	5
O 2,1	3	5	2	6	4
O 3,1	6	7	1	4	3
O 1,2	1	4	5	3	8
O 2,2	2	8	4	9	3
O 3,2	9	5	1	2	4
O 1,3	1	5	9	3	2
O 2,3	5	9	2	4	3

Table 1 Processing time of the operations.

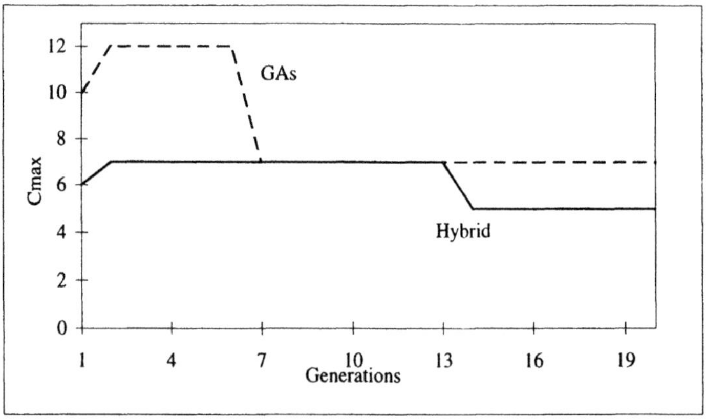

Figure 2 : Represent the Cmax for still only GAs and the Cmax for our hybrid method.

For the still only GAs, we proceed as following, we take one solution given by other methods (in this case by temporal decomposition method) and we apply the previously mutation operators to extend the population. We apply the GAs with the following operators :

- Size of population = 20
- Crossover probability (Pc) = 0.65
- Mutation probability (Pm) = 0.05
- Number of generation (NB gen) = 20.

For the CLP only, we use the procedure described in the section 4.3, in the figure 2 we represent the different cmax associated to set of scheduling

For the hybridization results, we take the set of solutions given by the CLP, we use it like the first population and we apply the GAs with the follow operators :

- Size of population = 12
- Crossover probability (Pc) = 0.65
- Mutation probability (Pm) = 0.05
- Number of generation (NB gen) = 20.

In this case the solution given by the hybrid method is better then of the solution gives by the GAs only and by the CLP only. But in order to prove the effectiveness of our hybrid method we should apply it for a big size problem.

5.2 Big size problem

We consider a big size problem with 10 jobs and 10 machines, each job has 3 operations and each operation can be performed by any machine(we have a total flexibility). In order to solve this problem by our hybrid method, we propose to proceed as in the following flow chart :

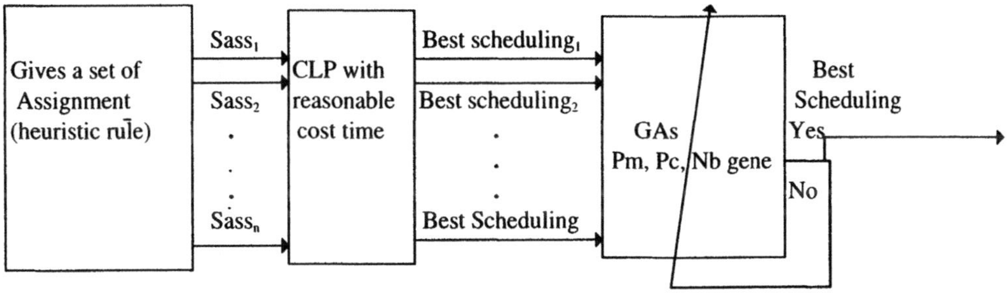

Figure 3 : Flow chart of hybrid method for big size problem.

With Sassi : a set of assignment of the operations to the machines.
i = 1 to n

This problem is very difficult to solve due to the very large combinatory process.

In this context, only applying CLP requires considerable computation time or complex mathematical formulation. Only applying GAs with a random initial population leads the 'genetic' operators to work in an inappropriate search space. So using an exact method such as branch-and-bound and dynamic programming or CLP method to create a good initial population is an important starting point for the evolution GAs.

6 CONCLUSION

The application of a new method to a flexible job-shop scheduling problem with real-word constraints has been defined. In this paper and for a little size problem the hybrid method is efficient and gives a good solution satisfying all of the constraints (precedence and resources constraints) and confirming the effectiveness of the proposed approach. This very good results gives by little size problem encourage for using this method like a figure 3 for a big size problem (10 jobs and 10 machines), for instance.

7 REFERENCES

Boronad-Thierry C. (1993) Planification et ordonnancement multi-site : une approche par satisfaction de contraintes, Thèse de doctorat, Université Paul SABATIER, Toulouse.

Carlier, J and Chretienne, P. (1988) Problèmes d'ordonnancement : Modélisation / complexité / algorithmes. Masson, Paris

Esquirol, P., Lopez, P. (1995) Programmation logique avec contraintes et ordonnancement : Automatique-Productique-Informatique Industrielle, Hermès, Vol. 29, n°4-5, pp379-407.

Fron A. (1994) Programmation par contraintes, Addison-Wesley.

Goldberg, D. E. (1989) Genetic algorithms in search, optimization, and machine learning. Addison-wesley.

Hammadi, S .Castelin, E. Borne, P.(1991) Efficient and feasible job-shop scheduling in flexible manufacturing systems : IMACS international symposium, Toulouse, France, March 13-15.

LLIA (1991) La programmation par contraintes, LLIA, n°73, July-august .

Mesghouni, K. Hammadi, S. Borne, P (1996). Production job-shop scheduling using Genetic Algorithms. proceedings of IEEE /SMC Conference, Beijing, China, Vol 2, October 14-17,. 1519-1524.

Mesghouni, K. Hammadi, S. Borne, P (1997) , Parallel Genetic Operators For flexible Job-Shop Scheduling, to appear in 1st International Conference on Engineering Design And Automation, Bangkok, Thailand, March 18-21.

Michalewicz, Z. (1992) Genetic Algorithms + Data Structures = Evolution programs: Springer Verlag.

Syswerda, G. (1990) Schedule optimization using genetic algorithm: in handbook of genetic algorithm (ed. Van Nostrand Reinhold), New-York.

PART FIVE

Mechatronics

13

Kinematic study of biped locomotion systems

F.M. Pereira da Silva
Modern University, Dept. of Control and Automation
Rua Augusto Rosa 24, 4000 Porto, Portugal
Tel: 351 2 2026938, Fax: 351 2 2026939, E.mail: fpsilva@fe.up.pt

J.A. Tenreiro Machado
Faculty of Engineering of the University of Porto, Dept. of Electrical and Computer Engineering
Rua dos Bragas, 4099 Porto, Portugal
Tel: 351 2 317105, Fax: 351 2 319280, E.mail: jtm@fe.up.pt

Abstract

This paper presents the kinematic study of robotic biped locomotion systems. The main purpose is to determine the kinematic characteristics and the system performance during walking. For that objective, the prescribed motion of the biped is completely characterised in terms of five locomotion variables: step length, hip height, maximum hip ripple, maximum foot clearance and link lengths. In this work, we propose four methods to quantitatively measure the performance of the walking robot: energy analysis, perturbation analysis, lowpass frequency response and locomobility measure. These performance measures are discussed and compared in determining the robustness and effectiveness of the resulting locomotion.

Keywords

Biped locomotion, locomotion variables, kinematics, motion planning, performance measure

1 INTRODUCTION

In recent years a growing field of research in biped locomotion has resulted in a variety of prototypes whose characteristics, in limited sense, resemble the biological systems (Golden, 1990; Kajita, 1994). This new generation of innovative robotic systems, for real-world applications, is bringing up fascinating perspectives in research and development work. The control of legged vehicles is a difficult and multidisciplinary problem and, even today, is regarded as the most crucial aspect of locomotion (Hemami, 1983). Vukobratovic *et al.* (1990) have proposed models and mechanisms to explain biped locomotion. On the other hand, Raibert and his colleagues built various biped walking machines. Extensive considerations of the major issues with dynamic balance have been presented by Raibert (1986) and Todd (1985). In another perspective, Galhano *et al.* (1992) demonstrated the appeal of implementing biological-inspired schemes in the analysis of mechanical systems. The theories and algorithms guiding the biological research suggested specific models to apply in biped robots.

In this line of thought, several researchers are in pursuit of better walking robots and continue to refine their models of locomotion (Kun, 1996; Yamaguchi, 1996). The main objective is to design and control an optimum leg geometry that provides good energy efficiency, the required working volume and simplicity in the structure. The movement of the robotic biped will be produced by joint actuators and will be moderated by environmental aspects. Therefore, the resultant motion depends on those factors influencing the structural and functional characteristics of the robot (kinematics problem) and those related to physical phenomena, such as gravity, friction and reaction forces (dynamics problem).

This paper concentrates in the kinematic study of a planar biped model with six degrees of freedom and rotational joints. The main purposes are threefold:

- To gain insight into the phenomenon of biped walking.
- To characterise the biped motion in terms of a set of locomotion variables.
- To establish the correlation among these locomotion variables and the system performance.

A wide variety of gait patterns are developed by prescribing the cartesian trajectories of the hip and the lower extremities of the leg. For the analysis of the resulting motion, various kinematic performance measures are proposed and discussed. These kinematic indices can be regarded as quantitative measures of the biped's ability in transporting the section of the body from an initial position to a desired position through the action of the lower limbs. The performance measures and the associated graphical results are used for evaluating the different locomotion patterns.

The remainder of the paper is organised as follows. A short description of the biped model is given in section 2. In section 3 we describe the method used to plan the kinematic trajectories of the biped robot. In section 4 various kinematic performance measures are proposed and discussed mathematically. Given this background, several simulation examples are presented in section 5 to illustrate the application of these methods. Finally, in section 6 we outline the main conclusions and the perspectives for future research.

2 BIPED MODEL

Figure 1 shows the planar biped model and the conventions used throughout this paper.

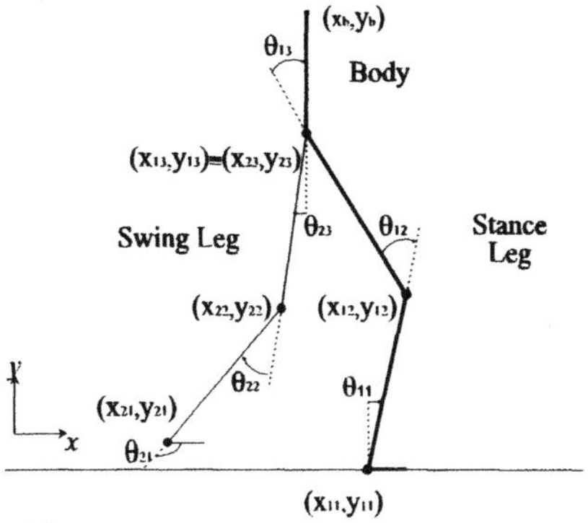

Figure 1 Planar biped model.

The kinematic model consists of seven links in order to approximate locomotion characteristics similar to those of the lower extremities of the human body (*i.e.*, body, thigh, shank and foot). In the present study, we consider a planar biped model with rigid links interconnected through rotational joints. During locomotion, the stance leg (*i.e.*, the limb that is on the ground) has a degree of freedom formed in the contact between the foot and the ground. A step in the gait of a biped consists of two phases:

- Single support phase in which one leg is in contact with the ground and the other leg swings forward.
- Exchange of support in which the legs trade role.

In the single support phase, the stance leg is in contact with the ground and carries the weight of the body, while the swing leg moves forward in preparation for the next step. The impact of the swing leg is assumed to be perfectly inelastic and the friction is sufficient to ensure that no slippage occurs. Additionally, we consider that the support shifts instantaneously from one limb to the other.

3 FORWARD MOTION PLANNING

A fundamental problem in robotics is to determine the trajectories that allow the biped robot to walk more skilfully (Shih, 1993). In early work, the determination of the biped trajectories was made largely on the basis of experience and intuition (*e.g.*, recording kinematic data from human locomotion (Silva, 1996)). In this work, the motion planning is accomplished by prescribing the cartesian trajectories of the body and the lower extremities of the leg. Furthermore, the prescribed motion is completely characterised in terms of five locomotion variables. Simultaneously, we determine the relation between the locomotion variables and the trajectories physically admissible. In this context, we introduce the concept of locomotion workspace.

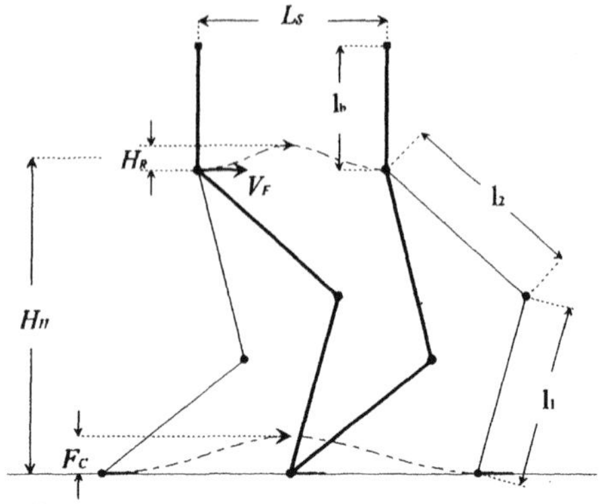

Figure 2 Locomotion variables.

3.1 Locomotion variables

The prescribed motion of the biped is completely characterised in terms of the following variables: length of the step L_S, hip height H_H, maximum hip ripple H_R, maximum foot clearance F_C and leg link lengths l_1 and l_2 (see Figure 2). The hip height is defined as the mean height of the hip along the cycle of walking. The hip ripple is measured as the peak-to-peak oscillation magnitude. During the experiments, we examine the role of the link lengths considering that $l_1 + l_2$ assumes a constant value equal to 1 meter.

3.2 Development of the kinematic trajectories

The proposed algorithm accepts the hip and feet's trajectories as inputs and, by means of an inverse kinematics algorithm, generates the related joint trajectories. To improve the smoothness of the motion we impose two additional restrictions: the body maintains an erect posture during locomotion and the forward velocity is constant. Therefore, the horizontal trajectory of the hip is implemented using a constant progression velocity V_F. One trajectory that undergoes smooth motion is the flat trajectory in which the stance leg adjusts itself so that the hip maintains a constant height. Simultaneously, it is of interest to exploit trajectories in which de hip translates with some vertical motion. In order to simplify the problem, we consider that such motions are produced based on sinusoidal functions.

In dynamic walking, at each footfall, the system may suffer impacts and incurs on additional accelerations that influence the forward velocity (Zheng, 1984). For this reason, we must impose a set of conditions (continuity conditions) on the leg velocities so that the feet are placed on the ground while avoiding the impacts. We denote the moment of exchange of support as time t_1, and by t_1^- and t_1^+ the times before and after the impact occurs, respectively. For smooth exchange-of-support, we require the angular velocities, before and after, to be identical, that is:

$$\dot{\theta}_{2i}\left(t_1^-\right) = \dot{\theta}_{1i}\left(t_1^+\right) \qquad 1 \le i \le 3 \tag{1}$$

The kinematic relations have been used and the differential problem solved to obtain the cartesian velocities immediately before and after contact. These equations constrain the biped to zero velocity foot falls. The foregoing derivation determined conditions for smooth exchange-of-support. In accordance, the equation of the tip of the swing leg along the x-axis is computed by summing a linear function with a sinusoidal function:

$$x_{21}(t) = 2V_F\left[t - \frac{1}{2\pi f}\sin(2\pi f t)\right] \tag{2}$$

where f is the step frequency (number of steps per unit of time). Moreover, the vertical motion, that allows the foot to be lifted off the ground, is implemented using the function:

$$y_{21}(t) = \frac{F_C}{2}\left[1 - \cos(2\pi f t)\right] \tag{3}$$

The trajectory generator is responsible for producing a motion that synchronises and coordinates the leg behaviour. In this perspective, we assure that the swing limb arrives at the contact point when the upper body is properly centred with respect to the two lower limbs.

3.3 Workspace analysis

One pertinent question involving the motion planning is to determine which trajectories are physically realisable. Upon this point, we define the locomotion workspace of the biped robot as the set of all physically possible hip heights and step lengths, relative to a given reference trajectory (see Figure 3). The kinematic workspace, as defined above, can be investigated by solving the kinematic equations given two kinds of restrictions.

- Sinusoidal restrictions due the type of velocity and acceleration trajectories used.
- Knee-foot restrictions since the legs are constrained by ground contact and the knee joints lock at certain positions (similar to human knee).

In this way, all the results derived ahead are capable of displaying the performance measure for interpretation over the entire stand-up workspace.

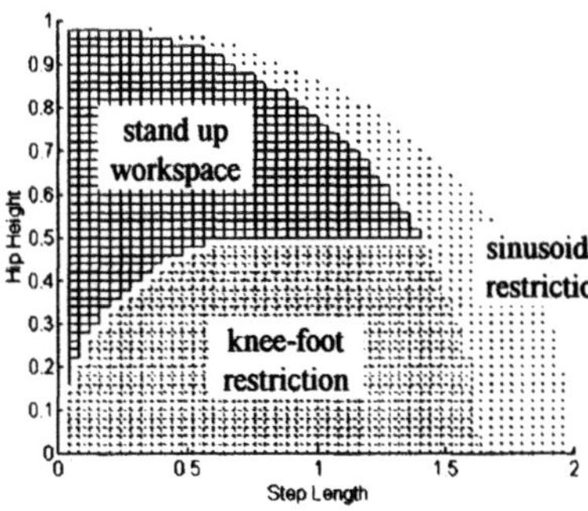

Figure 3 Workspace boundaries of the biped robot with $l_1 = l_2 = 0.5$.

4 PERFORMANCE EVALUATION

This section covers the implementation of different performance measures used in the evaluation of the biped's kinematic performance. The kinematic analysis is a challenging problem since the biped motion is characterised by gait patterns with complexity of large degrees of freedom (*i.e.*, multidimensional problem). In mathematical terms, we shall provide four global measures of the overall dexterity of the mechanism in some average sense. The aim is to verify whether a correlation between different viewpoints could be found in walking.

4.1 Energy analysis

The essence of locomotion is to move smoothly the section of the body from one place to another with some restrictions in terms of execution time. This makes the forward velocity V_F an important variable in determining the walking performance. On the other hand, the objective function should be expressed in the coordinate system on which movement control occurs. This appears to favour the use of joint coordinates to calculate the performance measure.

In this line of thought, we intent to determine the energy characteristics of the rotational-to-linear motion system. This motion conversion is concerned with taking the rotational motion from the joint actuators and producing a linear motion at the body. As a means of inferring the energy flow, we suppose that the translational system exhibits a fictional friction acting on the tip of the body. The linear relationship between the force and the velocity is given by,

$$F(t) = Bv_b(t) \tag{4}$$

where B is the viscous-friction coefficient. For purposes of analysis, suppose that the value of the fictitious friction coefficient is constant $B = 1$ Ns/m. The velocity at the body can be obtained, as function of the joint variables, by differentiating the forward kinematic equations:

$$\dot{x}_b = \Gamma_{X_1}(\theta)\cdot\dot{\theta}_{11} + \Gamma_{X_2}(\theta)\cdot\dot{\theta}_{12} + \Gamma_{X_3}(\theta)\cdot\dot{\theta}_{13} \tag{5}$$
$$\dot{y}_b = \Gamma_{Y_1}(\theta)\cdot\dot{\theta}_{11} + \Gamma_{Y_2}(\theta)\cdot\dot{\theta}_{12} + \Gamma_{Y_3}(\theta)\cdot\dot{\theta}_{13} \tag{6}$$

In order to assure an erect posture of the body, the transfer relationships between the joint space and the operational space, along the x and y-axis, are expressed by:

$$\Gamma_{X_1}(\theta) = l_1\cos\theta_{11} + l_2\cos(\theta_{11} + \theta_{12}) + l_b \tag{7}$$
$$\Gamma_{X_2}(\theta) = l_2\cos(\theta_{11} + \theta_{12}) + l_b \tag{8}$$
$$\Gamma_{X_3}(\theta) = l_b \tag{9}$$

and

$$\Gamma_{Y_1}(\theta) = -l_1\sin\theta_{11} - l_2\sin(\theta_{11} + \theta_{12}) - l_b \tag{10}$$
$$\Gamma_{Y_2}(\theta) = -l_2\sin(\theta_{11} + \theta_{12}) - l_b \tag{11}$$
$$\Gamma_{Y_3}(\theta) = 0 \tag{12}$$

For the case of a force acting in the same direction as the velocity of the body, the average power may be defined as the product of the force and the average velocity of the body, that is,

$$P_x = \frac{1}{T}\int_0^T \dot{x}_b^2(t)dt \tag{13}$$

$$P_y = \frac{1}{T}\int_0^T \dot{y}_b^2(t)dt \tag{14}$$

where $\dot{x}_b^2$ and $\dot{y}_b^2$ are related with the joint angular velocities by the following equations:

$$\dot{x}_b^2 = \sum_{i=1}^{3} \Gamma_{X_i}^2(\theta)\cdot\dot{\theta}_{1i}^2 + \sum_{j=1}^{3}\sum_{k=1,k\neq j}^{3} \Gamma_{X_j}(\theta)\cdot\Gamma_{X_k}(\theta)\cdot\dot{\theta}_{1j}\cdot\dot{\theta}_{1k} \tag{15}$$

$$\dot{y}_b^2 = \sum_{i=1}^{3} \Gamma_{Y_i}^2(\theta)\cdot\dot{\theta}_{1i}^2 + \sum_{j=1}^{3}\sum_{k=1,k\neq j}^{3} \Gamma_{Y_j}(\theta)\cdot\Gamma_{Y_k}(\theta)\cdot\dot{\theta}_{1j}\cdot\dot{\theta}_{1k} \tag{16}$$

The biped structure presents a nonlinear relationship between the input joint's rotation and the output's linear displacements. Consequently, some terms in the final equations (13) and (14) are not of simple physical interpretation. The first term in summation represents the energy transferred by each individual joint (P_{x_i} and P_{y_i}). The second term corresponds to product velocity terms associated with energy coupling ($P_{x_{jk}}$ and $P_{y_{jk}}$). The purpose of the energy analysis method is to determine the input-output characteristics of the biped system and, if possible, to unmask its detailed structure. This involves finding the locomotion variables that move the locomotion system in a desired manner and to minimise the energy expended.

4.2 Perturbation analysis

In many practical cases the robotic system is noisy, that is, has internal random disturbing forces. As such, an approach called "perturbation analysis" was implemented to determine how the biped model has enough flexibility to match these distortions. First, the joint trajectories are computed by the inverse kinematics algorithm. Afterwards, the angular acceleration vectors are 'corrupted' by additive noise. For simplicity reasons, it is used a uniform distribution with zero mean added to the acceleration signal. As result, the trajectories of the body suffer some distortion and can only approximate the desired one. By regarding the forward kinematics of the mechanism, we can determine a two-dimensional index based on the statistical average of the well-known mean-square-error and defined by:

$$\xi_x = \frac{1}{N_S}\sum_{i=1}^{N_S}\sqrt{\frac{1}{T}\int_0^T \left[\dot{x}_b^i(t) - \dot{x}_b^d(t)\right]^2 dt} \tag{17}$$

$$\xi_y = \frac{1}{N_S}\sum_{i=1}^{N_S}\sqrt{\frac{1}{T}\int_0^T \left[y_b^i(t) - y_b^d(t)\right]^2 dt} \tag{18}$$

Here, N_S is the total number of steps for averaging purposes, $\dot{x}_b^i$ and $\dot{x}_b^d$ are the *i*th sample and the desired forward velocities at the body section, respectively, and y_b^i and y_b^d are the *i*th sample and the desired vertical oscillations at the body. In this method, the stochastic perturbation penalises the system's performance and we shall be concerned with minimising both indices ξ_x and ξ_y.

4.3 Lowpass frequency response

In robotics, the electro-mechanical system that regulates the movement (*e.g.*, actuators and drives) is constrained by its bandwidth. An important point to remember is that the bandwidth limits the maximum speed with which a system can respond to an input command. Hence, the practical conditions of motor control should also be considered when evaluating the system's performance.

In this perspective, the joint velocities are expanded in Fourier series and the time domain signals are described by the coefficients of their harmonic terms. Afterwards, these series are truncated to the lower harmonic components by a lowpass filter, which passes a certain portion of the frequency spectrum while neglecting all the higher harmonics. In doing so, the response of the system will lag behind the input command, thereby producing a tracking error. Once again, we can compare the real and the desired trajectories produced at the body using, as our figure of merit, the mean-square-error as expressed in equations (17) e (18) for $N_S = 1$. This method, based on frequency domain analysis, has also been successfully employed. However, it is a less general method than the perturbation analysis, because it requires the *a priori* knowledge about the nature of the robotic actuators.

4.4 Locomobility measure

The motivation that led to the development of the locomobility index was to apply the concepts of kinematic dexterity to biped walking. This performance measure can be expressed, in mathematical terms, using the geometrical representation of the kinematics through the Jacobian matrix (Park, 1994). In our case, we selected a two-dimensional index of the global directional uniformity of the motion based on the projections of the major axis of the ellipsoid. The global index is obtained by averaging the longest projections of the major axis along the *x*-axis (E_{P_x}) and the *y*-axis (E_{P_y}), over a complete cycle:

$$L_x = \frac{1}{T}\int_0^T E_{P_x}(t)dt \tag{19}$$

$$L_y = \frac{1}{T}\int_0^T E_{P_y}(t)dt \tag{20}$$

The more suitable trajectory is that which maximises L_x (direction of movement) and minimises L_y.

5 SIMULATION RESULTS

In this section, we perform a set of experiments to estimate how the biped robot adapts to variations in the locomotion variables. The aim is to determine the main factors that optimise the leg motion and to compare the formulated dexterity measures. To evaluate the system's operating features, the simulations are carried out considering a forward velocity $V_F = 1$ m/sec.

5.1 Step Length and Hip Height

To illustrate the use of the preceding results, the biped locomotion was simulated. At this phase, a major simplification is introduced by allowing the swing foot stays on the ground for all time without violating any constraint. The individual indices are computed while maintaining a constant hip height during the stride. As shown in Figure 4, the performance function is evaluated with respect to the step length and the hip height (for equal link lengths). In this case, we are plotting the average power in terms of x and y coordinates. For convenience, the surface portion corresponding to hip height values less than 0.5 m is not represented.

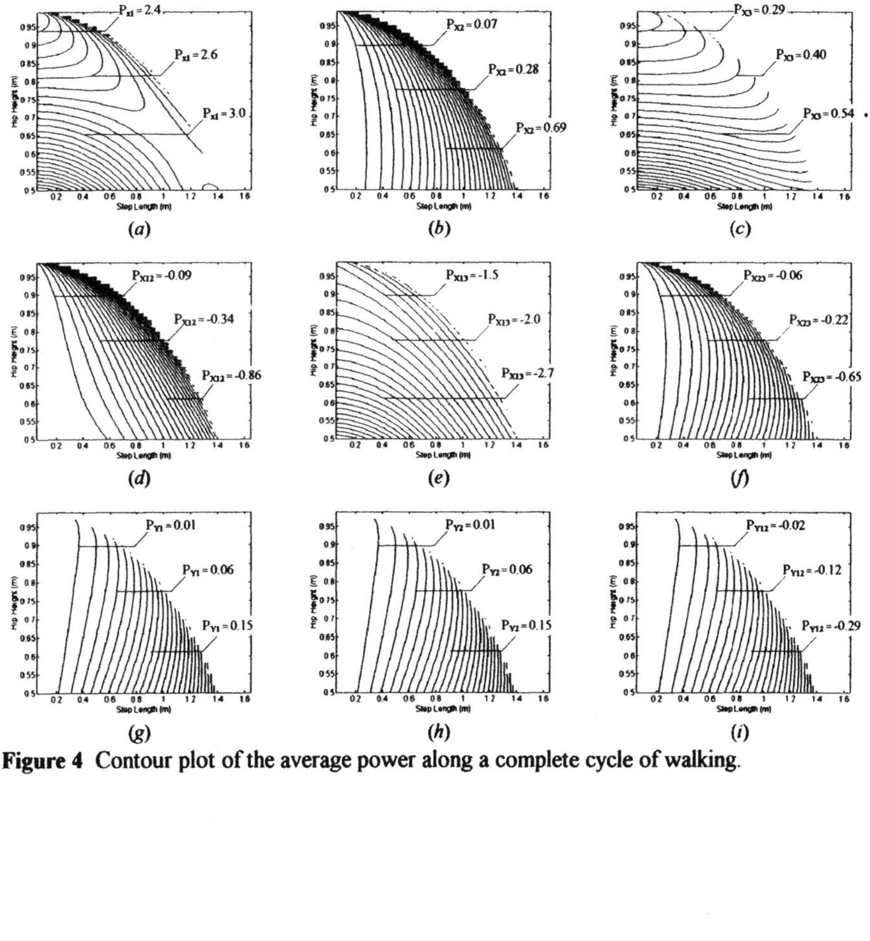

Figure 4 Contour plot of the average power along a complete cycle of walking.

The ankle joint is believed to dominate the process of walking, as seen from the energy values involved in every joint (Figure 4(*a-c*)). The simulation, also, indicates that the knee joint may be a major factor due a significant range of performance variation. As shown in Figures 4(*b*) and 4(*h*), an important degradation occurs as the step length value increases. On the other side, the hip seems to be particularly sensitive to the conditions of hip height (Figure 4(*c*)). In what concerns the cross coupling terms (Figures 4(*d-f*) and 4(*i*)), the average power along a complete cycle resulted in negative values.

Figure 5(*a*) corresponds to the perturbation analysis method. Considering these plots, we conclude that an important degradation occurs as the step length increases. The hip height variable also affects the measure, however, has a smaller influence. In Figure 5(*b*), we depict the results of applying the locomobility index. The basic idea to retain is that to maximise L_x and, additionally, to minimise L_y, the hip height is expected to round about 95% of the maximum height.

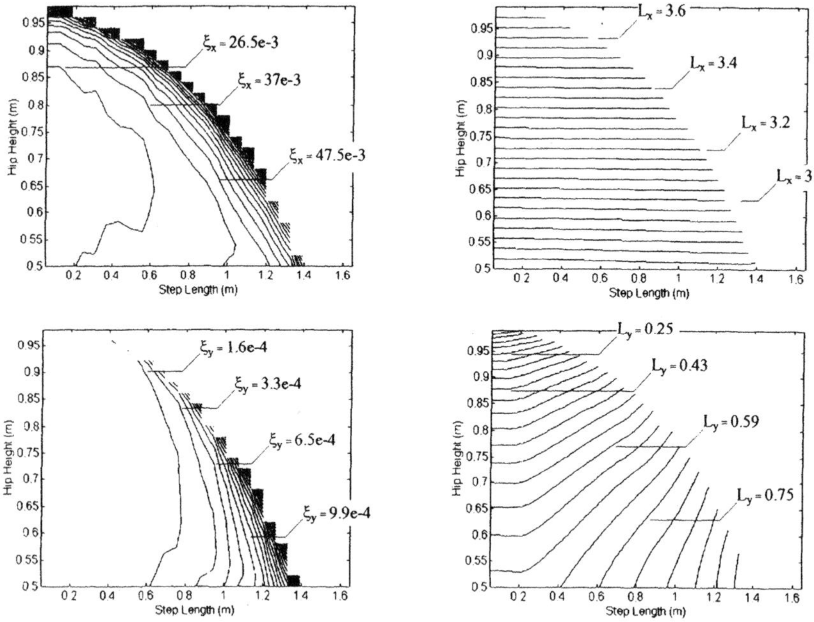

(*a*) (*b*)

Figure 5 Contour plot of the performance surface along the *x* and *y*-axis: (*a*) perturbation analysis; (*b*) locomobility index.

Finally, let us examine the "lowpass frequency response" method. The fundamental harmonic f_0 of the waveforms at the joints is given by,

$$f_0 = \frac{V_F}{2L_S} \tag{21}$$

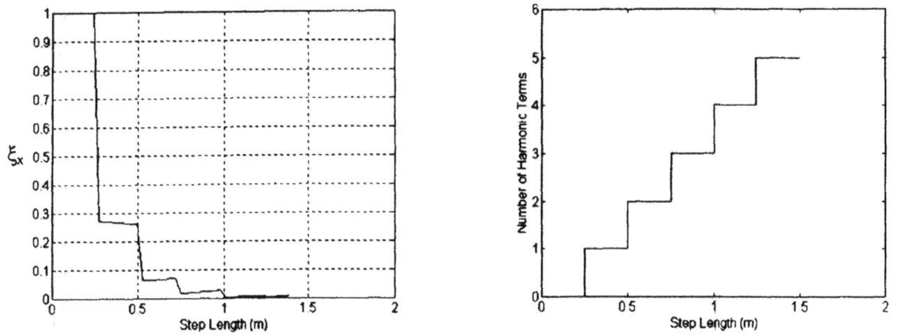

(*a*) (*b*)

Figure 6 Lowpass frequency response: (*a*) mean-square error ξ_x *versus* step length; (*b*) number of harmonic terms.

If, for example, we assume an ideal lowpass filter whose cutoff frequency is $f_c = 2$ Hz, the performance surface presents a set of discontinuities, as illustrated in Figure 6(*a*). These discontinuity points solely depend on the number of harmonic terms up to f_c. The number of lower harmonics passing the ideal filter *versus* the step length is shown in Figure 6(*b*). In contrast to the other methods, the above results suggest that there is an optimum solution for higher step length values, while the performance remains almost unchanged to hip height variations. In order to achieve broader applicability, we must provide information about the robotic actuators by means of a mathematical model (*e.g.*, frequency response).

5.2 Hip Ripple

Upon this point, we consider a hip trajectory with a sinusoidal oscillation. In this case, the foot of the swing leg remains at the ground for all cycle. The contour plot in Figure 7 displays the performance values using the perturbation analysis method. From this simulation, we can observe that the optimum hip ripple tends to a zero value and, furthermore, it remains almost unchanged to hip height variations.

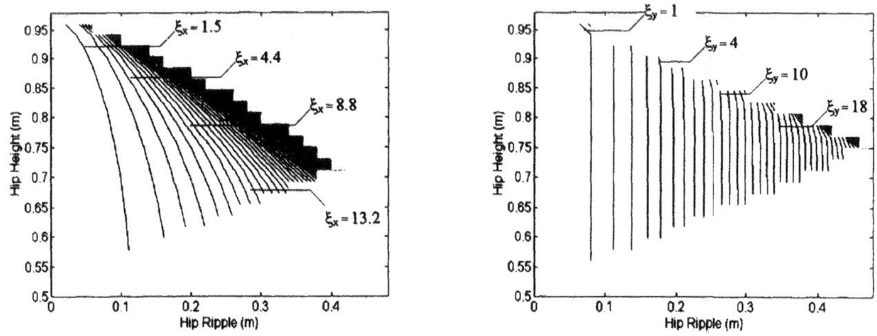

Figure 7 Contour plot: H_H *versus* H_R $(L_S = 40 \text{ cm})$.

5.3 Foot Clearance

Another variable that should be tested is the effect of foot clearance. Until now, all the experiences considered that the foot stays on the ground without any restriction. Next, the goal is to analyse the situation in which the foot can be lifted off the ground. In Figure 8 we study the influence of F_C for different hip height conditions. The results can be expressed as contours of constant performance measure values in terms of H_H. We confirmed this invariance property of the foot clearance using the locomobility analysis.

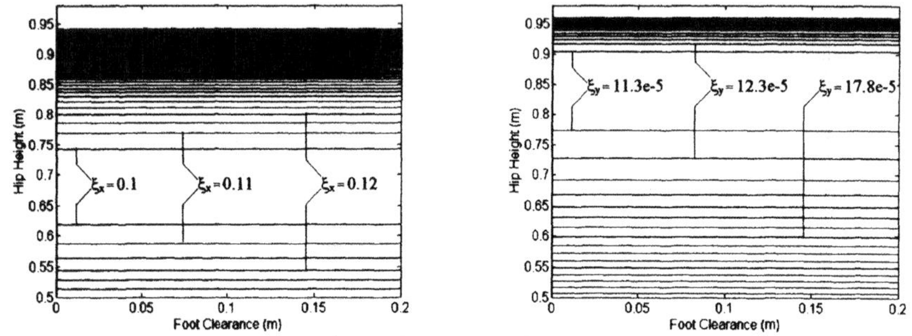

Figure 8 Contour plot: H_H *versus* F_C $(L_S = 40\text{ cm})$.

5.4 Relative Link Lengths

We now investigate the role of the link lengths, l_1 and l_2, in the system's performance. The choice of the leg lengths is a relevant aspect at the design process and affects the robot's mobility, as well. Figure 9 depicts the curve of ξ_x given the link length l_1, when employing the perturbation method. As can be observed, the optimum solution occurs when the knee-ankle length is higher ($l_1 = 0.74$ cm) than the hip-knee length ($l_2 = 0.26$ cm). The simulation also indicates that for l_1 in the range from 40 to 90 cm the performance index remains almost constant. The variation found along the y-axis is insignificant and is not represented.

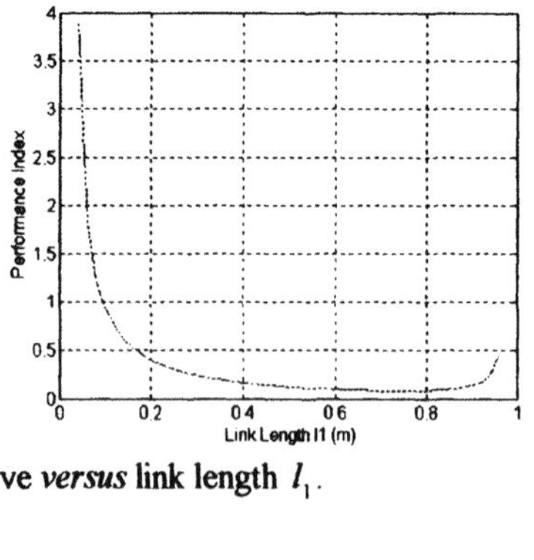

Figure 9 Performance curve *versus* link length l_1.

6 CONCLUSIONS

Nowadays, robots are widely used in the manufacture industry. However, the applications of these industrial robots are relatively simple and, for the most part, involving repetitive tasks previously performed by human beings. Bringing robots out of engineered environments and striving for increased autonomy, reliability and dexterity, is both an enormous challenge and a socio-economic necessity in the near future. One key component required to build such intelligent systems is the locomotor system. Accordingly, the biped locomotion aspects of robotics are gaining more and more attention in both research and applications.

In this paper, we have compared various kinematic aspects of biped locomotion. By implementing different motion patterns, we estimated how the robot responds to a variety of locomotion variables such as hip ripple, hip offset, foot clearance and relative link lengths. Various quantitative measures were formulated for analysing the kinematic performance of such systems. Given these goals, the graphical results provide a more concrete illustration of the system's properties. The energy analysis method has the advantage that is very amenable to computer implementation and, furthermore, it is computationally more efficient. However, this approach has the drawback that the resulting equations are not in a completely intuitive form. In general, the perturbation analysis method tends to be more elegant although computationally more exigent. Its random characteristic seems to be particularly tailored for examining the role of the different variables on the locomotion process. On the other hand, the results obtained using the locomobility measure present a correlation with that of the perturbation analysis. The lowpass frequency response produces basically different effects allowing a different perspective of the problem. Moreover, some knowledge of the actuator's characteristics will always be helpful when evaluating the system's performance.

While our focus has been on kinematic dexterity, certain aspects of locomotion are not necessarily captured by the measures proposed. Thus, future work in this area will address the refinement of our models to incorporate dynamics, as well as exploring human-like walking principles. In contrast to rotational actuators used in robotics, the skeletal is completely controlled by linear actuators organised in agonist-antagonist pairs. A complementary analysis could reveal some important properties of both approaches giving some hints to understand the performance differences between robotic and biological motion.

7 REFERENCES

Basmajian, J. and Luca, C. (1978) *Muscles Alive: Their Functions Revealed by Electromyography*. Williams & Wilkins.

Galhano, A.M., Tenreiro Machado, J.A. and Martins de Carvalho, J.L. (1992) Statistical Analysis of Muscle-Actuated Manipulators. *IEEE International Conference on Robotics and Automation*, Nice.

Golden, J.A. and Zheng, Y.F. (1990) Gait Synthesis for the SD-2 Biped Robot to Climb Stairs. *International Journal of Robotics and Automation*, vol. 5, n. 4.

Hemami, H. and Stokes, B. (1983) A Qualitative Discussion of Mechanisms of Feedback and Feedforward in the Control of Locomotion. *IEEE Transactions on Biomedical Engineering*, vol. 30, n. 11, 681-688.

Kajita, S. and Tani, K. (1994) Study of Dynamic Biped Locomotion on Rugged Terrain: Theory and Basic Experiments. *Proceedings of '94 International Conference on Advanced Robotics*, 741-746.

Kun, A. and Miller, W.T. (1996) Adaptive Dynamic Balance of a Biped Robot using Neural Networks. *Proceedings of the IEEE International Conference on Robotics and Automation*, 240-245.

Park, F.C. and Brockett, R.W. (1994) Kinematic Dexterity of Robotic Mechanisms. The *International Journal of Robotics Research*, vol. 13, n. 1, 1-15.

Raibert, M.H. (1986) *Legged Robots that Balance*. The MIT Press.

Shih, C-L., Gruver W.A. and Lee T-T. (1993) Inverse Kinematics and Inverse Dynamics for Control of a Biped Walking Machine. *Journal of Robotic Systems*, vol. 10, n. 4, 531-555.

Silva, F.M. and Tenreiro Machado, J.A. (1996) Research Issues in Natural and Artificial Biped Locomotion Systems. *Proceedings of 2nd. Portuguese Conference on Automatic Control*, 211-216, Controlo'96.

Todd, D.J. (1985) *Walking Machines: An Introduction to Legged Robots*. Kogan Page.

Vukobratovic, M., Botovac, B., Surla, D. and Stokic D. (1990) *Biped Locomotion: Dynamics, Stability, Control and Applications*. Springer-Verlag.

Yamaguchi, J. *et al.* (1996) Development of a Dynamic Biped Walking System for Humanoid-Development of a Biped Walking Robot Adapting to the Humans' Living Floor. *Proceedings of the IEEE International Conference on Robotics and Automation*, 232-239.

Zheng, Y.F. and Hemami, H. (1984) Impact Effects of Biped Contact with the Environment. *IEEE Transactions on Systems. Man and Cybernetics*, vol. 14, n. 3, 437-443.

14
Adaptive hybrid impedance control of robot manipulators: A comparative study

Luís F. Baptista, José M. G. Sá da Costa
Technical University of Lisbon, Instituto Superior Técnico
Department of Mechanical Engineering, GCAR/IDMEC,
Avenida Rovisco Pais, 1096 Lisboa Codex, Portugal.
Telephone: 351-1-8418190. Fax: 351-1-8498097.
email: baptista@gcar.ist.utl.pt / msc@gcar.ist.utl.pt

Abstract

This article presents an adaptive hybrid impedance control approach for robot manipulators. The proposed scheme, consists of an outer hybrid impedance control loop that generates a target acceleration to an inner adaptive controller in order to linearize the robot dynamics and track the desired force/position. In order to analyse the performance of the proposed control scheme, two distinct adaptive control algorithms in the inner control loop are tested. The performance of the hybrid impedance strategy is illustrated by computer simulations with a 2-DOF PUMA 560 robot, in which the end-effector is forced to move along a frictionless surface in the vertical plane. The results obtained for both adaptive algorithms, reveal an accurate force tracking and impedance control in robotic tasks that require force tracking capability in an rigid environment with a time varying stiffness profile.

Keywords

Robotics, hybrid impedance control, direct adaptive control, passivity-based adaptive control

1 INTRODUCTION

Control schemes for a robot manipulator in contact with an environment can be classified as open-loop force control or direct closed-loop force control. The first approach classified as impedance control (Hogan, 1985), do not attempt to control force explicitly but rather to control the relationship between force and position of the end-effector in contact with the environment. Controlling the position, therefore leads to an implicit control of force. The second approach, hybrid control (Raibert, 1982), separates a robotic force task into two subspaces, a force controlled subspace and a position controlled subspace. Two independent controllers are then designed for each subspace. In 1988, Anderson and Spong (Anderson, 1988), proposed a new method termed Hybrid Impedance Controller which combines the hybrid control and impedance control strategies, into a unified approach. It separates the task space into two subspaces, an explicit force controlled subspace and an impedance controlled position subspace.

Many researchers have studied the application of adaptive control in robotics, as in the paper by Craig and Sastry (Craig, 1986), where adaptive control is applied to control a robot in free motion, while Lu and Meng (Lu, 1991) and Colbaugh et al. (Colbaugh, 1995) applied adaptive impedance control to a robot in constrained motion.

In this article, the performance of the proposed adaptive hybrid impedance strategy with the algorithms proposed by Colbaugh et al. (Colbaugh, 1995) and by Siciliano and Villani (Siciliano, 1996) respectively, is analysed in order to show the closed-loop capability of the scheme in constrained motion.

This paper is organized into 6 sections. Section 2 summarises the modelling of the manipulator and the environment. Section 3 presents the hybrid impedance approach. Section 4 discusses the adaptive algorithms and the overall control schemes. Simulation results are presented and analyzed in Section 5. Finally, conclusions are given in Section 6.

2 MANIPULATOR AND ENVIRONMENT DYNAMICS

A rigid robot in constrained motion is governed by a set of nonlinear dynamic equations, given by

$$\boldsymbol{M}(q)\ddot{q} + \boldsymbol{c}(q,\dot{q})\dot{q} + \boldsymbol{g}(q) + \boldsymbol{d}(\dot{q}) = \tau - \boldsymbol{J}^T(q)\boldsymbol{f}_e \tag{1}$$

where q, $\dot{q}$ and $\ddot{q}$ correspond to the $(n\times 1)$ vectors of joint angular positions, velocities and accelerations, respectively. $\boldsymbol{M}(q)$ represents the $(n\times n)$ symmetric positive definite inertia matrix, $\boldsymbol{c}(q,\dot{q})\dot{q}$, describes the $(n\times 1)$ vector of Coriolis and centrifugal effects, $\boldsymbol{g}(q)$ accounts for gravitational terms and $\boldsymbol{d}(\dot{q})$ for the frictional terms in joint coordinates, respectively. The term $\boldsymbol{J}$ represent the Jacobian matrix that relates the joint velocity to the linear and angular velocities of the end-effector, the term τ represents the $(n\times 1)$ vector of applied joint torques and $\boldsymbol{f}_e$ denotes the $(n\times 1)$ vector of generalised forces exerted by the end-effector on the environment and measured by the wrist force sensor.

For non-redundant robots (1) can be written in cartesian space as:

$$\boldsymbol{M}_x(x)\ddot{x} + \boldsymbol{c}_x(x,\dot{x})\dot{x} + \boldsymbol{g}_x(x) + \boldsymbol{d}_x(\dot{x}) = \boldsymbol{f}_c - \boldsymbol{f}_e \tag{2}$$

where x is a six-dimensional vector representing the position and orientation of the manipulator end-effector, and the terms of (2) are well known and are given in the literature as in Spong and Vidyasagar (Spong, 1989).

Throughout this paper, a linear spring-damper is adopted for the environment model, according to:

$$\boldsymbol{f}_e = \boldsymbol{K}_e(x - x_e) + \boldsymbol{B}_e\dot{x} \tag{3}$$

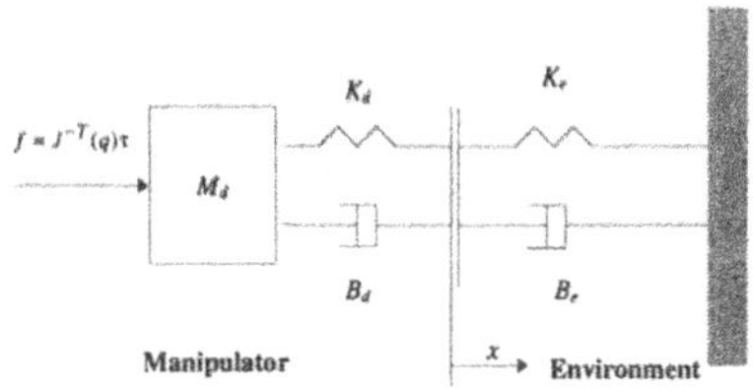

Figure 1 1 D.O.F. cartesian manipulator: B_e, K_e - environment parameters. M_d, B_d, K_d - Desired robot impedance parameters.

where K_e and B_e are the environment stiffness and damping matrices, respectively, and x_e corresponds to the environment position.

3 HYBRID IMPEDANCE APPROACH

The design of the hybrid impedance controller is accomplished in the task space. The task space is splited into two independently controlled subspaces, an impedance controlled subspace and a force controlled subspace. The object of the hybrid impedance controller is to design an algorithm for each subspace such that the dynamics exhibited by the system are replaced by better desirable dynamics. Figure 1 represents the 1-DOF cartesian manipulator case.

In the force controlled subspace, a desired force trajectory must be followed (Bickel, 1995). Additionally, in order to reduce impact forces when the manipulator loses and then regains contact with the environment, low velocities are desirable even in the presence of large force errors. According to Figure 1 the target dynamics are given by:

$$\boldsymbol{M}_d\ddot{x} + \boldsymbol{B}_d\dot{x} - \boldsymbol{f}_d = -\boldsymbol{f}_e \tag{4}$$

The desired reference acceleration for the force controlled subspace which guarantees the desired force following is taken by solving (4) in respect to the acceleration, leading to:

$$\ddot{x}_{tf} = \boldsymbol{M}_d^{-1}(-\boldsymbol{B}_d\dot{x} + \boldsymbol{f}_d - \boldsymbol{f}_e) \tag{5}$$

Therefore, if this acceleration can be tracked then the desired dynamics given in (4) can be accomplished.

The objective of the controller in the impedance controlled subspace is to track a desired position trajectory in the absence of any forces implied by the environment and, moreover, to react with a desired impedance whenever the presence of such environmental forces exist (Bickel, 1995), as shown in Figure 1. The dynamics in this subspace can be represented by

$$\boldsymbol{M}_d(\ddot{x}_d - \ddot{x}) + \boldsymbol{B}_d(\dot{x}_d - \dot{x}) + \boldsymbol{K}_d(x_d - x) = \boldsymbol{f}_e \tag{6}$$

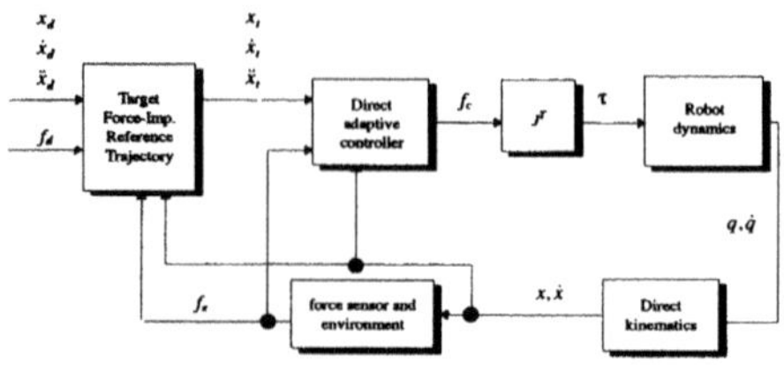

Figure 2 Direct adaptive control block diagram.

which leads to the target reference acceleration for the impedance controlled subspace

$$\ddot{x}_{ti} = \ddot{x}_d + \boldsymbol{M}_d^{-1}[\boldsymbol{B}_d(\dot{x}_d - \dot{x}) + \boldsymbol{K}_d(x_d - x) - \boldsymbol{f}_e] \tag{7}$$

The outer hybrid impedance control loop provides the target acceleration $\ddot{x}_t = [\ddot{x}_{tf} \;\; \ddot{x}_{ti}]^T$ to the inner control loop which is designed to linearize the robot dynamics and track $\ddot{x}_t$. Given the reference acceleration, we can obtain a target force/impedance reference trajectory, by numerical integration of $\ddot{x}_t$, with a proper set of initial conditions, i.e. $x_d(0) = x_t(0)$ and $\dot{x}_d(0) = \dot{x}_t(0)$. The target force/impedance trajectory $x_{ref} = [x_t \;\; \dot{x}_t \;\; \ddot{x}_t]^T$ coincides with the desired tracking trajectory over the unconstrained motion duration, because $\boldsymbol{f}_e$ is zero. However, when the robot end-effector reaches the surface it may differ considerably from the desired trajectory due to the environmental forces.

4 ADAPTIVE CONTROLLERS

4.1 Direct adaptive controller

In this control approach, a non-model based cartesian algorithm generates the control action $\boldsymbol{f}_c$ and the required torques to the robot, in order to achieve the desired force/impedance characteristics (Colbaugh, 1995). A block diagram of the hybrid impedance control structure with a direct adaptive controller (DAC), is given in Figure 2.
Consider the following auxiliary vectors:

$$\sigma = \dot{e}_t + \Lambda e_t \tag{8}$$

$$\dot{x}_r = \dot{x}_t + \Lambda e_t \tag{9}$$

where $\dot{x}_r$ is a "modified velocity", σ is a "weighted position-velocity error" where $e_t = x_t - x$, $\dot{e}_t = \dot{x}_t - \dot{x}$ are the target position and velocity errors, respectively, and Λ is a positive scalar constant. Consider the following control law:

$$\boldsymbol{f}_c = \boldsymbol{A}(t)\ddot{x}_r + \boldsymbol{B}(t)\dot{x}_r + \boldsymbol{f}_a(t) + \boldsymbol{f}_e + [\boldsymbol{K}_1 + \boldsymbol{K}_a(t)]\sigma \tag{10}$$

where $\boldsymbol{K}_1$ is a positive diagonal gain matrix. The adaptive gains $\boldsymbol{f}_a(t) \in \Re^{\eta}$, $\boldsymbol{A}(t) \in \Re^{n\times\eta}$,

$\boldsymbol{B}(t) \in \Re^{n\times\eta}$ and $\boldsymbol{K}_a(t) \in \Re^{n\times\eta}$, where η is the number of coordinates in task space, are updated with the following laws:

$$\boldsymbol{f}_a(t) = \boldsymbol{f}_a(0) - \int_0^t \alpha_1 \boldsymbol{f}_a dt + \int_0^t \beta_1 \sigma dt \tag{11}$$

$$\boldsymbol{A}(t) = \boldsymbol{A}(0) - \int_0^t \alpha_2 \boldsymbol{A} dt + \int_0^t \beta_2 \sigma \ddot{x}_r^T dt \tag{12}$$

$$\boldsymbol{B}(t) = \boldsymbol{B}(0) - \int_0^t \alpha_3 \boldsymbol{B} dt + \int_0^t \beta_3 \sigma \dot{x}_r^T dt \tag{13}$$

$$\boldsymbol{K}_a(t) = \boldsymbol{K}_a(0) - \int_0^t \alpha_4 \boldsymbol{K}_a dt + \int_0^t \beta_4 \sigma \sigma^T dt \tag{14}$$

In (11-14), the β_i are positive scalar constants and the α_i are functions of the form $\alpha_i = \alpha_{i0} + \alpha_{i1} \parallel \dot{x} \parallel$, where α_{i0}, α_{i1} are positive scalar constants (Colbaugh, 1995).

4.2 Passivity-based adaptive controller

In this control approach (Siciliano, 1996), the following model-based control law is considered

$$\boldsymbol{f}_c = \hat{\boldsymbol{M}}_x(x)\ddot{x}_r + \hat{\boldsymbol{c}}_x(x,\dot{x})\dot{x}_r + \hat{\boldsymbol{g}}_x(x) + \boldsymbol{f}_e + \boldsymbol{K}_D(\dot{x}_r - \dot{x}) \tag{15}$$

where $\hat{\boldsymbol{M}}_x, \hat{\boldsymbol{c}}_x, \hat{\boldsymbol{g}}_x$ are the estimates of $\boldsymbol{M}_x, \boldsymbol{c}_x, \boldsymbol{g}_x$, respectively, $\dot{x}_r, \ddot{x}_r$ are $(n \times 1)$ reference vectors and $\boldsymbol{K}_D$ is a positive definite matrix of proportional-derivative type. It is assumed that $\hat{\boldsymbol{c}}_x$ is skew-symmetric and that $\hat{\boldsymbol{M}}_x, \hat{\boldsymbol{c}}_x, \hat{\boldsymbol{g}}_x$ have the same functional form of $\boldsymbol{M}_x, \boldsymbol{c}_x, \boldsymbol{g}_x$, with a $(m \times 1)$ vector of estimated parameters $\hat{p}$. Then, using the property of linearity of the dynamic model (2) in terms of a proper set of manipulator and load constant parameters, i.e.

$$\boldsymbol{M}_x(x)\ddot{x} + \boldsymbol{c}_x(x,\dot{x})\dot{x} + \boldsymbol{g}_x(x) = \boldsymbol{W}_x(x,\dot{x},\ddot{x})p \tag{16}$$

where $\boldsymbol{W}_x(x,\dot{x},\ddot{x})$ is an $(n \times m)$ regressor matrix, the control law (15) can be given by

$$\boldsymbol{f}_c = \boldsymbol{W}_x(x,\dot{x},\dot{x}_r,\ddot{x}_r)\hat{p} + \boldsymbol{f}_e + \boldsymbol{K}_D(\dot{x}_r - \dot{x}) \tag{17}$$

Combining (2) and (17) leads to

$$\boldsymbol{M}_x(x)\dot{e} + \boldsymbol{c}_x(x,\dot{x})e + \boldsymbol{K}_D e = \boldsymbol{W}_x(x,\dot{x},\dot{x}_r,\ddot{x}_r)\tilde{p} \tag{18}$$

where $\tilde{p} = \hat{p} - p$, $\dot{x}_r = \dot{x}_t + e$ and $\ddot{x}_r = \ddot{x}_t + \dot{e}$. In this approach, the error e is related to the force and position errors (Siciliano, 1996) by the following equation:

$$e = \Delta\dot{x} + \lambda_1 \Delta x + \lambda_2 \int_0^t \Delta \boldsymbol{f} d\delta \tag{19}$$

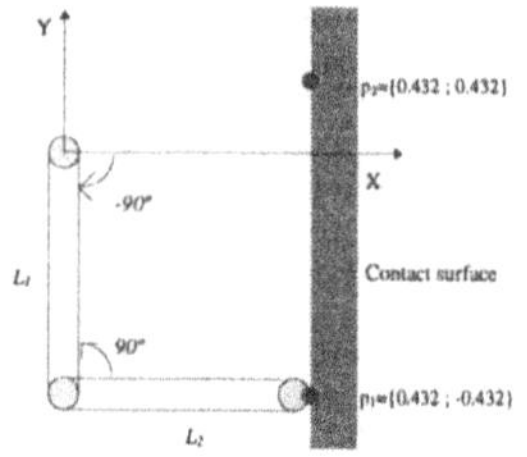

Figure 3 Two link manipulator in vertical plane and environment.

where $\Delta x = x_t - x$, $\Delta \boldsymbol{f} = \boldsymbol{f}_d - \boldsymbol{f}_e$ and $\lambda_1, \lambda_2 > 0$. Then, we obtain

$$\dot{x}_r = \dot{x}_t + \lambda_1 \Delta x + \lambda_2 \int_0^t \Delta \boldsymbol{f} d\delta \tag{20}$$

$$\ddot{x}_r = \ddot{x}_t + \lambda_1 \Delta \dot{x} + \lambda_2 \Delta \boldsymbol{f} \tag{21}$$

Applying the equations that relate the dynamics in the joint space and in the task space for non-redundant manipulators (Spong, 1989), and from (15), finally leads to the control law in joint space

$$\tau = \boldsymbol{W}(q, \dot{q}, \dot{q}_r, \ddot{q}_r)\hat{p} + \boldsymbol{K}_D \boldsymbol{J}^T(q)\boldsymbol{J}(q)(\dot{q}_r - \dot{q}) + \boldsymbol{J}^T \boldsymbol{f}_e \tag{22}$$

The parameter estimate update law in the joint space (Siciliano, 1996), is given by

$$\dot{\hat{p}} = \boldsymbol{\Gamma}^{-1} \boldsymbol{W}^T(q, \dot{q}, \dot{q}_r, \ddot{q}_r)(\dot{q}_r - \dot{q}) \tag{23}$$

5 SIMULATION RESULTS

In this section, the adaptive hybrid impedance control schemes presented in Section 4 are applied to an industrial robot through computer simulation. The robot chosen for this simulation study is the 2-DOF PUMA 560 planar robot which is acting in the vertical plane according to Figure 3. The numerical values for the link parameters of the robot under study are $m_1 = 15.91$ Kg, $m_2 = 11.36$ Kg, $l_1 = l_2 = 0.432$ m, which resembles the original parameters of links 2 and 3 of the Unimation PUMA 560 arm. Throughout the simulations, the constant time steps of 1 and 2 ms were used in both controller implementations, while the dynamic model of the robot is simulated in the MATLAB/SIMULINK environment using the Runge-Kutta fourth order integration method.

In this study a frictionless contact surface modelled as in (3) is placed in a vertical position of the robot workspace at x_e =0.432 m, as shown in Figure 3. It is considered that in all the simulations the end-effector maintains the contact with the surface during the execution of the complete task.

Both control schemes are tested considering an end-effector reference position trajectory from $p_1 = [0.432 \;\; -0.432]$ to $p_2 = [0.432 \;\; 0.432]$ with a cycloidal reference force profile. The

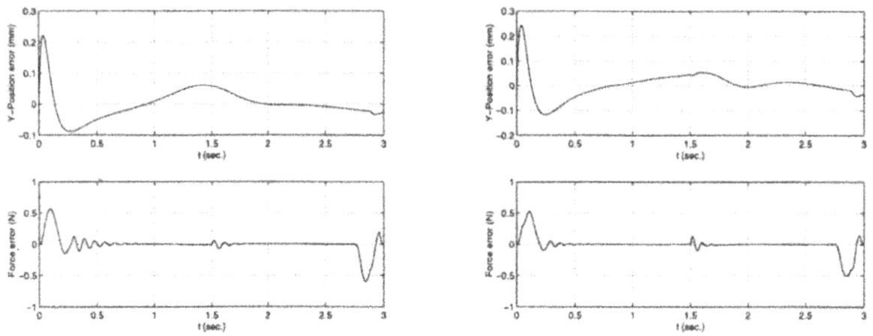

(a) *Position and contact force errors* (b) *Position and contact force errors*

Figure 4 a) Position and contact force errors (DAC). b) Position and contact force errors (IAC). (Note: Sampling time T_s= 2 ms).

impedance parameters in (6) are defined as $\boldsymbol{M}_d = diag[2.5 \;\; 2.5]$, $\boldsymbol{K}_d = diag[250 \;\; 250]$ and the elements of $\boldsymbol{B}_d$ are set to provide critical damping in the impedance model. In order to test the robustness of the force control schemes to environmental model uncertainties, a following time varying stiffness profile, was considered:

$$\boldsymbol{K}_e = \begin{cases} 10000 + 1000sin(\pi t/2) & \text{for } 0 < t < 2 \\ 10000e^{(-0.25(t-2))} & \text{for } 2 \leq t < 3 \end{cases} \tag{24}$$

and $\boldsymbol{B}_e = diag[5 \;\; 5]$. For the direct adaptive controller, the constant parameters are set as $\boldsymbol{K}_1 = diag[3500 \;\; 3500]$, $\Lambda = 30$ and the adaptive gains $\boldsymbol{f}_a$, $\boldsymbol{A}$, $\boldsymbol{B}$ and $\boldsymbol{K}_a$ as defined in Section 4 are set initially to zero, with $\alpha_{i0} = 0.001$, $\beta_i = 1$, $i = 1, ..., 4$. For the passivity-based adaptive controller (IAC), the values of the matrices $\boldsymbol{K}_D$, $\boldsymbol{\Gamma}$ and the initial values of the estimated parameter vector $\hat{p}$, are:

$\boldsymbol{K}_D = diag[1500 \quad 1500]$

$\boldsymbol{\Gamma} = diag[0.02 \quad 0.02 \quad 0.02 \quad 0.002 \quad 0.002]$

$\hat{p}(0) = [3.0 \quad 1.6 \quad 0.4 \quad 70.0 \quad 12.0]$

with $\lambda_1 = 30$, $\lambda_2 = 0.15$. The results presented in Figure 4, show that both adaptive controllers have good position and force tracking capability for the given trajectory. The results in Table 1 show also that the hybrid impedance approach with passivity-based adaptive algorithm (IAC) has a better performance, with position and force errors less than 0.3 mm and 0.6 N, respectively, at a cost of a higher number of MATLAB floating-point operations (FLOPS). Moreover, the IAC algorithm reveal to be more robust to a moderate increase in the sampling time (Ex: T_s= 4 ms).

Table 1 Sum squared errors of y coordinate and contact force $\boldsymbol{f}_e$.

	DAC	IAC	DAC	IAC
	$T_s = 1$ ms	$T_s = 1$ ms	$T_s = 2$ ms	$T_s = 2$ ms
SSE($\boldsymbol{y}$)	5.45×10^{-6}	6.68×10^{-6}	3.48×10^{-6}	4.04×10^{-6}
SSE($\dot{\boldsymbol{y}}$)	0.0028	0.0034	0.0024	0.0018
SSE($\boldsymbol{f}_e$)	56.11	45.61	27.65	22.71
# FLOPS	1.55×10^6	2.93×10^6	0.77×10^6	1.46×10^6

6 CONCLUSIONS

In this article an adaptive hybrid impedance control strategy for robot manipulators is presented. In order to test the performance of the proposed control scheme two distinct adaptive control algorithms were tested in the inner control loop. The study has shown good results and demonstrated some of the potential applications of the adaptive hybrid impedance control in robotic tasks that require explicit force tracking capability in an environment with a time varying stiffness.

Future research will cover the improvement of the proposed scheme, stability analysis, and an accurate study of the impact situation in the context of the hybrid impedance control approach.

REFERENCES

Neville Hogan. Impedance control: an approach to manipulation: Part I-II-III. *Journal of Dynamic Systems, Measurement, and Control*, 107:1–24, March 1985.

M. Raibert and J. Craig. Hybrid position/force control of manipulators. *Journal of Dynamic Systems, Measurement, and Control*, 102:126–133, June 1982.

R. J. Anderson and M. W. Spong. Hybrid impedance control of robotic manipulators. *IEEE Transactions on Robotics and Automation*, 4:549–556, October 1988.

J. Craig and S. S. Sastry. Adaptive control of mechanical manipulators. *Proc. IEEE International Conference on Robotics and Automation*, pages 190–195, 1986.

W. S. Lu and Q. H. Meng. Impedance control with adaptation for robotic manipulators. *IEEE Transactions on Robotics and Automation*, 7:408–412, June 1991.

R. Colbaugh, H. Seraji, and K. Glass. Adaptive compliant motion control for dextrous manipulators. *Journal of Robotic Systems*, 14:270–280, June 1995.

Bruno Siciliano and Luigi Villani. A passivity-based approach to force regulation and motion control of robot manipulators. *Automatica*, 36, 1996.

Mark W. Spong and M. Vidyasagar. *Robot dynamics and control.* John Wiley and Sons, Inc., Republic of Singapure, 1989.

R. J. Bickel and M. Tomizuka. Disturbance observer based hybrid impedance control of robotic manipulators. *American Control Conference*, June 1995.

PART SIX

FMS Control and Tooling

15
A Cell Controller Architecture Solution: Description and Analysis of Performance and Costs

António Quintas [1], Paulo Leitão [2]
[1]Professor Associado do DEEC da Faculdade de Engenharia da Universidade do Porto, Rua dos Bragas, 4099 Porto Codex; Tel: 351.2.2041844; email: aquintas@fe.up.pt
[2]Assistente da ESTG do Instituto Politécnico de Bragança/CCP, Quinta Sta Apolónia, Apartado 134, 5300 Bragança; Tel:351.073.3303083; email: pleitao@ipb.up.pt

Abstract

Nowadays, with the globalization of the markets, the companies must to implement new manufacturing technologies and concepts for the improvement of manufacturing systems productivity. The integration is one of the keys to solve the problems in manufacturing systems. The arrangement of the resources in cells and the implementation of control systems allows the improvement of productivity, flexibility, quality and the reduction of production costs.

This paper describes a Cell Controller architecture solution, implemented at the flexible manufacturing cell integrated into the demonstration pilot of an ESPRIT Program Project; special attention to the control structure and the system communication is taken. The last part of the paper focuses the costs involved in this solution and presents the alternative platforms and technologies to implement the Cell Controller.

Keywords

Cell Controller Architectures, Production Technologies, Industrial Communications, MMS.

1. INTRODUCTION

A Flexible Manufacturing Cell is a group of processing resources (PLC, NC machines), interconnected by means of an automated material handling (AGV, robots) and storage system, and controlled by a control system [Groover]. This control system is called Cell Controller, and it's a module of software that performs the integration of the several cell devices, allowing more flexibility and the improvement of production. The Cell Controller should be able to perform the following functions [Rembold, Groover]:

- **production control** - management of the cell production;
- **real time scheduling of production** - reaction in real time to the cell capacities;
- **resource management** - optimal use of materials and devices;
- **NC and RC programs management** - storage and download of NC and RC programs;
- **device monitoring** - visualization of cell and devices status;

• **error monitoring** - reaction to fault conditions.

The design and the implementation of a Cell Controller is a complex task, involving real time control restrictions and the manipulation of different operating systems, different communication protocols and machines supplied from different vendors.
The specification of a Cell Controller architecture results into the definition of three structures: control structure, communication system and information system.

The control system is very important for the final performance of the Cell Controller and it's structure can be performed with some basic control architectures: centralized, hierarchical and heterarchical [Diltis]. With the improvement of the performance/price relation for computing processing power, the architectures evolued from centralized architecture to the heterarchical architecture, allowing the distribution of the cell controllers functionalities in several hierarchical control levels.

The communication system is crucial to implement the integration of the cell resources, as to transfer the NC and RC programs to the cell resources and also to control these devices. There are several protocols available to implement the communication system: MAP (Manufacturing Automation Protocol)/MMS (Manufacturing Message Specification), FieldBus, serial link, TCP/IP. The choice of a communication protocol depends upon the costs, the standardization, and the control degree.

Following the paper, a Cell Controller architecture solution is described, with the analysis of his performance, and then, other platforms for the Cell Controller are discussed and a comparative performance evaluation is presented.

2. FLEXIBLE MANUFACTURING CELL DESCRIPTION

The manufacturing cell has two CNC machines and an anthropomorphic robot for the load/unload of the machines. One of these machines is a turning center *Lealde TCN10*, with a SIEMENS *Sinumerik 880T* controller; the other machine is a milling center *Kondia B500* model, with a FANUC *16MA* numerical control.

The robot is a KUKA *IR163/30.1* with a SIEMENS *RC3051* controller. The manufacturing cell has two transfer tables for the containers loading and unloading. These containers bring the material to be operated into the cell and take away the pieces produced.

The Cell Controller was developed and implemented in a Sun SparcStation 10 workstation with Solaris 2.4 Operating System. This workstation has a network card with two stacks:

- **TCP/IP stack**, for the communication with the Shop Floor Controller and the Project Department. This network allows the transmission of NC and RC programs from fileserver to the industrial machines;
- **OSI MAP stack**, for the communication with industrial machines, like NC machines and robots.

The manufacturing cell is connected to the controlling room by a LAN with a bus structure topology, based on a base band transfer media (10Mb/s). The LLC protocol used is 802.3 (Ethernet CSMA/CD). All the machines have MAP interface boards. These interfaces are: **CP 1476 MAP** for Siemens Sinumerik 880T machine controller, **CP 1475 MAP** for Siemens Sirotec robot controller and **GE FANUC OSI-Ethernet Interface** for GE Fanuc 16MA numerical controller.

Flexibility and open systems concept lead us to the application of an open systems communication standard protocol at the manufacturing process control level. MMS is a standardized message system for exchanging real-time data and supervisory control information between networked devices and/or computer applications in such a manner that it is independent from the application function to be performed and from the developer of the device or application. MMS is the international standard ISO 9506 [ISO/IEC 9506-1], based upon the Open Systems Interconnection (OSI) networking model - it's a protocol of the application layer of the OSI model. It runs on the top of an OSI stack providing a set of 80 services distributed for ten MMS functional classes.

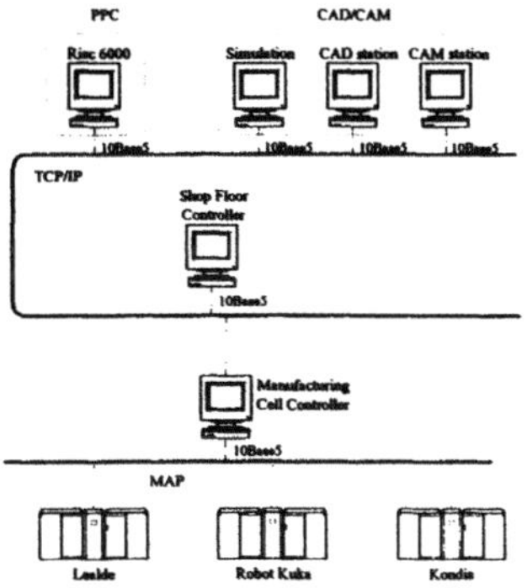

Figure 2 - Shop Floor Communication Infrastruture

3. MANUFACTURING CELL CONTROLLER ARCHITECTURE

The definition of control system structure for Cell Controller was done keeping in mind two important aspects:

- **modular structure**, which allows the future expansion of the control system, for instance, if the number of machines grows up;
- **real time requirements**, to guarantee that control system is able to execute all required functions performing the time restrictions, for example the cooperation between machines.

The specified control system for this Cell Controller is a blend of centralized and hierarchical architecture. It's a hierarchical architecture because the functions are distributed by several control levels; however, it's not completely hierarchical because the lowest levels may execute jobs and services but they are not able to execute the complete manufacturing orders.

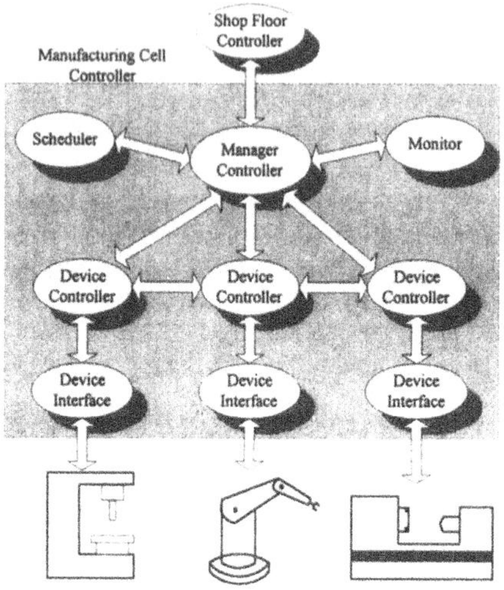

Figure 3 - Cell Controller Structure

This structure has three hierarchical levels, being each one responsible for the execution of control functions. The first level is designated by Manager Module, and it's responsible for the planning and the control of the cell production. In the second level, there are several modules to control industrial machines. Each of these modules, designated by Device Controller, is customized to the industrial machine and it has the responsibility for the execution of the jobs in the machines. Finally, the last level, the Device Interface, contains the interface between the Cell Controller and each of the industrial machines. The communication protocol implemented in these three modules is the MMS protocol.

This control structure is modular and flexible, allowing easily the expansion of the Cell Controller if the number of machines in the cell grows up. When a new machine is added to the manufacturing cell, it's only necessary to add a Device Controller module and a Device Interface module, customized to the new machine.

All Cell Controller code is written in C language and the communication between modules, inside the Cell Controller, is implemented with a Unix Operating System functionality, called pipes, which is made up of files to exchange the messages.

3.1 Manager Module

This module is the brain of the manufacturing cell controller, and it's responsible for the control and the supervision of the production process of the manufacturing cell and also for the management of cell resources. The main functions of this module are:

- Start the Cell Controller, verifying if it's a normal or abnormal start;

- Receive and process the messages from the Shop Floor Controller;
- Receive and process the results of services executed by the Device Controllers;
- Determine for each order, the next operation to be executed;
- Dispatch the orders to the Device Controllers;
- Notify the Shop Floor about the evolution of the orders and whenever an alarm occurs;

The management of the manufacturing cell is based upon information stored in the cell database. This database contains information about orders, resources, cell buffers and tools stored in the industrial machines of the cell.

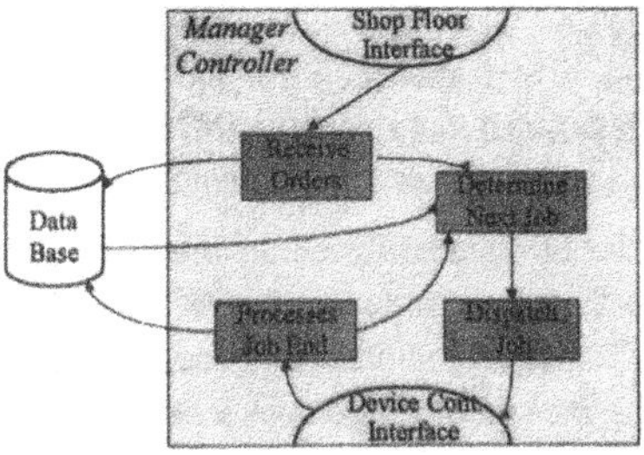

Figure- 4 Manager Module Structure

The Manager Module is a set of several small modules, each one responsible for the execution of different functions. The Shop Floor Interface and the Device Controller Interface send and receive messages to the Shop Floor Controller and from the Device Controller, respectively. The other four small modules execute the management of the manufacturing orders.

3.2 Device Controller Module

The Device Controller is responsible for the local control of the machine, and for the execution of the jobs requested by the Manager Module. This module receives jobs from the Manager Module and separates them into elementary services, to be executed by the Device Interface module in the industrial devices. For example, the execution of a machining program requires the execution of two services: the transmission of a NC program from fileserver to the NC machine, and the start of a NC program.

When all services are completed, the required job is concluded and this module sends a message to the Manager Module indicating the end of the job. If an error occurs in the industrial device, this module sends also an error message to the Manager Module.

3.3 Device Interface Module

The MMS protocol implements the interface communication between the Cell Controller and the industrial devices. The key features of MMS are:

- **the "Virtual Manufacturing Device" (VMD) model**. The VMD model specifies the objects contained in the server, and with MMS services it's possible to access and manipulate these objects;

- **the client / server model.** The client is a device or application that requests data from the server; the server responds with data requested by the client.

There is a distinction between a real device (e.g. PLC, CNC, Robot) and the real objects contained on it (e.g. variables, programs, etc.), and the virtual device and objects defined by the VMD model. Each developer of a MMS server device or MMS server application is responsible for "hiding" the details of their real devices and objects. The executive function translates the real devices and objects into the virtual ones defined by the VMD model.

In many cases, the relationship between the real objects and the virtual ones can be standardized for a particular class of device or application (PLC, CNC, Robots, PCs). Developers and users of these real devices may decide to define more precisely how MMS is applied to a particular class of device or application - the result is a Companion Standard.

The Device Interface module is responsable for the execution of services, required by the Device Controller, on the industrial devices. The MMS objects implemented in the server application (industrial device) and accessed by the client module (Device Interface module) are:

- **MMS Domains**, used for the standardized transmission of data or programs;
- **MMS Program Invocation**, used to manipulate NC/RC programs;
- **MMS Variables**, used to map typed data (e.g. Tool offsets, NC data, PLC data, etc.);
- **MMS Events**, reporting the end of program execution.

When the service is finished, this module sends a message to the Device Controller indicating the end of service execution. An error message is sent if a fault occurs in the machine, for example, a tool collision.

4. SOLUTIONS FOR SOME PROBLEMS

4.1 Execution of a Machining Program

The execution of a machining program requires the execution of a set of services: decoding a NC program, download the program to the machine and start the program by the machine. When the project team designs a product and generates the NC programs for the manufacturing of the product, it's not possible to know where the tools are stored in the machine. Therefore, the NC programs must contain the indication of the tool's type. However, the CAD postprocessor can only insert a code with two digits in the number or type of tool. Thus, it's necessary an auxiliary file to associate the tool types and the tool codes (two digits) that appears in the NC machine.

Original Program

MPF 1000
. . .
T22 D22
. . .
T4 D4
. . .
M30

Auxiliary File

4 fresa20
. . .
22 roca20

Tools Table

Tool Type	Position
drill10	1
fresa20	2
drill4	6
probe	10
roca20	12

Final Program

MPF1000
. . .
T12 D12
. . .
T2 D2
. . .
M30

Figure 5 - Decoding of a Machining Program

This limitation of CAD post-processor requires a decoding of a NC program before the download of the program to the machine. This function is implemented by a filter algorithm.

The filter algorithm searches the character "T" in the NC program; the next two digits are the tool code. It's necessary to know what is the tool type for this tool code, by reading the auxiliary file. After, it's necessary to verify if there is a tool inside the machine with the type required, by seeking the tools table. If there is a tool stored in the machine with the type of tool required for the operation, the tool code of NC program is replaced by the position of the tool inside the NC machine. When the filter algorithm finishes the changes in all lines of the NC program, this NC program is ready to be downloaded to the NC machine.

4.2 Execution of Setups

The execution of a new production requires the update of machine configurations. This operation is designated by setup and can contain the following tasks:

- change the tool stored inside the machines;
- change the gripper of robot;
- execute the machine maintenance;
- test the RC programs;
- change the machine parameters.

During the setup, the Cell Controller must execute a set of functions: transfer the tools offset to the machines and update the tools table with the parameters of the new tools. The transfer process is different for each machine:

- **Lealde machine** - the Device Controller makes the upload of tools offset from machine to a file. For each position, the old parameters are replaced by the new tools offset. When the file is completed, this file is downloaded to the machine through MMS Download service.
- **Kondia machine** - it's necessary to build a data structure with the following information: offset number, geometrical dimensions, radius, wear and tool life. This structure is transferred to the machine through MMS Write service.

5. PERFORMANCE ANALYSIS

5.1 Data Transmission rates

The execution of a program in a industrial device requires the existence of this program inside the device memory. Due to the memory capacity limitation, it's necessary to transfer the program to the device whenever it will be necessary. This operation is performed by the Cell Controller with the MMS Download service.

During the Cell Controller implementation tests, and with a network analyzer, it was possible to work out the transfer speed: 10 kbits/s. This speed is very slow and origines the Cell Controller lost of efficiency. For example, the Cell Controller spends approximately 30 seconds to download a NC program with 50 kbytes! This time is not profitable in the processing tasks, and causes the lost of productivity in the manufacturing cell.

With this slow transfer speed, it was necessary to know where the problem was located. After several tests, it could be concluded that the problem was not associated with MMS protocol, because two MMS applications, running at SUN worksation, could communicate with 10 Mbits/s transfer speed; thus, the problem is related to the internal machine problems. In fact, when a machine receives a program, this is analised inside the machine, line by line, to detect errors inside the program. This procedure causes the reduction of data transmission from 10 Mbits/s to 10 kbits/s.

For increasing the Cell Controller efficiency, it´s necessary to do the management of the NC programs inside the machines memory and download the needed NC programs into the Cell Controller, making use of it´s dead times. This is a complex solution to be implemented, and requires the introduction of Artificial Intelligence techniques inside the system control. The use of multi-agents systems in the Cell Controller architecture can be a good solution, to solve this problem and, in addition, to help the fault recovery tasks.

5.2 Costs of the solution

The performance of an integrated solution is analyzed into two points: operationality and cost. The cost is an important factor that will be decisive for the viability of a solution. It is possible to quantify the cost associated with this Cell Controller architecture solution, by analyzing the costs of the several platforms and interface boards used to implement the Cell Controller. These costs can be divided into two main areas: the communication protocols and the developing platform.

To implement a communication system with MMS protocol, the cost is obtained by adding the cost related to the MAP/MMS interface boards and the cost of the communication software. It´s possible to calculate the cost of the communication system implemented in the Cell Controller. The cost of the MAP/MMS interface boards is around \$6000 (US dolars) each one and the MMS software (MMSEASE from SISCO in this case) around \$6700. In this application, the number of MAP/MMS interface boards is 4 (3 servers/machines and 1 client), so it´s necessary to spend \$30700 to implement the communication system. For a better perception of this cost value, it is possible to make a comparation with the cost associated with the resources of the cell (two NC machines and one robot). The cost of these three machines is around \$333300. So, the communication system cost value is approximately 9,2% of the costs associated with the resources of the cell.

With this scenario, it can be concluded that it´s very expensive to use a standardized communication protocol like MAP/MMS, basically because the price of the interface boards are very expensive. In alternative, other communications protocols could be implemented for the communication system: FieldBus and Serial link, for instance.

The FieldBus interface boards are cheaper, because they only implement three levels of OSI the stack. The FieldBus is a standardized protocol, like MAP/MMS, and uses the RS485 protocol to implement the physical layer. This protocol is used to interconnect the sensor and actuator devices. The Serial link is very cheap, but it requires the development of an API (Application Program Interface) for each industrial device, customized to the device funcionalities. For a higher number of devices, this task is complex due to the different funcionality of each device. Another disadvantage of the Serial link is the connection point to point between devices and the controller station, that is worst for the expansion of the system and also for the machines which are distant from the controller station. Solutions which only transfer programs to the devices required from device operator, is a good answer for applications with low control degree or with manual production, which only requires the download of NC or RC programs.

Another disadvantage of this Cell Controller solution is the developing platform. The cost of the developing platform is \$15340, due to the cost of the SUN SparcStation 10 and the developing software SUN SparcWorks. These costs can not be reached by SME (Small and

Medium Enterprises), which also demands for new platforms like Windows NT. Actually, the companies need cheap control systems solutions and short developing time. The Windows NT platform allows the reduction of hardware costs and the use of more userfriendly programming tools, like Visual Basic or Visual C++.

An important problem with the Unix platforms is the monitoring or graphical interface that is very complex to implement. The use of Visual C++, for instance, allows the reduction of development costs and a more quick implementation of the solution for the desired application, using all the potentialities of these programming tools.

6. CONCLUSIONS

The cell controller architecture presented in this paper is now working well. The two major aspects of a Cell Controller specification were focused: the control structure and the communication system. In the control structure, the use of hierarchical architectures is growing up, mainly due to the increasing of computing system capacities and the reduction of the prices. The control structure presented in the Cell Controller solution is a blend of a centralized architeture and a hierarquical architecture, allowing the flexibility and modularity of the application. The communication system could be implemented with MAP/MMS, FieldBus, Serial link, between others. The protocol presented is the MAP/MMS due basically to it´s standardization and the simplicity of it´s implementation.

With the analysis of the MMS implementation, it was possible to conclude that the costs of the MMS interface boards are the limiting factor to the implementation of MMS protocol in the integrated applications. Nevertheless, the MMS protocol is a good solution for applications which require a total control of resources or when the integration involves a high number of resources with high degree of complexity. For cheaper solutions required by SME´s, it´s necessary to use other protocols, like serial link and field bus. These solutions are a valid alternative to implement the communication system of Cell Controller.

The Cell Controller architecture described in this paper, is developed in an Unix platform, and despite the good results of his operationality, it´s necessary, in the future, to build the Cell Controller on a cheaper development platform, like Windows NT, and also it is wise to use the Artificial Intelligence methods to perform the control and the faults recovery tasks.

7. REFERENCES

Diltis, D., Boyd, N., Whorms, H. (1991) *The evolution of control architectures for automated manufacturing systems*, in Journal of Manufacturing Systems, Vol 10 N°1, pp 63-79

Lee, K., Sen, S. (1994) *ICOSS:A two-layer object-based intelligent cell control architecture*, in Computer Integrated Manufacturing Systems, Vol 7 N°2, pp 100-112

Pimentel, J. (1990) *Communication Networks for Manufacturing*, Prentice Hall

Groover, M. P. (1987) *Automation, Production Systems and CIM*, Prentice-Hall

ISO/IEC 9506-1, (1992) *Industrial Automation Systems - Manufacturing Message Specification, Part 1 - Service Definition*

Rembold, U., B.O. Nnaji, B.O. (1993) *Computer Integrated Manufacturing and Engineering*, Addison-Wesley

CCE-CNMA 2, (1995) *MMS: A Communication Language for Manufacturing*, ESPRIT Project CCE-CNMA 7096, Volume 2, Springer

16

Strategies and fundamental structures for FMS tool flow systems

S. C. Silva
Universidade do Minho,
Campus de Gualtar, 4710 - Braga, Portugal, phone:+351+53 60 4455, fax: +351+53 60 4456, email scarmo@ci.uminho.pt

Abstract

A Tool Flow System, TFS, is an integral part of a wider system within a Flexible Manufacturing System, FMS: the Material Flow System. This also includes the Work Flow System.

Tool Flow Systems deal essentially with the storage, transport and handling of tools which are exchanged at the machine spindles. Work Flow Systems are particularly concerned with the storage, transport and handling of parts and auxiliary handling devices, such as fixtures and pallets.

This article focuses attention on the building blocks of TFS's and presents a classification, discussion of application and advantages of tool flow fundamental structures. It is argued that the configuration of TFS's and the level of tool replication, may greatly influence the performance and flexibility of production in FMS's.

Keywords

Flexible manufacturing systems, FMS, tool flow systems, multiple stage systems, MS, combined stage systems, CS and single stage systems, SS.

1 INTRODUCTION

Tool storage and flow in manufacturing systems, in general, and in Flexible Manufacturing Systems, FMS, in particular, requires careful planning. According to Stute (1974), in 1967

Dolezalek used the term Flexible Manufacturing System to refer to a number of machines interlinked, through common control and transport systems, in such a way that automatic manufacture of different workpieces, requiring a variety of different operations, could be carried out. This definition still applies today as a general definition of a Flexible Manufacturing System, FMS.

When machines are very versatile they are able to perform a large range of operations during the manufacturing period conditioned, however, by those tools which a machine can access. If the access is restricted to a few tools, the machine might only be able to function as a special purpose machine. Therefore its versatility is not used completely during the manufacturing period with probable loss of efficiency.

To take advantage of FMS machine versatility, short term scheduling may be prepared off-line. In this case a machine loading plan can be prepared for the manufacturing period, i.e. shift or day, where part processing sequence and part and tool assignment to machines can be established in advance of production. This assignment is known as the FMS loading problem (Stecke, 1981).

On the assumption that no machine breakdown happens, during the planned manufacturing period, then manufacturing according to the plan can be completely carried out by providing the machines with only the required tools. If big disturbances do occur during the planned manufacturing period then rescheduling the parts and tools is likely to be necessary. Small variances may be coped with by adequate on-line reloading control (Stute,1978).

Tool variety reduction

In FMS there is a predominance of multipurpose machines, i.e. machines which are capable of performing a large number of different operations. This simplifies the control of tools and parts flow.

So, there is interest in establishing efficient tool flow systems with a coordinated tool supply to the different work stations. This is particularly relevant when tool variety and quantity can become large. In this case, usually, there are both economical and organisational advantages in reducing the number of tools in the system. This can be achieved on the one hand by adopting an off-line preparation of FMS loading plans based on the tools available, as referred to above, and on the other hand through a variety of standardisation and rationalisation measures directed at tool variety reduction (Hoop 1982, Zeleny 1982, Hankins 1984), figure 1.

Tool availability

One major aspect in the selection of the TFS configuration is the need to improve machine readiness as it is affected by tooling. Consequently, there is interest in separating, as much as possible, the tool set-up function from the machine cycle. To achieve this, new tool system configurations can be applied as it is discussed below. The best configuration to choose is influenced by the degree of automation in connection with the autonomous period desired for unattended manufacture and by other operational and organisational aspects as well as by economical reasons.

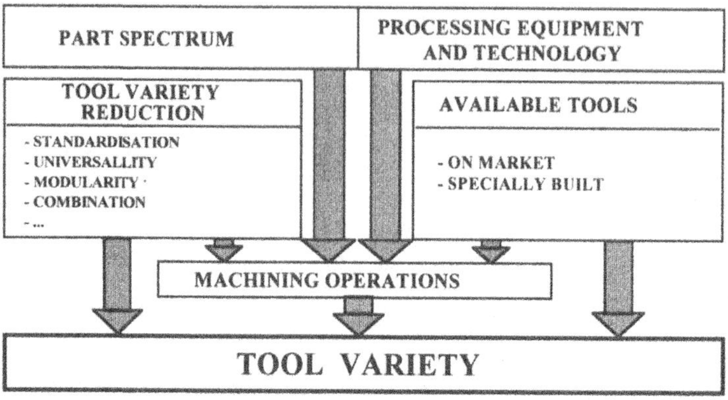

Figure 1 Intervening factors for tool variety determination.

2 TOOL FLOW STRUCTURES

A classification of fundamental tool flow structures for TFS's configuration design is presented in figure 4. These structures can be developed from pertinent combinations of basic tooling systems elements as classified and shown in figures 2 and 3. These elements include local and central tool stores, tool magazines, tool exchangers and tool and magazine transport and handling elements, namely vehicles, conveyors and automatic manipulators, such as industrial robots, widely used in TFS's of many FMS's.

Practical application of the classified tool flow structures were reported by Chmielnicki 1980, Hammer 1983, Binder 1983, Zeleny 1985, FMS Magazine 1985, Tomek 1986 and other authors.

When tools cannot be automatically accessed for tool change at the machine spindle, *manual tool change* has to be used, figure 4-case 1. Such a solution allows some degree of flexible automation, since in FMS's machines are computer controlled, but unattended working is not possible.

By providing the machines with an *automatic tool change* system, performed from a local and/or central tool store, different degrees of unmanned work are possible.

2.1 Tooling structures with central tool storage

Unattended work

The solution of automatically accessing a central tool store permits a high degree of automation and, for large central stores, can allow long periods of unattended work. This solution can lead, also, to a good level of utilisation of tooling resources. This can be achieved through a continuous flow of tools from and into the tool magazines of the machines. This configuration is likely to require constant computerised supervision and control of the tool requirements, tool flow and tool life.

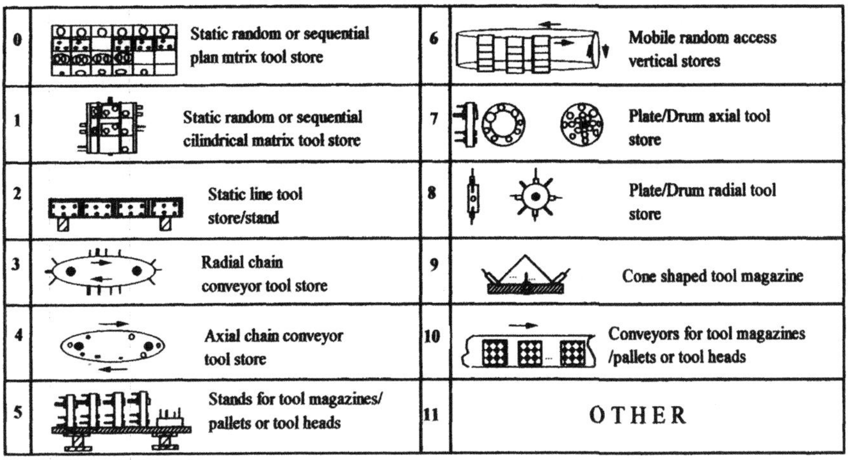

Figure 2 Basic tool buffer and storage elements for FMS Tool Flow Systems.

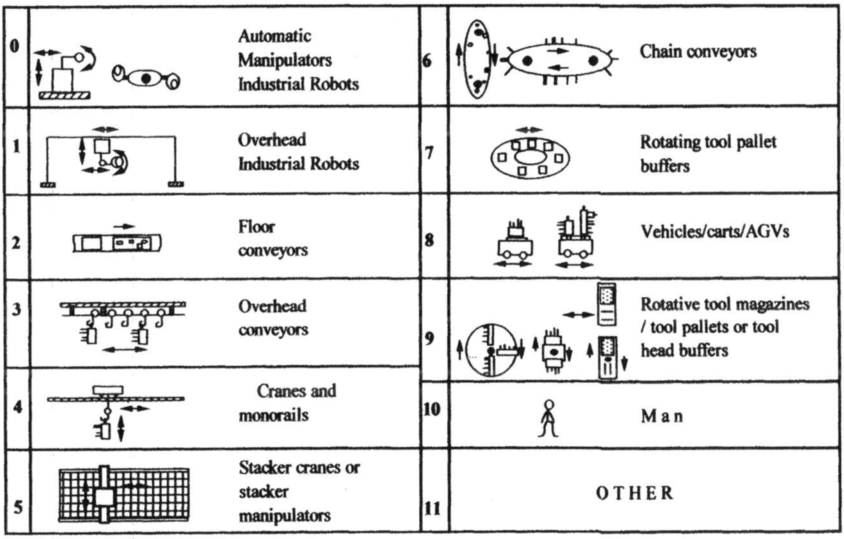

Figure 3 Basic tool transport and handling elementsfor FMS Tool Flow Systems.

Tooling reduction

A further advantage of this system is that it may allow a reduction in the required number of identical tools in the manufacturing system. However, with centralised tool storage configurations, processing interference among machines may result when real-time machine loading and operations scheduling or sequencing is used. This is due to the fact that all machines are simultaneously sharing the same resources, in this case the same tool central store and, possibly, the same stored tools. Therefore, at some instants, it may happen that different machines are "fighting" for the same tools or at least simultaneously requiring the service of the tool exchange mechanism for tool change. This problem can be partially solved if off-line manufacturing loading plans are prepared in advance and tools are accordingly and adequately provided. This, may require some degree of tool duplication in the store.

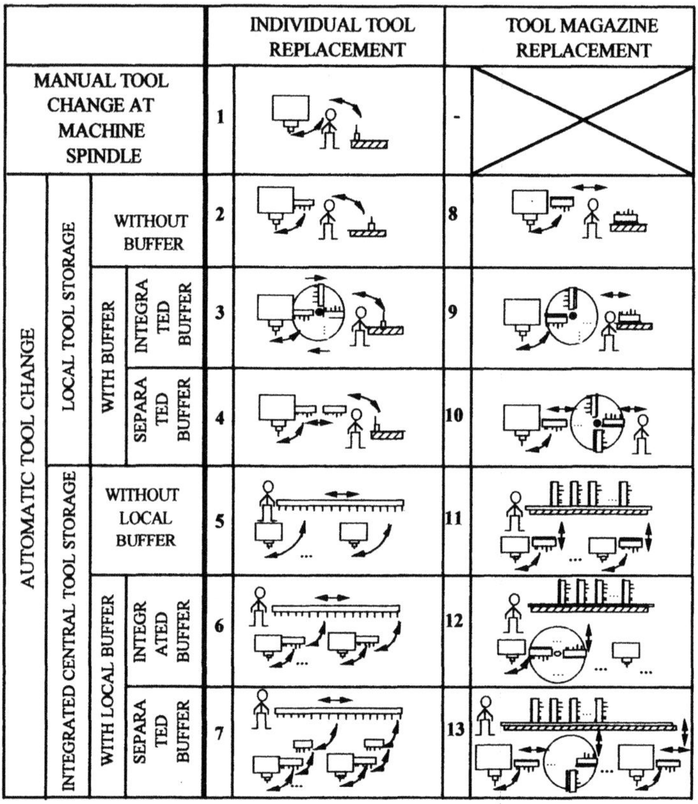

Figure 4 Tool flow strutuctures for FMS tool flow systems.

Although some action can be taken for avoiding simultaneous requirements of the same tools, through careful sequencing and part loading, in practice, applying this off-line loading strategy, may be unsatisfactory. This is due to the unpredictable behaviour of the tool requirements dynamics during part processing.

A disadvantage of the centralised tool storage configuration is the increased risk of complete system stoppage due to breakdown of the tooling system. To reduce this problem, machines can be designed and prepared to also work standing alone and tools manually changed while the tooling system recovers.

Minimum tooling and machining systems configuration

Part routing flexibility advantages can be obtained applying centralised tool storage even under minimum tooling requirements, figure 5.

The situation of minimum tooling requirements, in the extreme, is associated with the existence, within the system, at any time, of a single tool of each required type. This is theoretically enough for carrying out the processing of a chosen part mix, during a planned period of production as long as dynamic tool loading, (Silva, 1997), is allowed.

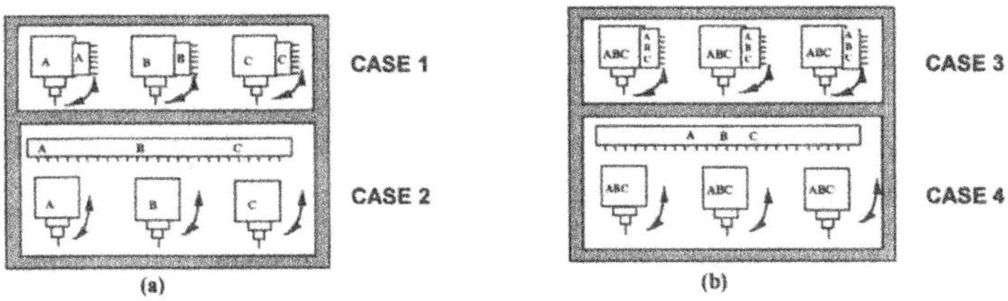

CASE	CASE 1	CASE 2	CASE 3	CASE 4
Number of Tools used	T	T	T	T
Routing flexibility	None	None	None	Good

Legend:
(a) - Single purpose machines/Multi-stage Systems
(b) - Multiple purpose machines/Single Stage Systems
T - Minimum number of tools to process a part mix
Tool Store A B C
Machinetool

Figure 5 The use of a central store for direct tool change into the spindle of the machines works best for single stage systems.

Although some flexibility advantages exist from using centralised tool storage in multiple stage systems, MS, it is in the combined stage systems, CS, and particularly in the single stage systems, SS, (Opitz, 1972), that full advantage can be taken from such a tool flow structure.

In MS systems part processing requires the use of a different machine type for carrying out each particular machining process. This, usually, requires a set of different tools. Unless duplication of machines of some type exists, part routing flexibility does not exist in these systems whatever the nature of the tool flow structure. In CS systems machines can carry out a few different processes. In SS systems it is considered that full part processing, of any loaded part, can be carried out in any of the machines in the system. In this case the full range of centrally stored tools are potentially available to all machines. Maximum routing flexibility can be achieved with this integrated concept, i.e. central tool storage and SS systems. These systems can be configured from very versatile identical machining centres.

For minimum tooling requirements, routing flexibility is not available, under local storage, whatever the type of system.

It should be stresses that, under minimum tooling and centralised tool storage approach, interference among working units may occur, for example, due to tooling and handling requirements. Therefore, this tool flow solution, in terms of reliability and flexibility, may not be a particularly good solution. Tool replacement can be enhanced to allow tools to be replaced during processing as a means to increase machine availability and, possibly, system performance.

2.2 Individual tool replacement

When tool magazines are an integral part of the machine, tool replacement at the machine is normally done on a tool by tool basis, figure 4, cases 1 to 7.
In addition, or as an alternative, to a tool central store a replacement back-up tool buffer, at each machine, can also be provided. The tools in the buffer may be handled and accessed either manual or automatically. An operator or a handling device could be applied.

A few versions of the back-up tool buffer arrangement can be found in practice and three structures fitting this approach, under single tool replacement, are shown in figure 4, namely cases 3,4, and 7. They are designed to accommodate the tools needed for a large parts processing autonomy period and to reduce waiting time of the machines. Typically the tools in the tool magazine are replaced with the tools in the back-up store to cope with next parts processing requirements (Marsh 1981, FMS Magazine 1985). This buffering function is, in many cases, provided by the machine tool magazine itself, figure 4-case 6. The tool buffer also functions as a means of extending the tool storage capacity near the machines.

2.3 Exchanging magazines of tools

The exchangeable magazine replacement approach, in opposition to the tool by tool replacement, can be associated with two basic arrangements as shown in figure 4:

A - Local tool exchangeable magazine store, cases 8 to 10.

B - Central tool exchangeable magazine store, cases 11 to 13

In the situation A the tool store is local with one or more magazines waiting to be replaced with the ones being used.

In the situation B the bulk of tool magazines or pallets with tools, destined for more than one machine, are located in a central tool magazine store. They are transferred to the machine areas through mechanised or automated transport and handling equipment. This concept integrates the local magazine storage cases.

By using the exchangeable magazines approach, in a manner which is similar to part pallets replacement at machines, many operational benefits can be explored. In particular, tool magazines can be associated with the processing requirements of the parts on a pallet or pallet group and routed together to the machines. When the flow of tool magazines is "linked" to that of part carriers or pallets the control of the flow of tools and parts is simplified and synchronised. In this case the same parts pallet carrier may also simultaneously carry the tool magazine for processing them.

When, as frequently it happens, the tool magazine or buffer represents an increased capacity of the local tool storage, longer periods of unmanned manufacture can be achieved.

Magazine replacement at a back-up tool buffer, performed during processing, together with an effective swapping magazines mechanism, in the processing area, tends to increase considerably the availability time of machines contributing, in this way, for improved performance of production.

The exchangeable magazine solution is likely to require considerable investment in tooling, tool magazines and handling equipment. To reduce this, not only tools but also tool magazines should be simple, standardised, universal, modular and flexible, Bullinger (1986), Hoop (1982), Zeleny (1982), Hankins (1984).

3 LEVEL OF TOOL REPLICATION

Although, as was referred above, flexibility of production may be achieved under minimum tooling, it is important to understand the impact of different levels of tool replication in system performance, under different TFS configurations.

The need for tool replication in FMS's, in a situation of tool sets replacement, has been studied by Silva (1988). The study does not centres on tool flow systems but solely, for a given tool flow system configuration, evaluates the influence of different levels of tool replication in CS and SS FMS system configurations. One of the main findings of this work is that, under adequate production control, only a very reduced level of tool replication is likely to be necessary for achieving very high levels of FMS performance. However, this work does not take into account the need for replication due to tool wear.

In a particular FMS case study Carrie, (1986) concluded that, the need for tool changes due to part variety was only a small part of the total number of changes and that tool wear can be highly influential of the need for tool changes.

The interrelationships and interdependence between tool flow structures and the level of tool replication and their articulation with other factors such as tool wear, part variety and operating strategies, seem to be so great that it is wise to study the full influence of tools replication and tool flow structures if good design and good operation of FMS's is to be achieved.

4 CONCLUSION

The number of replicated tools and the configuration of Tool Flow Systems, TFS's, which integrate both the function of storage and that of movement and handling of tools, is critical to the efficiency of manufacturing systems and in particular to FMS's.

Many arrangements can be designed considering both the handling of tools either in sets or individually and the storage of tools centrally in the system and/or locally to machines.

It was shown that, in FMS's, the combination of machining system configurations with some TFS solutions, for the same number of tools, can be very restrictive to part routing flexibility while others can provide good flexibility for even minimum tooling.

The TFS solutions which can be generated by combining basic arrangements of tool flow structures for adequate level of tool replication, greatly influence the design of FMS's as well as the manner how FMS's must be run. Therefore for good design and operation of FMS thorough analyses of the influence of TFS configurations and levels of tool replication, for different operation strategies, must be conducted.

5 REFERENCES

Binder, R. and Hammer, H. (1983) Flexible Manufacturing Cell for Vertical 5 side Machining of Large Parts. *WT-Zeitshrift für industrielle Fertigung*, **73**, No.4.

Bullinger, H.-J., Warnecke, H.J.and Lentes, H.-P. (1986) Toward the Factory of the Future. *International Journal of Production Research*, **24**, No. 4, 697-741.

Carrie, A. S. and Perera, D. T. S. (1986) Work Scheduling in FMS under Tool Availability Constraints. *International Journal of Production Research*, **24**, No.6, 1299-1308.

Chmielnicki, S. and Stute G. (1980) Simulation - ein Hilfsnittel bei der Auslegung des Werkzeugflusses in Flexiblen Fertigungssystemen. *Industrie Anzeiger*, **102**,. No. 20, 28-9.

FMS Magazine, (1985) British Aerospace Aims Sky High. *FMS Magazine*, April.

Hammer, H. (1983) Verbesserung der Wirtshaftlichkeit durch flexible Automatisierung beim Bohren unf Fräsen. *ZwF-Zeitschrift*, **78**, No.2, 77-96.

Hankins, S.L. and Rovito, V.P.(1984) The Impact of Tooling in Flexible Manufacturing Systems. *2nd. Biennial Int. Machine Tool Technical Conf., National Machine Tool Builders' Association.*

Hoop, P. (1982) Einsatz von kombinationswerkzeugen in FFS. *Zeitschrift für Metalbearbeitung*, **76**, No 7.

Marsh, P. (1981) Britain advances in computerized factories. *New Scientist*, **89**, No.1245.

Opitz, H. and Hormann, D.(1972) Planning and Utilization of FMS. *Annals of the CIRP* **21/1/72**, 29- 30.

Silva, S. C. (1988) An Investigation into tooling requirements and strategies for FMS operation. *PhD Thesis*, Loughborough University of Technology, U.K..

Silva, S.C. (1997) Analytical assessment of tooling requirements for FMS design and operation. Submited to the *OE/IFIP/IEEE International Conference on Integrated ans Sustainable Industrial Production, ISIP'97*

Stecke, K. and Solberg, J.J.(1981) Loading and Control Policies for FMS. *International Journal of Production Research*, **19**, No.5, 481-90.

Stute, G., (1974) Flexible Fertigungssysteme. *WT-Zeitshrift für industrielle Fertigung*, **64** No.3.

Stute, G. (1978) Organizational Control in an FMS FMS *Workshop conference, Milwaukee, USA.*

Tomek, P. (1986) Tooling Concepts for FMS, Proc. of the 5th Int.Conf. on FMS.

Zeleny, J.(1985) Unmanned Technology and Control Strategies in Flexible Manufacturing Systems with Random Flow of Parts and Tools, Annals of the CIRP Vol. 34/1/1985

Zeleny, J. (1982) Manufacturing Cells with Automatic Tool Flow for Unmanned Machining of Box-like Workpieces Manufacturing Systems, Vol.11 No.4.

BIOGRAPHY

Dr. S. Carmo Silva is a member of the Departamento de Produção e Sistemas at the Universidade do Minho, in Braga, Portugal. At present Dr. Carmo Silva is the head of the Industrial Management and Systems group. He got his PhD, in 1988, from Loughborough University and his MSc degree from UWIST in U.K, both in the area of Production Systems and Management. He is also graduated in Mechanical Engineering by Oporto University, Portugal. His main academic and research intestests fall into the general concept of Manufacturing Systems Engineering and, in particular into the area of design and operation of integrated manufacturing and assembly systems.

17

Analytical Assessment of Tooling Requirements for FMS Design and Operation

S. C. Silva
Universidade do Minho,
Campus de Gualtar, 4710 - Braga, Portugal, phone:+351+53 60 4455, fax: +351+53 60 4456, email scarmo@ci.uminho.pt

Abstract

FMS's do not work well if the required amount of tools to carry out processing, during a given manufacturing period, are not available. High operating efficiency can be ensured, once the right number of tools of each type is available, by conveniently solving the problem of assigning tools and parts to machines. This has been identified as the FMS loading problem (Stecke, 1981). Moreover it is essential that good manufacturing control of part processing priority is realised.

In this paper an analytical method is presented for initial assessment of the total number of required tools for efficiently running an FMS. The method is based on tool life duration and on a concept referred to as tool cycle time, as well as on the planned manufacturing period. This is usually taken as the time frame for the solution of the FMS loading problem.

Two fundamental situations are considered, namely that of static tool loading and dynamic tool loading.

Keywords

Flexible manufacturing system, static tool loading, dynamic tool loading.

1 INTRODUCTION

In simplified terms a Flexible Manufacturing System, FMS, may be defined as a computer aided manufacturing system able to simultaneously manufacture a variety of parts through computer control of processing and material flow.

Much research has been carried out to solve the FMS loading problem identified by Stecke (1981). The problem and a few variations of it, for a number of different constraints, have been addressed by Sodhi 1994, Sarin 1987, Rajagopalan 1986, Stecke 1983 and other authors. In most cases considerable simplification of reality is assumed.

The development of a FMS, requires technical, economical and performance analysis of the alternatives for control, machining and material flow systems, (Warnecke 1982, Eversheim 1980, Storr 1979).

A tool flow system, TFS, is an integral part of the material flow system within a FMS. TFS's deal with the storage, transport and handling of tools which are exchanged at the machine spindles, and can be configured in many different ways (Silva, 1997). Therefore tool flow systems are conceived to handle a set of important manufacturing aids, i.e., the machining tools to be used by the machines to carry out the processing of parts.

This paper addresses the analytical estimation, of the number of tools to be loaded in the machines to run an FMS.

Two fundamental situations are considered, namely that of static tool loading and dynamic tool loading. These situations are closely related to the concepts of static tool change and dynamic tool change presented by Sodhi (1994). In *static tool loading* tools are only loaded at the start of each planned manufacturing period in such a way that a complete tooling autonomy exists during that period. Here no tool recycling is allowed In *dynamic tool loading* tools can be removed from, and fed into, the system during the planned manufacturing period.

The analytical methodology used can easily be extended to the determination of other manufacturing aids in FMS or other manufacturing and assembly systems.

2 NUMBER OF TOOLS TO RUN A FMS

2.1 Number of tools under static tool loading

While tool variety is primarily determined by the variety of part operations, the total number of tools required is dependent on the length of the manufacturing period for which tooling autonomy is desired, figure 1. Under static tool loading tools are loaded only at the beginning of each planned manufacturing period.

Tools with identical tool lives

The number, f, of tools required to operate a single machine during a single time unit is given by:

$$f = 1 / t_l \tag{1}$$

t_l is the tool life time.

Any fraction of tool, should be interpreted as a tool which is required only for an equivalent fraction of tool life duration.

Under static tool loading having to run an FMS with a number of machines m for an autonomous period of duration T the total number of tools, F, required is given by:

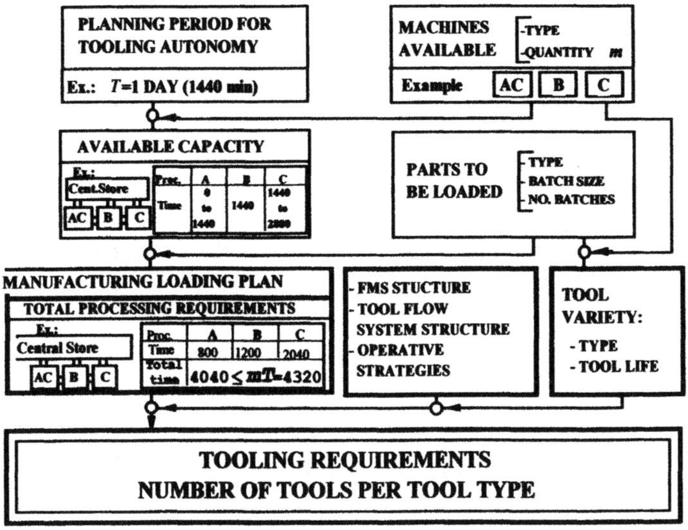

Figure 1 Tool requirements for a manufacturing planned period

$$F = mTf \tag{2}$$

or

$$F = mT\frac{1}{t_l} \tag{3'}$$

As an example, considering a manufacturing period of a 3 eight hour shifts, i.e., a 24 hour day work, and 10 machines in the system, for tools with tool life of 20 minutes, applying equation (3'), the number of tools required is:

$$F = 10*24*60*\frac{1}{20}$$

$$F = 720 \text{ tools}$$

Therefore, for complete tooling autonomy, during the 24 hour manufacturing period, a total of 720 tools would be necessary. The result assumes that machines are fully and constantly utilised in actual machining during the planned manufacturing period T and that all tools are used to their full lives.

When expected average machine utilisation in actual machining is U, this would mean that the machining time per machine is U times T. On the other hand, due to technical, organisational or economical reasons resulting from system operation, usually tool life is not used up completely. Only a percentage, say β, of tool life duration is used. In this case, a generic expression for the number F of required tools, which also takes into account the average machine utilisation, can be written as:

$$F = UmT\frac{1}{\beta t_l} \quad (3)$$

Taking account of tool life differences

When more than a single type of tool, with different tool lives, are going to be used, the number $\bar{f}$ of tools required, to continuously operate a machine during a single unit time, is the average number of tools, over all tool types, and is given by:

$$\bar{f} = \frac{1}{n}\sum_{i=1}^{n} f_i \quad (4)$$

- $\bar{f}$ is the number of tools that, on average, will be necessary for running a machine during a single time unit;
- n is the number of different tools to be used;
- f_i is the number of tools of type i which would be required to operate continuously a machine, during a single time unit, and is given by: $f_i = 1/t_{l_i}$.

Therefore, equation (4) can be rewritten as:

$$\bar{f} = \frac{1}{n}\sum_{i=1}^{n} \frac{1}{t_{l_i}} \quad (5)$$

which allows, to write the average number, $\bar{F}$, of tools necessary to run an FMS with m machines for an autonomous period T, as:

$$\bar{F} = mT\frac{1}{n}\sum_{i=1}^{n} \frac{1}{t_{l_i}} \quad (6')$$

and, generalising for taking average machine utilisation U into account and the proportion β of tool life used, it will become:

$$\bar{F} = UmT\frac{1}{n}\sum_{i=1}^{n} \frac{1}{\beta t_{l_i}} \quad (6)$$

- $\bar{F}$ is the average number of required tools for running a FMS under static tool loading;
- m is the number of machines in the FMS;
- T is length of the manufacturing planned period under which tooling autonomy exists;
- n is the number of different tools to be used;

- β is the percentage of used tool life;
- t_{l_i} is the tool life of tool type *i*.

2.2 Number of required tools to run an FMS under dynamic tool loading

Dynamic tool loading is characterised by dynamic tool changing (Sodhi, 1994), i.e. the possibility of dynamically changing tools at the magazines of the machines, and, also, feeding the manufacturing system with new tools to replace those which wear out and are removed, during the planned working period. This means that full tool recycling in adopted.

Under dynamic tool loading the number of required tools, at any time, to run a FMS, can be considerably reduced. To determine this number let us consider first that t_f is the average tool flow time, inside the system, until tool life ends. This time will be called *tool cycle time*, figure 2. Tool cycle time includes not only the time the tool is involved in processing but also the time the tool is delayed in the system due to waiting, transport, handling and set-up or preparation.

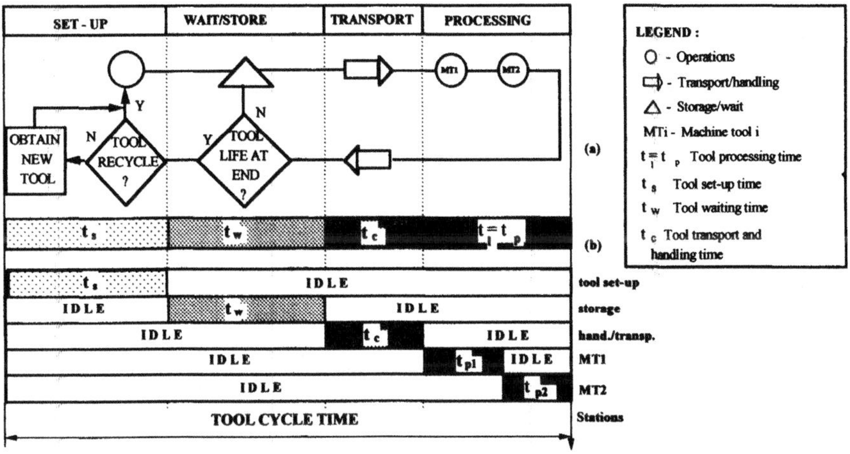

Figure 2 (a) - Schematic representation of the tool cycle time in an FMS
(b) - Multiple activity cycle diagram for a single tool.

After this time t_f has elapsed a tool tip may be replaced and the tool reused. Alternatively, for non reusable tools, tool replacement is required. In either case the average number of tools in the system, at any time, is kept constant. It is, in fact, as if the tools were recycled with new tool lives.

A dummy usage rate, R, per tool can, therefore, be established as a function of the manufacturing planned period T and average tool cycle time t_f.

This usage rate, R, is, by definition, given by

$$R = \frac{T}{t_f} \tag{7}$$

- R can be interpreted as the number of times that, on average, each tool is reused or substituted after tool life ends during a planned manufacturing period, with first time use also included.
- t_f is the average tool cycle time.

The number of required tools in a FMS at any time, under dynamic tool loading is, therefore, scaled down, in relation to the number of tools under static loading, by the R factor.

Tools with identical tool lives

If tools are identical or have the same tool life duration, the number F_D of required tools in a FMS, at any time, under dynamic tool loading, as referred above, can be obtained from scaling down the number of tools F under static tool loading:

$$F_D = \frac{F}{R} \tag{8}$$

or, from equations (3') and (7):

$$F_D = \left(m\frac{T}{t_l}\right) \Big/ \left(\frac{T}{t_f}\right)$$

and therefore:

$$F_D = m\frac{t_f}{t_l} \tag{9}$$

Equation (9) shows that the number of tools required in a FMS, at any time, is not dependent on the length of the manufacturing period, T, but solely dependent on tool cycle time t_f, on tool life, t_l, and on the number of machines, m in the system. Moreover we can conclude that, on average, for each machine in the system a number t_f/t_l of tools is required. As figure 2 suggests, t_f can be expressed as:

$$t_f = t_l + t_s + t_c + t_w \tag{10}$$

- t_l is the tool life time ;
- t_s is the tool set-up time per tool cycle;
- t_c is the tool transport and handling times per tool cycle;
- t_w is the tool waiting time, due to storage and buffering during manufacturing, per tool cycle.

Therefore F_D can be rewritten as:

$$F_D = m\frac{(t_l + t_s + t_c + t_w)}{t_l} \tag{11}$$

or

$$F_D = m + m\left(\frac{t_s}{t_l}\right) + m\left(\frac{t_c}{t_l}\right) + m\left(\frac{t_w}{t_l}\right) \tag{12}$$

or

$$F_D = F_m + F_s + F_c + F_w \tag{12'}$$

- F_m are the tools used at machining, given by m;
- F_s are the tools used at preparation and set-up, given by $m(t_s/t_l)$;
- F_c are the tools being carried or handled, given by $m(t_c/t_l)$;
- F_w are the tools waiting in system, given by $m(t_w/t_l)$;

From equation (12) we can see that reducing tool waiting and the time of some activities such as transport and tool set-up can mean a reduction in tools required in the system. Dependent upon the amount of time, this could be a substantial reduction. In the extreme, the minimum number of tools is equal to one per machine and corresponds to the permanent use of a single tool at each machine.

Equation (11), for the number F_D of required tools in a FMS, can be adjusted to take account of tool replacement before tool life ends, with β as the percentage of used tool life:

$$F_D = m\frac{(\beta t_l + t_s + t_c + t_w)}{\beta t_l} \tag{13}$$

Minimum number of required tools in an FMS

The minimum number $F_{D\min}$ of tools to run a FMS is obtained when tools do not wait, that is, they are constantly used for part processing, being transported or set-up. In this case:

$$F_{D\min} = m\frac{(\beta t_l + t_s + t_c)}{\beta t_l} \tag{14}$$

This assumes that tools flow continuously between machines and tool preparation area and, therefore, theoretically, no tool stores are necessary. This tooling situation is similar to the flow of parts in a pure part flow line without intermediate storage buffers between stations.

Tool requirements considering tool cycle time and tool life duration differences

The number of required tools in the system has been determined considering that all tools in the system have identical tool lives and tool cycle times.

Taking account of equation (9) and differences on tool life and tool cycle times the average number $\overline{F}_D$ of required tools for running a FMS under dynamic tool loading is:

$$\overline{F}_D = m\frac{1}{n}\sum_{i=1}^{n}\frac{t_{f_i}}{t_{l_i}} \tag{15}$$

and, by decomposing t_{f_i} in its elements and taking the percentage β of tool life used, it will become:

$$\overline{F}_D = m\frac{1}{n}\sum_{i=1}^{n}\frac{\left(\beta t_{l_i} + t_{s_i} + t_{c_i} + t_{w_i}\right)}{\beta t_{l_i}} \tag{16}$$

- $\overline{F}_D$ is the average number of required tools for running a FMS under dynamic tool loading;
- m is the number of machines in the FMS;
- n is the number of different tools to be used;
- β is the percentage of used tool life;
- t_{l_i} is the tool life of tool type i;
- t_{s_i} is the tool set-up time per tool cycle;
- t_{c_i} is the tool transport and handling times per tool cycle;
- t_{w_i} is the tool waiting time, due to storage and buffering during manufacturing, per tool cycle.

The equation (16) takes account of differences in individual tool life, tool set-up. tool transport and handling and tool waiting times. However, the waiting time, in the tool cycle time, can be seen as a safety factor which can be adjusted according to the tooling flow system response time, i.e., the readiness for tool set-up and tool transport and handling. Therefore, the waiting time is dependent on the strategy of FMS operation, but particularly dependent on the tool flow system configuration (Silva, 1997). Thus, for example, the existence of a single tool central store, dynamically feeding the machines of the FMS which have each a single position tool buffer for the tool to process the next operation, is likely to reduce the overall waiting time of tools with consequent reduction of the number of tools required, at any time, in the system.

3 DISCUSSION AND CONCLUSIONS

The method presented here is analytical and does not take into account the operation dynamics of FMS's. As a consequence, it offers values which can only be considered as a first approximation to the required number of tools necessary to operate a FMS. This number is determined under two situations, namely static tool loading and as dynamic tool loading.

The precise number of each tool required is highly dependent on many aspects of system flexibility. This is related to system design and dynamics of system operation. Such dynamics results, for example, from time variability of processing and handling as well as from operative strategies and sequencing priority rules. A rigorous approach to the determination of tools required for a particular FMS would, most probably, require the use of a detailed simulation model of the FMS, run every time a FMS loading solution was required.

The number of tools required cannot be smaller than the number of different tools necessary to process the selected parts in the FMS. This number is, therefore, a lower bound to the total number of tools necessary. Further, as seen, under static tool loading, the length of the manufacturing period is determinant of the amount of tools required. Under dynamic tool loading this period is not important.

Manufacturing systems which are running for some time could take particular advantage of the analytical assessment presented. In fact these systems are in a good position for providing the data necessary by the method. Thus an evaluation of shortage or excess of tooling resources could be carried out. Moreover, the method can be useful in FMS planning or replanning because, by establishing estimations of the total number of tools, it helps in the estimation of requirements for tool storage and handling within a FMS. This, is relevant for the design of the tool flow system. Further, the estimation of the number of tools can help to make an initial and rough evaluation of performance of running FMS's under static or dynamic tool loading. This evaluation would be based on the number of tools really available as compared to the number which is required, estimated by the method.

The points discussed show that the analytical method presented can help reengineering manufacturing systems towards better design and operation.

4 REFERENCES

Eversheim, W and Herrmann, P., (1980) "Technical and Economical Planning of Automated Manufacturing Systems." *Proc. 4th Int. Conf. on Production Research*, Tokyo.

Rajagopalan, S. (1986) "Formulation and heuristic solutions for parts grouping and tool loading in Flexible Manufacturing Systems." *Procs. of the 2nd ORSA/TIMS Conf. on Flexible Manufacturing Systems*, Elsevier.

Sarin, S. C. and Chen, C. S. (1987) "The machine loading and tool allocation problem in a flexible manufacturing system." *International Journal of Production Research*, **25,** No 7, 1081-94.

Silva, S.C. (1997) "Strategies and Fundamental Structures for FMS Tool Flow Systems." accepted to the *OE/IFIP/IEEE Int. Conf. on Integrated and Sustainable Industrial Production*"

Sodhi, M. S., Askin, R. G. and Sen, S. 1994" Multiperiod tool and production assignment in flexible manufacturing systems". *International Journal of Production Research*, **32**, No.6 1281-94.

Stecke, K E. (1983) "Formulation and solution of non-linear integer production planning problems of FMS". *Management Science*, **29**, 273-88.

Stecke, K. E. (1981) "Production Planning Problems for Flexible Manufacturing Systems". PhD Thesis, *Purdue University*.

Storr, A. (1979) Plannung und realisierung flexibler Fertigungssysteme. *Wt Zeitschrift für industriell Fertigung* **60** No.11.

Warnecke, H.J. and Vettin G. (1982) Technical Investment Planning of Flexible Manufacturing Systems. *Journal of Manufacturing Systems*, **1** No.1.

BIOGRAPHY

Dr. S. Carmo Silva is a member of the Departamento de Produção e Sistemas at the Universidade do Minho, in Braga, Portugal. At present Dr. Carmo Silva is the head of the Industrial Management and Systems group. He got his PhD, in 1988, from Loughborough University and his MSc degree from UWIST in U.K, both in the area of Production Systems and Management. He is also graduated in Mechanical Engineering by Oporto University, Portugal. His main academic and research intestests fall into the general concept of Manufacturing Systems Engineering and, in particular into the area of design and operation of integrated manufacturing and assembly systems.

PART SEVEN

Systems Design and Control

18

Sustainable software design with design patterns

Adelino R. F. da Silva
Departamento de Engenharia Electrotécnica
Universidade Nova de Lisboa
2825 Monte de Caparica, Portugal
T: +351.1.294 83 38; Fax: +351.1.294 85 32
Email: asilva@acm.org

Abstract

Researchers in software engineering and information engineering have been struggling to modernize software life-cycle processes and to improve the quality of software. Recently, several methodologies and organizational changes have been put forward to improve systematic software specification and reuse. Design patterns offer a broader view helping us think at the architectural level. In this paper, several architectural abstractions guiding the development and reuse cycle in the production of complex software systems are introduced. We follow the design patterns methodology in the description and presentation of domain specific patterns taking the open distributed processing domain as reference. Several design patterns illustrating the mechanisms used for capturing design discipline knowledge are presented.

Keywords

Design patterns, Object-oriented methodologies, Software engineering, Distributed processing, Architecture description languages

1 INTRODUCTION

Researchers in software engineering and information engineering have been struggling to modernize software life-cycle processes and to improve the quality of software. Recently, several methodologies and organizational changes have been put forward to improve systematic software specification and reuse. These methodologies are aimed at developing adaptable software components from which application software can be constructed. Object technology, for instance, has been heralded as a promising solution to software complexity. However, a growing number of researchers, managers and practitioners have begun to realize that the object-oriented approach alone does not provide the right level of abstraction. Design patterns (Gamma, 1995) offer a broader view helping us think at the architectural level. The idea of using patterns and pattern languages is borrowed from the work done in building architecture by the architect Christopher Alexander (Alexander, 1977). In Alexander's terms, "each pattern describes a problem which occurs over and over again in our environment, and then describes the core of the solution to that problem, in such a way that you can use this solution a million times over, without ever doing it the same twice". A pattern is then a "rule which describes what you have to do to generate the entity which it defines". Software system designers are faced with an analogous situation. Expert designers know how to reuse solutions that have worked for them in the past.

In order to simplify the process of building increasingly large and complex computing systems, international standards bodies have addressed the problem of coordinating the standardization of architectural frameworks for the integrated support of distributed applications. In particular, the Reference Model of Open Distributed Processing (RM-ODP) (ITU-T X.901, 1995) is a joint effort by the international standards bodies ISO and ITU-T to develop a coordinating framework for the standardization of open distributed processing (ODP). Open distributed processing describes systems that support heterogeneous distributed processing, both within and between organizations through the use of a common interaction model. RM-ODP is an architectural framework for the integrated support distribution, inter-working, inter-operability, and portability of distributed applications. It provides an object-oriented reference model for building open distributed systems. The general goal of ODP standardization is the development of a framework for system specification.

As a meta-standard, RM-ODP requires a supporting architecture to coordinate and guide the development of application-specific ODP standards. In this paper, several architectural abstractions guiding the development and reuse cycle in the production of complex software systems are introduced. We follow the design patterns methodology in the description and presentation of domain specific patterns, taking the open distributed processing domain as reference. Several design patterns illustrating the mechanisms used for capturing design discipline knowledge are presented. We propose the utilization of the design patterns approach as the appropriate methodology to support the development of a framework for ODP system specification. By referencing approaches that have been followed in multiple design domains, we show how design patterns abstract knowledge representation.

2 THE DESIGN PATTERNS APPROACH

Recently, software designers have discovered strong analogies between software architecture patterns, and the building architectural patterns of C. Alexander. He found recurring themes in architecture, and captured them into descriptions and instructions. He used the term 'pattern' to express the replicated similarity in a design. Patterns provide a way to share design expertise in an application-independent way, and emphasizes their potential for reuse. The pattern characterization makes room for variability and customization of the intervening elements. Alexander's pattern language (Alexander, 1977) contains over 250 patterns, organized from high level to low level. The goal in documenting patterns is to build a system of patterns. A coherent system of patterns in a design domain is often referred to as a pattern language. A pattern language in this context is not a programming language in the ordinary sense of the term, but is a structured collection of patterns that build on each other to guide and inform the designer toward well-designed architectures. A pattern language is an architectural technique because it tries to characterize relationships within and between the parts. The goal is not merely to apply a divide-and-conquer approach, but rather make a complex work understandable by organizing it in a system of well-known patterns. In a given situation there may be multiple patterns to apply. The chosen relation between the patterns is said to form a pattern language in a given problem domain. Knowledge of development design and process management has typically been kept unstructured in terms of the underlying architectural patterns for reuse. Much of this knowledge can be captured using the patterns methodology.

3 EXPRESSING DESIGN PATTERNS

Patterns are well suited to documenting design techniques, expressing architectural considerations independent of language and design methodology. A common way of expressing patterns is to use the *Alexandrian form,* which tells what is contained in a pattern. An Alexandrian-like pattern consists of the following components:

- *Name.* The pattern name is the name by which the pattern is called.
- *Problem.* This is a concise problem statement It describes briefly the problem the pattern is trying to solve.
- *Context.* It describes the situations in which the solution is more or less applicable. A pattern is expected to solve a problem in a given context.
- *Forces or tradeoffs.* In a given context, conflicting constraints and requirements may affect the solution. The pattern finds a reasonable compromise between the conflicting requirements and 'resolves the forces'.
- *Solution* The solution component describes the elements that make up the design, their relationships, responsibilities and collaborations. The solution component describes behavior as well. It documents the actions to take or structures to build.
- *Example.* Examples show how to implement the pattern to solve typical problems in given contexts.
- *Discussion.* This component describes how the pattern relates to other patterns in the language. Related patterns may provide alternatives to the proposed solution.

4 ARCHITECTURES FOR INFORMATION PROCESSING SYSTEMS

Many architectures for information processing systems, guiding complex development projects, have been proposed in the past. These architectures often emerge and are elaborated on the context of a specific development process. Knowledge gained from previous projects is typically kept in one of the forms: human expertise and project-oriented documentation. The goal of abstracting from the peculiarities of particular designs is rarely considered. What is new in the patterns approach is its commitment to produce a 'handbook of useful solutions'. Collecting a patterns-catalogue systematically becomes part of the quality program of a development team or organization. We present below three design patterns for the specification of software architectures.

4.1 Pattern 1: abstraction levels

- *Problem*: How to specify the abstraction levels relevant to carry out a design discipline ?
- *Context*: A design and implementation discipline needs to be outlined for a well-understood domain. The analysis phase of the system development process is often called conceptual modeling, since it results in a conceptual model of the system to be implemented.
- *Forces*: A design discipline may have several levels of abstraction. Different design methodologies are used to manage design problems at different levels of abstraction. Although design is simpler at the higher abstraction levels, difficulties arise because it is necessary to manage object interactions between levels. Design at higher levels of abstraction becomes dependent on performance characteristics only observable at lower levels of abstraction.
- *Solution*: For each design object in a design domain name and characterize the target abstraction levels more suitable to carry out the design objectives. List the names of the abstraction levels by order of abstraction
- *Example*: RM-ODP defines a division of ODP system specification into five abstractions or viewpoints, in order to simplify the description of complex systems. Each ODP viewpoint provides a representation of the system with emphasis on a specific set of concerns (see Figure 1). The five ODP viewpoints are: the enterprise viewpoint (EntV), the information viewpoint (InfV), the computational viewpoint (ComV), the engineering viewpoint (EngV), and the technology viewpoint (TecV). EntV focuses on the purpose, scope and policies of the system. InfV focuses on the semantics of the information and information processing performed by the system. ComV is concerned with the description of the system as a set of objects that interact at interfaces, enabling distribution through functional decomposition. EngV focuses on the mechanisms and functions required to support distributed interaction between objects in the system. TecV focuses on the choice of technology in the system.
- *Discussion*: This pattern is common in multiple design domains. In (Kleinfeltd, 1994), for instance, the pattern has been applied to electronic design (CAD) framework systems. Fishwick (Fishwick, 1995), has described a related pattern in the context of simulation model design, under the name 'multimodeling'. He defines multimodeling in the following terms: "Multimodeling is a modeling process in which we model a system at multiple levels of abstraction. ... With abstraction we are concerned with many levels - each of which has the potential to be independently simulated". Kerth (Coplien, 1995), uses the Caterpillar's

Fate pattern to guide the transformation of a system from the analysis stage into the design stage. The pattern suggests a methodology consisting of an analysis phase and a design phase.

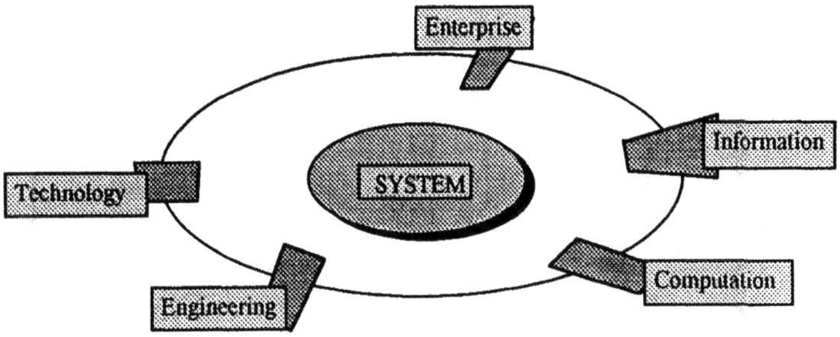

Figure 1 ODP viewpoints.

4.2 Pattern 2: design domains architecture

- *Problem*: How to create a design architecture in terms of design domains and their specializations, so that the system can meet its objectives and the requirements satisfied.
- *Context*: In this pattern we are concerned with many levels of abstraction, each of which has the potential to be independently designed and implemented. It is well known that small projects are easier to manage than big projects. A layered architecture is beneficial when many design objects are involved. A design domain is a set of design objects. These objects exist at different levels of abstraction. The design domains in a given design discipline can be arranged in a design domain hierarchy. When moving down a hierarchy, one refines the design domain. When moving up, one abstracts the design domain.
- *Forces*: The question of how to create a layered architecture emerges in this situation. A good architecture simplifies the implementation of the system. Some design methodologies are more appropriate to characterize structure than others. A design domain may be represented by different abstractions at the same level, corresponding to different perspectives of the system. In this case, a mapping relationship may be required to integrate both perspectives in a coherent system.
- *Solution*: Arrange the design domains for a given design discipline in a design domain hierarchy. The top level domain corresponds to the design discipline itself. Lower level design domains correspond to specializations of the parent level design domain. This specification must be done at the appropriate target abstraction level for the specific domain. The objective is to separate the concepts and notation used for different scopes, or levels of granularity. Follow the object-oriented approach, in order to identify key roles and interactions between levels.
- *Example*: Two main structuring approaches are used in the ODP architecture: viewpoints and distributed transparencies. These two approaches provide orthogonal representations. A viewpoint is a subdivision of the specification of a complete system relevant to some particular area of concern. On the other hand, distributed transparencies provide standard

mechanisms to access heterogeneous systems and components. The standard mechanism is 33said to provide a transparency. With transparencies the application designer works in a world of independent concerns. For instance, access transparency masks differences in data representation and invocation mechanisms to enable interworking between objects. Eight ODP transparencies have been defined (see (ITU-T X.901, 1995) for details).

- *Discussion*: The five ODP viewpoints are not organized in a hierarchy or in layers but represent the complete set of related viewpoints of a system architecture. However, they effectively provide different levels of abstraction. Rettig et al. (Rettig, 1993), in the context of the PADRE project planning and development process, uses a similar pattern based on modules: "Planning proceeds top-down until the work has been divided into many small pieces called 'modules'. Each team member uses the module-level process definition (...) to manage the progress of his or her current module, while the manager guides the progress of the whole series of modules toward the final goal". The 'Pedestal' pattern (Coplien, 1995), addresses the problem of how to create a layered architecture for a mechanical-process control system. He considers that domains should the selected and ordered based on the domain's purpose in the system.

4.3 Pattern 3: design problem decomposition

- *Problem*: How to decompose complex design objects to meet the domain requirements specifications.
- *Context*: A design discipline encompasses complex design objects of some broader type. A given domain decomposition works within a given abstraction level, which is adequate for carrying out the design objectives. The interactions between sub-domains are well-understood.
- *Forces*: It is possible that a design domain may have several possible decompositions Design domains may decompose into others of the same type or domain. The iterative nature of the design process may require a refinement of the modular decomposition, both in terms of abstract levels and in terms of domain decomposition.
- *Solution*: Choose the design strategy most appropriate to decompose complex design objects to meet the domain requirements specifications. Additional considerations, such as risk analysis, time to market, facility of reuse and easy maintenance, may strongly impact this choice as well. Several methodologies may be followed to partition a domain, namely object-oriented, object-based, structured, data flow diagrams, control flow diagrams, etc. Traditional process-oriented models stress sequencing of activities to accommodate chronological dependencies. For simple decomposition problems, a strict functional decomposition may be sufficient. In this case, apply design objective decomposition to break up the design goal into sub-goals which apply to the same whole design object. In general, however, more powerful methodologies are required. Object-oriented design emphasizes the discipline of identifying key roles and interactions in a design domain. Once the principal roles have been identified, responsibilities are assigned to each of these roles, and the circumstances under which roles interact with one another are defined.
- *Example*: In order to cope with the complexity of large ODP systems, it is necessary to provide a common framework for partitioning overall management. Management domains are used in RM-ODP to specify boundaries of management responsibility and authority.

They permit a set of managed objects to be controlled under a common policy. Each management domain specifies the management policy for a group of managed objects. On the other hand, for each abstraction it is necessary to define a structured set of concepts in terms of which that representation (or specification) can be expressed. This set of concepts provides a language for writing specifications of systems from that abstraction point of view. The RM-ODP enterprise language, for instance, introduces basic concepts necessary to represent an ODP system in the context of the enterprise in which it operates. In order to express the objectives and the policy constraints, the system may be represented by one or more enterprise objects, and by the roles in which these objects are involved. These roles, represent, for instance, the owners, developers, customers, and users involved in the enterprise activities. Moreover, in order to link the performers of the various roles and to express their mutual obligations, the enterprise language introduces the notion of contract. For example, the enterprise language for a distributed client-server system with one operator and two users might be specified in the following terms (see Figure 2): (i) Two management domains; (ii) Roles of the participating agents in terms of the system objectives (human-operator role, system-provider role, system-accessor role, and human-user role); (iii) Interactions between agents (represented by links in Figure 2); (iv) Contracts representing the policies and requirements governing the interactions between agents. Similar decompositions may be specified for the each RM-ODP viewpoint, in terms of its specific language.

- *Discussion*: Rettig et al. (Rettig, 1993), after breaking the planning and deveplopment process into modules, applies a *micromanagement* technique to each module based on five sequential milestones: plan, approve, do, review and revise, and evaluate. Fishwick (Fishwick ,1995), decomposes the process of making tea in three sequential stages: heating water, inserting bag in water, and steeping tea. This process is the refinement of a top-level abstraction concerned with the 'Making tea' objective. This pattern has similarities with the 'Divide and Conquer' pattern proposed by Coplien for organizational patterns (Coplien, 1995). The patterns 'Decouple Stages' and 'Hub, Spoke, and Rim' (Coplien, 1995), provide indications on how to decompose a domain. Several methodologies have been proposed to address the question of how to design behavioral composition in object-oriented systems (Rumbaugh, 1991). They emphasize the elaboration of a semantic net of domain objects in terms of interacting entities.

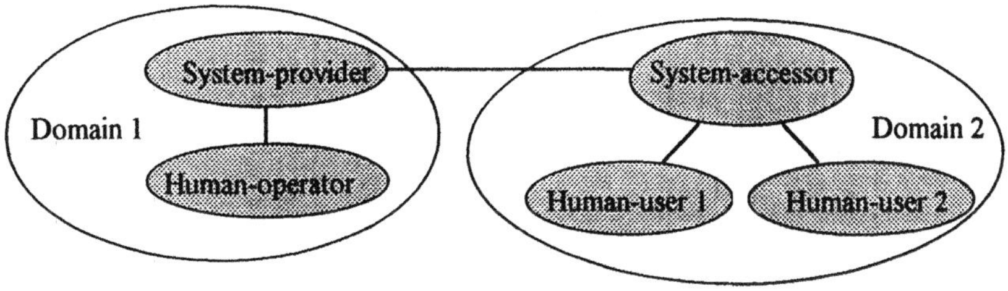

Figure 2 Enterprise viewpoint for a client-server system.

5 CONCLUSION

Design patterns are an emerging methodology for guiding and documenting system design. A large range of patterns for software architecture, design and implementation has been produced by the pattern community. Patterns covering the activities of design process management and project development are beginning to emerge as well. As the repository of such patterns grows, it will be possible to produce a pattern system for the organization and categorization of related patterns. By relying on architectural abstractions to capture existing design knowledge, patterns are a promising approach to guide the development and reuse cycle in the production of complex systems.

6 REFERENCES

Alexander, C., Ishikawa, S., and Silverstein, M. (1977) *Pattern Language - Towns-Buldings-Construction.* Oxford University Press.

Coplien, J. and Schmidt, D. (1995) *Pattern Languages of Program Design*, Addison-Wesley.

ITU-T X.901 (1995) *ODP Reference Model Part 1: Overview*, ISO/IEC 10746-1.

Fishwick, P. (1995) *Simulation Model Design and Execution - Building Digital Worlds,* Prentice-Hall.

Gamma, E., Helm, R., Johnson, R. and Vlissides, J. (1995) *Design Patterns: Elements of Reusable Object-Oriented Software*, Addison -Wesley, Reading, MA.

Kleinfeltd, S., Guiney, M., Miller, J.K. and Barnes, M. (1994) Design Methodology Management, *Proc. IEEE*, vol. 82, nº 2, pp. 231-250.

Rettig, M. and Simons, G. (1993) A project Planning and Development Process for Small Teams, *Comm ACM,* vol 36, nº 10, pp. 45-55.

Rumbaugh, J., Blaha, M., Premerlani, W., Eddy, F. and Lorensen, W. (1991) *Object-Oriented Modeling and Design*, Englewood Cliffs, Prentice Hall.

7 BIOGRAPHY

Adelino R. F. da Silva is an assistant professor in the Electrical Engineering Department at Universidade Nova de Lisboa, Lisbon. He received his PhD in Systems Engineering in 1989 from the Universidade Nova de Lisboa. Previously, he had been a Fulbright teaching and research assistant with the University of Pennsylvania. His main interests are Object-oriented methodologies, User-interfaces, Interactive systems and Human-computer interaction. Dr. Adelino Silva is a member of ACM and IEEE-CS.

19
SimBa: A Simulation and Balancing System for Manual Production Lines

Isabel C. Praça, Adriano S. Carvalho
Faculdade de Engenharia da Universidade do Porto
Instituto de Sistemas e Robótica - Grupo de Automação Industrial
R. dos Bragas 4099 Porto Codex - Portugal
Tel:02-2041619 email isp@tormentas.fe.up.pt, asc@garfield.fe.up.pt

Abstract
The balancing and the comparison between different solutions through the simulation of operation of man based production lines is an important tool to help to decide properly at the level of production planning. This tool becomes much more important in the planning of lines based on variable performance work stations, due to needs of learning, fast changes of operations, which cause frequent bottlenecks and difficulties in previewing the future performance of the production lines.
In the paper the authors analyse this problem and present a validated solution to perform the line balance and to provide a tool to simulate posterior production to help the production manager configure the line according to the objective of balancing it.

Keywords
flow lines, balancing of manual flow lines, simulation, manufacturing systems

1 INTRODUCTION

The production planning is performed based on decisions on the shop floor at different levels. In manual manufacturing systems the production lines balancing becomes an important matter. The balancing, an issue classifiable as NP-hard, presents a practical solution based essentially on the application of heuristic rules.

Some of the questions to give answer are:

- How are the manufacturing operations to be allocated to work stations?
- What should the cycle time (production rate) be?
- Should any work stations be paralleled, i.e. duplicated, and should multiple manning be employed?

The dependency on the human performance causes great complexity in the analysis of manual manufacturing lines. In fact, taking into account standard times for the operations - times estimated during the product design step - would allow lines balancing that do not guarantee a balanced operation in run time. The strong dependency of human performance in the operation times for this type of lines, influenced by factors such as absenteeism, the learning capacity and predisposition, the physical and emotional conditions of the operators, presents an additional requirement for evaluation of different factors and its role in the production planning.

In this context, simulation of the lines operation seems to become the appropriate tool for analysis and comparison of different line configurations according to predefined criteria. The simulation provides quantitative and qualitative information to the planner that allows him to take the right decision. So, it becomes an useful contribution to the planning system.

The tool presented in this paper was developed concerning the requirements of a production system organized as a set of manual flow lines. There was studied a set of three heuristic rules that allow the production manager to get different line balances in order to simulate them and evaluate their performance according to well defined and meaningful parameters.

This paper presents the characteristics and functionalities of a system, called SimBa, developed to balance and simulate manual production lines, according to appropriate rules. Some discussion on the role of simulation tools to help the production manager to take planning decisions is also presented.

2 MANUAL FLOW LINES BALANCING AND SIMULATION

The problem of balancing lines is a very important and complex one. Since the 50 decade several techniques and procedures have been developed, each having its own advantages and disadvantages. Those methods can be divided into optimal and heuristic procedures.

Optimal procedures require integer and linear programming algorithms and branch-and-bound algorithms. In (Talbot, 1984), (Held, 1962), and (Bowman, 1959), some of them are described.

The line balancing is an NP-hard problem. That means that the computacional power to obtain an optimal solution increases exponentially with the increase in problem dimensions. Therefore, for large problems it is usual to employ heuristic methods capable of achieving good solutions in shorter time. The heuristic procedures can also be easily adapted to the caracteristics of each problem, such as restrictions on the number of components to be worked in each work station, mutually exclusive operations, multiple objectives and variable operation durations.

Several criterion can be used to compare different line configurations. The most important are:

- number of work stations (number of equipment and operators) needed;
- balancing loss;
- levels of stocks between work stations;
- bottleneck stations;
- work stations utilization;
- number of days to complete the production of all items.

The last four criteria can only be evaluated by means of simulating each line configuration in analysis, so it is noticeable the importance of simulation in this field.

Simulation is a technique used in the analysis of complex systems in which it is too expensive, difficult or even impossible to do experiences with the real system. Some of the application field of simulation is the study of production and economic systems, distribution networks, and scheduling of services and personnel.

Simulation is used essentially to acquire knowledge about a system and to do experiences that can help to evaluate future performance and evolution. The information provided by a simulation study becomes valuable in the support of decision making.

The simulation of manufacturing systems is frequently a discrete event simulation. So it is the simulation of flow lines, since what is relevant is the changes in system state after occurrence of well defined events. Some of the most relevant events during line operation are:

- the end of batch processing in a work station;
- the need of a new work station by a batch to be processed on;
- breakdown of equipment;
- lack of raw materials or components;
- absence of operator.

The possibility of using information provided by a system capable of obtaining and analysing different line configurations, allows the production manager to take better decisions.

3 SIMBA SYSTEM

The analysis of a production system organized as a set of several production lines and of the difficulties that the production manager has to deal with was the starting point for the development of this system.

It implements three heuristic rules, that allows the user to obtain different line configurations. These different line configurations can be compared by means of a simulation module which gives the user qualitative and quantitative information about the future performance of the system.

The general architecture of the SimBa system is described in the next figure.

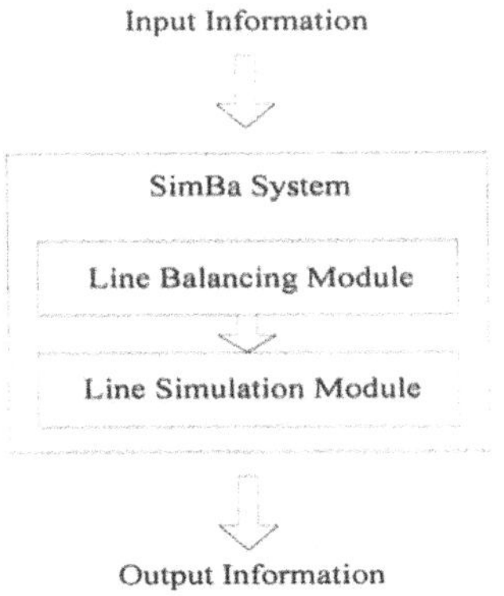

Figure 1 Architecture of SimBa system.

The system input information is related to:

- number, duration and operations precedence;
- number of items to produce;

• learning curves of the operators;
• desired cycle time.

The SimBa system is constituted of two modules: the line balancing module and the line simulation module.

3.1 Line Balancing Module

The Line Balancing Module implements three heuristic rules which allow the user to obtain different line configurations. The criteria used, computational time involved in and probability of solving optimal solution, to select the heuristic rules were based on the performance of the rules in a comparative study developed by Boctor (Boctor, 1995). Boctor evaluates the performance of 16 heuristic rules in 67 problems taken from the open literature and in 700 randomly generated problems. The good performance of the rules, evaluated in the mentioned study before, assures they are capable of finding good solutions.

It must be taken into account the simplicity of the rules, since their simplicity allows to obtain fast results, an important factor when considering future improvements on the rules to adapt them to fit each particular case.

The selected rules were the Largest Candidate Rule, the Ranked Positional Weight Method and the Multiple Heuristic Rule.

Largest Candidate Rule

Using the Largest Candidate Rule, the operations are allocated to stations, beginning with the first station, by selecting whose operations are feasible in descending order of size. It is a very simple rule, future alterations can be easily made, and has proved good performance on the vast study realized by Boctor.

Ranked Positional Weights Method

With the Ranked Positional Weights Method it is necessary to find the positional weight of each operation. It is calculated by summing the operation's own standard time and the standard time for all following operations. The positional weight is, therefore, a measure of the operation time and its position in the sequence of operations. The operations are allocated to work stations in order of decreasing positional weight and without violating precedence constraints. It is more complex than the Largest Candidate Rule, however it is a simple rule that proved good performance on the mentioned study before.

Multiple Heuristic Rule

The Multiple Heuristic Rule was proposed by Boctor. It is a composite line balancing method were the operations are assigned to work stations in decreasing order of its priority. The priority of the operations is determined by applying several rules to the set of schedulable operations (operations for which all predecessors are already assigned to a work station). To choose among the schedulable operations, the one being assigned to the current work station select:

1. the operation having a duration equal to the remaining time.
2. the severe operation (operation having a processing time greater than or equal to one half of the cycle time) having the largest number of subsequent candidates (number of operations that become schedulable after assigning the operation in question to the work station).

3. the combination of two operations having a duration equal to the remaining time.
4. the operation having the largest number of subsequent candidates.

To break ties in the four mentioned rules there are other defined rules.

The Multiple Heuristic Rule is the more complex of the three heuristic rules selected, however it was the one that prove better performance in the referred comparative study.

The information provided by the Line Balancing Module of the SimBa system is the number and the constitution of the work stations to be used in the production of the items. The user can obtain a line configuration for each of the three heuristic rules implemented, which can after be simulated in order to evaluate the future performance of the system and compare the solutions obtained.

3.2 Line Simulation Module

The simulation module allows the comparison of different line balances obtained by means of the heuristic rules implemented, or another suggested by the user.

It is possible to simulate the operating of lines with standard durations of the operations and also taking into account the learning curves of operators, which allows an analysis that reflects the variability of human performance. Using a simulation considering the learning curves of the operators allows the user to test different operators allocation in order to optimize the mapping between work places and operators.

It is a discrete event simulation based on a queuing model, where each work station is represented as having one or more servers and an associated waiting queue that represents the input buffer. The events are:

- Kickoff: start the production of the items, which can be based on batches, i.e. to start the production of the first batch or item at the work station where the first operation is assigned.
- Release_workstation: when an item or batch finishes processing in a work station.
- Request_workstation: when an item or batch request a work station in order to continue its production. If the next work station is occupied, the batch has to wait in the input buffer, otherwise it starts to be produced in it.
- NewBatch_type: when an assembly operation occurs it originates a new type of item.
- Update_operation_times: due to the learning curves of the operators the processing time of each operation will change along the time. This event deals with the necessity of updating the processing times.

The Line Simulation Module provides information about:

- levels of stocks between work stations;
- work stations utilization times;
- number of days to complete the production of all the items.

With this information it is possible to compare different line configurations. A line configuration with lower levels of stocks between the work stations, high utilization times and fewer days to complete the production of all the items is the most suitable.

The information concerning the levels of stocks in the input buffer of each work station allows the detection of bottleneck stations, which can be detected by the high level of stock in the input buffer and high utilization time.

Based on the output information of the SimBa system, the user can support the decision of choosing the line configuration that better fits the production.

The SimBa system was developed according to an object oriented methodology, specifically the "Object Modelling Technique" by Rumbaugh (Rumbaugh, 1991). It was developed in C++, being the Line Simulation Module also supported by Sim++ libraries of the SimPack package developed by Prof. Fishwick team at the University of Florida (Fishwick, 1992).

4 EXAMPLE OF APPLICATION OF THE SIMBA SYSTEM

In this section an example of the application of this system is described, in order to illustrate how it can help the managers for the production lines in taking decisions about the line configuration to implement.

The example shown is taken from a case study of a real production system organized as a manual production line. This and other examples can be found in (Praça, 1996).

The graph with the precedence restrictions of the operations necessary to produce one type of product made in that industry, as well as the operations durations, is shown in Figure 2.

Input information

The input information should be presented in the format:

Product Code 3830
Number of batches 50
Number of products in each batch 100
Number of operations 18
Number of branches of the operations graph 3
Cycle time 1.70
Operations:

code; duration; next operation; number of following operations; precedent operations; branch; initial = 1

0-> 0.15 1 0 0 1
1-> 0.40 2 1 0 0 0
2-> 0.50 3 1 1 0 0
3-> 0.15 4 1 2 0 0
4-> 0.80 5 1 3 0 0
5-> 0.25 6 1 4 0 0
6-> 0.20 10 1 5 0 0
7-> 0.65 8 0 1 1
8-> 0.30 9 1 7 1 0
9-> 0.40 10 1 8 1 0
10-> 1.70 11 2 6 9 2 0
11-> 1.65 12 1 10 2 0
12-> 0.45 13 1 11 2 0
13-> 0.35 14 1 12 2 0
14-> 0.95 15 1 13 2 0
15-> 0.15 16 1 14 2 0
16-> 0.45 17 1 15 2 0
17-> 0.90 -1 1 16 2 0

With this information the user can select which of the heuristic rules to use to obtain a line balance. It is possible to obtain three different line configurations. For each line configuration the results obtained with the Line Balancing Module show the constitution of the line, i.e. the number of work stations and the operations that were assigned to each one.

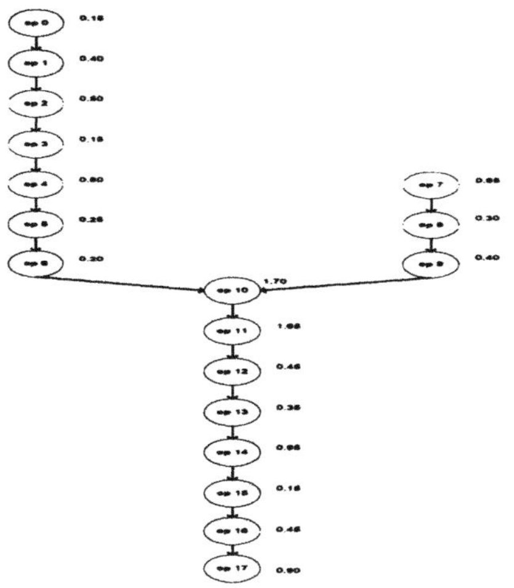

Figure 2 Graph of operations preference.

4.1 Largest Candidate Rule

Line configuration:
Number of work stations: 8

Work station: 1
Number of operations: 4
Operation: 7
Operation: 8
Operation: 9
Operation: 0
Work station: 2
Number of operations: 3
Operation: 1
Operation: 2
Operation: 3
Operation: 3
Work station: 3
Number of operations: 3
Operation: 4
Operation: 5
Operation: 6
Work station: 4
Number of operations: 1
Operation: 10
Work station: 5
Number of operations: 1
Operation: 11
Work station: 6
Number of operations: 2
Operation: 12
Operation: 13
Work station: 7
Number of operations: 3
Operation: 14
Operation: 15
Operation: 16
Work station: 8
Number of operations: 1
Operation: 17

For this and other line configurations, obtained by the Line Balancing Module or suggested by the user, the line can be simulated with standard duration of the operations or taking into account the variability introduced by the operators learning curve.

In this example the results obtained in both cases are illustrated.

Results obtained with the standard durations:

Days to produce all items:18

System Utilization: 70.4%
Arrival Rate: 0.010828, Throughput: 0.005414
Work stations Statistics

F 1 (work station): Idle: 18.8%, Util: 81.2%, Preemptions: 0, Longest Q: 2
F 2 (work station): Idle: 43.2%, Util: 56.8%, Preemptions: 0, Longest Q: 0
F 3 (work station): Idle: 32.3%, Util: 67.7%, Preemptions: 0, Longest Q: 0
F 4 (work station): Idle: 8.0%, Util: 92.0%, Preemptions: 0, Longest Q: 6
F 5 (work station): Idle: 10.7%, Util: 89.3%, Preemptions: 0, Longest Q: 0
F 6 (work station): Idle: 56.7%, Util: 43.3%, Preemptions: 0, Longest Q: 0
F 7 (work station): Idle: 16.1%, Util: 83.9%, Preemptions: 0, Longest Q: 0
F 8 (work station): Idle: 51.3%, Util: 48.7%, Preemptions: 0, Longest Q: 0

Results considering the learning curves of the operators:

Days to produce all items:15

System Utilization: 71.2%
Arrival Rate: 0.013508, Throughput: 0.006754
Work station Statistics

F 1 (work station): Idle: 19.0%, Util: 81.0%, Preemptions: 0, Longest Q: 2
F 2 (work station): Idle: 44.4%, Util: 55.6%, Preemptions: 0, Longest Q: 0
F 3 (work station): Idle: 35.0%, Util: 65.0%, Preemptions: 0, Longest Q: 0
F 4 (work station): Idle: 8.1%, Util: 91.9%, Preemptions: 0, Longest Q: 6
F 5 (work station): Idle: 16.4%, Util: 83.6%, Preemptions: 0, Longest Q: 0
F 6 (work station): Idle: 48.4%, Util: 51.6%, Preemptions: 0, Longest Q: 0
F 7 (work station): Idle: 17.4%, Util: 82.6%, Preemptions: 0, Longest Q: 0
F 8 (work station): Idle: 41.7%, Util: 58.3%, Preemptions: 0, Longest Q: 0

4.2 Ranked Positional Weight Method

Line configuration obtained through the Ranked Positional Weight Method:
Number of work stations: 8

Work station: 1
Number of operations: 4
Operation: 0
Operation: 1
Operation: 2
Operation: 3
Work station: 2
Number of operations: 3
Operation: 7
Operation: 4
Operation: 5
Work station: 3
Number of operations: 3
Operation: 8
Operation: 9
Operation: 6
Work station: 4
Number of operations: 1
Operation: 10
Work station: 5
Number of operations: 1
Operation: 11
Work station: 6
Number of operations: 2
Operation: 12
Operation: 13
Work station: 7
Number of operations: 3
Operation: 14
Operation: 15
Operation: 16
Work station: 8
Number of operations: 1
Operation: 17

Results obtained with the standard durations:

Days to produce all items:24

System Utilization: 52.5%
Arrival Rate: 0.008078, Throughput: 0.004039
Work station Statistics

F 1 (work station): Idle: 51.5%, Util: 48.5%, Preemptions: 0, Longest Q: 1
F 2 (work station): Idle: 31.3%, Util: 68.7%, Preemptions: 0, Longest Q: 5
F 3 (work station): Idle: 63.7%, Util: 36.3%, Preemptions: 0, Longest Q: 3
F 4 (work station): Idle: 31.3%, Util: 68.7%, Preemptions: 0, Longest Q: 18
F 5 (work station): Idle: 33.4%, Util: 66.6%, Preemptions: 0, Longest Q: 0
F 6 (work station): Idle: 67.7%, Util: 32.3%, Preemptions: 0, Longest Q: 0
F 7 (work station): Idle: 37.4%, Util: 62.6%, Preemptions: 0, Longest Q: 0
F 8 (work station): Idle: 63.7%, Util: 36.3%, Preemptions: 0, Longest Q: 0

Results considering the learning curves of the operators:

Days to produce all items:19

System Utilization: 54.9%
Arrival Rate: 0.010369, Throughput: 0.005184
Work station Statistics

F 1 (work station): Idle: 50.2%, Util: 49.8%, Preemptions: 0, Longest Q: 1
F 2 (work station): Idle: 30.8%, Util: 69.2%, Preemptions: 0, Longest Q: 4
F 3 (work station): Idle: 63.8%, Util: 36.2%, Preemptions: 0, Longest Q: 2
F 4 (work station): Idle: 29.5%, Util: 70.5%, Preemptions: 0, Longest Q: 17
F 5 (work station): Idle: 34.9%, Util: 65.1%, Preemptions: 0, Longest Q: 0
F 6 (work station): Idle: 60.0%, Util: 40.0%, Preemptions: 0, Longest Q: 0
F 7 (work station): Idle: 36.4%, Util: 63.6%, Preemptions: 0, Longest Q: 0
F 8 (work station): Idle: 54.9%, Util: 45.1%, Preemptions: 0, Longest Q: 0

Discussion of the results

With the information provided by the Line Balancing Module of the SimBa system it becomes possible to compare different line configurations which gives support to the decision of choosing the best one to implement in the production of the items.

In the example given, both configurations have the same number of work stations, and so the quantity of human and material resources necessary to the production will not affect the choice between them.

In the line configuration obtained through the Ranked Positional Weight Method the work station 4 has a huge accumulation of items in the input buffer. It is probably a bottleneck station, so the manager must take this into account in the decision making. Also, the number of days to complete the work is less with the configuration obtained with the Largest Candidate Rule and the levels of stocks are shorter. These factors and the higher utilization times of the work stations suggest the performance of this line configuration will be better.

But, the user can also compare it with other solutions and do some changes in order to find a better one.

The SimBa application in the real context has validated the results showed, through comparison with lines performance obtained by the methods used. In (Praça, 1996) the system is described with enough detail and more examples of its application to the real system are given and discussed.

5 CONCLUSIONS

This paper presents an approach to achieve good line balances, namely lines that involve a great component of manual work.

Simulation is an important framework in the analysis of flows global performance, and in the evaluation of alternative configurations. With simulation it is possible to obtain information on the performance evolution of lines, namely the detection of bottleneck stations and high levels of stocks.

So the use of simulation in a system capable of giving the user quantitative and qualitative information about alternative line configurations helps the production manager to compare different solutions, and so helps him to make decisions.

With SimBa, a developed and presented manufacturing application, becomes possible to obtain different line balances based on different heuristic rules implemented. The evaluation of this balances through simulation is very important in order to take correct decisions. The well performed detection of high levels of stocks, and so of bottlenecks, is a strong advantage of this system.

With this system different operator distributions can be tested, in order to achieve the best solution. It is possible then to establish a correct allocation of operators to the work stations, based on the results obtained when considering their learning curves during the simulation.

6 REFERENCES

Boctor, Fayez F.(1995) A Multiple-rule Heuristic for Assembly Line Balancing. *Journal of the Operational Research Society*

Bowman, E. H. (1959) Assembly-Line Balancing by Linear Programming. *Massachusetts Institute of Technology, Cambridge*

Fishwick, Paul A.(1992) Simpack: Getting Started width Simulation Programming In C And C++. *Dept. of Computer & Information Science, University of Florida*

Held, Michael, Karp, Richard M. and Shareshian, Richard (1962) Assembly-Line Balancing-Dynamic Programming with Precedence Constraints. *International Business Machines Corporation, New York"*

Praça, Isabel C. (1996) *Simulação e Balanceamento de Linhas Manuais de Fabrico*. Masters Thesis. Faculdade de Engenharia da Universidade do Porto.

Rumbaugh, James(1991) *Object-Oriented Modelling and Design*. Prentice-Hall Internacional

Talbot, F. Brian and Patterson, James H. (1984) An Integer Programming Algorithm with Network Cuts for Solving the Assembly Line Balancing Problem. *Management Science vol. 30*

20

SOMA: an organic and autonomous approach to manufacturing planning and control

A. A. Queiroz, Ph.D. and H. A. Lepikson, M.Eng.
UFSC/GRUCON- Federal University at Santa Catarina State,
Research Group on Numerical Control and Industrial Automation
Address: UFSC/EMC/GRUCON
88040-900, Florianopolis-SC, Brazil
Phone: +55-48-3319387 Fax: +55-48-2341519
E-Mail: hal@grucon.ufsc.br

Abstract

The complexity of manufacturing systems overload their structure and impose hard conditions for their routine operation. This paper first analyses this problem and proposes a new approach for planning and operating manufacturing systems, based on the dynamic aggregation of autonomous manufacturing units working in a co-operative, negotiated way based on the attendance of the present and prospective market needs. Then it sets out a system architecture and introduces a methodology for the shop floor planning and control which is capable both of meeting the objectives and constraints of these autonomous units with enhanced synergy and of keeping the system's integrity.

Keywords

Distributed manufacturing, shop floor planning and control

1 INTRODUCTION

Today organisations tend to become more and more sophisticated so as to cope with the challenge of the frequent change to attend market demands, including the introduction of new products. Complexity is a problem faced by the companies on every organisational level, with particular and important consequences for the workshop structure.

Complexity imposed on the big companies lacks agility and, to the small ones, lacks capital as well as technological and managerial support to play the game in this new and competitive arena, as observed by Fernandes (1995) and Hamel (1994). Complexity also brings an enormous struggle to the manufacturing system, since constancy, discipline and learning with experience are important for its performance. Operation of manufacturing routines is highly sensitive to disturbances, due to interwoven flows of material and information that lead to mutual dependence on resources. As a consequence, the planning and operation of the manufacturing system is truly a hard task. Scheduling, for instance, is often obsolete even before it is about to be carried out. Rescheduling comes to be more the rule than the exception.

The traditional approach to solve these problems is based on the deploying functions in a hierarchically way in order to divide big issues into smaller ones, that are easier to deal with. (See, for instance, Scheer, 1993 or Williams, 1989). The culture of the company is structured by this way of thinking, and, by extension, so are management issues, personnel training, software development, resource planning, and so on. This has been historically a valid concept, inherited from Taylor's postulates, with the following main characteristics (Bullinger, 1993):

- separation between planning and execution activities;
- hierarchical, centralised, management and control;
- organisation by functions (e.g., design, dispatching, assembly functions).

As long as the manufacturing companies have evolved, this concept has led to a more and more complex decision structure, with numerous interfaces and relations and, as a consequence, with a greater struggle for co-ordination and control of resources. The results of research on manufacturing that has sought to understand and deal with this complexity have shown that it can be a very hard work, and not always successful (e.g., Bienert, 1993).

This present reality brings new threats, but also new opportunities. Among the threats, the exponential growth in complexity and awkward relations with the market deserve special attention. On the other hand, among the opportunities, there are the technical and organisational resources potentially able to help in managing the situation, such as the optimum qualification of human resources, the sophistication of information systems and the flexibility achieved by new equipment.

To successfully take advantage of this situation, companies must rethink their vision of business, their strategy and, as a consequence, their way of planning the manufacturing process. The approach here proposed is an attempt to help industries plan their workshop in this direction and to simplify their manufacturing structures. It is based on the concept of Autonomous Units working with high synergy (organicity). First, the concept and its architecture are introduced. Then the paper addresses the mechanisms which make the concept possible in the particular case of manufacturing planning and control.

2 THE ORGANIC AUTONOMOUS APPROACH

The Organic Autonomous Approach, called here **SOMA** (an acronym for the Portuguese translation of Organic Autonomous Manufacturing System, which is **S**istema **O**rgânico de **M**anufatura **A**utônoma) is somewhere between the traditional, hierarchical and centralised way of thinking and the new research towards fully decentralised structures (see, for instance,

Mathews, 1995, Tharumarajah, 1996, Ueda, 1994, Warnecke,1993). **SOMA** is particularly fitted to repetitive manufacturing kind of industries which, at the same time, frequently need to introduce new products.

Compared with the hierarchical organisation (figure 1), the **SOMA** Approach breaks with the traditional manufacturing paradigms. In this sense, its main characteristics are:

- aggregation of planning and execution in focused, independent, units;
- decentralised management and control, mainly supported by negotiation strategies;
- organisation by process (manufacturing process, maintenance process, etc.).

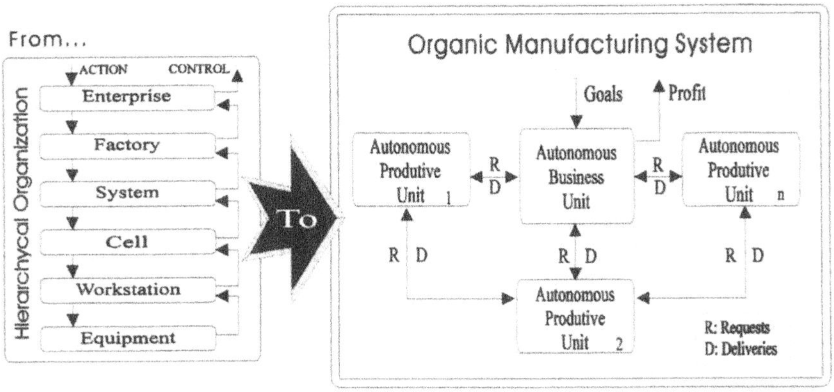

Figure 1 Comparison between traditional and Autonomous approach to manufacturing organisation.

On the other hand, compared with the fully decentralised systems, the main differences are twofold: first, the Autonomous Approach does not set out to build a fully decentralised system from the machine level on. Although this could be done, by concept it is transparent to the system. **SOMA** envisages the Unit as its smallest and only instance to be managed. Second, it does not pursue a totally automated control system. Instead, it takes advantage of the human skill, capacity of judgement and experience to give leverage to the competencies on each Unit and, by means of co-operation and negotiation among them, to also provide leverage for the whole system.

The AUs are focused on product. Hence, they have only one mainstream process to maintain and develop. This means that every Autonomous Unit (AU) has a very clear view regarding which core competencies they must pursue and, as a consequence, the strengths and weaknesses related to this competencies that they are to deal with in order to enhance their competitiveness and, in the end, their prospective "market share" (it must be previously understood that every AU sells most of its products to its preferential clients - probably another AU - and that its price and quality are under constant evaluation by benchmarking comparison). Focus on product also helps to organise the AUs by process: material, resources and information flow are co-ordinated by process objectives.

The basic architecture of **SOMA** is based on two kinds of AUs: The Autonomous Business Unit (ABU) and the Autonomous Production Unit (APU). An ABU is mainly market-oriented. It has the responsibility of observing, analysing, and acting according to the prospective market desire. Figure 2 illustrates how **SOMA** works, observed by the information flow perspective.

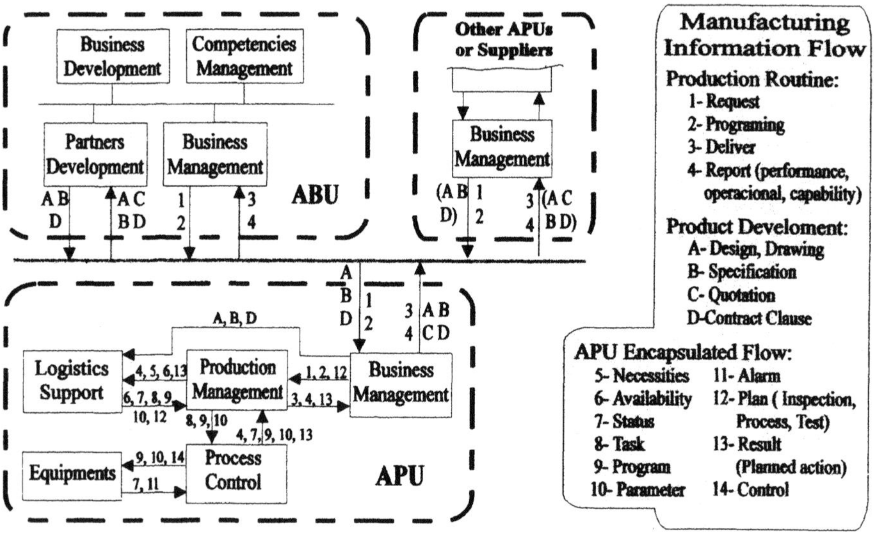

Figure 2 Basic architecture and the information flow on the Autonomous Manufacturing.

There is one ABU for each group of focused core competence. This is an important characteristic, for competencies are not always directly related to only one family of products. It is assumed that, in the long run, competitiveness derives from the ability to build these competencies. They, in turn, may generate unanticipated products to attend these prospective market desires (Prahalad, 1990) The ABU clearly assumes, in this sense, a differentiation strategy under Porters' perspective. But, on the short term basis, it also gives special attention to costs, since it is based on negotiation to operate with the APUs. This possibility partially solves a contradiction in Porters' strategy model (where cost and differentiation are excluding strategies) and comes nearer the perspective of building synergy by mutual compromise between quality and cost, as analysed by Corsten, 1993 and Fernandes, 1995.

The interface between ABU and APUs is oriented by the Business Management Module (BMM), which is similar, to the communication viewpoint, at every Unit (figure 2). BMM co-ordinates the supply chain by request-deliver derived functions. This means that ABU translates market needs into requests and orders for products and their related parts, assembly and services among AUs. To cope with this task, BMM must have a complete register of suppliers (mainly APUs) and their scope of competence. The functionality of the system will be explained in advance. At the ABU, BMM is supported by the Partners Development Module (PDM) in the activities aimed at the development of new products or new partners. PDM is necessary because of its long term perspective, quite different from the routine relation on a day-by-day basis to the existing products which are manufactured. PDP also has the important mission of managing and articulating product know-how from the organisation standpoint (which is not necessarily the same as product parts know-how, which are under APUs dominion).

Despite the importance of PDM, its functions are beyond the objectives of this paper, which is focused mainly on the APUs management and its relation to the whole system.

It is important observe that APUs are conceived as internal suppliers within a company. However, due to the **SOMA** approach, there is no hindrance to external suppliers when the situation calls for them. Additionally, the system can be potentially fitted for Virtual or Extended Enterprise compositions (NIST, 1994 and Browne, 1995).

APUs have a strong interrelation, since they are recognised and specialised by their core competencies. Hence, they are stimulated to develop alternative capacities and "markets" (i.e., new product ranges) within their core competencies to better occupy their resources. It is not important for the system to know how each APU works. As it can be seen in figure 2, the real-time control activities are totally encapsulated within the APU. The Partners Manager Module is the only one which has any relation with the outside world, considering the APU. This is a very useful characteristic, since the system can support working together AUs with different abilities or level of automation. Hence, this paper will focus on the relations between the AUs.

Two kinds of APUs may be distinguished: Product mainstream APU and Service APU. The first is directly dedicated to product manufacturing or assembling. The second is focused on specialised services to support production. Maintenance, testing and inspection, tooling, quality assurance, transport and handling or NC programming are some examples.

There could exist a third kind of APU, though it is not directly related to production. It is oriented to general services necessary within the organisation, such as accounting, finances, training, as well as juridical and other services. They are not considered here as APUs, since they are out of production scope. Nevertheless, nothing hinders them from being treated as APUs which sell services to other APUs/ABU, and using the same request-deliver functions. This extension of the approach simplifies cost management, since it allows the use of the same basic accounting functions to balance incomes and expenses among APUs as well as between APUs and ABU. This alternative strongly simplifies the manufacturing cost account, a problem still unsolved today (see, for example, Harmon).

APUs are managed like small enterprises. They extend the concept of cellular manufacturing to provide the necessary flexibility in size, lay-out, resources and even profits (in this case, the final result of primary performance measurements). The limit for its growth is defined by the Unit's self-management capacity. As a reference, the team should not exceed 15 people. This is necessary to keep the APUs' agility and focus. On the other hand, it is also important to maintain the mutual dependence and equilibrium among APUs.

3 THE SOMA MANAGEMENT AND CONTROL SYSTEM

SOMA management and control in the Autonomous Unit concept is based upon two key-words: negotiation (among units) and encapsulation (inside each unit). This means that control exists only inside each AU. And, among AUs, only requests and delivery-derived functions are sufficient to maintain the defined client-vendor relation (figure 1). The objective is to assure complete modularization of the manufacturing structure (which also has consequences in standardisation).

Figure 3 gives an overview of the concept. Negotiation is based on request-deliver functions which are applied through messages to objects which, then, internally invoke the necessary operations. When necessary, the invoked object returns (delivers) an answer. Figure 2 introduces some of the most important functions and where they apply.

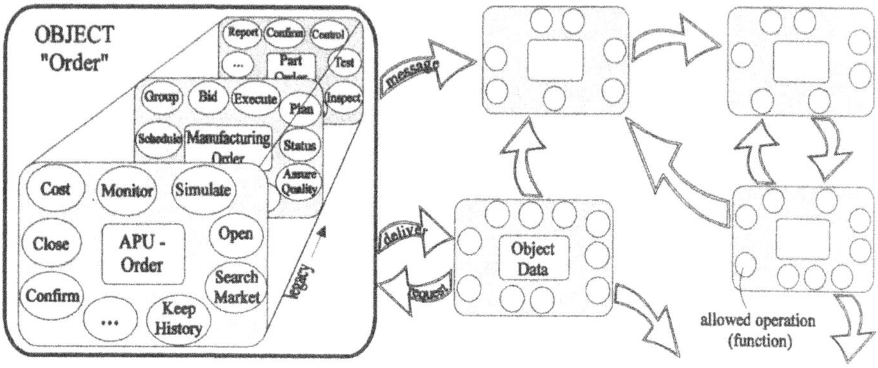

Figure 3 Control encapsulation and negotiation flow between objects.

Functions are assynchronically managed, which means that no control actions are involved. An object is any entity about which information must be manipulated and stored, including its manipulation methods. Objects (for instance, a manufacturing order for a specific part) are instances of object types (orders in the example). In this sense, a manufacturing order, a service order or an assembly order are all sub-objects of object Order. They inherit the properties and methods of the type order and add others, specific to their necessities (Martin)

Encapsulation is the consequence of hiding from the users details of how objects are manipulated. Its extended meaning makes it possible to shelter control actions within the AUs. In this context, AUs can be seen as object types which enclose sub-objects such as people, machines, software, tools, inventories, clients, suppliers, accounting, capability and so on.

As it can be noted, internal management and control are concealed from the external system (the AUs are objects themselves). This characteristic permits each AU to develop its own culture and personality, just like any enterprise. This brings enormous advantages to the SOMA concept, even when compared with the other decentralised approaches.

This characteristic also brings organicity to the system and imposes a common behaviour on every AU that is:

- market-driven, including:
  - ◊ performance concern;
  - ◊ inherent benchmarking with partners and competitors (which can be similar to AUs);
  - ◊ product life cycle attendance;
  - ◊ costumer audits;

- product-driven, by the capacity of:
  - ◊ continuous tuning in to the costumer needs;
  - ◊ fast flaw correction;
  - ◊ continuous quality, cost and lead time evolution;
  - ◊ fast response to new product development;
- process-driven, since it is able to:
  - ◊ to be in tune with the minor changes or evolution of the market;
  - ◊ rapidly take advantage of the experience curve;
  - ◊ be sensitive to changes in capacity or capability when demanded by the market.

4 THE UNIT PLANNING AND CONTROL SYSTEM

Considering its smooth flow and the partnership built gradually, APUs can be co-ordinated in a kanban-like style, with some concessions to the **Soma**'s specificity. From the organisational viewpoint, the APUs have a structure similar to Flexible Manufacturing Cells (as defined by Lepikson: product-driven and with self-control capacity), with three important differences:

- local planning and management capacity;
- free decision about resource allocation and opportunity use to advantage;
- APUs are supposed to address and manage the resources by themselves (including information and people) to accomplish their products.

As already stated, the relations between AUs are based on negotiation mechanisms. The system management and control is based on a reward-penalty system, oriented by market values. This concept adapts, for this scheduling purpose, the similar alternatives presented and tested by Márkus, 1996 and Iwata, 1994. It works as follows:

1. ABUs keep a performance portfolio of the other ABUs and APUs with which they maintain a close relation (besides their capacity to access other AUs within the **SOMA**).
2. When an ABU starts negotiation of a new feasible order, it is technically and commercially evaluated as a whole and, when necessary, in every critical sub-product (parts of the product).
3. The product is then exploded into its sub-products and a call for tenders is placed for them among the APUs, including its specifications, process plans, drawings and so forth. It also includes the due date, the reward which ABU is ready to pay and the anticipated penalties for quality and delay problems.
4. APUs which are somehow interested in the business, negotiate the tender contents with ABU and then formally answer with a standard-format bid, with the reward and penalties they accept.
5. The ABU evaluates the best alternatives and places the orders (always in standard-format). It is considered for evaluation, along with the proposed revenues, the reliability index of the proponents (translated from their ranking on the reward-penalty board, which is also used for wage and profit share among the units).
6. If any invitation is not answered after the specified time, it is interpreted that the sub-product is it too risky for the anticipated reward. The ABU then reassesses its technical and commercial conditions and places another invitation for a tender, in newer basis. This is a

simple and effective filter for the manufacture against those unfeasible designs, with the additional advantage of stimulating partnership from the earliest product development stages (to prevent those unfeasible designs).

7. APUs process the orders and deliver according the plan, while ABU applies rewards and penalties as contracted.
8. The ranking is continuously updated on the reward-penalty board and then published among the AUs.

Its worth noting that this process takes place for every new product that is to be introduced by SOMA or in any exceptional case (e.g., a reliability problems in an APU). Once the product is introduced and the AUs are already tuned in, the ABU converts the process into a pull order system, based on electronic kanbans, in the place of the tender invitation. APUs then will proceed to improve their position in the ranking, and, as a consequence, their revenue. Enhancement in productivity brings increase in capacity and a new interest in applying it by looking for new tender calls. On the other hand, AUs which are low-ranking will deserve special attention because they probably will fit into one of the three situations:

- they are overloaded, which is an opportunity for enhancement or duplication;
- they are inefficient (opportunity for improvement);
- they are frequently running idle (opportunity for deactivation).

This concept also brings some other interesting characteristics:

- When an urgent order (or special opportunity) comes to **SOMA**, it may stimulate fast reaction of the system by simply raising the rewards and the penalties according to the necessity;
- it is supported on highly modularised and standardised patterns, which allows, for example, an APU to use the same approach to search for partners to complement resources or competencies towards competing for a new product;
- it unifies most of the productivity measurements and greatly simplifies the system management;
- it makes it possible to optimise the system for short and long terms simultaneously, without losing any of the perspectives

5 CONCLUSION

The Autonomous approach focuses on the modularization of the manufacturing system. This brings the manufacturing planning and control problem to a manageable size and introduces new perspectives. The main aspects to highlight are:

- the capacity to deal with the current technologies and concepts (JIT, kanban, MRP, TQC, etc.) and to gradually migrate to others without cultural breakage;
- the inherent characteristic to protect itself from the tendency to introduce complicated, hierarchical management and control tools;
- the ease in expanding or reconfiguring manufacturing system;
- business orientation, high level of individual responsibility and quality understanding;
- clear identification and separation of value- and non-value-added activities;

- transparency: traditional puzzle-like problem solving, such as hindered inventories, bottlenecks or production inefficiency are exposed before everybody's eyes;
- the use of the same primary performance measurement indicators to evaluate (with the same "currency") the various productive units of the enterprise.

Another aspect to pinpoint: in the Autonomous approach small and medium enterprises have the same opportunity to compete as the big ones. They have the opportunity to put together the competence and resources that they need in many different ways and with great flexibility and agility. This assures them a new competitive advantage even in an environment which favours concentration by capital capacity.
The concept embedded in the SOMA is now under test by simulation tools to evaluate its feasibility. The first results are encouraging. It is also planned to handle further tests with a real system in a medium size manufacturing company which currently works in a MRP-like style. Another test is planned in a JIT environment. The objective of these tests is to compare the SOMA performance against these two more traditional systems.

6 REFERENCES

Bienert, A. Schönenberger, D. (1993) Computer Aided Enterprise Modeling - A Global Approach, in *Proceedings of. Advances in Production Management Systems (B-13)*, Elsevier,. 323-30.

Browne, J. (1995) The Extended Enterprise Manufacturing and the Value Chain, in *Proceedings of IEEE/ECLA/IFIP BASYS'95*, Chapman & Hall,. 05-16.

Bullinger, H. J. et al (1993) Integrated Management: How to Combine Lean Production and CIM in *Proceedings. of 2nd International. Conference on Computer Integrated Manufacturing*, Singapore.

Corsten, H. Will, T. (1993) Simultaneous Achievement of Cost Leadership and Differentiation by Strategic Production Organization, in. *Proceedings of Advances in Production Management Systems (B-13)*, Elsevier, 47-55.

Fernandes, D. Lepikson, H. A. (1995) Product Differentiation as Basis for Competitive Strategy, in *Proceedings of. 11th ISPE/IEE/IFAC CARS & FOF'95 International Conference*, Pereira, Colombia, 562-67.

Hamel, G. Prahalad, C K (1994) *Competindo pelo Futuro*. Ed. Campus, Rio de Janeiro.

Harmon, R L. Peterson, L D. (1991) *Reinventando a Fábrica*. Ed. Campus, Rio de Janeiro.

Iwata, K. Onosato, M. (1994) Random Manufacturing System: A New Concept of Manufacturing Systems for Production to Order, *Annals of the CIRP*, **43** (1), 379-383.

Lepikson, H.A. (1990) Padronização e Interação das Unidades de Fabricação, Manipulação e Inspeção de uma Célula Flexível de Manufatura. Universidade Federal de Santa Catarina, Dissertação de Mestrado.

Márkus, A. Váncza, T.K. Monostori, L. (1996) A Market Approach to Holonic Manufacturing, *Annals of the CIRP*, **45** (1), 433-36.

Martin, J. Odell, J.J. (1996) *Análise e Projeto Orientados a Objeto*. Makron Books, Rio de Janeiro.

Mathews, J. (1995) Organizational Foundations of Intelligent Manufacturing Systems - The Holonic Viewpoint, *Computer Integrated manufacturing Systems*, **8** (4) 237-43.
NIST (1994) *Workshop on Virtual Enterprise*, Ann Arbor, MI, March 29-30, NIST- National Institute of Standards and Technology, Gaithersburg, MD.
Porter, M. E. (1991) *Estratégia Competitiva.* Ed. Campus, Rio de Janeiro.
Prahalad, C.K. Hamel, G. (1990) The Core Competence of the Corporation, *Harvard Business Review*, May-June, 79-91.
Scheer, A.-W. Jost, W. (1993) Knowledge-Based Optimization of CIM-Systems Using Industry Specific Reference Models, in *Proceedings on Advances in Production Management Systems (B-13)*, Elsevier, 21-28.
Tharumarajah, A. Wells, A.J. Nemes, L. (1996) Comparison of the Bionic, Fractal and Holonic Manufacturing System Concepts, *International Journal of Computer Integrated Manufacturing*, **9**, (3), 217-26.
Ueda, K. (1994) Biological-Oriented Paradigm for Artifactual Systems, *Japan-USA Symposium on Flexible Automation- A Pacific Rim Conference*, 1263-66, July 11-18.
Warnecke, H.-J. (1993) *The Fractal Company,* Springer-Verlag, Berlin.
Williams, T.J. (1989) *A Reference Model for Computer Integrated Manufacturing.* ISA-Instrumentation Society of America / Purdue Research Foundation.

7 BIOGRAPHIES

Abelardo Alves de Queiroz, Ph.D.: Senior Professor at UFSC - Universidade Federal de Santa Catarina (Federal University at Santa Catarina State), Department of Mechanical Engineering. Responsible, for the manufacturing strategies research subgroup at the GRUCON - Research Group on Numerical Control and Industrial Automation. At the present, he is also Head of the Graduate Programs in Mechanical Engineering. Areas of interest: manufacturing strategies, manufacturing logistics, operation management.

Herman Augusto Lepikson, M.Eng.: Assistant Professor at UFBA - Universidade Federal da Bahia (Federal University at Bahia State), Department of Mechanical Engineering. At the present, joining the GRUCON - Research Group on Numerical Control and Industrial Automation in the UFSC for the studies and research towards his Doctorate degree. Areas of interest: manufacturing logistics, shop floor management and control.

PART EIGHT

Simulation

21

Life Cycle simulation for verifying sustainable model of products

M. Johansen
Department of Production Technology, Narvik Institute of Technology
Lodve Langes gate 2, 8500 Narvik, Norway

Y. Umeda and T. Tomiyama
Department of Precision Machinery Engineering, Graduate School of Engineering, The University of Tokyo
Hongo 7-3-1, Bunkyo-ku, Tokyo 113, Japan
Telephone: + 81-3-3812 2111(ext. 7776), Fax: +81-3-3812-8849
e-mail: {umeda, tomiyama}@zzz.pe.u-tokyo.ac.jp

Abstract

This paper presents a life cycle model with a simulation tool which can be used at a early stage of the design process. The main objective is to evaluate the possibility of reducing the total volume of waste related to mass production and mass consumption. In order to verify the objective, an example is presented.

Keywords

Life cycle simulation, post mass production paradigm, closed loop life cycle, modular products

1 INTRODUCTION

Some decades ago industry alone was more or less considered responsible for environmental damage. However, it is not industry but the industrial society with mass production that creates main part of the environmental damage. Due to the technological development, consumers have frequently been offered products with better quality to a lower price based on the principles as " the bigger, the better" or "the more, the better." As a result of this development, quantity of discarded products is rapidly growing.

One of the most evident effects is that the landfill capacity is being used up. In Tokyo they have landfill capacity only for 15 more years. Figure 1 shows that this landfill problem is caused mainly by cars, household appliances and consumer electronic goods.

We only have this planet to survive on, and the resources on the earth are limited, as well as its capacity of coping with pollution and waste. These problems coupled with expected exponential grow in world's consumption and force us to react promptly.

Based on the fact that a modern life cycle economy can only function if the polluter takes the responsibility and the cost for recovering and disposing the waste, the Germany government has adopted a new act regarding "environment policy concept of the closed substance cycle and waste management," which forces the producers to take fully responsibility for their products based on the principles "from the cradle to the tomb."

ISO has started to develop a new 14000-standard regarding the environment. The new ISO 14000 Environmental Management Systems (EMS) specifies requirements for an environmental management system. This type of regulations and standards force the producers to take the responsibility for their products "from the cradle to the tomb." As a consequence, the producers have to deal with issues such as disassembly, reuse and recycling.

The purpose of implementing a "closed life cycle loop" is to minimize the volume of waste by reusing and recycling all parts of the product, with keeping quality of life and profits of producers in the same level.

In this paper, first, we the environmental background and a contents of Post Mass Production Paradigm (PMPP) which has a strong relationship to the environmental issue in Section 2. Then, Section 3 illustrates a simulation model of closed life cycle loop we have developed and an example of simulation result is shown in Section 4. After describing some pending factors in Section 5, Section 7 concludes this paper.

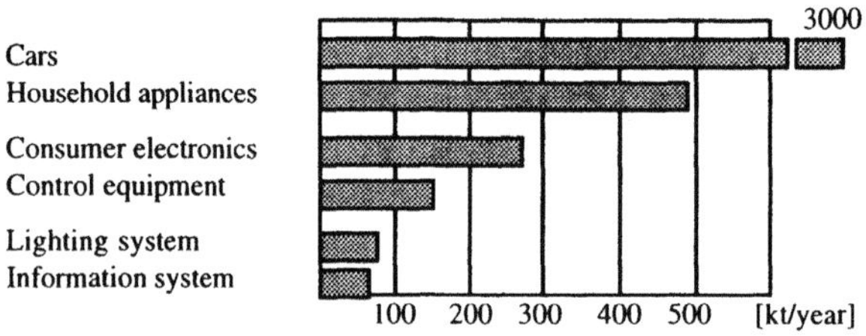

Figure 1: Yearly scrap rate for consumer products in Germany (Hentschel, 1993).

2 POST MASS PRODUCTION PARADIGM

In order to achieve the aim of reducing the volume of waste, we have proposed Post Mass Production Paradigm (PMPP) (Tomiyama, et. al., 1996). PMPP represents a new production paradigm which aims at reducing production of artifacts in terms of volume, but at the same time keeping up the same standard of living. PMPP suggests a way to overcome the growing environmental issues by decoupling economic growth from resource and energy consumption and waste creation. It is a system of economic activity capable of encouraging and sustaining economic growth without depending on mass production and mass consumption that inevitablly involves the problems mentioned in Section 1. PMPP includes shift of objective of production of artifacts from today's quantitative sufficiency to more qualitative satisfaction.

In order to implement PMPP, a new methodology to reduce the production of artifacts in terms of volume, while maintaining and improving living standard has to be developed as well as a new type of artifacts. A "knowledge intensive society" is a society where economic development is based on the creation of high value products that depend on intellectual recourses rather then natural recourses. In this concept, knowledge plays a decisive role and a major change will take place from " more, cheaper and better products" to " quickly deliver innovative products that can offer better and more responsive service to customers' demand."

2.1 Redefining manufacturing industry as life cycle industry.

The present manufacturing industry provides services from marketing through sales, however reclamation, recycling and discarding of waste are normally outside of its domain. Due to the limitations imposed by the environment, manufacturing industry will be forced to gradually provide services in the later stages of the product's life cycle. When the industry has to consider the total life cycle, the life cycle price of a product has to be defined in order to include costs currently ignored in the market. For example, the price of "single-use camera" is determined in such a manner. Because the single-use camera is in part reused and the rest recycled, it can be understood that its price consists of film price, unlimited rental fee, and insurance fee against, e.g., loss and damage.

Under PMPP, manufacturing industry will clearly be recognized as and expected to serve as *life cycle industry* (see Figure 2). In the craft society, reclamation and recycling were carried out by nature, while in the industrial society local communities provide these services paid mainly by taxes. In the knowledge intensive society, reuse and recycling should be performed by manufacturing industries. In other words, the focus of the manufacturing industry expands into all stages of product life cycle.

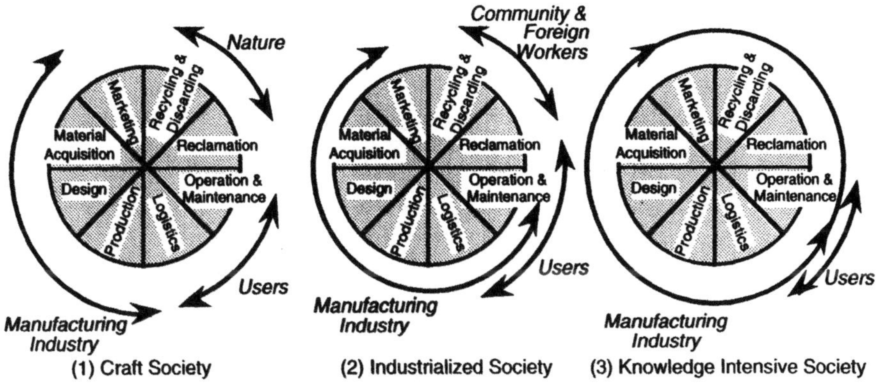

Figure 2: Craft, industrialized, and knowledge intensive societies (Tomiyama, et. al., 1996).

3 LIFE CYCLE SIMULATION

The aim of the life cycle simulation is to verify how the volume of waste can be reduced and what effect it will have on energy and material consumption. We have constructed a model for this simulation based on some assumptions in order to verify the aim of PMPP. Details of this model are described below.

This model is a general model, and is not built for any specific product. The aim is to make a "template" which, with some small modifications, can be used for several products such as household applicants and consumer electronics.

3.1 Module based closed life cycle

There are several ways to reduce the volume of waste. One is to encourage reuse and recycling of the entire product. The other is to divide the product into modules. whenever a module is broken, it is replaced by a new module. In other words, some modules can live "forever" and others have limitations. Therefore, if we make a module based product, we can increase life time of the product by replacing a broken module with a new module instead of throwing away the entire product whenever the product is broken. The broken module will then be reused and recycled.

In this simulation, we introduce a module based product into a closed life cycle loop. In order to close the life cycle loop, processes such as disassembly, reuse and recycling should be implemented. Not only broken products are thrown away, we often want new and upgraded products even though the old one is still working. One of the advantages of the module based product is that a product can be upgraded without throwing away the entire product by replacing some modules.

Figure 3 depicts an example of the closed loop life cycle with the involved processes. In this closed loop of product life cycle, whenever a product is broken, the product is either maintained or dismissed for disassembly. From disassembly some parts will be reused and the rest goes to recycling either as material or energy. Material and reusable parts are then either sold or sent back to the manufacturing process. From use, maintenance, disassembly and recycling there will always be some waste. But the objectives of this model include to try to get this volume as small as possible.

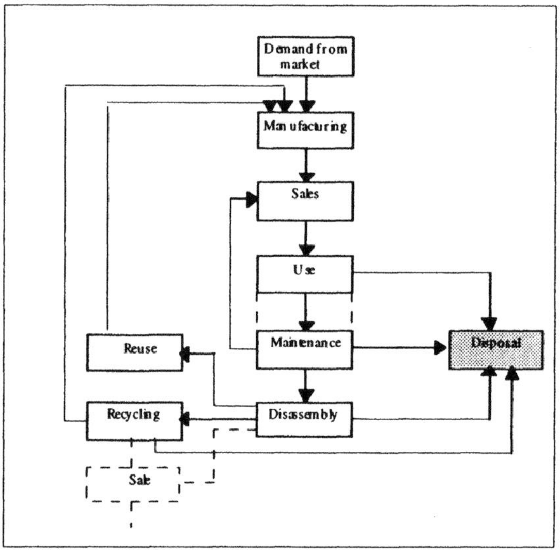

Figure 3: A closed life cycle loop.

3.2 Parameters

The model is built in order to examine influents of different numbers of modules and their life time.

Evaluation Parameters

In order to verify feasibility of the Post Mass Production Paradigm, we evaluate the following parameters in this model ;

- volume of disposal (EP1),
- energy consumption (EP2), and
- material consumption (EP3).

Here, we evaluate these parameters relatively by dividing these values in the module based product life cycle by those values in the traditional product life cycle.

Given Parameters

For executing the simulation, values of some parameters should be given by the user. On of the objectives of this simulation includes to support a designer to design and optimize a product life cycle based on results of this simulation by varying these given parameters. The given parameters should be determined based on real figures from the production. The given parameters considered here are as follows. The user should specify values of these parameters before executing the simulation.

- Cost
  all expenses involved in each processes such as labor, material, and assembling.
- Profit
  total demand of profit by the producer in each month.
- Distribution coefficients
  the distribution coefficients has to be decided in various outputs. For example, a distribution coefficient represents how large percentages of disassembled modules go to reuse, recycling and disposal.
- Size of market
  this parameter denotes a fixed demand from the market in each month.
- Energy used
  energy used for, e.g., manufacturing, operation and recycling, in each process.
- Raw material
  after subtracting reused and recycled material which can be used in the manufacturing process, some new material are compensated for material that became waste and was used for energy recycling.
- Number of modules
  number of modules in each product.
- Life time
  the estimated life time of each module.

3.3 Processes in the life cycle model

In this model, we have chosen to divide the life cycle into nine different processes. Each process is represented by given parameters, output parameters, input parameters and equations.

Given parameters requires user's input before executing simulation. In other words, the user can examine optimal evaluation of each given parameter by executing a lot of simulations with this life cycle simulation tool. Each process calculates values of the output parameters with the equations from the given parameters and the input parameters of which values are propagated from other processes (see Figure 4). Links between the processes represent these propagation paths.

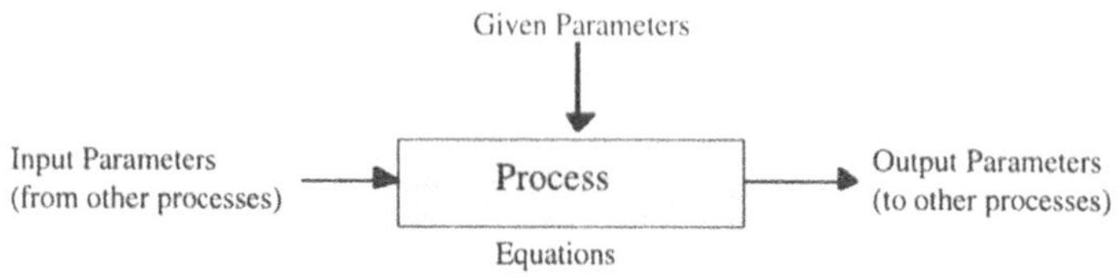

Figure 4: Example of a process in the model.

The following processes are considered in the simulation model:

1) Market

This is the fundamental process of all production. This process provides data regarding number of modules in a product and the life time of each module, which are determined by design.

Given parameters:

* size of market
* number of modules
* life time of each module
* needed material for modules

2) Manufacturing

A product is assembled in the manufacturing process. This process calculates how much new raw material is needed by comparing the demand from the market with available reused and recycled components. This process also calculates the manufacturing cost by using the information given to the model together with the feedback values gained from the disassembly and recycling processes.

Given parameters:

* cost of raw material
* assembly cost
* cost of energy
* amount of energy used in the process

Input parameters:

* amount of reusable modules and recycled material provided from the disassembly and recycling processes.
* cost of reusable modules and recycled material provided from the disassembly and recycling processes.

3) Sales

The product is brought out to the consumer in this process. The profit is divided into each module and are added to the manufacturing cost before the total price of the product is calculated.

Given parameters :

* profit demand

4) Maintenance

Whenever a module is broken, it needs to be replaced with a new module, and this modification is what is called *maintenance*.

Given parameters :

* cost of maintenance
* cost of energy

* amount of energy needed for replacing modules
* distribution coefficient that denotes percentage of replaced modules which go to disassembly. The rest goes to disposal.

Input parameters :

* number of broken modules in a month

5) Disassembly

Given parameters:

* cost of disassembling a module
* cost of energy
* amount of energy needed for disassembling a module
* distribution coefficients representing percentage of modules that go to reuse, recycling and disposal

Input parameters:

* number of modules to be disassembled in a month

6) Reuse

Some modules can be reused without any additional work. In this model, the reuse process is represented as a path of modules from the disassembly process to the manufacturing process.

7) Recycling

The recycling process consists of material recycling and energy recycling. In general, it is difficult to determine how large part supposed to be recycled as material and how large part goes to energy recycling. There is no certain rule for it and it has to be determined depending on kinds of material used in the module with the help of this simulation tool.

Given parameters :

* cost of recycling
* cost of energy
* amount of energy needed in this process
* distribution coefficient representing percentages of modules that go to material recycling, energy recycling and disposal

Input parameters :

* number of modules sent to this process in a month

8) Disposal

This process summarizes the volume of disposal sent from the maintenance, disassembly and recycling processes.

Given parameter:

* tax

Input parameters:

* number of modules sent to this process in a month

3.4 Simulation tool

The simulation tool is used for planing the life cycle of a product at the very early stages of design in order to support to designers to design the product and its life cycle at the same time. Figure 5 depicts an example of the life cycle model in this simulation tool.

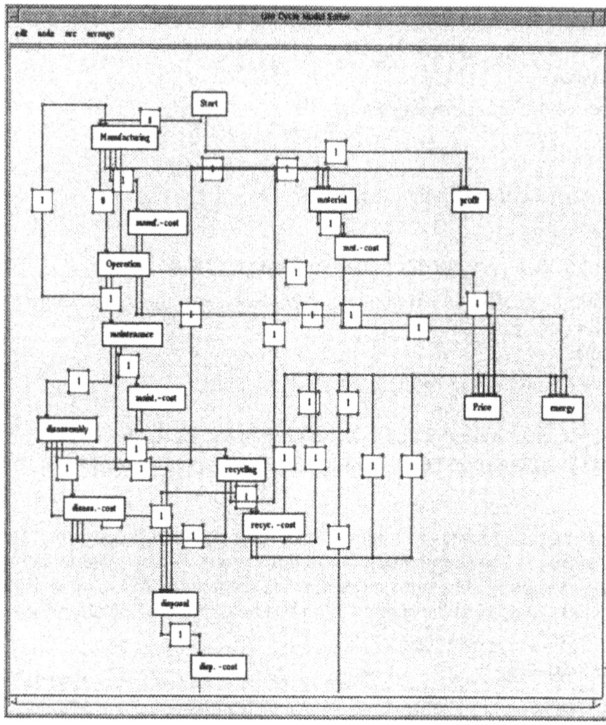

Figure 5: Simulation model.

4 SIMULATION EXAMPLE

4.1 Assumptions

In order to verify advantages of the module based product in the closed life cycle loop, we here compare its simulation result with the result gained from the module based product in an open life cycle loop. In the open loop, processes like reuse and recycling are not introduced. Moreover, in order to see influences of difference in number of modules in a product, the following assumptions have been made :

- One product consists of ten components
  This assumption means that total size, weight and functionality of a product are the same regardless how many modules the product has.
- One product can consists of any number of modules

- All modules in a product contain the same number of components
  This assumption is introduced for simplifying the simulation. Namely, each module consists of 10/*N* components, where *N* denotes number of modules in a product.

Table 1 shows values of main given parameters in this simulation example. Moreover, in some processes, the distribution coefficients should be determined. For instance, from the process of disassembly some modules are reused, others go to recycling and the rest is just thrown away. For this simulation, the distribution coefficients are chosen as shown in Table 2. For a real life situation, the distribution coefficients should be determined based on experiences and facts.

For executing the simulation, number of modules and their life time have to be determined. In order to verify influences of module based production, we perform the simulation with different numbers of modules; namely 1,3,5 and 7 modules in a product. The reason of selecting 1 module is to verify influents of recycling and reuse of non modular products.

Table 1: Given parameters

Size of Market	30 [product / month]
It takes 50 months to introduce the new product into the market.	
Material Cost	30 [money / product]
Assembly Cost	50 [money / product]
Energy for Assembly	10 [energy / module]
Maintenance Cost	20 [money / module]
Energy for Maintenance	8 [energy / module]
Disassembly Cost	50 [money / module]
Energy for Disassembly	8 [energy / module]
Recycling Cost	100 [money / product]
Energy for Recycling	20 [energy / module]
Disposal Tax	50 [money / product]

Table 2: Distribution Coefficients

Process	Reuse	Material Recycling	Energy Recycling	Disposal
Disassembly	25 %	55 %	-------	20 %
Recycling	--------	45 %	25 %	30 %

The minimum life time of a module is set to 50 months, and the maximum to 200 months. The average life time of each product is between 60 - 150 months.

4.2 Simulation Results

Disposal

If the average life time of a module based product is lower then that of a non module based product, the amount of disposal will be larger. Figure 6 shows the result of simulation in terms of volume of disposal evaluated by the evaluation parameter EP1. In this simulation, the traditional products, of which life time is 70 months, are submitted for disposal when the life time is gone. A increasing life time is one of the benefits of a module based product.

There is no doubt that the most significant reduction in the volume of disposal is gained by introducing recycling and reuse. However, introduction of recycling and reuse is not longer a question, but *only a matter of time* because all producers are forced to implement and perform it sooner or later. A guideline could be to firstly sort out components that are suitable for reuse and have highest value. Then the components that can not be reused directly should go to material recycling. The rest should be used for energy recycling.

If we look upon the results from the closed life cycle and compare one module based products with the seven modules based products, the big gain in reduction of disposal volume appears when the module based product is fully introduced to the market. After this time, the market is mainly based on upgrading and changing modules with only some few "new sell."

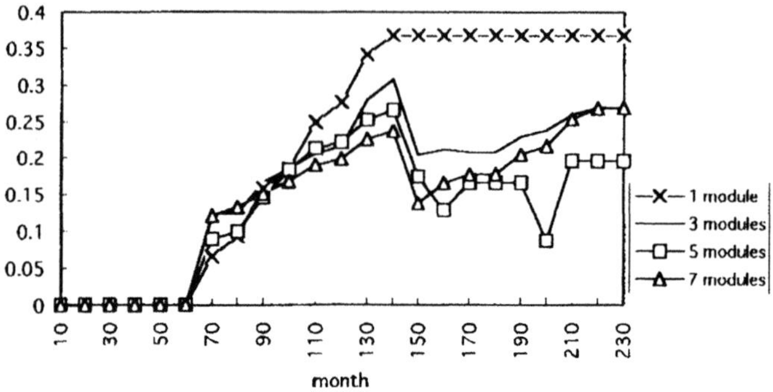

Figure 6: Evaluation of disposal.

Material consumption

By introducing recycling and reuse, reduction of material consumption will not appear before the end of life time of modules. According to the simulation result, the major gain appears when module based products are fully introduced to the market.

Energy consumption

The simulation result shows no large difference in terms of the energy consumption. Even though introduction of maintenance, disassembly and recycling might increase the energy consumption, the gain from the energy recycling compensates this increase. Figure 7 shows the simulation result based on the evaluation parameter EP2.

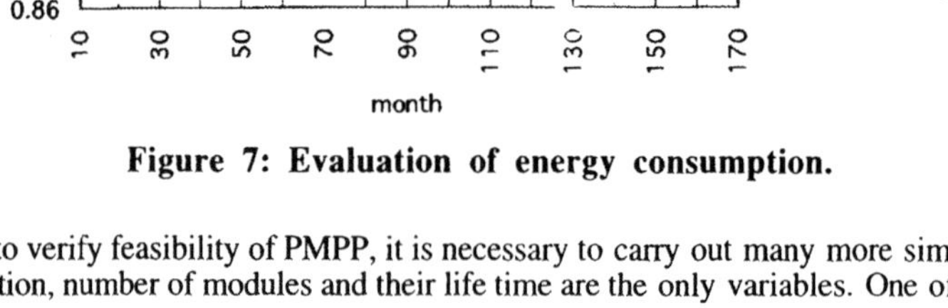

Figure 7: Evaluation of energy consumption.

In order to verify feasibility of PMPP, it is necessary to carry out many more simulations. In this simulation, number of modules and their life time are the only variables. One of the effects of changes of these parameters denote that with a increasing number of modules the volume of disposal decreases. Another effect is that longer average life time of product decreases the volume of disposal. Therefore, we conclude that the simulation result shown here implies its feasibility in term of module based products. Some other interesting parameters to study would be the distribution coefficients and the consumption of energy as well as the product price.

5. THE PENDING FACTORS

One of our future works is to include the following factors into the simulation model.

Economy

- Profit
  It is important for producers to keep their profit unchanged. According to decrease in production volume, we will have to increase the profit gained from each product.
- Product price
  By introducing disassembly, recycling and reuse, the production cost will increase due to more production time spent on each product. But, if the producers are regarded as fully responsible for their products and heavy taxes are introduced on disposal, the price of a traditional product will also increase.

Customers

- Habit
  Today behavior of consumers is closely related to fashion and controlled by the producers. The custom of buying new products is artificially created and, therefore, can be changed. New product are mainly sold not because the old ones do not work, but simply because the new ones are looking different either in color or shape.
- Satisfaction
  From the customers' point of view, there is basic desire to satisfy their needs. Since maybe 80 % of the needs is artificially created, it can also be changed through consciousness and education. With a growing concern in the environmental issues it will be advantages for producers to have an "environmental stamp." But, to change our habits will take both time and a big effort.

Employment

Other points include the problem of unemployment due to the less production. But, even there will be less production new areas of work will emerge regarding disassembly, recycling and reuse. Therefore, we do not see unemployment as a pending factor.

Laws and regulations

In order to meet new demand from authorities and standardization organizations, implementation of the closed life cycle will be a indispensable. It do not mean that every factory has to deal with their own disposal, but that the producer is fully responsible for it. In order to make the recycling business profitable, some companies has to be specialized in this area. In Europe the most fast growing business is the environmental business.

Implementation

On of the big challenges in sustainable development is to educate engineers for a new way of thinking. Achieving global sustainable production and consumption require sustainable structural changes over a long term. There is no doubt that the related changes have to take part already in the designing process, and that may be the most pending factor for implementation.

6. CONCLUSION

In this paper, we have presented a life cycle simulation tool. The result, gained so far, indicates that there is a significant reduction in the volume of waste and consumption of energy and material by introducing module based products. Future works include to add environmental effects to each process in order to support the inventory analysis in the life cycle assessment. This will lead us to develop indisputable tool to support designers to plan product life cycle at the very early stage of design.

REFERENCES

Hentshel, C. (1993) The greening of products and production: a new challenge for engineering, *Proceedings of International conference on advances in Production Management system.*

Tomiyama, T., Sakao, T., Umeda, Y. and Baba, Y. (1996), The post mass production paradigm, knowledge intensive engineering, and soft machines, *Life cycle modeling for innovative products and processes*, Chapman & Hall, London, pp. 367-379.

22

Manufacturing System Simulation Model Synthesis: Towards Application of Inductive Inference

Goran D. Putnik, João Almeida das Rosas
University of Minho, Production Systems Engineering Centre
Azurem, 4800 Guimaraes, Portugal, fax.:+351-53-510268
e-mail: {putnikgd, mpic_jar}@eng.uminho.pt

Abstract

A hypothesis of an inductive inference approach as a paradigm for automation of a simulation model synthesis process is presented. It differs from a standard approach in the sense that it does not refer the simulation model specification by some simulation programming language. The inductive inference algorithm/system infers a probably good approximation to a target system simulation model. It is expecting that the inductive inference approach could contribute to a productivity, correctness and independence of a human expert's experience in the process of the simulation model synthesis. An example, of the "standard" approach and of the inductive inference approach applied to a simple hypothetical production system, is given. Elements of the software tool developed, for inductive inference, are presented.

Keywords

Simulation, simulation model, simulation model synthesis, learning, inductive inference

1 INTRODUCTION

In this paper we are concerned with the simulation system model building issue. A hypothesis of the **inductive inference** approach **as a paradigm for automation of a simulation model synthesis process** is presented.

In the first part of the paper we present the concept of automation of the simulation concept. The objective of this part is to locate the problem we elaborate. In the second part of the paper we present a standard (conventional?) approach to the simulation model specification, based on an intuitive model of the simulation model synthesis process, applied to one hypothetical production system. In the third part of the paper we present the *automated inductive inference application* for a simulation model synthesis and elements of the software tool developed (for inductive inference).

2 THE CONCEPT OF SIMULATION AND ITS AUTOMATION

Today a simulation is probably the most used tool for manufacturing systems design. Different authors specify a concept of simulation in different ways, for example (Guimarães Rodrigues, 1988), (Tempelmeier, Kuhn, 1993). Generally, we could say that the simulation, as a concept, is defined by processes performed on two levels:

1) The level of a model building, i.e. **simulation model synthesis process** and
2) The level of experimentation over the model, i.e. **the simulation as a process** (the simulation itself).

To be implemented, these processes need to have:

1) The **simulation concept logical (abstract, mathematical) definition**, specifying its 2 levels structure (referred above), and
2) The corresponded **physical implementation**.

The physical implementation of the simulation concept is a critical issue. It depends of the language type we use. Of the greatest importance is use of some formal language. It provides us with a possibility for **algorithmisation** and, further, for **automation** of simulation processes, or of the simulation model synthesis process itself. Automation of these processes implies their physical implementation in a computerised environment, defining specific software and hardware resources. This type of implementation we will call **machine implementation**.

Otherwise, it means that we have only an intuitive model, which is usually known, and performed, by a human. In this case, processes, within the concept of the simulation, are *not implementable in a computer environment*. We do not have a knowledge representation of the simulation processes and, thus, necessarily, they must be performed by a human.

Logical definition and corresponded implementations graphical symbols, we will use in the further text, are given in Figure 1.

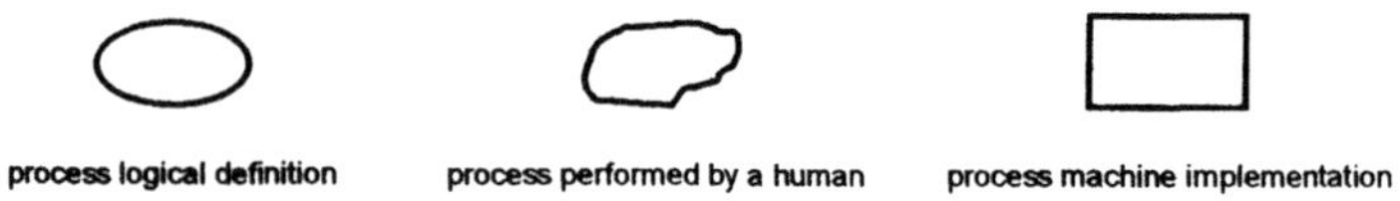

Figure 1. Graphical symbols used for the simulation processes representation.

Two level simulation concept logical structure definition, is given in Figure 2. **The simulation model synthesis process outputs the simulation model.**

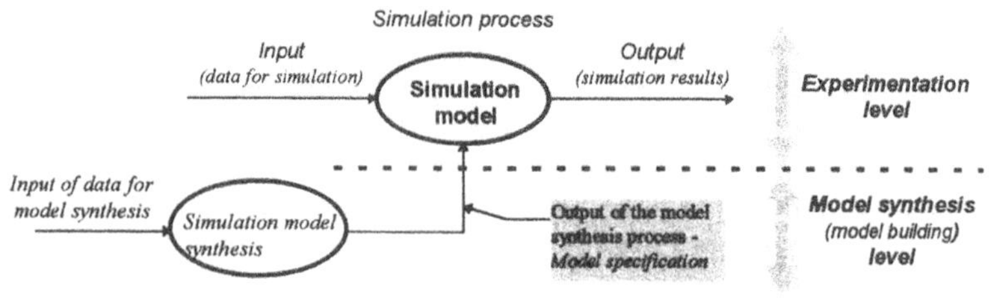

Figure 2. The simulation concept logical definition.

Our objective is to achieve a certain level of automation of all processes within the simulation concept which, is represented by the *implementation model 2*, Figure 3.

Unfortunately, due to enormous difficulties in a general design theory developments, application of the formal specification language was limited to the experimentation level, i.e. to the simulation model specification. It means that the model of the simulation model synthesis process is based on an intuitive model, *performed by a human*. This (intuitive) model **outputs a simulation model specification** in some *formal language* which is *implementable in a computer environment*, in some specific computer programming language, for example ECSL - Extended Control and Simulation Language.

The structure described above is represented by the *implementation model 1*, Figure 3. The simulation process, i.e. the simulation model specification, is implemented in a computer environment. This is *the output of an intuitive simulation model synthesis process on the model building level*. ***The model synthesis process itself is not specified !***

We will call this approach the **standard (conventional?) approach**.

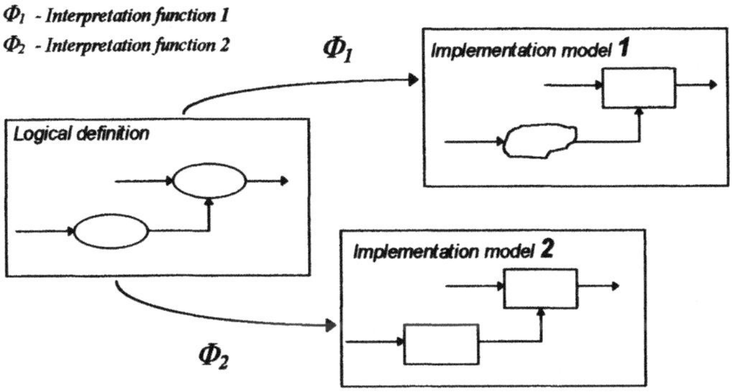

Figure 3. The simulation concept logical definition and its implementations.

Today, for the simulation model specification and implementation, a number of programming languages, and corresponded interfaces for data input and output, are developed. For example, GPSS, SIMULA, SIMSCRIPT, GASP, SLAM, ECSL, etc., ***but still not for the model synthesis process !***

An example of a higher level interface language for the simulation model specification is Activity Cycle Diagram (ACD), which is a kind of specific graph language. A specific software tool, in this case it is CAPS-ECSL (CAPS - Computer Aided Programming of Simulation) simulation software tool, compiles a computer readable format of an ACD description and outputs a simulation model specification in the simulation model specification "original", "lower" level, language, in this case in ECSL language. Improved "*implementation model 1*" of the simulation model specification is represented in Figure 4.

The problem is that the **simulation model synthesis process** itself is still performed intuitively by the human, which is very a serious limitation from some points of view.

Thus, we came to the problem of the **simulation model synthesis process automation**. Implementation of the simulation concept with the **automated simulation model synthesis process** is shown in Figure 8, chapter 5.2.

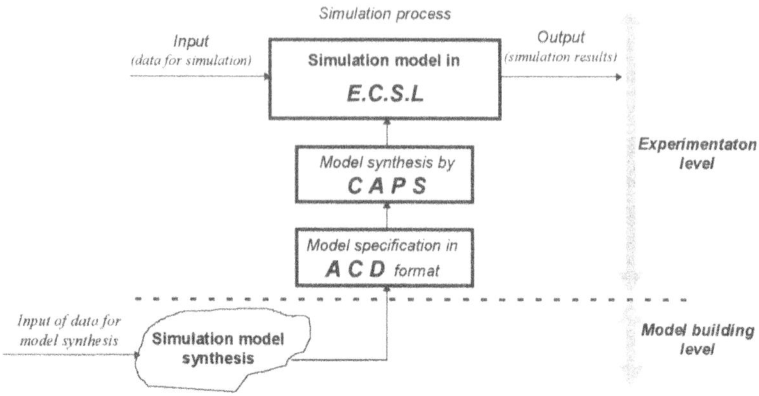

Figure 4. Standard approach to a simulation model synthesis.

As this approach requires automation of the process usually considered as a creative one, or, as an intelligent one, we will call this approach the **"intelligent" approach**.

3 PRODUCTION SYSTEM EXAMPLE - PROBLEM DEFINITION

A production system to be analysed represents an assembly line for car-radios, Figure 5. The line has six workstations: 3 workstation for chassis, boards and mechanisms assembly, 1 workstation for wiring, 1 workstations for quality control and 1 workstations for packaging.

The production process is characterised by a high level rejection rate of finished products caused by defects on chassis, boards, mechanisms and wiring. These defects are identified by the quality control performed on the corresponded workstation. Car-radios with defects are sent to reparation and after the reparation car-radios are introduced again on the assembly line to be refined. Defects originated by the wiring are repaired by the new wiring.

The objective is to design and to simulate the system. The system simulation should determine the impact of the strategies applied for resources management with the objective to optimise the system performances, for example, to obtain higher efficiency, higher production rate, optimisation of human resources, etc.

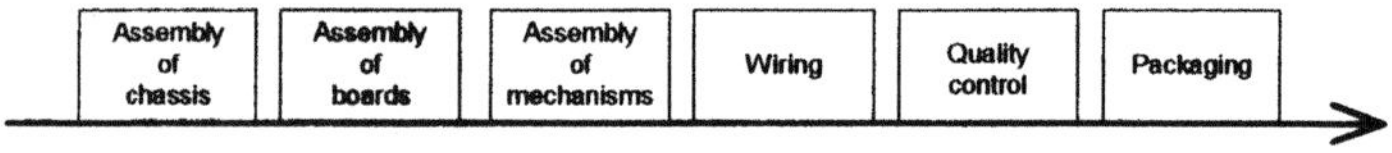

Figure 5. Car-radios assembly line.

4 STANDARD APPROACH TO A SIMULATION MODEL SYNTHESIS

The model synthesis is performed through the phases (Figure 4.):

1. Preparation of a complete detailed Activity Cycle Diagram (ACD). Preparation of the simulation model attributes: the maximum number of each type of entity, the queue discipline, the length of simulation, activity priorities, etc.

2. Describing the Activity Cycle Diagram in a format readable by the computer - Input and editing of cycles and their attributes in CAPS;
3. CAPS generates a file containing the ECSL program for simulation.

Simulation model concept specification - Activity Cycle Diagram

The specification of the model concept by the ACD diagram is conceived as follows. A cycle shows the sequence of states through which the typical entity (product, task) will pass. There are two types of states: activities and queues.

Three main conventions for an ACD diagram drawing, which we will use, are:

1. For any particular type of entity there will be always an alternation of activities and queues;
2. Activities are represented by rectangles and queues by circles inside which is written the name of the state;
3. The states are connected by arrows to show the valid sequences to show the valid sequence in which states can occur;

In our experiment, activities (processes) and queues, on the assembly line, are named as:

ACTIVITIES		***QUEUES***	
CHRD	- radios arrival (fictive activity - the *task* is entered the line)	EXT	- external world
		RDESP	- task radio is waiting for a chassis
CHMONT	- assembly of chassis	ESPPL	- task radio is waiting for a board
PLMONT	- assembly of boards	MESPM	- task radio is waiting for a mechanism
MMONTA	- assembly of mechanisms	AFIESP	- task radio is waiting for a wiring
TEAFIN	- wiring	RDECQ	- task radio is waiting for a quality control
RDCQ	- quality control	ERM	- task radio is waiting for a chassis reparat.
RDRMEC	- mechanism reparation	ERP	- task radio is waiting for a board reparation
RDRPL	- boards reparation	ERC	- task radio is waiting for a mechanism reparation
RDRCH	- chassis reparation		
REEMB	- packaging	EMBRD	- task radio is waiting for packaging

Corresponded ACD diagram of the system given is represented in Figure 6.

The second phase of the simulation model synthesis by the CAPS software tool is input and editing of the simulation model ACD diagram defined above. Description of the ACD diagram, for the system given, in the CAPS readable format, for the entity RADIO is:

RADIO,600,QEXT,ACHRD,QRDESP,ACHMONT,QESPPL,APLMONT,QMESPM,AMMONTA,QAFIESP,
ATEAFIN,QRDECQ,ARDCQ,QERM,RSA,40,QERP,RSA100,QERC,RSA,10,QAFIESP,RSA,50,QEMBRD,
AREEMB,QEXT,+,QERM,ARDRMEC,QAFIESP,+,QERP,ARDRPL,QAFIESP,+,QERC,ARDRCH,QAFIESP

For example, the cycle presented in Figure 7a, is represented in the input file as (in bold):

RADIO,*600*,**QEXT,ACHRD,QRDESP,ACHMONT,QESPPL,APLMONT,QMESPM,AMMONTA,**
QAFIESP,ATEAFIN,QRDECQ,ARDCQ,*QERM,RSA,40,QERP,RSA100,QERC,RSA,10,QAFIESP,RSA,50,*
QEMB
RD,AREEMB,QEXT,*+,QERM,ARDRMEC,QAFIESP,+,QERP,ARDRPL,QAFIESP,+,QERC,ARDRCH,QAFIESP*

and the cycle presented in Figure 7b, is represented in the input file as (in bold):

RADIO,*600,QEXT,ACHRD,QRDESP,ACHMONT,QESPPL,APLMONT,QMESPM,AMMONTA,* **QAFIESP,TEAFIN,Q**
RDECQ,ARDCQ,*QERM,RSA,40,* **QERP,RSA100**,*QERC,RSA,10,QAFIESP,RSA,50,QEMBRD,REEMB,QEXT,+,*
QERM,ARDRMEC,QAFIESP,+,**QERP,ARDRPL,QAFIESP**,*+,QERC,ARDRCH,QAFIESP*

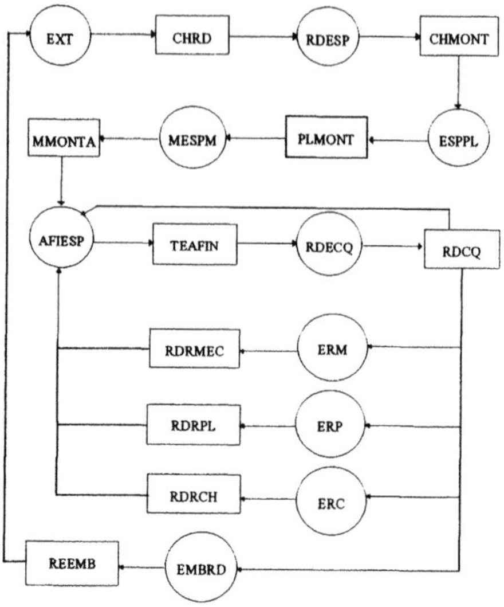

Figure 6. Activity Cycle Diagram for the manufacturing system analysed.

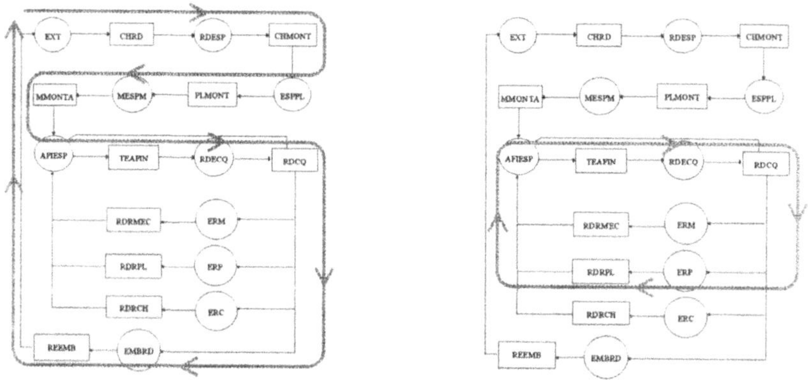

a) *b)*

Figure 7. Activity cycles examples.

5 INDUCTIVE INFERENCE APPROACH TO AUTOMATED SYNTHESIS OF THE SIMULATION MODEL

5.1 Inductive inference

The inductive inference as a learning paradigm means that a learning algorithm takes, in an iterative process, as input, a set of particular examples (one by one) by taking into account all given examples and by adding some new rules to build some hypothesis about correct general rule. Inductive inference can be defined on many ways. For example (Angluin, Smith, 1983):

"Inductive inference is a system which try to infer a general rules from examples", or

"Inductive inference is a process of hypothesising a general rule from examples"

The set of examples given, denoted by I, is the set of sentences which belong to the language we want to learn. Thus, the inductive inference is attempt to find such grammar that $I \subset L(G)$. The methodological definition is (Miclet, 1990):

Definition 1: *"Inductive inference is automatic learning of formal grammars from finite subset of the languages they generate"*

Learning is provided by the **separated** learning/design algorithm for which the object transformation system - a general rule or a machine - is output (it means, we learn about the rule or we design the machine - both are *transformation systems*), accordingly to the given set of examples. Learning/design algorithm is the **meta-transformation system** for the object transformation system.

Besides the learning algorithm itself, the most important problems is quantification of the learning process attributes and of the learned concept attributes.

The paradigm of Probably Approximately Correct (PAC) learning algorithm provides us with a quantitative measure for evaluation of the learning process and of the concept learned. This paradigm is introduced by L. Valiant (Valiant, 1984).

The PAC learning algorithm can be described as follows. Let the correct output concept, or the target concept, be denoted by f, and the hypothesis, or learned concept, be denoted by g. Both belong to a corresponded class of concepts F. The learning algorithm will output, with probability at least (1-δ), good approximation g of the target concept f, where by "good approximation" is meant that the probability that f and g differ on a randomly chosen string (example, instance) is at most ε , i.e., $P(f \Delta g) \leq \varepsilon$. Parameters δ and ε are called "confidence parameter" and "error of learning" respectively, and they can be controlled by the number of examples and by the number of hypothesis changing. Thus, "the output of the learning algorithm is *probably* a good *approximation* to the target concept" (Natarajan, 1991).

Time complexity of the learning algorithm is directly connected with a representation class by which we represent a problem, or, in other words, with a knowledge representation system.

5.2 Synthesis of the simulation model by inductive inference

The inductive inference approach refers **the simulation model synthesis, not the simulation model description compilation**, as it is by the standard approach (compilation of the system specification from one language to another language).

The inductive inference algorithm infers the model **based on some incomplete information about the objective system (which, information, is different than the system description itself**. For example, this information could be examples of the system behaviour.

Information, about the systems behaviour, given to the learning algorithm does not present all possible system's behaviours, only some of them. It means that the system does not infer the "ideal" model, but **the inductive inference algorithm/system infers *probably* a good *approximation* to the target model**, in our case, **to the system simulation model**.

The inductive inference approach to the system simulation model synthesis is conceived as a structure shown in Figure 8. The inductive inference algorithm will take as an input description of the system functionality in some "very high level" language. Output of the inductive inference algorithm will be the system simulation model specification in the "ACD language". This is from the very practical reasons: 1. the "pipeline" ACD-CAPS-ECSL (the sequence of the compilations) is already developed and 2. the ACD diagram can be represented in some "graph language", which, in general, belongs to the class of Context-Free Languages and in some special cases to the class of Regular Languages, for which we already have developed inductive inference algorithms.

In our experiment, we use a *Finite State Automata* (FSA) as a representation class.

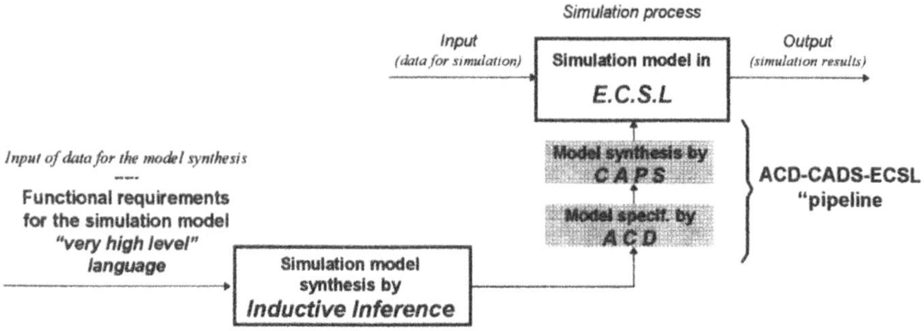

Figure 8. Synthesis of the simulation model - "intelligent" or inductive inference approach.

5.2.1 <u>Problem representation by the functional requirements for the system : examples of production processes required as examples for learning</u>

As it can be seen on the ACD diagram, of the system under consideration, each activity is preceded by a corresponded queue. Thus, we will attach to each pair "queue-activity" a particular name as bellow, and as it is presented in Figure 9.:

- EXT - CHRD - **o01**
- RDESP - CHMONT - **o02**
- ESPPL - PLMONT - **o03**
- MESPM - MMONTA - **o04**
- AFIESP - TEAFIN - **o05**
- RDECQ - RDCQ - **o06**
- ERM - RDRMEC - **o08**
- ERP - RDRPL - **o09**
- ERC - RDRCH - **o10**
- EMBRD - REEMB - **o07**

The **<u>system functionality description is introduced by giving the process sequences</u>**, which reflect the system functionality.

This type of the functionality representation represents a kind of *a <u>"very high level" language for the system specification</u>*.

We distinguish two types of the system functionality, i.e. examples of the system behaviour. These will be represented by a pair (x, y), where x will be a string, a sequence of process/operation symbols representing the system functionality, and the $y=f(x)$, where $y \in \{1,0\}$. If $y=f(x)=1$, then (x, y) is a *positive* example, representing the required system functionality, and if $y=f(x)=0$, then (x, y) is a *negative* example, representing the system functionality to be avoided.

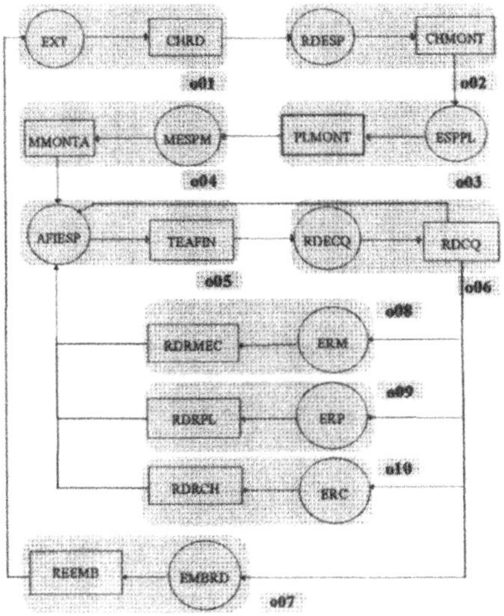

Figure 9. Renaming Activity Cycle Diagram for the manufacturing system analysed.

For example, one positive example of the system functionality can be represented as:

"o01o02o03o04o05o06o07" 1

One negative example, of the system functionality, to be avoided, can be represented as:

"o02o01o02o03o04o05o06o07" 0

5.2.2 Synthesis of the model in the FSA representation class, or, *Learning Finite State Automata*

Learning Finite State Automata (FSA) in general, is a process which belongs to the NP class. So, it is necessary to apply nondeterministic (ND) algorithm for learning with polynomial time complexity, in order to be efficient. Realisation of the ND learning algorithms is performed by introducing an additional source of information, carried on by a routine called ORACLE. In a real system, the oracle can be a human expert, database, deduction system, which will implement corresponded expertise, heuristics, helping in searching for the target concept. Of course, we are interested in outside help, during the learning process, to be minimal.

The second condition for the learning process efficiency is related to the sample size (number of the examples for learning), which is necessary to be the polynomial-sample size.

Algorithm defined by Angluin D. (Angluin, 1987), presented bellow, *satisfies* above mentioned conditions for efficiency (for the proof see (Angluin, 1987) or (Natarajan, 1991)).

The algorithm constructs a hypothesis about FSA for the target concept and then tests the hypothesis over examples. Algorithm halts if it is found the hypothesis that is valid with high probability. Otherwise it uses counterexample for construction a new and improved hypothesis.

To algorithm is permitted to make checking of membership of each sentence to the target concept. This is realised by a subroutine called MEMBER (performing ORACLE function).

<u>*A nondeterministic algorithm for FSA learning*</u> [1] :

```
input :   ε, δ, n .
begin
    set  S = { λ }  and  E = { λ } ;
    set  i = 1 ;
    call MEMBER on λ and each a ∈ Σ ;
    build the initial observation table (S, E, T) ;
    repeat forever
        while (S, E, T) is not closed or not self-consistent  do
        if (S, E, T)  is not self-consistent  then
                find  x, y ∈ S ,  a ∈ Σ ,  and  z ∈ E  such that
                    row(x) = row(y)  but  T(xaz) ≠ T(yaz) ;
                E = E ∪ {az} ;
                extend T to (S ∪ S·Σ)·E  by calling MEMBER ;
        end
        if (S, E, T)  is not closed  then
                find  x ∈ S ,  a ∈ Σ  such that
                    for all  y ∈ S ,  row(y) ≠ row(xa) ;
                S = S ∪ {xa} ;
                extend T to (S ∪ S·Σ)·E  by calling MEMBER ;
        end
        end
        let M = M(S, E, T) ;
        make m = (1/ε)[2·ln(i+1) + ln(1/δ)]  calls of EXAMPLE ;
        let Z be the set of examples obtained ;
        if M is consistent with Z  then
        output M and halt;
        else
        let (x,y) be an example in Z with which M is
                not consistent ;
        add x and all prefixes of x to S ;
        extend T to (S ∪ S·Σ)·E  by calling MEMBER ;
        end
        i = i + 1
    end
end
```

$$m = \frac{1}{\varepsilon}\left[2\cdot\ln(i+1) + \ln\left(\frac{1}{\delta}\right)\right]$$

---

[1] For detailed explanation of the algorithm. see references. Also, for the purpose of our experiment, it is not considered the control of parameters ε and δ.

5.2.3 Synthesis of the model

Synthesis of the simulation model is performed through the input of the system functional requirements - *examples for learning*, and running the inductive inference algorithm which outputs the simulation model in FSA representation class. During the algorithm run, the designer communicates with the MEMBER routine defining *implicitly* the heuristic which helps the model synthesis.

The first part of the input is input of the alphabet used:

```
..define "o01"
  define "o02"
  define "o03"
  define "o04"
  define "o05"
  define "o06"
  define "o07"
  define "o08"
  define "o09"
  define "o10"
```

In the second part of input, examples (for learning/synthesis) are given.

In *the first experiment* it is given the description of the system functionality by the positive example of a process sequence (which reflects the system functionality):

```
"o01o02o03o04o05o06o07"          1
```

The system, generated over this example of the system functionality was not satisfactory.

In *the second experiment*, to the first (positive) example, it is added the second example, the negative one, giving the information about the system functionality to be avoided:

```
"o01o02o03o04o05o06o07"         1
"o02o01o02o03o04o05o06o07"      0
```

The output, i.e. the system generated over these examples of the system functionality, in the FSA representation class, is given:

```
Recognizer = RADIOS

Initial_state = {0}
Final_states = {7}
Alphabet = {o01,o02,o03,o04,o05,o06,o07,o08,o09,o10}
Productions
{
   S0 -> o01S1          0
   S1 -> o02S2          0
   S2 -> o03S3          0
   S3 -> o04S4          0
   S4 -> o05S5          0
   S5 -> o06S6          0
   S6 -> o07S7          0
   S6 -> o08S4          0
   S6 -> o09S4          0
   S6 -> o10S4          0
}
```

The system generated can be evaluated as satisfactory. The graphical representation of the output generated is given in Figure 10. It is evident the similarity with the ACD representation of the system simulation model defined in the chapter 5.2.1, Figure 9.

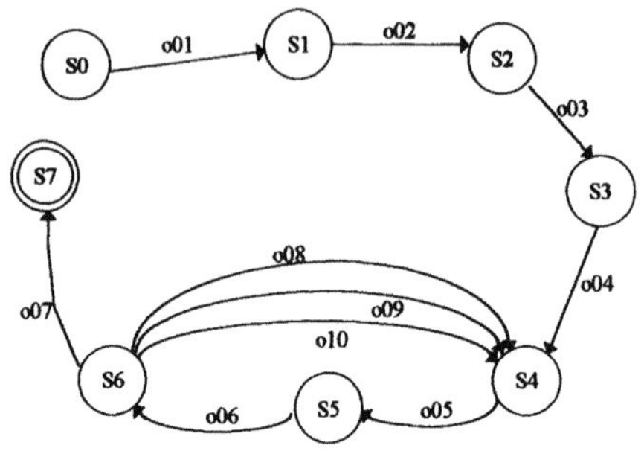

Figure 10. System simulation model generated in the FSA representation class.

6 SOFTWARE TOOL DEVELOPED

The inductive inference algorithm presented is implemented in the C language in the PC environment. In Figures 11a,b,c,d. are given, respectively: the screen of the introductory menu, the screen of the data input, the screen of the dialogue with the MEMBER (oracle) routine and the screen of the output FSA.

7 CONCLUSION

The *inductive inference* approach presented in the paper, represents *a paradigm for automation of the simulation model synthesis process*, and *does not refer the simulation model description compilation*, as it is by the standard approach. Inference algorithm takes as input an incomplete information about the system functionality (examples for learning), and outputs the system simulation model, which is the hypothesis about the targeted "ideal" manufacturing system/enterprise model. The *inductive inference algorithm/system infers probably a good approximation to the target model, in our case, to the system simulation model*. We call this approach **the "intelligent" approach**.

The simulation model generated is a **hypothetical model** about the target manufacturing system/enterprise model. This is the real case we have in manufacturing system/enterprise re-engineering problems. The first step should be to define the target model. The best practise is if we can *test different hypothesis, i.e. different hypothetical manufacturing system/enterprise models*, candidates to satisfy our objectives.

The approach, we have proposed, is believed to be highly applicable and useful tool for an **manufacturing system/enterprise organisation re-engineering**. This expectation is based on very high intelligence built in the concept of inductive inference, together with implemented capabilities for quantitative measuring of learning/synthesis process **enabling the best engineering approach**. We propose the inductive inference approach expecting that it could contribute to a manufacturing system/enterprise re-engineering by:

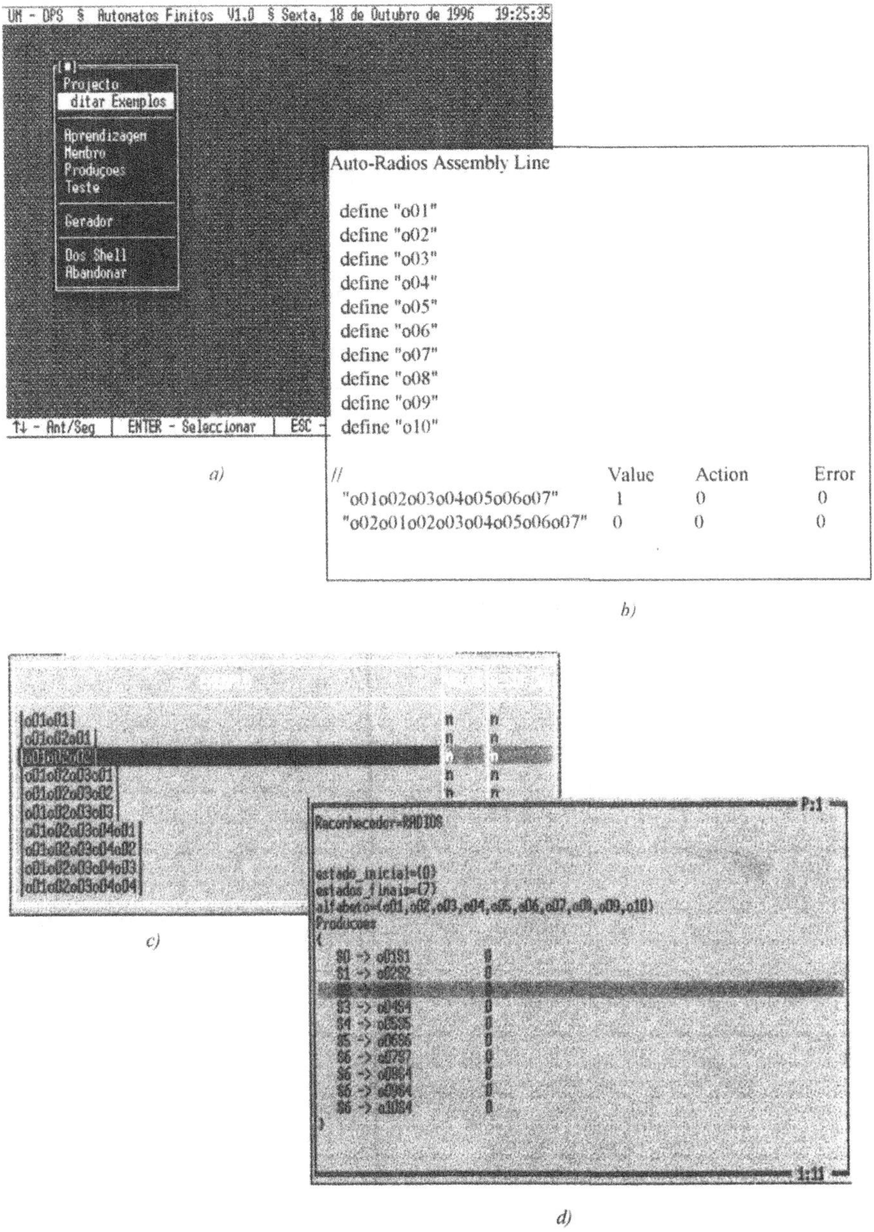

Figure 11. a) Introductory menu, b) Data input, c) Dialogue with the MEMBER routine, d) Output FSA.

1. **productivity** in building enterprise simulation models to simulate hypothetical organisational models, which is *of the greatest importance in re-engineering processes*;
2. **correctness** of the simulation model synthesis process and its *controllability*;
3. **independence** of a high skilled expert's experience, which enables company to use own resources in the best way.

The experiment, we have presented in the paper, is performed over one simple hypothetical production system in order *to strengthen the initial hypothesis of the inductive inference applicability to the simulation model synthesis process automation*. The model generated by the inductive inference algorithm differs from the reference model in few details (cycle AFIESP-TEAFIN-RDECQ-RDCQ-AFIESP, figure 6. and in the arc which closes the main cycle REEMB-EXT) but we think this is not a serious obstacle.

The future research direction identified are: 1) implementing inductive inference algorithm for more powerful representation classes, for example, for the class of Context-Free Languages; 2) development of a tool for a hypothesis generation control strategies, that is, heuristics on hypothesis generation process management, 3) implementing attribute concepts, in order to permit real simulation model automated synthesis, 4) developing post-processors for chosen simulation programming languages.

8 REFERENCES

Angluin D. (1987) Learning Regular Sets from Queries and Counterexamples, *Information and Control*, Vol. 75, pp 87-106.

Angluin D., Smith C.H. (1983) Inductive Inference: Theory and Methods, *ACM Computing Surveys*, Vol. **15**, No. 3, pp 237-269.

E.C.S.L. System - *User's Manual*

Guimarães Rodrigues A. (1988) *Simulation*, University of Minho (on Portuguese).

Miclet L. (1990) Grammatical Inference, in *Syntactic and Structural Pattern Recognition Theory and Applications* (ed. Bunke H., Sanfeliu A), World Scientific Publ. Co.

Natarajan B. K. (1991) *Machine Learning: A Theoretical Approach*, Morgan Kaufmann.

Tempelmeier H., Kuhn H. (1993) *Flexible Manufacturing Systems - Decision Support for Design and Operation*, John Wley & Sons Inc.

Valiant L.G. (1984) A Theory of the Learnable, *Communication of the ACM*, Vol. **27**., Number 11.

9 BIOGRAPHIES

Dr. Goran D. Putnik received his MSc and Ph.D. from Belgrade University, both in domain of Intelligent Manufacturing Systems. Dr. Putnik's current position is assistant professor in the Department for Production and Systems Engineering, University of Minho, Portugal, for subjects CAD/CAPP, Intelligent Manufacturing Systems and Design Theory. His interests are machine learning and manufacturing system design theory and implementations.

João Almeida das Rosas received his Eng. Lic. diploma from the University of Minho, in the domain of industrial electronic engineering. Eng. Rosas' current position is lecturer on Escola Superior de Tecnologia e Gestão. He is also a MSc student at the University of Minho, Course on Computer Integrated Manufacturing. His interests are CIM systems, design theory and process control and automation.

PART NINE

Multi-Agent Systems

23

Multi-Agent Collaboration in Competitive Scenarios

Florian Fuchs
Technische Universität München, 80290 München, Germany
phone: ++49 -89 48095 254, fax: ++49 -89 48095 250
email: fuchsf@informatik.tu-muenchen.de

Abstract

For many multi-agent scenarios one can assume that the agents behave cooperatively and contribute to a common goal according to their design. However, our work focuses on competitive scenarios which are characterized by the agents' strong local interests, their high degree of autonomy, and the lack of global goals. Therefore, two agents will cooperate if, and only if, both will gain — or at least expect to gain — from that cooperation.

This paper presents a conflict resolution mechanism which is appropriate for competitive resource allocation in dynamic environments which is based on compromising. It integrates a goal relaxation mechanism, negotiation histories, and multilateral negotiations.

Keywords

Agents, non-cooperative, collaboration, compromise

1 INTRODUCTION

In contrast to multi-agent settings which follow a fully cooperative model, there are scenarios with quite different qualities. These are settings without a global goal function and with agents which follow exclusively local goals. Cooperation between two agents takes place only if both sides expect to profit. We denote such scenarios as competitive.

A restricted view to the global system state and agents pursuing different local goals are typical qualities in multi-agent systems, whereas in traditional centralized structured approaches these qualities hardly play any role. As a consequence of these problem characteristics conflicts arise among the agents which have to be resolved. I.e. mechanisms for conflict resolution have to be integrated into a multi-agent approach in order to obtain global consistent solutions.

In competitive scenarios some problems make the conflict resolution more difficult than in fully cooperative ones. First of all, the amount of common knowledge is very small. In general, agents do not know the other agents' plans, goals, strategies, etc. An agent will not provide the others with that kind of information to avoid that they can take advantage of this knowledge. Thus, all the agents can do is to extract information out of the negotiation processes and build models of its competitors which are largely uncertain.

Furthermore, if an agent provides some information, it may not tell the truth. Interesting questions are: When is it profitable for agents to lie, and how can lies be discouraged. Palatnik and Rosenschein investigated some problems arising in that context [PR94].

The resolution of a conflict is achieved by the application of a conflict resolution strategy. There are different situations in which conflicts occur, e.g. the conflict may concern a part of the schedule which is either still coarse grained or where already much fine planning was involved. Furthermore, we have different scenarios, e.g scenarios with many alternative resources or such without any alternatives. Different situations and different problem scenarios require varying strategies to resolve a conflict successfully. An agent therefore needs conflict resolution knowledge that enables it to react appropriately when it is faced with a conflict.

This paper presents a conflict resolution mechanism for distributed resource allocation which is suitable for competitive scenarios. The rest of the paper is structured as follows: In section 2 requirements for an appropriate conflict resolution mechanism are pointed out. Section 3 is dedicated to our application scenario and section 4 introduces our compromising concept, and finally, in section 5 some conclusive remarks are made.

2 COMPETITIVE AGENTS

Competitive scenarios show some specific characteristics which require appropriate conflict resolution strategies. Most methods suitable for fully cooperative settings are not applicable.

Many approaches to conflict resolution among cooperative agents employ a more or less static set of strategies, one of which is chosen as a common strategy in the case a conflict occurs. This strategy might be either chosen through a negotiation between the agents (see for example the Cooperative Experts Framework [LLC91]), or might be agreed upon directly because the agents have identical conflict resolution units (see for example the Cooperative Design Engine [KB91]). With strategies which are common to all agents it is virtually impossible to model competitive scenarios where the agents are not willing to share the knowledge about their employed strategies. They have to be enabled to apply arbitrary strategies, however they believe their local interests are supported efficiently with.

Another popular method in cooperative domains is to let one of the conflicting parties play a special role in the conflict resolution process. In the work of Polat and Guvenir, for instance, the agents have different roles in the negotiation process according to their knowledge and problem-solving capabilities [PG92]. As far as competitive scenarios are concerned, this is not a feasible way. Here, agents will refuse to take part in negotiations in which they do not have equal rights, and therefore, may be put at a disadvantage or may be cheated.

Quite similar drawbacks have approaches in which the conflict resolution is done by an arbitrator agent (see for example [Wer90] or [SRSF90]). The existence of such an arbitrator is a quite unrealistic assumption in most cases. Conflicts rather have to be resolved by direct negotiation between the conflicting parties.

Let us briefly summarize the requirements for a conflict resolution mechanism suitable for our scenario which we have found so far:

- **Symmetry**
  None of the conflicting parties must play an extraordinary role in the conflict resolution process, they must be given equal rights.

- **Private Strategies**
  Strategies have to be private in the sense that they are freely choosable by each of the conflicting parties, and they are not accessible by the other agents.

- **No Arbitrators**
  The agents have to resolve conflicts without help from some higher level instance.

Furthermore, there are several requirements which are not specific for competitive scenarios, but have to be considered in other settings as well. The most important ones are listed in the following:

- **Stability**
  The conflict resolution should not yield an unstable behavior of the overall system, since instability results in undesired unpredictability of future market trends.

- **Adaptability**
  An important characteristic of a market scenario is its dynamic, the agent's goals are subject to constant change. Therefore, an agent must have the capability to choose appropriate strategies, taking into consideration its own goals, the model it has about its negotiation partners, as well as the needs of its clients.

3 APPLICATION SCENARIO

The environment in which our agents are situated is a scenario from the field of distributed resource allocation. Orders which are brought into the system from extern clients have to be assigned to resources under consideration of temporal aspects. Therefore, the scenario includes a distributed scheduling problem, which is to be solved by competitive agents. The participating agents try to influence those parts of the global schedule positively with respect to their local optimality criteria which are interacting with their local plans.

We are treating only worth-oriented domains which allow for a partial satisfaction of goals, and therefore, for a compromise in the case of goal conflicts. This stands in contrast to task-oriented or state-oriented domains, where one of the conflicting parties has to refrain from its goal totally in the case of a conflict.

3.1 Scheduling model

The employed scheduling model is similar to what is proposed in [Win93]. This extended job shop comprises a set of *resources* which is divided into *resource groups* corresponding to certain *activities*. Then there is a set of *tasks* aggregated to *jobs* by means of a partial order. The model allows that jobs are brought into the system dynamically (online scheduling). This model is characterized in detail in [Fuc96].

Schedules are not represented by means of exact time intervals or time points. Despite, density functions over time are used for each resource. A density function serves as a coarse grained representation of the capacity usage of a resource and provides flexibility with respect to the actual start points of intervals. They allow for an efficient evaluation of resource requests. Dealing with exact schedules would be virtually impossible considering the early planning state, and therefore, the high planning uncertainty at this planning level.

3.2 Agent model

This section gives an overview of the applied agent model. Therefore, we will describe our agent design and how the agent is situated within its environment. Figure 1 shows the conceptual structure of the agent model which is used in our cooperation scenario. The agent is embedded in its environment which consists of three parts: The client environment from where jobs are brought into the system, the execution environment functioning as an outlet for jobs, and finally the agent environment where jobs may be distributed via some cooperation mechanisms.

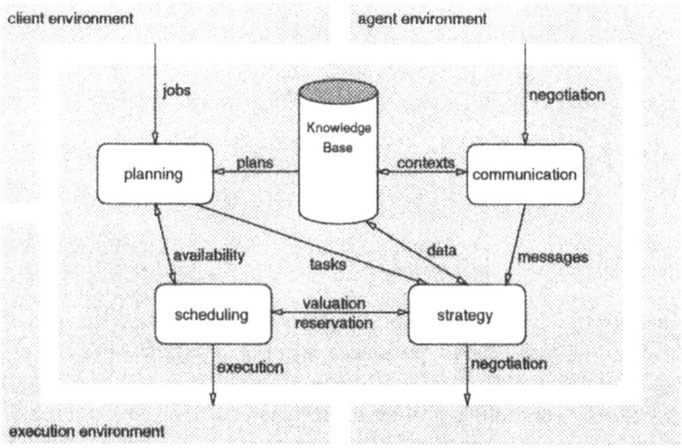

Figure 1: agent architecture

An agent encapsulates all the knowledge it needs to realize its goals and also the appropriate processing mechanisms for that knowledge. The knowledge is kept explicitly represented in a local knowledge base which contains in the first place knowledge about other agents, the own internal state, and skeleton plans that are instructions on how to process jobs.

In addition to the knowledge base there are four agent components which operate quite independently and are connected via some internal interfaces.

The planning unit maintains a set of jobs it receives in irregular intervals from some clients. The planning unit has to build a production plan for a job — mostly by getting a prebuilt plan from the knowledge base — and determines a task which is to be scheduled next.

An agent is able to communicate through its communication unit with other agents. From a received message it extracts information and updates its local knowledge base accordingly. Furthermore, the message is put into a certain context according to its type and contents. The agent then changes to the role intended for this context.

The strategy unit contributes the strategic part to the conflict management. This unit is responsible for generating an appropriate response to a received message (see section 3.3).

The scheduling unit manages schedules and evaluates proposals for resource allocations from the strategy unit. This unit is also the one which is connected to the execution environment and makes the necessary steps for the execution of tasks.

3.3 Market System

The basis of our cooperation model is a *market-oriented system* (cf. the notions *computational ecology* [HH88], *agoric open systems* [MD88], and *market-oriented programming* [Wel93]). With market-oriented system we understand a system which provides market mechanism like a pricing system or negotiation platforms which can be used by the participants of the market.

Those participant are agents. They own resources which are necessary for processing the jobs. Generally, not all resources necessary for all steps of a job are available locally. Therefore, an appropriate decomposition and distribution of jobs is essential.

For this reason, a pricing mechanism is introduced as follows. Each resource causes costs. This is especially true for the time a certain activity is processed. But also resources lying idle are not free of costs. All those costs have to be payed by the respective owner of a resource. Whenever a job is given out to some supplier this supplier will pass on the costs to its client. An additional fee is charged for the usage of a resource what allows for making some profit.

4 CONFLICT RESOLUTION

Different local interests yield inevitably goal conflicts which are eliminated by mutual goal relaxation — a compromise. This goal relaxation and the necessary negotiation is discussed in detail in the following two sections.

4.1 Negotiation

The allocation of resources is subject to negotiation. A negotiation is characterized by a negotiation protocol describing a set of negotiation primitives as communicable messages and all valid sequences of those primitives of a negotiation.

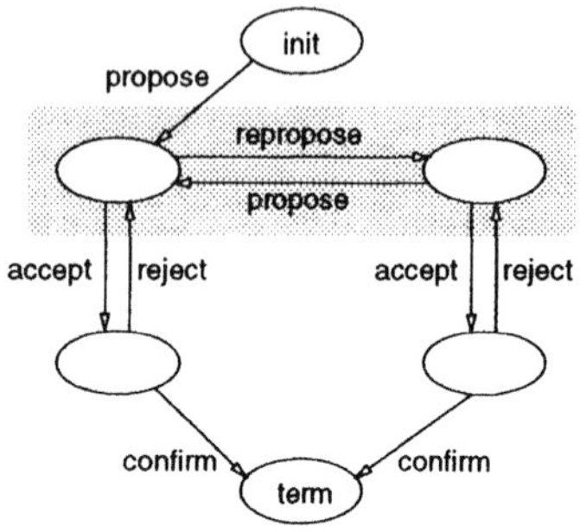

Figure 2: protocol

The negotiation protocol which is made use of is an extended version of the well-known contract net protocol [Smi88]. The four phases *request*, *offer*, *order*, and *confirmation* are replaced with an embedding of *propose* and *repropose* (see figure 2) — necessary to enable the compromising mechanism.

The contents of these primitives is taken as a proposal for a contract (*draft contract*), which includes two parts: First, a specification of the object of the contract — mainly the resource in question — and second, negotiable terms of the contract, for example the allocation time interval or the price for the allocation. During a negotiation the agents may either agree on the terms and thus make a contract, or the negotiation fails because the local interests are incompatible.

At each negotiation at least one *supplier* and one *client* takes part. Supplier and client are *roles* the agents play temporarily.

4.2 Compromising

The agents are characterized by a rational behavior. Rationality is given when an agent adopts the goal to increase its own profit which is defined by means of its local utility function.

The *utility* an agent gets from a contract is influenced mainly by two aspects. On the one hand *costs* arise from what an agent is obliged to by a contract. The client is obliged to pay the negotiated price and the supplier has to provide resources. On the other hand a contract has a certain *worth* for both parties. The supplier gets the negotiation price and the client is given resources and thus the possibility to fulfill its due dates. The resulting utility is the difference of worth and costs.

There are two more points, which are important for an agent's utility. Resources do not only cause costs when they are allocated but also when they lye idle. And furthermore, for exceeding due dates some kind of punishment is introduced. Both points serve as an incentive for the agents to take part in negotiations and show willingness to compromise.

Costs and worth of a draft contract depend on the current schedule and the quality of the resulting schedule with respect to the preferences of an agent, i.e. with respect to capacity usage, flow time, job tardiness, etc. If an agent was alone in the society, it could build the schedule that has maximum worth and therefore, the maximum utility. Otherwise, if it is in a society of competitors, goal conflicts are inevitable and compromises concerning the schedule have to be found. The market-oriented system serves as a tool to reach those compromises.

For achieving a compromise between two agents a mutual modification of a draft contract takes place. Such a modification is like the search in a search space which is defined by the terms of the contract. Each point in the search space corresponds exactly to one possible draft contract. The search for a counterproposal is guided by a evaluation function which assigns an evaluation to every draft proposal. This evaluation depends on the current schedule and state of the negotiation, i.e. especially the current terms of the contract.

The following three aspects are considered through the evaluation function and are weighted appropriately:

- **local goals**
  An agent's local goals slip into the evaluation by means of the agent's utility.

- **negotiation history**
  An agent tries to extract information out of its encounters with other agents and to use this information in future encounters. Therefore, it finds out the flexibility of its competitors in negotiation concerning each single term of a contract. It uses these flexibility values to guide the search for a counter proposal into a direction which is acceptable for its competitor.

- **time pressure**
  An agent is faced with two kinds of time pressure during a negotiation. First, it has to prevail over its competitors and is therefore forced to come to a acceptable proposal in reasonable time. Second, there are time constraints which result from given due dates. For these reasons, the evaluation of a draft contract is corrected appropriately depending on the expired negotiation time.

In the resulting search space one of the standard search algorithms — e.g. simulated annealing — can be applied to generate a counter proposal. The termination of this process is guaranteed because the utility of a contract increases when the probability of exceeding a due date becomes higher. The convergence of this process can be controlled by the agents by means of giving the time pressure aspect more or less weight in their evaluation.

Figure 3 shows an example for typical evaluations which are extracted from a test run with a given planning state. For visualization reasons just two terms of a contract are displayed. One can see clearly the contradictionary interests of supplier and client concerning the price.

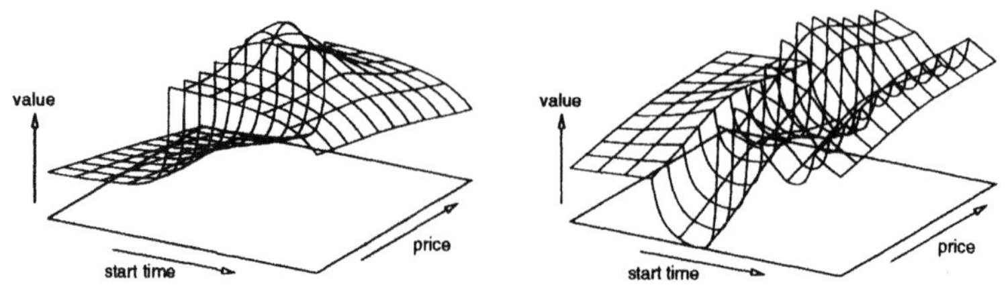

Figure 3: example for proposal evaluations of client (left) and supplier (right)

5 CONCLUSION AND FUTURE WORK

This paper presented a conflict resolution concept for multi-agent systems which is suitable for competitive scenarios which is based on compromising.

The most important feature of this concept is that it follows strictly the non-cooperation paradigm. It integrates aspects which are realistic for that kind of scenarios, goal relaxation, uncertain knowledge, negotiation histories, and multilateral negotiations. The agents' conflict resolution strategies are private, i.e. not accessible by other agents. Therefore, an agent is enabled to preserve its local interests. Of course the possibility to share a common strategy remains.

Currently effort is spent on finishing a prototypical implementation of the described system. This prototype will also include a simulation environment which serves as a testbed for the efficiency of different strategies in different manufacturing scenarios. The next goal is to investigate the effects of some strategies on the scheduling results — especially the impact on job lateness, in-process inventory, job idle time, and machine utilization. We hope to finish the first prototype implementation in a few months and will come out with the first experimental results soon afterwards.

References

[Fuc96] F. Fuchs. Multi-Agent Collaboration in Competitive Scenarios. In *Proc. of the* 15th *Workshop of the UK Planning and Scheduling Special Interest Group*, Liverpool, 1996.

[HFL96] S. Hahndel, F. Fuchs, and P. Levi. Distributed Negotiation-based Task Planning for a Flexible Manufacturing Environment. In J. W. Perram and J.-P. Müller, editors, *Distributed Software Agents and Applications*, number 1069 in Lecture Notes in Artificial Intelligence. Springer, 1996.

[HH88] B. A. Huberman and T. Hogg. The Behaviour of Computational Ecologies. In B. A. Huberman, editor, *The Ecology of Computation*, number 2 in Studies in Computer Science and Artificial Intelligence, pages 77–115. North-Holland, 1988.

[KB91] M. Klein and A. B. Baskin. A Computational Model for Conflict Resolution in Cooperative Design Systems. In S. M. Deen, editor, *CKBS'90: Proc. of the International Working Conference on Cooperating Knowledge Based Systems, October 1990, Univ. of Keele, UK*, pages 201–219. Springer, Berlin, 1991.

[LLC91] S. Lander, V. R. Lesser, and M. E. Connell. Conflict Resolution Strategies for Cooperating Expert Agents. In S. M. Deen, editor, *CKBS'90: Proc. of the International Working Conference on Cooperating Knowledge Based Systems, October 1990, Univ. of Keele, UK*, pages 183–200. Springer, Berlin, 1991.

[MD88] M. S. Miller and K. E. Drexler. Markets and Computation: Agoric Open Systems. In B. A. Huberman, editor, *The Ecology of Computation*, number 2 in Studies in Computer Science and Artificial Intelligence, pages 133–176. North-Holland, 1988.

[PG92] F. Polat and H. A. Guvenir. A Conflict Resolution Based Cooperative Distributed Problem Solving Model. In *Proc. of AAAI-92*, pages 106–115, 1992.

[PR94] M. Palatnik and J. S. Rosenschein. Long Term Constraints in Multiagent Negotiation. In 13th *International DAI Workshop*, pages 265–279, Seattle, Washington, 1994.

[Smi88] R. G. Smith. The Contract Net Protocol: High-Level Communication and Control in a Distributed Problem Solver. In A. H. Bond and L. Gasser, editors, *Readings in Artificial Intelligence*. Morgan Kaufmann Publishers Inc., San Mateo, California, 1988.

[SRSF90] K. Sycara, S. Roth, N. Sadeh, and M. Fox. Managing Resource Allocation in Multi-Agent Time-Constrained Domains. In *Proc. of a Workshop on Innovative Approaches to Planning, Scheduling, Control.* Kaufmann, 1990.

[Wel93] M. P. Wellman. A Market-Oriented Programming Environment and its Application to Distributed Multicommodity Flow Problems. *Journal of Artificial Intelligence Research*, 1:1–23, 1993.

[Wer90] K. Werkman. Design and fabrication problem solving through cooperative agents. Technical report, Lehigh University Bethlehem, 1990.

[Win93] A. Winkelhofer. *Zeitrepräsentation und merkmalsgesteuerte Suche zur Terminplanung.* PhD thesis, Technische Universität München, 1993.

24
The Establishment of Partnerships to Create Virtual Organizations: A Multiagent Approach

M. A. Hochuli Shmeil, E. Oliveira
Faculdade de Engenharia da Universidade do Porto
Rua dos Bragas, 4099 Porto Codex, Portugal
Tel +351-2-2004527 E-mail: {shmeil, eco}@fe.up.pt

Abstract

Virtual Organizations are aggregations of independent organizations or individuals aiming to contribute to a common goal. Two basic steps are needed to assemble a virtual organization: (i) a business process partitioning, and (ii) a partners' selection process. Broadly speaking, in this paper we present a computational approach to model organizations based on Distributed Artificial Intelligence - Multiagent Systems (DAI-MAS) as well as Symbolic Learning (SL) paradigms. Each organization, which is seen as an agent, is provided with the needed observation, planning, coordination, execution, communication and learning capabilities to perform its social roles. In particular, we present a specific inter-organization relationship: the selection process that leads to the automatic establishment of contracts between organizations. This selection process is composed of a bid evaluation phase followed by a negotiation phase as a mean for agents conflicts resolution. Through negotiation interactions, a set of offer and counter-offer values which are seen as instances (positives and negatives) are supplied for further analysis in order to support the learning activities. The contribution of our work lies, not only, on the computational model proposed for the society of organizations, but also in some extent, on the learning methodologies applied to the established partnerships, in particular, and to the community, in general.

Keywords

Multiagent system, Organization Modeling, Application.

1 INTRODUCTION

The goods and services market is composed of two types of economic agents, the organizations and the individuals, related by their interactions. They can, dynamically, assume productive, distributive and consumptive basic roles. The interactions play a decisive part in the behavior of the organizations and individuals, and are used to: (i) decide what and how to produce, (ii) indicate where and how to distribute the production and (iii) determine how to promote and validate the products. In our scenario proposal (Figure 1), the involved organizations are conceived as goods and services producers, consumers or distributors (suppliers), depending on the role they play in each interaction. In this scenario individuals are consumers. To model this domain, an under developing computational system called ARTOR (ARTificial ORganizations) uses Distributed Artificial Intelligence-Multi Agent System (DAI-MAS) and Symbolic Learning

(SL). Its goal is, to create, maintain and provide an adequate setting where organizations and their interactions are simulated (Figure 2). In our first approach we are focusing the hierarchical form as an organizational model.

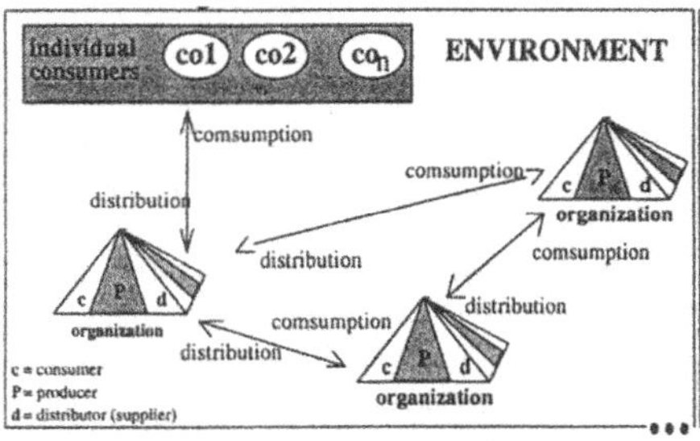

Figure 1 - The scenario proposal.

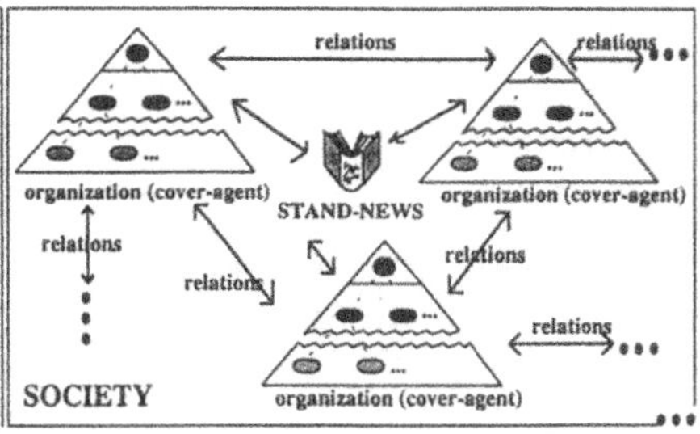

Figure 2 - The Multiagent society.

In this setting each organization acts as a cognitive agent with the capability of performing any of the following economic roles: production, distribution (supplying), and consumption. These roles or behaviors are dynamically enabled/disabled by the established interactions. The agents that map such organizations are referred as *cover-agents* (Figure 3). Cover-agents are semi-autonomous agents that need to establish and maintain interactions in order to "survive". Cover-agents are implemented as a set of finer grain agents. The inter-cover-agent communication can be of three types: (i) point to point, when a cover-agent addresses its message or observation to another cover-agent, (ii) point to multi-point, when a cover-agent addresses its message or observation to more than one cover-agent, and (iii) newspaper-like, when a cover-agent wishes to make public some data. The latter is built using a blackboard [1] called *stand-news*, where the data is posted and kept for public consultation. From the society point of view, the cover-agents are distributed agents, that act in conformity with their cognitive capabilities[1].

2 THE INNER OF A COVER-AGENT

The adopted cover-agent architecture (related to our actual focus - the hierarchical organizations) is based on the three layered hierarchical structure decomposition [2]. The top and intermediate layers are composed of cognitive, homogeneous, semi-autonomous agents, while the bottom layer contains cognitive, heterogeneous, semi-autonomous agents. The number of agents per layer depends on the layer's hierarchical level:

(i) the top level layer has exactly one agent;

(ii) the intermediate layer number of agents is problem-domain dependent - there can be further in layer hierarchical decomposition, both horizontal and vertical, according to the complexity of the domain being mapped;

(iii) the bottom layer number of agents is also problem-domain dependent - there is no hierarchical decomposition, and the amount of agents varies horizontally, based on the needs imposed by the problem-domain.

Time-sharing, agent performance, resources similarity and process nature are some of the criteria used for establishing the composition of the intermediate and bottom layers.

The required capabilities of the agents of the top and intermediate layers are planning, action coordination, communication and learning. These agents are called *administrator-agents*. The required capabilities of the agents of the bottom layer are task execution (domain expertise),

[1]Perception, memory and thinking.

communications and learning. These agents are called *executor-agents*. From the cover-agent point of view, the intermediate and bottom set of inner agents are distributed and centralized.

There are two different sets of knowledge present in an organization: the corporate knowledge (CK) or the cover-agent knowledge, and the individual knowledge (IK) or the internal agent's knowledge. The CK and IK are acquired, transformed and learnt by the inner agents during their interactions. The CK can be divided in two classes [3]. The first knowledge class, known as self model (SM), represents the styles, missions, plans, resources, processes, etc., of the inner world. The second knowledge class, denominated acquaintance model (AM) represents the outer or external world. The AM is divided in two sub-models: (i) the cover-agents sub-model (CS) containing a local representation of the remaining cover-agents, and (ii) the society sub-model (SS) containing a local representation of the society. The continuous analysis of the agents' interactions plays an essential role: it acts as the knowledge acquisition interface that enables learning. Consequently, both sub-models (CS and SS) are dynamic.

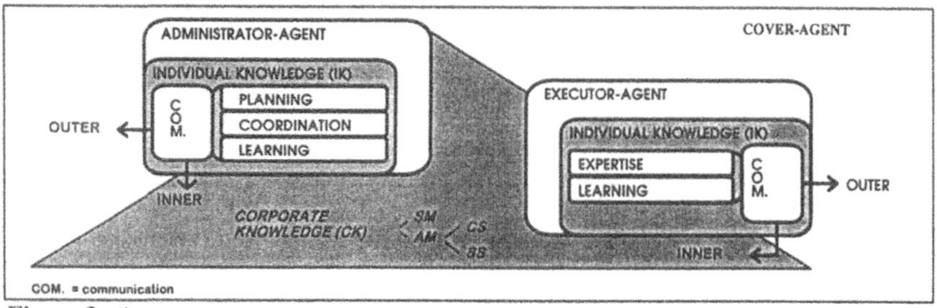

Figure 3 - A cover-agent.

The corporate knowledge is a shared knowledge base which is maintained and updated by the organization's inner agents. This corporate knowledge base contains (Figure 4) not only the organization's functional model but also encompasses the corresponding model of competencies. The functional model is supported by a general *descriptor* of the functions performed by the organization. The descriptor is an abstract representation of the functional objects and their relations, and provides an integrated model of the concepts, processes and resources of the organization. The acquisition, transformation and learning of new corporate knowledge performed during the execution of the organization tasks is oriented by this abstract descriptor of the corporate knowledge base.

The model of competencies defines the duties and the rights of each inner agent. The competencies of an inner agent can be viewed as a projection over the corporate knowledge base descriptor. The inner agent is responsible for maintaining, updating and learning about the concepts of the corresponding corporate knowledge subset defined by the adequate projection.

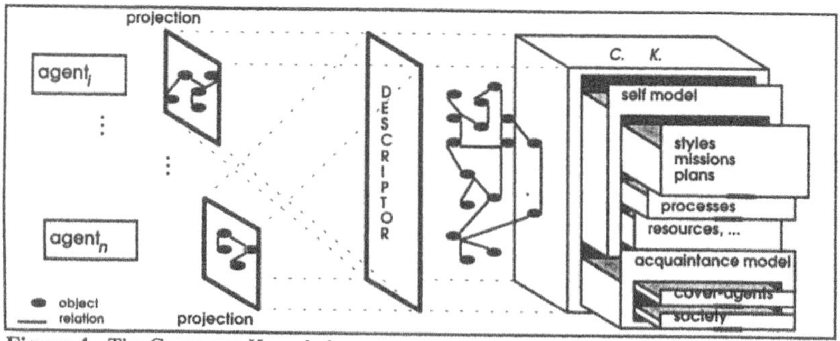

Figure 4 - The Corporate Knowledge.

3 THE PARTNER COVER-AGENTS' SELECTION PROCESS

Our work is focused on a well-defined type of interaction: the interactions that occur between cover-agents acting as producers and cover-agents acting as suppliers. The producer's intent is to find the most adequate supplier[2] of a good or service. A cooperative action is established when a supplier is contracted by a producer. This contracting is the result of a selection process. The selection process consists in sending an invitation (posted in the stand-news by the producer) to the society, followed by the evaluation of the received offers. Negotiation is invoked when conflicts arise. The contract will be celebrated with the supplier that satisfies the producer's constraints. In the specific setting of our approach we assume that the selection process is undertaken by the bottom layer executor-agents of the involved cover-agents. The cover-agents engaged in a cooperative mission have well-defined roles [3]: (i) the producers, or *organizers* are the cover-agents which are searching for cooperative partners, and (ii) the suppliers, or *respondents*, are the cover-agents that wish to participate in the proposed cooperative activity.

The evaluation process for each good or service to be contracted is based on the analysis of the existing local constraint list (CL). A CL is a conjunction of units denominated evaluation constraints (EC). They are used by the organizer to validate the respondents' proposals according to three criteria: (i) an organizational criterion that validates organizational attributes (structural, financial , etc.), (ii) a technical criterion that validates the characteristics of the good or service (color, dimension, etc.) and (iii) a commercial criterion that validates commercial attributes (price, quantity, etc.).

Each EC is a 5-tuple where:

$EC_i = (id_i, vd_i, va_i, pr_i, sm_i)$ where:

- id_i is the constraint identification
- vd_i is the constraint default value
- va_i is represented by a 2-tuple (vac_i, fed_i) where:
  - vac_i represents the EC_i domain
  - fed_i is the most favorable evolution direction within the domain:
    - right — when it is advantageous to move towards the right;
    - left — when it is advantageous to move towards the left;
    - none — when there is no preference among the any of the possible values (vac_i).
- pr_i is the value of the agent's utility($pr_i \in$ Dcardinal)
- sm_i represents the agent's state of mind (open or close minded):
  - grounded — the first EC_i offer will be posted specifying a defined value;
  - free — the first EC_i offer will be posted without any value specification.

This sort of CL is called *default_CL*. As an example, is presented a default_CL, involving three constraints: color, price and payment_period.

{ (color, blue, ((black, green, blue), none), 2, ground),
(price, 30, ([0, 33], left), 5, free),
(payment_period, 60, ([30, 60, 90], right), 2, ground) }.

Alternative CLs can be provided representing, either alternative constraint values or pre-defined constraint value's relaxation, which define the alternative bids. These new CLs are called *alternative_CLs*. As an example, is presented an alternative_CL, involving the same three constraints: color, price and payment_period.

{ (color, blue, ((black, green, blue), none), 2, ground),

---

[2] The most adequate supplier is the one that satisfies the constraints imposed by the producers.

(price, 34, ([34, 36], left), 4, ground),
(payment_period, 120, ([120, 150], right), 3, ground) }.

The union of the cartesian product of the domains of the ECs that belong to the default_CL with the cartesian product of the domains of the ECs that are specified in the alternative_CLs, define the plausible negotiation space state for a good or service. Every EC has two different utility parameters: (i) the first utility parameter denominated pr_i represents the EC's importance within the CL; and (ii) the second utility parameter expresses the most favorable evolution direction (fed_i) of the concept within its domain (left, right, none). The global utility of a CL is given by the sum of the individual ECs utility parameters (pr_i and fed_i).

The respondents have, internally, the same CL representation data format. There is, however, one exception: the respondents do not communicate their states of mind. The selection process is started and ended by the organizers. It starts when the organizer sends the invitation to the stand-news using the following data type:

⟨ *idorg*, *datli*, *list_of_constraints*, *qty* ⟩,
where:
idorg is the identification of the organizer;
datli is the answering deadline;
list_of_constraints is a list of constraints represented by (*id*, *vd*, *negotiation*)
where:
id is the constraint identification;
vd is the constraint default value;
negotiation is y when vac contains other values than vd;
negotiation is n, otherwise;
qty is the maximum number of iterations per negotiation

Example:

⟨ $agent_n$, 12, ((color, blue, y), (price, _, y), (payment_period, 60, y)), 10 ⟩

In our setting, the economic principles that guide the behavior of the organizations make them eager to cooperate.

Every cover-agent that has available resources continuously, poles the stand-news in search of new business opportunities. The invitations posted in the stand-news are analyzed by every one of these potential respondents. Those who have the necessary resources send their first offer to the organizer.

3.1 The Evaluation Process

The evaluation process is exclusively performed by the organizers. The goal of this process is to measure the distance between the specified demand and the received offers. In order to calculate these distances' values two filters are applied: a) Filter 1 is applied to the ECs that contain only default values, b) Filter 2 is applied to the remaining types of ECs. It computes the distance between the default and answered values for each remaining EC and for each respondent (*dist = default_value - answered_value*).
Three situations may arise:

(i) there is only one respondent such that dist ≤ 0 (this means that either the respondent offer satisfies (dist = 0) the CL specified by the organizer, or that it may contain more attractive values (dist < 0). Negative distances (dist < 0) occur when the offered values are more advantageous than the proposed ones;

(ii) more than one respondent has presented attractive offers (dist ≤ 0);

(iii) there are no respondents that, totally or partially, fit into the initially specified default values (dist > 0) for any or all ECs.

3.2 The Negotiation Process

The negotiation process is a decision making process, which helps the agents to satisfy their goals. The organizers look for respondents such that, $\forall$ EC, *dist* $\rightarrow$ Z, ($Z \leq 0$). The respondents' aim, while trying to satisfy the organizer's CL, is to be contracted. In their effort to satisfy the organizer bids they make use of the available EC utilities. Organizers use the conflict resolution further negotiation. Conflicts occur when the distance calculated in the evaluation phase is either negative or positive, giving rise to the positive and negative conflict types [3].

The selection process has diverse halting conditions: (i) Through the celebration of a contract when only one respondent has met the bid's requisites, (ii) After a pre-defined number of evaluation-negotiation interactions. This value is defined by the organizer. This condition is met when: (a) more than one respondent has made tempting offers (dist $\leq$ 0), (b) no agent is making offers, and (c) none of the respondents' offers are sufficiently close to what the organizer has specified (dist > 0).

Before beginning the negotiation phase each involved agent selects one strategy (Figure 5), among those available in its strategies set. This choice is made based on the consultation of the following agent's knowledge: the self model (SM) and the acquaintance model (AM). In the case of the self model, the agent's style is what is more relevant. The agent's style is a meta-knowledge that characterizes the agent inner and outer behavior. This style may be classified as win/win or win/lose. A win/win agent governs its activities with reciprocal advantage concepts towards its partners. It is characterized for taking into account, totally or partially, the answered values. A win/lose agent directs its activities in an egocentric fashion, solely moved by his own advantage. It is characterized for disregarding the answered values. The acquaintance model helps the agent to select its strategy based on what it knows about the involved cover-agent (CS) as well as about the society (SS) as a whole. At startup, ARTOR's cover-agents have empty acquaintance models. As a result of the established interactions, the AMs are, incrementally, built and refined.

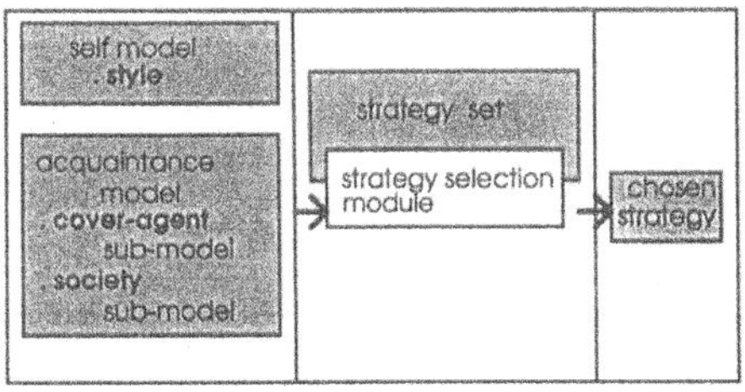

Figure 5 - Strategy choice.

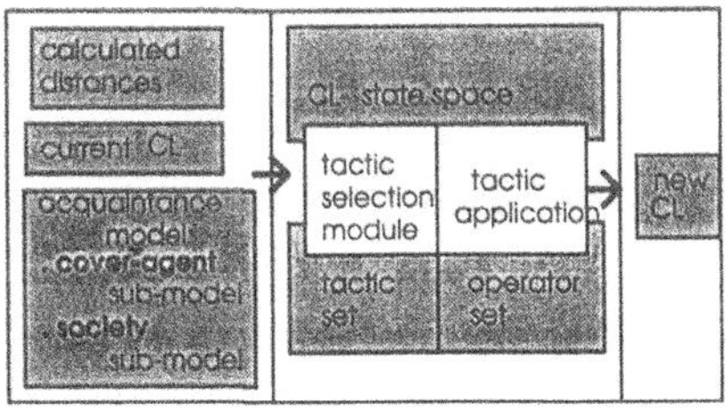

Figure 6 - Counter-proposal generation.

Once chosen the strategy, organizer and respondent, have to select the most adequate tactics. Each strategy has an associated number of tactics and operators that are applied to the current CL in order to create the new CL (Figure 6). The organizer selects the most appropriate tactics according to: (i) the current CL (ECs values); (ii) the calculated distances - when the offer exceeds the values of the previous demand (*dist* $\leq$ 0), the organizer employs the tactics that maximize its benefits. Otherwise it applies tactics that provide alternative_CLs and (iii) acquaintance model.

The respondent chooses the tactics that try to meet the organizers demand, based on the current CL and acquaintance model.

Each iteration (Figure 7) is composed of an evaluation of the proposals (CL_n), followed by the generation of a counter-proposal (CL_{n+1}) and the subsequent sending of the new counter-proposal (CL_{n+1}) to the involved agent(s). The generated values or counter-proposal (e.g., CL_{n+1}) is the result of the application of the strategy, tactic and operator chosen.

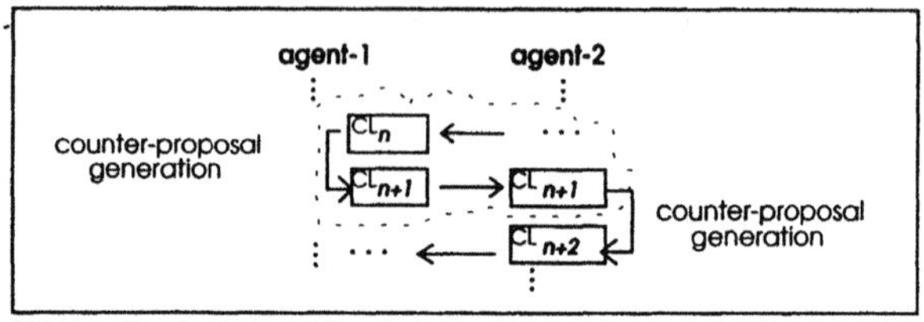

Figure 7 - An interaction.

The set of styles, the strategies, the tactics and the operators of an agent are represented as frames (Figure 8). Each style has an associated set of concept frames defining it. The set of applicable strategies is defined by the style. Associated to each strategy there are frames representing tactics which support the strategy enforcement via specific operators. Operators can increase, decrease or maintain the EC values.

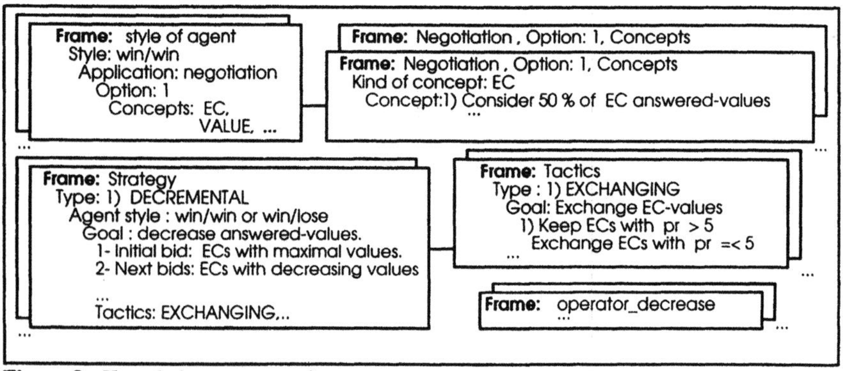

Figure 8 - Knowledge representation.

3.3 Reasoning about Interactions

The model chosen for the study of the discussed segment of the goods and services market is being implemented according to the DAI-MAS methodologies [4]. In this context, it is possible to detect a number of learning opportunities, based on the analysis of the interactions established among the cover-agents (organizations). In particular, ARTOR's cover-agents are focused on two specific learning opportunities:

(i) learning organizational concepts - The conceptualization of the behavior of other cover-agents or the construction and refinement of the cover-agents sub-model (CS);

(ii) learning social concepts - The conceptualization of the behavior of the society, as a whole, or simply, the construction and refinement of the society sub-model (SS).

In the ARTOR's model, the learning capabilities lay on the individual inner agents. The learning capability of a cover-agent, as a whole, is the sum of the individual inner agents' capabilities. The inter-agent exchanges that occur during the evaluation and negotiation steps, constitute the set of examples used for learning. Each example is a triplet (P_{r1}, CL_{n+1}, CL_{n+2}) (Figure 9). P_{r1} is the internal transformation process used to generate the new constraint list from the current CL (CL_n). The CL_{n+1} is the new internal constraint list that will be generated once P_{r1} is applied to CL_n. The CL_{n+2} is the constraint list that will be externally produced upon the reception of CL_{n+1}. The learning will be concentrated on the virtual consequence dependency that can be established between the internally generated CL_{n+1} and the externally generated CL_{n+2}.

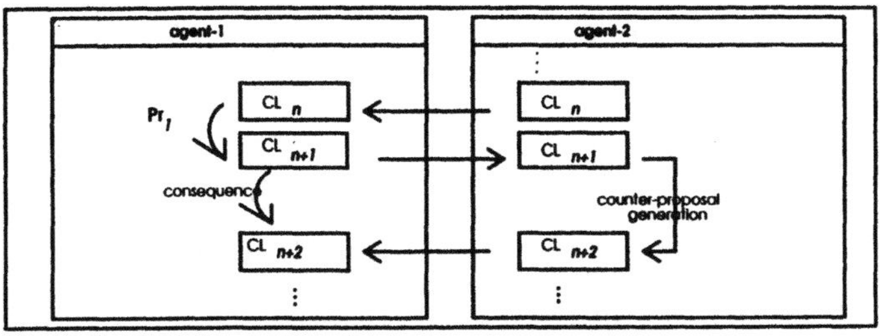

Figure 9 - An inter-agent exchange.

The purpose of the individual learning activity (within different layers- top, intermediate and low) is to hypothesize assumptions regarding the behavior of others. This type of learning is based on the analysis of the CLs changes that exist during the selection process. From the comparative analysis of consecutive CLs (CL_{n+1} and CL_{n+2}) a classification of the experienced EC changes emerges. The calculated deltas, or variations are cataloged in three possible classes:

(i) null delta or = class - when the ECs have remained unchanged;

(ii) positive delta or ↑ class - when the ECs have experienced an incremental alteration;

(iii) negative delta or ↓ class - when the ECs have suffered a decremental variation.

The calculated deltas represent the difference between the desired and offered EC values. In (Figure 10) a default_CL which contains three constraints (EC) denoted by a, b, c is presented. The default ECs values, expected by the organizer, are in the central line of the first column (CL_i). The other columns represent posterior states which were generated by respondent and organizer during the selection process.

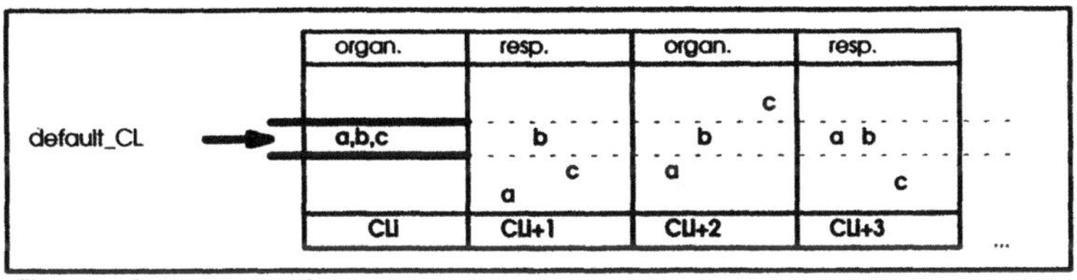

Figure 10 - An example.

In this case the calculated deltas are:

CL	a	b	c		a	b	c
CL_i	initial	initial	initial	CL_{i+1}	↓,[value]	=	↓,[value]
CL_{i+2}	↑, [value]	=	↑,[value]	CL_{i+3}	↑,[value]	=	↓,[value]

⇒ organizer ⇐⇒ respondent ⇐

The learning process has two phases: (i) example analysis and (ii) submission of the new example set to the learning algorithms The analysis of the example consists in establishing for each constraint (EC) the extremes of the negotiated values. The applied algorithm uses both inductive and deductive reasoning. The agents involved in a selection process (the organizer and the respondent agents) will have, by the time it finishes, a more accurate internal model of its counterpart. The quality of the model created depends on the quantity of information made available during the selection process, or in other words, the richness of the set of examples provided. The inductive and deductive learning methodologies are applied over the same set of examples but distinct complementary concepts are obtained. Through the application of: (i) the inductive learning paradigm [5], higher level knowledge about the behavior of the CLs during the selection process is produced, (ii) the deductive learning algorithm based on (EBL, mEBG) [6], higher level knowledge concerning the style of the agents involved in the selection process, is generated.

In the first case (i), the organizer and the respondent not only learn new organizational concepts concerning the evolution and relative importance[3] of the negotiated constraints, but also learn new local social concepts regarding their respective social roles/behaviors. In the second case, the agents use as background knowledge (BK) the knowledge present in each agent style, strategies and tactics slots to deduce new local social perspectives. However, when the background knowledge does not support deductive reasoning, deductive learning is not possible. In such circumstance specialize concepts are hypothesized.

Generically, the learnt assumptions concerning the organizational concepts are represented in the cover agent sub-model (CS) by a ⟨*id_ag*, *assumptions*⟩ slot, while the social concepts are represented in the society sub-model (SS) using a ⟨*assumptions*⟩ slot, where *id_ag* is the identification of the counterpart agent, and *assumptions* are the learnt assumptions (rules, propositions, etc.).

As an example, and according to the interaction described in Figure 10, the organizer ($agent_1$) has learnt about the respondent ($agent_2$), that in those specific circumstances EC_b remains constant:

$$< agent_2\ EC_b = default\_value \text{ when } EC_a \subset [a_1, a_n] \text{ and } EC_c \in \{c_1, c_i\} >.$$

Assuming that after the selection process of executor $agent_1$, a quality control process occurred at executor $agent_{11}$, and that both executor agents are managed by the same administrator $agent_i$ (intermediate layer agent), a higher level concept (Figure 11) about external $agent_2$ can be learnt:

$Agent_1$ < $agent_2$, was selected to provide the specified goods/services >.
$Agent_{11}$ < $agent_2$, ISO quality requirements = 92% >.
$Agent_i$ < $agent_2$, satisfaction = OK >.

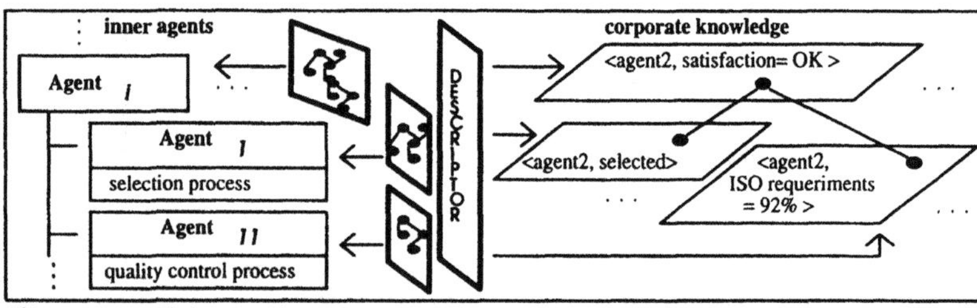

Figure 11 - Higher level concept.

In future selection and quality control processes involving this same good/service the cover-agent of executor $agents_1$ and $_{11}$, will take into account the satisfaction concept learnt about the cover-agent of executor $agent_2$.

[3] The importance of a constraint can be measured in terms of its weight in the final sucess or failure of the selection process: hard constraints are non relaxable and soft contraints are relaxable.

4 CONCLUSIONS

The work we have described in this paper, aims at contributing to the computer simulation of the interacting of organizations. Several different approaches for modeling of the goods and services market as well as its organizations have been attempted and implemented. The work of Wellman [7] and the research of Barbuceanu and Fox [8] are some good examples of the currently undergoing investigation on this subject. Conflict resolution via negotiation in multi-agent systems is also based on previous work by Oliveira [9] and by Sycara [10], while cooperative learning has been explored in Sian [11] and others.

The contribution of our work lies, not only, on the model proposed for the society of organizations, but also, on the distributed learning methodology applied to the established partnerships, in particular, and to the community, in general. The combination of the three layered hierarchical structure with the in-layer multiagent architecture provides a rich and adequate setting for the representation of complex organizations. The inherent modularity of the presented architecture allows the representation of organizations with different levels of complexity, by means of social grouping. In particular, our efforts are concentrated on the analysis of a specific type of interaction which is one of the basic step to assemble a virtual organization: the cooperative interaction between producers and suppliers. The producer's intent is to find the most adequate supplier, according to a set of specified constraints. The supplier's aim is to be contracted while trying to meet every producer's demand. The mechanism used to select the most adequate partner is based on a selection process which is composed of evaluation, negotiation and conflict resolution phases. Once an interaction is established, every exchanged constraint list is collected. The set of collected constraint lists constitutes a cooperative example, which can be used by the learning algorithms to generate assumptions about the behavior of the involved organizations, as well as about the society as a whole.

5 ACKNOWLEDGMENTS

The first author's current research is being supported by PUC-Pr and CNPq - Conselho Nacional de Pesquisa, Brazil, Grant Number 200413/92-9.

6 REFERENCES

[1] Engelmore, R., Morgan, T. (1988) Blackboard Systems. Addison - Wesley Publishing Company.

[2] Simon, H.A., Decision Making and Organizational Design. In: Pugh, D.S.ed. Organizational Theory. Penguin Books. p.189-212.

[3] Oliveira, et.al. (1993) Negotiation and Conflict Resolution within a Community of Cooperative Agents. In: Proceedings of The First International Symposium on Autonomous Decentralized Systems, Kawasaki, Japan.

[4] Gasser, L. Huhns, M.N. (1989) Distributed Artificial Intelligence, vol.II, Pitman Publishing, London.

[5] Michalski, R.S. (1990) Learning Flexible Concepts: Fundamental Ideas and a Method Based on Two-Tired Representation. In: Machine Learning - An Artificial Intelligence Approach, vol. III, Edited by Yves Kodratoff and Ryszard Michalsky, Morgan Kaufmann Publishers, Inc.

[6] Kodratoff, Y. (1990) Learning Expert Knowledge by Improving the Explanations Provided by the System. In: Machine Learning - An Artificial Intelligence Approach, vol. III, Edited by Yves Kodratoff and Ryszard Michalsky, Morgan Kaufmann Publishers, Inc.

[7] Wellman, M.P. (1993) A Market-Oriented Programming Environment and its Application to Distributed Multicommodity Flow Problems. In: Journal of Artificial Intelligence Research, 1 (1993) 1-23, AI Access Foundation and Morgan Kaufmann Publishers.

[8] Barbuceanu, M., Fox, M.S. (1994) The Information Agent: An Infrastructure for Collaboration in the Integrated Enterprise. In: Proceedings of the 2nd International Working Conference on Cooperating Knowledge Based Systems, Editor S.M.Deen, University of Keele.

[9] Oliveira, E., Mouta, F. (1993) Distributed AI Architecture Enabling Multi-Agent Cooperation. In: Industrial and Engineering Applications of Artificial Intelligence and Expert Systems, Edited by Paul W.H. Chung, Gillian Lovegrove and Moonis Ali, Gordon and Breach Science Publishers.

[10] Sycara, K.P. (1989) Multiagent Compromise via Negotiation. In: Distributed Artificial Intelligence, vol.II, Edited by Les Gasser and Michael N. Huhns, Pitman Publishing, London.

[11] Sian, S.S. (1991) Adaptation Based on Cooperative Learning in Multi-Agent Systems. In: Decentralize A.I. - 2, Edited by Yves Demazeau and Jean-Pierre Muller, Elsevier Science Publishers B.V.

PART TEN

Modeling and Integration

25

Business process modelling: Comparing IDEF3 and CIMOSA

F.B Vernadat
Laboratoire de Génie Industriel et Production Mécanique
Université de Metz and ENIM
Ile du Saulcy, F-57012 Metz cedex 1, France
Tel: +33 (0)3 87 34 67 23 Fax: +33 (0)3 87 34 69 35
vernadat@poncelet.univ-metz.fr

Abstract

Enterprise Integration (EI) and business process reengineering (BPR) both require business process modelling methods to assess and reengineer enterprise behaviour and functionality in order to create better synergy among business entities. IDEF3 and the CIMOSA process model are such methods. These two methods are reviewed and compared in this paper with regards to common business process modelling structures necessary to model industrial systems and manufacturing enterprises. Applications of the methods to the modelling of structured and semi-structured processes as well as to non-deterministic and concurrent processes are also discussed.

Keywords

Enterprise modelling, Enterprise Integration, BPR, Business processes, CIMOSA, IDEF3

1 INTRODUCTION

Enterprise Integration (EI) is concerned with braking down organisational barriers to improve synergetic effects and increase overall productivity within an enterprise or among cooperative enterprises (Bernus *et al.*, 1996; Vernadat, 1996a). Central to EI is the modelling of enterprise activities and business processes to assess and rationalise enterprise functionality and enterprise behaviour (Vernadat, 1996b). CIMOSA (AMICE, 1993) and IDEF3 (Mayer *et al.*, 1992) both promote a horizontal approach based on a modelling method enforcing the *business process* concept (as opposed to previous vertical approaches focusing on organisational functions). Both can be used in business process reengineering (BPR) or systems engineering projects. They are presented and compared in this paper as state of the art workflow methods to model manufacturing environments.

2 IDEF3

IDEF3 is a process description capture method for Concurrent Engineering (CE), Total Quality Management (TQM) and BPR initiatives (Mayer *et al.*, 1992). It has been developed as part of the IDEF suite of methods in the ICAM project supported by the US Air Force.

IDEF3 complements IDEF0, a functional modelling notation, and IDEF1x, an information system modelling technique based on the entity-relationship model. IDEF3 introduces a process-centred view to the IDEF method. It is itself complemented by IDEF4, a method for modelling enterprise objects in the form of Object State Transition Networks (OSTNs) and IDEF5, a method for acquiring CIM ontologies (Benjamin *et al.*, 1995).

Basic elements of the IDEF3 process description method include (Mayer *et al.*, 1992):

- *Units of Behaviour* (UOBs) to define process steps (i.e. operations, activities or sub-processes)
- Three types of *links* (precedence, relational and object flow) to describe interrelations between two UOBs
- Three types of *junctions* (AND, XOR and OR junctions) which can be used as fan-in (convergent) and fan-out (divergent) junctions in either the synchronous or the asynchronous mode to formulate process flow description. This makes up to ten types of junction possibilities: four ANDs, four ORs and two XORs (the synchronous XOR has no meaning).

In addition, the concept of *referent* is used to refer to other models or previously defined parts of the model (e.g. a UOB, a junction, an object, a scenario, a note, an OSTN). Referents are also used to build "go-to" statements. Referents can be unconditional, asynchronous or synchronous.

Each of these constructs is defined in the IDEF3 formalism by means of a description template with predefined entries and by a graphical representation illustrated by Figure 1. Figures 2 and 3 provide examples of a process and one of its sub-processes expressed in IDEF3 about controlling items of an equipment to be assembled as a kit.

3 CIMOSA

CIMOSA is a reference architecture for enterprise integration in manufacturing developed by the ESPRIT Consortium AMICE (AMICE, 1993). It includes an enterprise modelling framework based on a process-centred approach, an enterprise engineering methodology based on the system life cycle concept and an integrating infrastructure for enterprise integration. In this paper, only its modelling framework, and especially its process model (Vernadat, 1993), is considered.

The CIMOSA process modelling language is a formal language and has no official graphical formalism because graphics are always ambiguous. It is made of a set of modelling constructs defined as object classes embedding declarative and procedural language statements. The language addresses the four CIMOSA views: function view, information view, resource view and organisation view. Thus, the set of constructs covers the concepts of domains, events, domain/business processes, enterprise activities, functional operations, functional entities (resources), enterprise objects, object views, organisation units and organisation cells (CIMOSA Association, 1996).

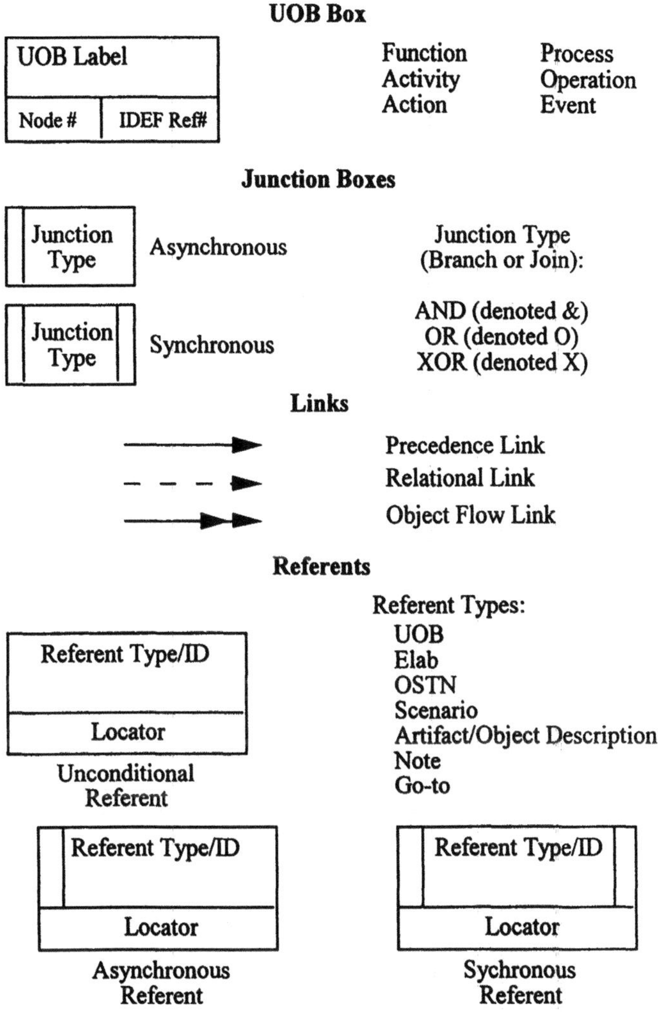

Figure 1 IDEF3 graphical formalism

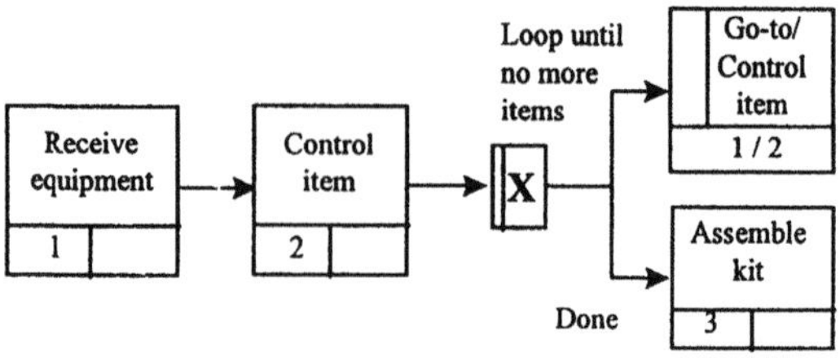

Figure 2 An IDEF3 process

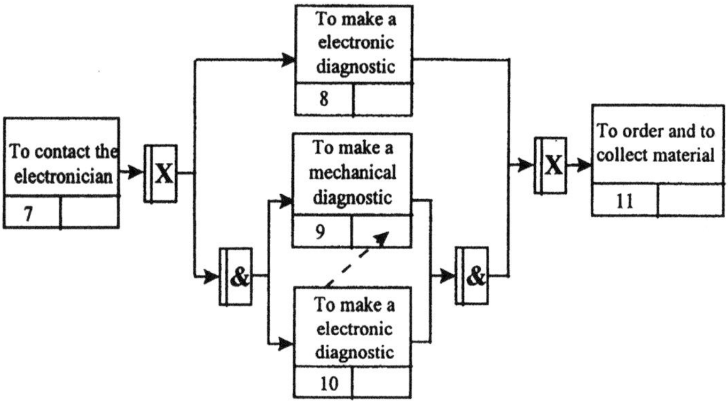

Figure 3 Control item sub-process

Activities, specifying the enterprise functionality, are defined by their name, their inputs (function, control and resource), their outputs (function, control, resource), the function transforming inputs into outputs and the set of all their possible termination statuses, called ending statuses. Function inputs and outputs are object states transformed by the activity, control inputs are objects used but not modified by the activity, control output gives the ending status of the activity upon its completion as well as events generated during the activity execution (if any), resource inputs are objects used as resources and resource outputs give information about the resource status at the end of the activity. Finally, the transformation function of an activity is in most cases defined by procedural statements involving functional operations, i.e. atomic commands (or requests) to be executed by functional entities (i.e. active resources or actors of the enterprise).

In CIMOSA, the control flow of a business process for describing enterprise behaviour is defined by means of a declarative language made of statements of the form:

WHEN (condition) DO action

called behavioural rules in which:

- "condition" is a Boolean expression on events (or triggers) and/or expressions on termination statuses of process steps (denoted by ES(F) where ES stands for "ending status" and F is either an activity or a sub-process)
- "action" is the active part and indicates what to do next (i.e. it indicates the name(s) of activity(ies)/process(es) to execute when the triggering condition is satisfied).

4 BUSINESS PROCESS MODELLING

In this section, common workflow structures for modelling manufacturing environments or industrial systems and suitable for BPR are discussed on the basis of examples of Figures 2 and 3.

4.1 Process start

IDEF3 does not explicitly model process starts. In CIMOSA, domain processes are triggered by nothing but events (one event or a logical combination of event occurrences). Events indicate that a fact has happened requiring some action. The process start rule for the example of Figure 2 is as follows:

WHEN (START WITH event_RE) DO Receive_Equipment

indicating that an occurrence of Receive_Equipment will be executed as soon as an occurrence of event_RE occurs. If the process needs to be started by two events, event_1 and event_2, we can write:

WHEN (START WITH event_1 AND event_2) DO Receive_Equipment

4.2 Sequential structures

Both IDEF3 and CIMOSA provide a sequential structure. In IDEF3, this is modelled by the precedence link, an oriented arrow as shown in Figure 2. CIMOSA provides (1) a conditional sequential rule and (2) a forced sequential rule using the reserved word "any" meaning that whatever the ending status ES(x) of process step x is, the process can go forward as soon as x is completed. Examples for Figure 2 are:

WHEN (ES(Receive_Equipment) = any) DO Control_Item
WHEN (ES(Control_item) = done) DO Assemble_Kit

4.3 Conditional structures (branching)

The conditional (or fork or branching) structure can be expressed either by a fan-out OR or by a fan-out XOR in IDEF3. The fan-out XOR indicates that one and only one alternative from several alternatives will be executed next (See, Figures 2 or 3). A fan-out OR indicates that one or several or even all of the possible alternatives can be executed next and in any order (i.e. k among n alternatives, with $1 \leq k \leq n$). In CIMOSA, only the XOR structure is provided. An example with two alternatives for Figure 2 is:

WHEN (ES(Control_item) = more-items) DO Control_Item
WHEN (ES(Control_item) = done) DO Assemble_Kit

4.4 Spawning structures

The spawning structure is used either in synchronous or asynchronous mode to start n parallel paths ($n \geq 2$) in a workflow. In IDEF3, the fan-out AND junction box is used (Figure 3). In CIMOSA, the parallel operator (&) is used in the action clause for asynchronous spawning as follows for the example of Figure 3:

WHEN (ES(To_Contact_Electronician) = E&Mdiag) DO
To_Make_Mechanical_Diagnostic & To_Make_Electronic_Diagnostic

For the synchronous mode, the reserved word SYNC must be used as follows:

WHEN (ES(To_Contact_Electronician) = E&Mdiag) DO
SYNC (To_Make_Mechanical_Diagnostic & To_Make_Electronic_Diagnostic)

4.5 Rendez-vous structures

The rendez-vous structure is the symmetric structure of the spawning one. It is modelled by fan-in AND in IDEF3 and in CIMOSA by rules of the following form:

WHEN (ES(EAi)=si AND ES(BPj)=sj AND ...) DO EAn)

For instance, for Figure 3 we can write:

WHEN (ES(To_Make_Mechanical_Diagnostic) = MOK AND
ES(To_Make_Electronic_Diagnostic)=EOK) DO To_Order_And_Collect_Material

IDEF3 allows the use of synchronous fan-in AND. However, since all parallel paths being executed must be finished to complete a rendez-vous and proceed further, there is no real need for such a structure for discrete processes because control must wait until the completion of the latest path in the set. The structure can be useful for continuous processes.

4.6 Loop structures

Most workflow languages do not explicitly include a loop statement. Both in IDEF3 and CIMOSA, the loop must be constructed using a branching structure.

In IDEF3, a loop can be constructed in two ways. The first one is to use an XOR junction box followed by two branches: one going back to the loop entry point (an upstream

UOB or junction box in the workflow) and the other one going to the rest of the workflow. The second way consists of using a go-to statement, i.e. using a referent from an XOR junction box onto a UOB or even a connector upstream in the workflow. IDEF3 recommends to use the second way because there is a rule in IDEF3 saying that it is not allowed to have arrows going from right to left in an IDEF3 diagram.

In CIMOSA, the loop is constructed using a branching rule as follows (assuming that EA-1 precedes EA-2 and EA-3 follows EA-2 in the workflow):

WHEN (ES(EA-2) = s2) DO EA-3
WHEN (ES(EA-2) = loop-condition) DO EA-1

4.7 Roll-back structures

Roll-back structures can be found in engineering processes in which the process can execute up to a certain point where it is realised that it goes to a dead-end and that control must be passed to an earlier point upstream in the process to come back to a previous state and resume from that state, ignoring what was previously done after it. Neither IDEF3 nor CIMOSA supports this structure.

4.8 Process end

IDEF3 does not explicitly model process ends. In CIMOSA, the end of a process is specified by rules containing only the reserved word FINISH in the action part. For instance, for the sub-process of Figure 3, we will write:

WHEN (ES(To_Order_And_Collect_Material) = Done) DO FINISH

4.9 Comparison

To compare a workflow description in IDEF3 and CIMOSA, the full process behaviour of the sub-process of Figure 3 follows (for sub-processes, the starting rule has no event):

```
WHEN (START) DO To_Contact_Electronician
WHEN (ES(To_Contact_Electronician) = Ediag) DO To_Make_Electronic_Diagnostic
WHEN (ES(To_Contact_Electronician) = E&Mdiag)
        DO To_Make_Mechanical_Diagnostic & To_Make_Electronic_Diagnostic
WHEN (ES(To_Make_Electronic_Diagnostic) = any)
        DO To_Order_And_Collect_Material
WHEN (ES(To_Make_Mechanical_Diagnostic) = MOK
        AND ES(To_Make_Electronic_Diagnostic)=EOK)
        DO To_Order_And_Collect_Material
WHEN (ES(To_Order_And_Collect_Material) = Done) DO FINISH
```

As the example shows, the complete process behaviour can be expressed in CIMOSA in a formal and non ambiguous way using WHEN DO declarative statements. Because these statements can be expressed as Horn clauses or as Prolog statements, this makes possible the definition of consistency checking procedures based on logics.

5 CONCURRENT PROCESSES

Manufacturing enterprises can be viewed as complex dynamic systems made of many communicating and/or concurrent processes which need to be synchronised.

5.1 Communicating processes

There are two basic mechanisms to handle communicating processes (Hoare, 1985). Communication can be made:

- by message passing
- by shared memory

IDEF3 only provides the relational link (dashed arrow) between two UOBs (of the same process) or the use of a referent (from a UOB of a process onto a UOB of another process) to model interaction between two processes. The concept of messages does not exist in IDEF3.

In CIMOSA, communication between two processes can be established by message passing (through identified channels) or by shared memory at the activity level. This is realised by means of the use of predefined functional operations in the activity behaviour description. Available functional operations are (CIMOSA Association, 1996):

- *request* (a, c, m) to ask for message or data m via channel c from activity a
- *receive* (a, c, m) to receive message or data m via channel c from activity a
- *send* (a, c, m) to send message or data m via channel c to activity a (a is undefined in the case of a *broadcast*)
- *acknowledge* (a) to let activity a know that the message/data sent has been received

The difficulty is to make sure that a given activity sends/receives the right message from the right receiver/sender activity (occurrence identification problem).

5.2 Process synchronisation

In many cases, concurrent processes need to be explicitly synchronised. They are implicitly synchronised in the model by resource access in the case of shared resources and by the flow of objects flowing from one process to another. However, other mechanisms influencing the control flow are required. IDEF3 only considers the flow of objects.

In CIMOSA, processes can be synchronised in three ways:

- *by events*: Events are used to define starting conditions of processes. Events can be timers (clock times), real world happenings (e.g. customer order arrivals, machine break-downs) or requests. Events can come from the outside world (external requests) or from the system itself (raised by resources or by activities). In CIMOSA, an activity can raise an event (issued as a control output) by means of the CreateEvent predefined functional operation. Therefore, a process can generate an event which will start another process which can in turn generate results for the calling process using synchronisation by messages.
- *by messages*: An activity of a running process can send a message (a request) to another activity of another running process using primitives presented in section 5.1.This caters for synchronous or asynchronous communications (e.g. message passing or hand-shaking).
- *by object flows*: Objects created by one activity of a process are used by other activities of the same process or of other processes. Availability of object inputs is defined as pre-conditions to the start of activities.

6 NON-DETERMISTIC PROCESSES

6.1 Non deterministic workflow

Most business process modelling languages and tools are well suited to describe well-structured processes, i.e. processes the control flow of which is completely known and deterministic. Few methods address ill-structured processes. CIMOSA provides two special behavioural rules to cope with these situations:

- The *run-time choice rule*: It is used in a process when there is an exclusive choice (XOR) among several alternatives at a given process step. Only one alternative will be executed as decided by the resource at run-time. Example:

WHEN (EF(EF1) = s1) DO XOR (EF2, EF3, EF4)

- The *unordered set rule*: It is used when a set of process steps, seen as a sub-process, must all be executed next by the same resource(s) but the exact order of execution is not known and will be decided by the resource(s) at run-time. Example (A is the name of the set seen as a sub-process):

WHEN (ES(EF1) = s1) DO A = AND (EF2, EF3, EF4)

IDEF3 only provides the OR junction box (which is semantically ill-defined) to cope with these situations. Extensions to IDEF3 based on these rules and called IDEF3x to deal with semi-structured processes and non-structured activities have been recently proposed (El Mhamedi *et al.*, 1996).

6.2 Exception handling

Another way to deal with non-determinism is to address exceptions. Only foreseeable exceptions can be detected in the model. Other types of exceptions require the intervention of a control system or supervisor monitoring the execution of the model.

CIMOSA provides two mechanisms for exception handling both at the process and activity description levels, namely:

- *time-out mechanism*: the activity or the process is stopped after a fixed elapsed time, its ending status forced to a pre-defined value and control is forced according to that value,
- *watch-dog mechanism*: an exception condition is defined as a predicate and, when it happens to become true during activity or process execution, an event is created and generated which triggers an exception handling process. In this case, the faulty process is stopped and its ending status forced to a pre-defined value.

No such mechanisms are defined in IDEF3. Exceptions can only be documented and recorded as comments in plain text in the template form attached to UOBs.

7 CONCLUSION AND RELEVANCE TO BPR

In this paper, two methods for modelling business processes and enterprise activities have been compared on the basis of the most common structures necessary to model manufacturing environments like in EI or BPR projects. The first one, IDEF3, is a diagrammatic method well-suited as a business user language to capture system requirements. The second one, derived from CIMOSA, is more complete and can be applied at the business level for requirements definition but also at the design specification level as a semi-formal description technique. Because of its graphical notation, IDEF3 has been found ambiguous in a number of situations which can be more properly handled by CIMOSA.

The paper has identified a number of typical situations that any workflow language should address for advanced business process modelling. The BPR analyst can therefore compare modelling languages of interest to him or her against these basic patterns and assess their suitability for his/her targeted BPR project.

8 REFERENCES

AMICE (1993) *CIMOSA: Open System Architecture for CIM*, 2nd edn, Springer-Verlag, Berlin.

Benjamin, P.C., Menzel, C.P. Mayer, R.J. and Padmanaban, N. (1995) Toward a method for acquiring CIM ontologies. *Int. J. of Computer-Integrated Manufacturing*, **8**(3): 225-234.

Bernus, P., Nemes, L. and Williams, T.J. (eds) (1996) *Architectures for Enterprise Integration*, Chapman & Hall, London.

CIMOSA Association (1996) CIMOSA Technical Baseline, Version 4.2. CIMOSA Association, Stockholmerstr. 7, D-71034 Böblingen, Germany, April.

El Mhamedi, A., Schroeder, V. and Vernadat, F. (1996) IDEF3x: A modelling and analysis method for semi-structured business processes and activities. Proc. Fourth Int. Conf. on Control, Automation, Robotics and Vision (ICARCV'96), Singapore, 3-6 December 1996.

Hoare, C.A.R. (1985) *Communicating Sequential Processes*. Prentice-Hall, Englewood Cliffs, NJ.

Mayer, R.J. *et al.* (1992) *Information Integration for Concurrent Engineering (IICE), IDEF3 Process Description Capture Method Report*, AL-TR-1992-0057, Air Force Systems Command, Wright-Patterson Air Force Base, Ohio, 45433.

Vernadat, F. (1993) CIMOSA: Enterprise modelling and integration using a process-based approach. In *Information Infrastructure Systems for Manufacturing* (H. Yoshikawa and J. Goossenaerts, eds.), North-Holland, Amsterdam. pp. 65-79.

Vernadat, F.B. (1996a) *Enterprise Modeling and Integration: Principles and Applications*, Chapman & Hall, London.

Vernadat, F.B. (1996b). Enterprise Integration: On business process and enterprise activity modelling. *Concurrent Engineering: Research and Applications*, **4** (September): 219-228.

François Vernadat is a French and Canadian citizen. He holds a Master degree in Electronics and Automatic Control and obtained the Ph.D. degree in 1981 from University of Clermont, France. From 1981 till 1988, he has been a research officer at the Division of Electrical Engineering of the National Research Council of Canada, Ottawa. In 1988, he joined INRIA, a French research institute in computer science and automatic control. He is currently a professor at University of Metz, France.

His research interests include CIM, database technology and information systems, enterprise modelling, enterprise integration, manufacturing system specification, formal description techniques, Petri nets and workflow enactment. Beside his work on the M* methodology for CIM information systems, he has headed the development of a main-memory database system (DBS/R) and was one of the chief architects of CIMOSA, an Open Systems Architecture for CIM initially developed as a series of ESPRIT projects (AMICE). He is the author of over 100 scientific papers in journals, conferences and books. He is the author of "*Enterprise Modeling and Integration: Principles and Applications*" (Chapman & Hall) and a co-author of the book "*Practice of Petri nets in Manufacturing*" (Chapman & Hall). He has been a co-chairman and invited speaker of several conferences on CIM. He is the European editor for the *International Journal of Computer-Integrated Manufacturing* and serves the scientific committee of other journals and conferences.

Dr. Vernadat served several times as a technical expert for the CIM programme of the Commission of the European Communities (CEC). He is the vice-chairman of the IFAC Technical Committee on Manufacturing Modelling, Management and Control and a member of the IFAC/IFIP Task Force on Architectures for Enterprise Integration. He is a member of the IEEE Computer Society, ACM, and SME.

26
Integrating Infrastructures for Manufacturing a Comparative Analysis

José P. O. Santos[1], João J. Ferreira[2], José M. Mendonça[3]
1University of Aveiro / Dept. of Mechanics / 3800 Aveiro, Portugal
Tel.: +351 34 370892, Fax: +351 34 370953, email: jps@inesca.pt
2,3FEUP-DEEC/INESC/ Rua José Falcão 110, 4000 Porto, Portugal
Tel.: +351 2 2094300, Fax: +351 2 2008487, email: jjpf@inescn.pt

Abstract

This paper presents an overview of the evolution of manufacturing systems integrating infrastructures and standards, ranging from industrial network standards[1] such as MAP[2] and CNMA[3], that essentially support hardware integration, distributed platforms like OSF/DCE[4] and OMG/CORBA[5] promoting industrial applications integration, up to integration frameworks like CIMOSA[6] which try to encompass all the aspects linked to business integration. This overview has a particular emphasis on the integrating infrastructure concepts towards enterprise integration.

The advantages and constraints of a CIMOSA integrating infrastructure implementation using OSF/DCE is also presented and discussed.

Keywords

Integrating Infrastructures, CIMOSA, DCE, CORBA, MAP, MMS

1 INTRODUCTION

Enterprise integration is becoming a must in competitive manufacturing, particularly when improving delivery performance, time-to-market reactivity and flexibility is the issue.

The enterprise's activity integration can be made at three levels:

- Physical systems integration: providing physical interconnection and reliable communication between enterprise resources, through the use of *industrial networks* standards, like MAP an CNMA. Despite industrial network's standardisation efforts, it is hard to promote physical systems integration at all levels.
- Applications integration:making use of distributed software for industrial automation overcoming: hardware, operating systems, networks, programming languages and data base heterogeneity. It is necessary to use a standard and portable packet of software (*middleware*)

---

[1] Open System Interconnection (OSI) based.
[2] Manufacturing Automation Protocol (MAP).
[3] Communication Networks for Manufacturing Application (CNMA), It is not yet a standard.
[4] Open Software Foundation/Distributed Computing Environment (OSF/DCE).
[5] Object Management Group/Common Object Request Broker Architecture (OMG/CORBA).
[6] Open System Architecture for CIM (CIMOSA).

that provides a common application program interface (API). This middleware must provide not only communication services but also a set of common distributed services, to be used in the development and operation of all distributed applications (planning, CAD, CAM, MRP, shop floor applications, sales, etc.).

- Business integration: The business integration necessity comes from the fast market evolution imposing a continuos enterprise adaptation to rapid changing product specifications and production strategies. Despite the technological advantages offered, by present distributed platforms advantages in hardware and applications integration it is important to have fast means to analyse, design and implement new functions and operations, as well as to have production process modelling and simulation [8] integrated with production control and monitoring, as defined by *CIMOSA* architecture.

Within the context above this paper presents and compares the advantages and constraints of, industrial networks (MAP, EPA), distributed platform (DCE), CIMOSA architecture, and attempts to answer to the following two questions:

- Which relationship, advantages, and constraints have distributed platforms and industrial networks?
- Keeping in mind that we want to find a general Information Technology (IT) platform able to support CIMOSA architecture, *which advantages and constraints have industrial networks and distributed platforms to do it ?*

2 INDUSTRIAL NETWORKS [1][2]

Keeping in mind the integration of activities we present some communications protocols [1,2,3], its relationships with the OSI reference model [4], and its advantages and limitations.

Since the 60's computers began to be used in industry for Accounting and others Management functions, Computer Aided Design (CAD), Computer Aided Manufacturing (CAM), Computer Aided Engineering (CAE), Computer Aided Production Planning (CAPP), Supervisory Control And Data Acquisition (SCADA), etc.

The exponential growth of the amounts of information, that had to be exchanged and stored, together with organisational changes imposed by the rapid market evolution (small series of many products), led to the unfeasibility of traditional methods (like diskette, tape, paper and informal communication).

In the 80's better integration of these enterprise functions was attempted (fig. 1) by several suppliers, by using proprietary networks and protocols (fig. 1).

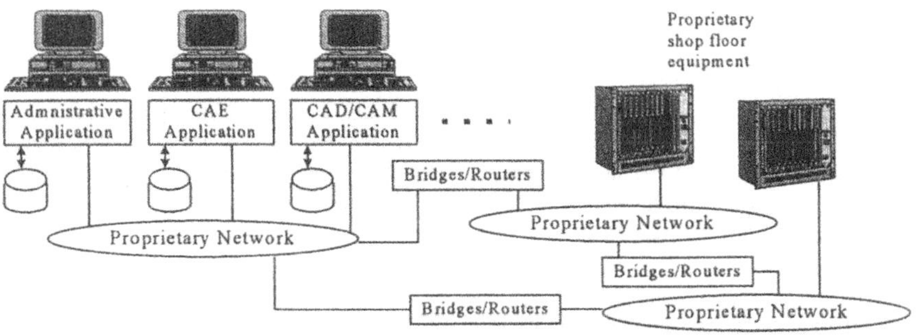

Figure 1 Networking: The first level of integration.

Despite the good results achieved to reach the actual integration of activities in an environment with multiple pieces of proprietary equipment and networks has been a difficult and expensive task. The only way to have a maximum level of integration is the adoption of communication (networks and protocols) standards, like MAP [5] (fig. 2) and CNMA [6], cost issue however still unresolved .

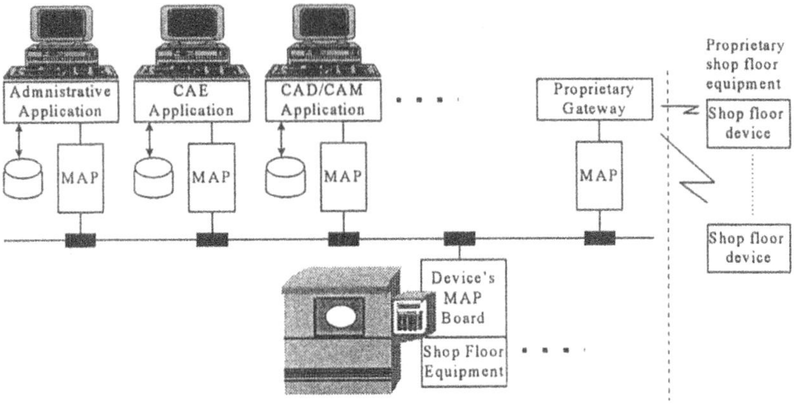

Figure 2 Using MAP interfaces in manufacturing integration.

Both MAP and CNMA standards selected a sub set of pre-existing communication protocols according his specifics needs to some or all, OSI layers. For example MAP (fig. 3) uses the following protocols.

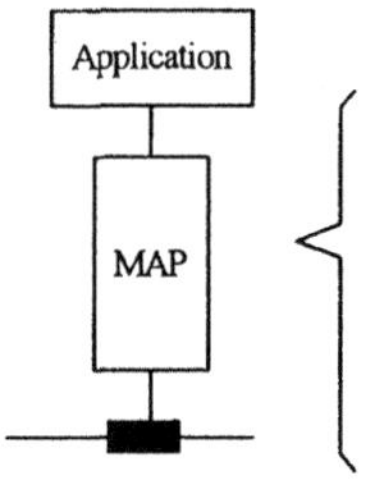

OSI Layer	MAP ver. 3.0
7- SASEs CASEs	ISO-8571(FTAM), ISO-9506/1/2(MMS),...
6- Presentation	ISO-8822(3)/ASN.1 ISO-8824(5)
5- Session	ISO-8327
4- Transport	ISO-8073 (class 4)
3- Network	ISO-8473 (Internet Protocol, connection less)
2- LLC MAC	ISO-8802/3 (type1, class1) / IEEE 802.2 ISO-8802/4 (Token bus) / IEEE 802.4
1- Physical	ISO-8802/4 (Bus, Broadband Cable-CATV)

Figure 3 MAP protocols.

As figure 3 portrays MAP uses many communication protocols which require some processing capacity on every network node. However, many of the manufacturing components have a limited processing capacity, and therefor on the third MAP revision a new communication standard (mini-MAP) called Enhanced Protocol Architecture (EPA) was defined. EPA architecture (fig. 4) is much more easily implemented than MAP because it has only three layers: application, link and physical layer.

Despite MAP and EPA resemblance, a bridge it is necessary to interconnect both industrial networks because they use different modulation signals on physical level and consequently different medium access control in the link layer.

The third level of EPA (application level) interfaces directly with the manufacturing applications.

Thus EPA communication protocols do not provide network and transport services thus lowering message transmission reliability only supports the transmission of a limited number of bits per message. EPA does not provide session services (ex. dialogue synchronism points) neither presentation services (ex. ASN.1).

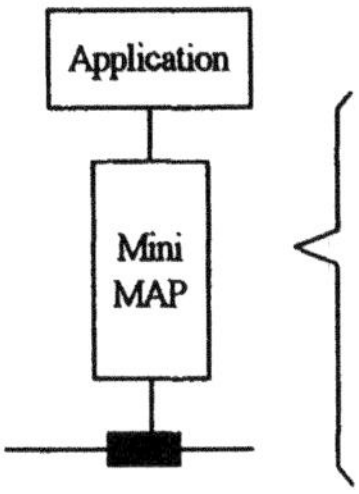

OSI Layer	MiniMAP ver. 3.0
7- SASEs CASEs	ISO-8571(FTAM), ISO-9506/1/2(MMS),...
6, 5, 4, 3	Null
2- LLC MAC	ISO-8802/3 (type1, class1) / IEEE 802.2 ISO-8802/4 (Token bus) / IEEE 802.4
1- Physical	ISO-8802/4 (Bus, Carrierband)

Figure 4 MiniMAP protocols.

From the figure 4, one can say that EPA defines an application layer directly on the top of a Local Area Network.

In order to allow the connecting a device into the network, a computerised controller with a network interface board must be used.

Usually the services provided by both the physical layer and the medium access control sub layer are implemented by the network board firmware. The others layers of MAP, EPA, CNMA, etc. are implemented by software routines which are ,in turn, used by the application software. These routines are executed, in the device motherboard, and they usually access the interface network board services using its I/O address, memory and interrupts.

One can say that the greatest contribution of MAP or EPA to manufacturing activities integration was the definition of MMS (ISO9506/1/2).

However, there are others ways: to transmit information through the factory with an higher security, to synchronise the enterprise activities, to co-ordinate factory resources and others ways to locate orders and products.

3 DISTRIBUTED PLATFORMS

Industrial network standards were just a first step within the scope of the global integrating infrastructure. Distributed platforms are another important step to promote a really distributed manufacturing integration infrastructure.

Nowadays, because of the enormous quantity and diversity of information, as well as hardware and software components generating processing and using it, not only at the shop floor level but also at all enterprise levels an encompassing enterprise integration framework is necessary beyond industrial network's standardisation.

In order to develop distributed systems for industrial automation overcoming hardware, operating systems, networks, programming languages and data base heterogeneity, is necessary to use standard and portable middleware providing a common application programming interface (API) that will easy the, development of applications (planning, CAD, CAM, MRP, shop floor control, sales,...). These manufacturing management applications must co-operate, share information and resources. A middleware structuring providing a set of common and distributed services to be used by all these distributed applications is therefor required.

These distributed platforms must overcome several problems:

- A CIM system is composed of several applications that must communicate and store information in a standard format.
- CIM applications must deal with high number of PCs, minicomputers, mainframes and shop floor devices produced by many different manufacturers.
- Global and distributed information access with appropriate security levels should be provided.
- Machines, computers and applications, may crash. On a distributed platform, if one equipment breaks, another one must replace it while providing exactly the same services (reliability and fault - tolerance).
- Development and maintenance costs of distributed manufacturing applications tend to be high.
- The hardware components of an enterprise must be located wherever they are needed (flexibility).
- An enterprise must be easily upgraded with more and new components to satisfy ever changing requirements (scalability, upgradability).

As to overcome the problems above referred, distributed platforms must provide a set of useful common distributed services, such as:

- Standard communication and distributed service protocols :
- Standard data representation:
- Directory service: a distributed system will have a high number of enterprise resources and the data associated with them to be stored in several places that can change dynamically from one place to another. A directory service can help providing the resources locations and properties.
- Security service providing several services to verify and control the access to shared resources.
- Distributed time service: Each enterprise has many resources with their own clock to synchronise their individual activities. To allow for an integration of enterprise activities one must have means to synchronise them. This is a problem, for example, for distributed applications that care about the sequence of events if each resource has a different notion of the current time.

Nowadays, several computing environments already use distributed platforms as middleware between the applications and specific hardware/networks. *One possibility is to use these platforms also in industrial environments*, alone or together with industrial network protocols (MAP,EPA,..), and in this later case *to choose which is the best distributed platform to reach integration.*

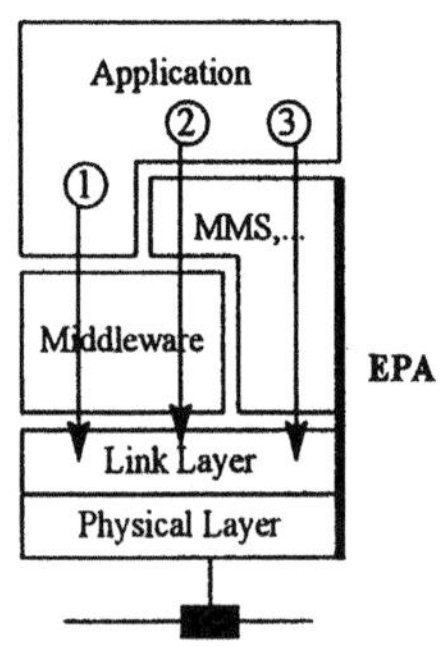

Figure 5 Middle ware and EPA integration.

As illustrated in the fig. 5 such middleware can be used together with existing industrial communication protocols. As the middleware needs some computational capacity we recommend to use, if possible, EPA industrial communication protocol. As figure 5 shows, an application can use the middle ware services alone (1), the EPA application layer services alone (3), or even MMS messages inside the middleware messages (2). In turn, both middleware and EPA application layer will use the EPA' Local Area Network (Physical layer and MAC sub layer) to send data through the physical transmission cable.

The middle ware provides a set of software tools and defined services, that make it much more easy to develop and operate distributed application. The middleware runs on the top of the operating system and networking software of the computerised equipment therefore hiding resources heterogeneity. A great number of funcionalities are provided, that are used by applications and other software, that in turn are used by human operators.

Basically there are two generic middle ware standards, which are not proprietary, available in the market with a set of services, commercial support, and distributed platform products. These are the OSF/DCE and the OMG/CORBA.

These platforms use a client server architecture. When a device makes a request, a server or several servers return responses. To each request made by a software module running on the client side there is an associated software module running on the server side, that processes and returns the right results to the client.

3.1 OSF/DCE

DCE [7,8,9] supports both portability and interoperability by providing the developer with capabilities that hide differences among the hardware, software and networking elements.

As illustrated in fig. 6 DCE has seven components (services), that are described bellow:

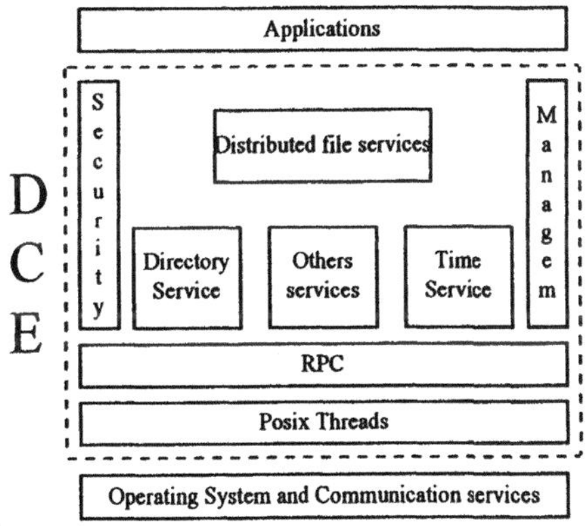

Figure 6 DCE components.

- The DCE Remote Procedure Call (RPC) provides a mechanism for communication between software modules running on different systems (figure 7), much simpler to code than older methods, like socket calls. RPC is probably the most important component of DCE, because all communication between clients and servers are performed using the DCE RPC protocol. On each call there is one or several parameters (data) that are transmitted to the server. The data transmission format used by DCE RPC is called Network Data Representation (NDR). The call parameter (data) are is first of all marshalled using client stubs (software routines). The client stub gathers together all of the arguments (in or out arguments), pack them into a message, and sends it to the server, via network services provided by client runtime software. The server that has the corresponding server side stub receiving the invocation message, pulls out the arguments (unmarshalling), and calls the server procedure. When this procedure returns, the returned parameters (data) are marshalled by the server side stub into another message, which is transmitted back to the client side stub, which pulls out the data and returns it to the original caller.

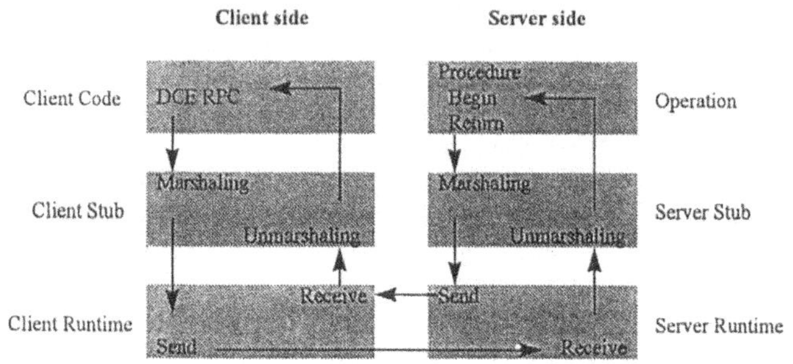

Figure 7 RPC execution steps.

- The DCE Directory Service, has three main components: Cell Directory Service (CDS), Global Directory Service (GDS), and Global Directory Agent (GDA).
- The DCE Security Service, is based on Kerberos (version V) and POSIX 1003.6 ACLs, it provides: user identification and authentication, access authorisation, secure RPC with several authentication options, data integrity and confidentiality.
- The DCE Distributed Time Service, provides services to periodically synchronise the clocks of different enterprise resources, and synchronises them with an external and correct time.
- The DCE Threads: Threads and Processes have a similar behaviour, although they are different. Each process has its own memory and CPU time. Each thread has its own CPU time, but *they share the same memory and resources.* It is possible to have several threads inside each UNIX process sharing the same memory and resources, but not CPU time. DCE middle ware provides the means to have one or several threads for each application on a single operating system task like MS-DOS or inside one UNIX process. DCE threads are based on the API interface for multithreaded programming draft 4 of POSIX 1003.4a.
- The DCE Distributed File Service provides a way to share remote files anywhere in the network. Using DFS remote files appear like local files to the applications.
- The DCE Management Service, can be used by all other services and applications providing services such as: event notification, software installation, distribution and maintenance services, and software licensing control.

Development process

There are several steps to develop a DCE application. First of all the application programmer, using the DCE Interface Definition Language (DCE-IDL), defines a RPC interface, and its associated data types. An interface is a group of operations (RPCs) that a server can perform. The definition of a RPC is similar to a local procedure with hers input and output parameters, except that in DCE-IDL only the calling interface is defined, not the implementation of the procedure side.

The DCE-IDL file is created with a text editor using IDL, a language that uses a C-like syntax to declare the attributes of the interface, its operations, and parameters.

Next, the programmer compiles the DCE-IDL file using the IDL compiler. The compiler produces output either in a conventional programming language, C language, or object code. The output of the compilation consists of a client stub, a server stub and a header file.

The #include (.h) file contains functions prototypes for all the operations and other data declarations defined in the interface, and must be used by both the client and the server.

The programmer then writes the client application code using the RPCs defined in the DCE-IDL file. Thus the client stub code must be linked with this application code, to perform a task behaving like a local procedure call, but that will in fact be performed on the server side of the application.

For the server side of the application, the programmer writes application routines that implement he operations defined in the DCE-IDL file. Thus, the server stub generated by the DCE-IDL compiler must be linked with the server application code. The server stub unpacks the RPC parameters and makes the call to the application as if the client program had called it directly.

3.2 OMG/CORBA

The Object Management Group (OMG) is defining a standard architecture for interoperable distributed software equipment, called Common Object Request Broker Architecture (CORBA). This architecture provides a high level of abstraction and multiple object oriented services to support distributed applications. Within this distributed object oriented platform client server architectures may be built, by allowing clients to issue requests through a local object interface accessing services supplied by the remote object implementation.

They are quite similar to C++ objects where each object's method definition is made on one header file and its implementation on other file. In this client server architecture (fig. 8) the object interface (method definition) is defined and used in the client side of one application. The object implementation (server side) providing the intended service, is used by the application server side, returning the results to the client side. This object interfaces are defined using the so-called OMG Interface Definition Language (IDL), that is a general object oriented language used to create application program interfaces. The OMG/IDL files look likes C++ header files and can be converted on specific languages like as C, C++, e Smaltalk.

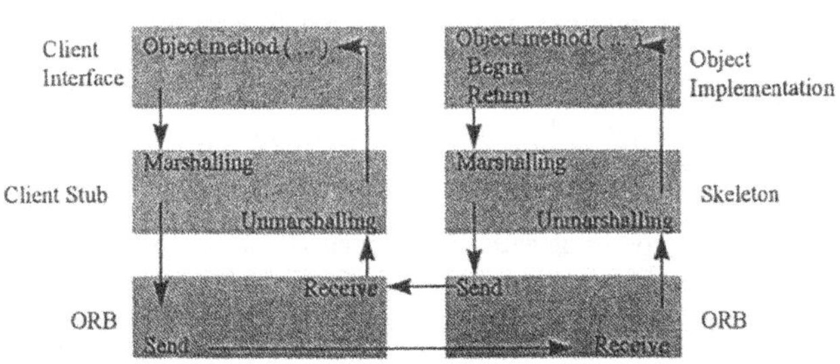

Figure 8 Object Interface / Object Implementation.

This architecture also defines how to locate (the right server) and to access (via ORB) the object implementation, but first of all the application client side must to know the object interface names.

- Interface Repository - IR: This repository keeps the objects interfaces names. The object's name is the only way for an application locate and access its services. This means that each object must write its interface's names on a IR.
- Object Request Broker - ORB: As presented, on a distributed client / server architecture, an application must have information about the interface name of the object's method (service) and a communication mechanism to access it. To this end, the OMG defined the Object Request Broker, which provides a set of high level communication services, being a CORBA main component. It is responsible for the transmission and routing of the client request through the distributed environment to a particular server, and for returning the results to the client. Through the ORB an application can remotely execute procedures as such as execute local procedures. The stub, the skeleton and the header files are generated by IDL specification compiler. The parameters of each object method are first of all marshalled using the client stub code. Each client stub is a piece of code that gathers together all of the arguments (in and out arguments), of its interface, and packages them into an adequate format message that it sends via ORB, to the server.
- Skeleton : On the server side there is a peace of code known by Skeleton Is the Skeleton, when the Skeleton receives a request, via ORB, from the Client, it calls the right object implementation. This object can already exist on a library or can be created by one user.
- Dynamic Interface Invocation - DII: There are two ways to call an interface (an object method): a static way, where the number and type of parameters pre defined at compile time, and a dynamic way where they can change dynamically at run time, namely the interface name it self.

Both DCE and CORBA architecture intended to be a middleware interface providing more or less similar services such as client/server communication, services to handle security and authentication, services to locate servers, and so on. A detailed comparison analysis between DCE and the CORBA, as well an evaluation of a CIMOSA integrating infrastructure implementation using the CORBA architecture are beyond of the scope of this paper.

4 CIMOSA REFERENCE ARCHITECTURE [5][9]

Keeping in mind that we wish to present the advantages and constraints of a CIMOSA integrating infrastructure implementation using either distributed platforms (DCE) or industrial networks (MAP) middleware. We will present a CIMOSA overview [10,11] with particular emphasis on the CIMOSA integration infrastructure requirements.

Despite distributed platforms advantages (hardware and applications integration) it is important to have fast means to analyse, design and implement new operations/processes. It's also important to have production and process simulation [8] integrated with production control and monitoring.

An ideal solution would be to be able to generate automatically from new market requirements for a new product all necessary programs to design and produce it.

An increasingly feasible approach consists of building on an enterprise computer representation (computer models) that would permit at first to simulate all daily enterprise activities. With such environment, simulation of the controlling components of a manufacturing system would be then possible. The new application programs would be first of all tested on the simulation environment and afterwards, using the same integrated environment, loaded on shop floor equipment to be used on the daily production.

To reach this goals CIMOSA provides, not only the integrating infrastructure concept, but also other two inter-related concepts: the Life Cycle (which deals with the development and maintenance of enterprise and product models) and the Modelling Framework (enterprise models development process and reference architectures). CIMOSA is an architecture that aims to integrate, not only industrial hardware and applications but also enterprise activities (business integration).

The need for business integration comes from the fast market evolution which imposes a continuous enterprise adaptation to quick changes in product as well as in process specification, evaluation and implementation.

4.1 CIMOSA System Life Cycle

CIMOSA sub divides this integrated environment into two parts: development and maintenance of enterprise models (*enterprise engineering environment*), and the day to day control and monitoring of the enterprise's activities (*enterprise operation environment*). CIMOSA *enterprise operation environment* provides support to execute and control the enterprise activity through the *particular enterprise operation implementation description model*. Both environments use the CIMOSA *integrating infrastructure*.

A clear separation between the enterprise engineering and the enterprise operations concepts help us to simulate and evaluate new enterprise models without affect the day-to-day production.

4.2 CIMOSA Modelling Framework

To define the enterprise models (*enterprise engineering environment*) many CIMOSA development steps are necessary. The model creation processes will now be presented in reverse order. We begin from the particular enterprise operation implementation description model, used to control and monitor the day to day operations. The development of particular enterprise models is then discussed, up to the reference architecture generic building blocks used to define the enterprise models.

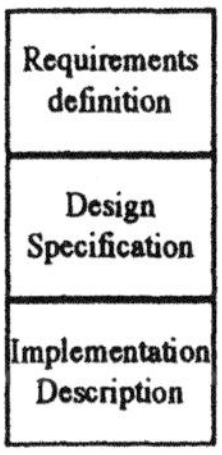

Figure 9 Particular enterprise model Levels.

The particular enterprise operation implementation model to be used in the enterprise operation environment is released from the enterprise engineering environment, more exactly from particular enterprise models.

Particular enterprise models (fig. 10) development is sub divided into three inter-related model levels which correspond to different life cycle phases.

At the so called third level (**analysis of requirements**) based on the market and enterprise goal's, functional specification is built using a user friendly language (particular requirement definition model). This model defines the main events, data and macro operations. However at this level the model does not yet consider issues like the implementation environment or technology constraints.

At the second level (**design**), one allocates the previously defined functional specifications to each ,implementation independent enterprise resource. Using a computer processable language one defines how to synchronise and allocate the several enterprise operations to each resource, how to recover from some resources fails, etc. (*particular design specification model*). This model can already be use for simulation purposes. However at this level the model does not consider yet aspects like specific characteristics of proprietary resources.

At the first level (**implementation**), one defines commercial enterprise hardware and information technology specific assets as well several specific implementation details to enterprise operation (*particular implementation description model*).

During the development of each of these three levels is important to consider several enterprise aspects (fig. 11) necessary to reach enterprise requirements, like:

Function view: which planning, control and monitoring function and system behaviour are required. To develop functional and behavioural enterprise models CIMOSA proposes techniques such as SADT, IDEF0.

Resources view: which equipment, application, databases, human and communication functional entities are needed.

Information view: which data and data bases are required. To develop information models, CIMOSA proposes semantic object-oriented modelling techniques, entity-relationship diagrams, and logical and physical data models. The global information modelling framework is compliant with the three-schema approach proposed by ANSI/X3/SPARC, 1976).

Organisation view: who is the responsible for each: function, resource, information and exception, definition and maintenance.

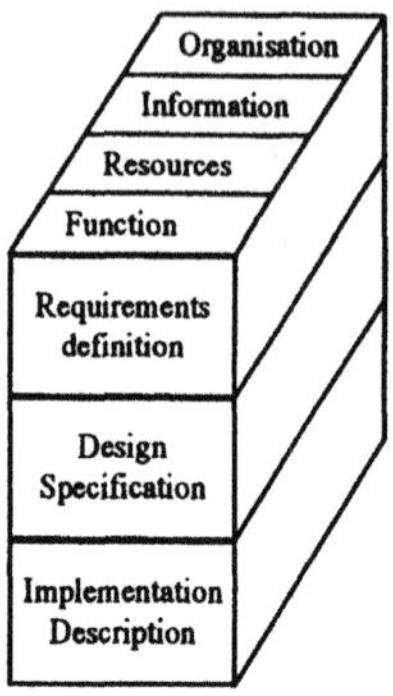

Figure 10
Particular enterprise model (Levels and View).

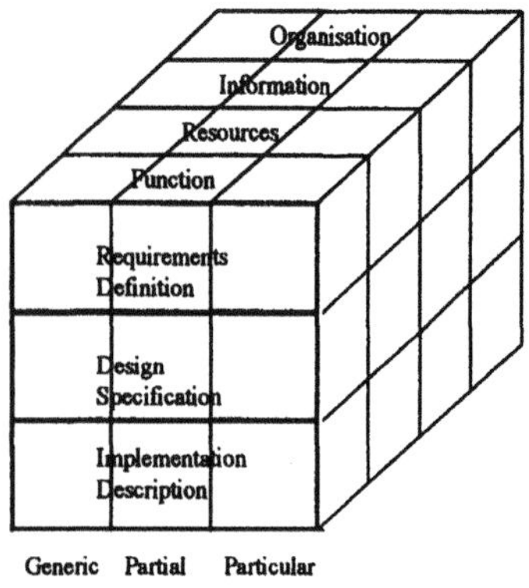

Figure 11 CIMOSA Modelling Framework (Levels, View, Generecity).

Until now we only (almost) speak about **particular enterprise models**, however CIMOSA provides (fig. 11) a partial and a generic set of building block (*reference architecture*) that we must use to get particular models.

CIMOSA **generic** building blocks :The generic building blocks are the basic blocks of CIMOSA architecture. These blocks are called "generic" in the sense that can be reused to build the models of all types of enterprises.

CIMOSA **partial** building blocks uses a sub set of generic blocks organised in set of incomplete models, each set to one type of enterprise. Each set of models can be then refined to get one set of particular enterprise models.

4.3 CIMOSA Integrating Infrastructure [12,13]

In order to allow enterprise implementation description models (and its functional operation - FO) simulation and execution *we need an integrating infrastructure that provides not only, a homogeneous and distributed access to all enterprise resources already achieved by industrial networks (communication infrastructure, abstract manufacturing services) and distributed platforms (global information infrastructure, resources location, data replication , access security, resources synchronisation)*, but also additional services required to execute enterprise CIMOSA models.

CIMOSA integrating infrastructure provides five entities (fig. 12) and each of them provides several inter related services:

- Business entity: is composed by the Business process control, resources, management and activity control part. They provide services to monitor and control enterprise operations. These services are used, for example, to execute and control the released Particular Implementation Description Model.
- Presentation entity:provides machine, functional and human dialogue services. These services are used by both distributed applications and the other entities.
- Information entity: provides data access, data integration and data manipulation services. These services are used by business and presentation entities. It's composed by system wide data services and data management services.
- System management entity: provides all network and system management services. These services are used by the other entities to perform tasks such as : to install and distribute new release of components, fault and performance management.
- Common services entity: It provides a set of common communication (exchange message, priority, transparency,...) and distributed services (naming, security, time,...). These services are used by the other entities.

Management	Information	Business	Presentation
Common services entity			
Information Technology base services (IT)			

Figure 12 Integrating Infrastructure - five entities.

However the better suited CIMOSA information technology base services (IT), must be able to satisfy as much as possible the CIMOSA IIS requirements[14]:

- Distribution: The IT base services ought to provide distributed processing and distributed data access to the manufacturing and information systems enterprise's resources, connecting them via communications networks. The IIS itself ought be distributed.
- Openness: The IT base services are built of many components that must be integrated, working together. To achieve an higher degree of integration each component must (be Open) to provide standard services through a well defined and documented interface.
- Portability and connectivity support: An IT base service should allow a software enterprise model to be easily ported from one device to another without any problem (portability) and easily connected to any machine (connectivity).
- Portability of an integrating infrastructure: The IT base services themselves must be portable to any device, regardless of the devices on which they are presently running: operating system, language, etc.
- Use of existing standards: It is desirable that an IT base service is defined in compliance to standards, as to allow for portability and openness.
- Reliability: The IT base services should provide the means to recovery from faults or failures in the distributed environment components, such as retrying operations, duplication of used components, etc.
- Performance: The IT base services should provide an high performance.
- Migration support: The IT base services itself and the upper distributed system must support a fast time-migration.
- Conformance/compliance with the framework for integrating infrastructure: The IT base services ought to overcome the business, presentation, information, management and common services entities needs referred to above.

5 CONCLUSION

A few "questionable issues " and their answers can be aligned under the form of concluding remarks.

Which relationships, advantages and constraints, have distributed platforms and industrial networks ?

As presented, one can use both distributed platforms and industrial networks, each of them separately or together, to reach a greater integration of enterprise's activities (both engineering and operation environments). Thus, the question is therefore which is the better suited to each environment.

Both distributed platforms and industrial networks provide network communication facilities. As presented in figure 5, one can separately use (1,3) both middleware on the same device or even sends MAP-MMS messages as DCE-RPC parameters (2). However, MAP industrial network is better suited to be used in the shop floor environment than distributed platforms because:

- Most of distributed platforms capabilities, such as dynamic resources location, shared resources, dynamic task allocation and others, will not be useful to the shop floor activities because shop floor resources, physical layout, as well as material flow specificities do not allow it.
- Only few and expensive shop floor resources are able to support (i.e. run) distributed platform middleware.
- The MMS messages and the file transfer protocol provided by MAP are the most useful facilities to the shop floor activity integration.

Distributed platforms are, on the other hand, obviously better suited to improve the performance of Computer Aided Engineering Environment (CAEE) activities than industrial networks.

So within the context above, one can say that DCE and MAP middleware are complementary.

Table 1 CIMOSA IIS requirements

CIMOSA IIS requirements	MAP	DCE
Distribution	poor	very good
Openness	good	good
Models portability	poor	good
IIS portability	good	very good
Standards	*de jure*	*de facto*
Reliability	poor	very good
Performance	good	heavy
Time-migration support	poor	good
IIS compliance		
-business	--	--
-presentation	ASN.1	--
-information	poor	poor
-management	--	combined
-common	poor	OSF/DME
services		very good

Keeping in mind our objective of finding a general IT platform able to support CIMOSA architecture, which advantages and constraints have industrial networks and distributed platforms to do it ?

MAP, DCE, and CIMOSA IIS requirements have been presented, thus we are in position to compare them and conclude, which advantages has DCE over MAP to support CIMOSA-IIS. To implement the CIMOSA IIS one could use distributed platforms or MAP. However, as presented on the table 1, MAP will not provide most of CIMOSA IIS requirements which implies that we would have to implement them on the top of MAP. On other hand, as DCE is much more complete it frees the CIMOSA IIS from having to implement most of IT basic services.

Some DCE constraints are nevertheless recognised:

- DCE is only a standard de facto not a standard de jure.
- DCE only provides a Network Data Representation (NDR), which is not enough to support the required presentation entity services.
- DCE itself does not provide management services; however OSF also defined another complementary standard (de facto), the Distributed Management Environment (DME), which combined with the DCE would provide most of the CIMOSA management entity requirements.

One can conclude that DCE provides most of the Common entity services and most of the general IIS requirements. However it does not provide any CIMOSA business or presentation services.

6 REFERENCES

[1] G. Messina, G. Tricomi, "Software standardisation integrating industrial automation systems",Computers in industry 25/(1994) 113-124.

[2] S. A. Koubias, G. D. Papadopoulos, "Modern Field bus communication architectures for real-time industrial applications", Computers in industry 26/(1995) 243-252.

[3] Fred Halsall, "*Data Communications, Computer Networks and OSI*", Addison-Wesley, 1988.

[4] Etienne Lapalus, Shu Guei Fang, Christian Rang, Ronald J. Van Gerwen, "Manufacturing Integration", Computers in industry 27 (1995) 155-165

[5] Zoubir Mammeri, Jean-Pierre Thomesse, "Résaux locaux industriels", Eyrolles, 1994.

[6] Esprit Project 7096, "CIME Computing Environment Integrating a communications network for manufacturing applications", (CCE-CNMA), 1994.

[7] OSF, "Introduction to OSF DCE", Prentice-Hall, 1994.

[8] Thomas W. Doeppner Jr., "A Technical Introduction to DCE for Managers", OSF/DCE Conference, London, UK, April 25-27, 1995.

[9] Harold W. Lockhart, "OSF DCE Guide to Developing Distributed Applications", McGraw-Hill, 1994.

[10] Esprit consortium AMICE, "CIMOSA Open System Architecture for CIM".

[11] Kurt Kosanke, "CIMOSA-Overview and status", Computers in Industry 27(1995)101-109.

[12] IFIP transactions Towards world class manufacturing, "Architectures for integrating manufacturing activities and enterprises", T.J.Williams, et al, 1993.

[13] IFIP transactions Towards world class manufacturing, "CIMOSA: Integrating the production", Jackob Vlietstra, 1993.

[14] CEN Report/TC 310, "CIM Systems Architecture- Enterprise model execution and integration services (EMEIS)"-"Annex E: CIMOSA 's Framework for integrating Infrastructure", CR 1832:1995, Feb.1995.

PART ELEVEN

Systems Integration I

27

Designing an Internet based Teleproductics System

João Paulo N. Firmeza, Fernando M. S. Ramos
Electronics and Telecommunications Dep. – University of Aveiro
Campus Universitário de Santiago, 3800 Aveiro, Portugal
Phone: +351 34 370500 Fax:+351 34 370545
E-mail: firmeza@inesca.pt & fmr@inesca.pt

Abstract

The improvement of development tools for the Internet potentates the appearance of new challenging fields of applications to the global network. One new promising field is Teleproductics that comprises the remote access and control of industrial facilities through the Internet.

This paper describes in detail how current Internet development tools can be used for this application, and results from a research project that is focused on the development of virtual learning laboratories for the teaching of mechanical engineering.

Keywords

Teleproductics, Internet, Client-Server architecture, WEB, Distance Learning

1. GENERAL CONCEPTS

Teleproductics is a recent concept that refers to the interaction among industrial facilities, telecommunication networks and computer systems.

Applications of Teleproductics are very wide; examples of applications are virtual laboratories for teaching mechanical engineering (enabling cost reductions due to the overcome of multiplication of similar facilities), and the remote control of self-contained highly automated industrial plants.

A teleproductics system enables the remote operation and control of one (or several) industrial process(es) by using a telecommunications network.

Figure 1 illustrates the concept of Teleproductics.

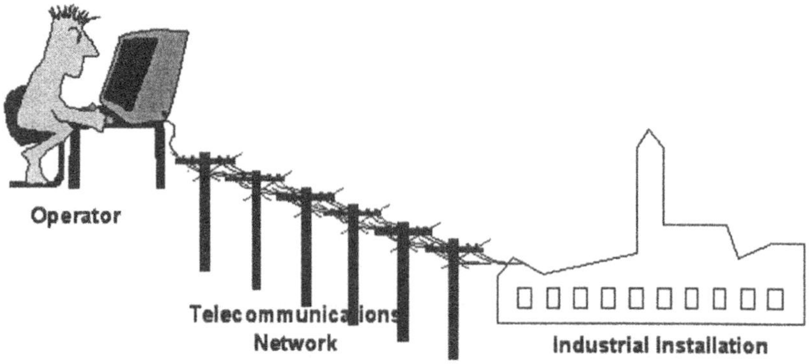

Figure 1 The Teleproductics concept.

This article describes an implementation approach for such type of system, and includes a discussion on the tools and strategies that will be used and a proposal of architecture.

2. IMPLEMENTATION APPROACH

The implementation foreseen is a prototype of a Teleproductics system, and is intended to demonstrate the capabilities and feasibility of the general concept.

The system to be designed will have the following main modules:

- User interface to the facilities of the industrial process.
- Telecommunications network.
- Industrial site interface.
- Industrial process (viewed in the perspective of the mechanical parts and the electro-mechanical control interfaces).
- Feedback circuits (return video channel and control data transfer).

Figure 2 Simplified diagram of the proposed system.

2.1 The control and management Interface

In this proposal the user interface will be built mostly over an HTTP (*HyperText Transfer Protocol)* client. This client acts like a container for the user graphical interface.

The HTTP client, an ordinary World Wide Web (WEB) browser, will be the support for the display of all the operations related to the direct control of the industrial process. The WEB browser should be upgraded with standard add-ons available as plug-in parts, to enable the implementation of the features required.

The use of a browser as the host for the user interface improves the platform independence of the management module. The system on the operator's side will work under any platform containing an installed browser compatible with the system's specifications.

The target environment for the client will be Microsoft Windows. This option is quite obvious, since it's one of the most common operating environments, with lots of existing commercial applications and development tools. Another strong point of this option is the low cost associated with the solution: Windows is a very cheap operating system that runs on relatively low cost platforms. We must also emphasise its good performance, good graphical user interface and good network component integration.

Resulting from these previous considerations, on the operator's side, we have the architecture represented on figure 3:

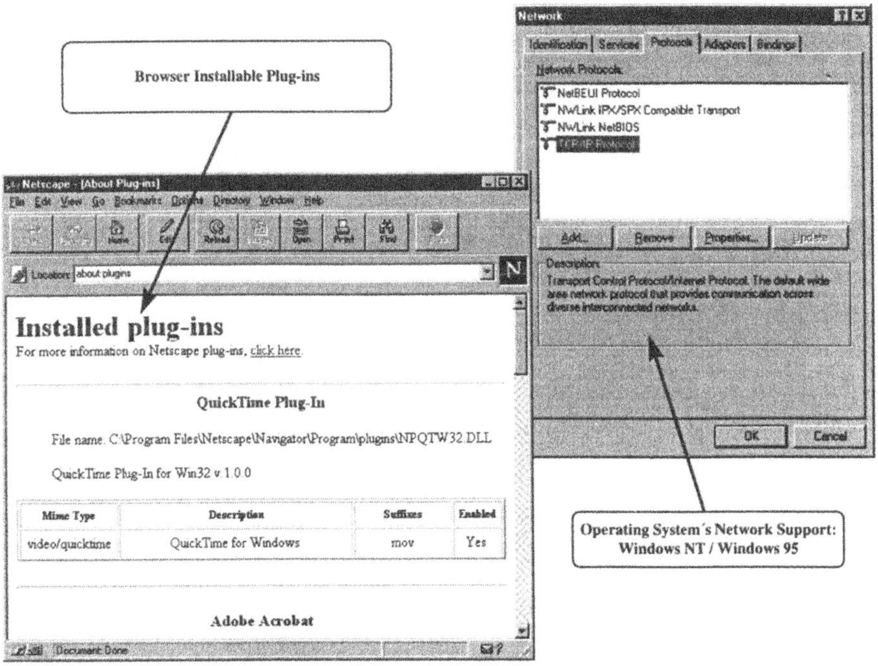

Figure 3 The user interface. Implementation over a WEB *browser*.

2.2. Network

On the proposed system, the network acts like one of the main components, since all the main modules (hardware & software) will relay on the presence of a network.

On the network physical layer, no restrictions will be imposed, allowing the use of any of the common physical topologies, as far as the TCP/IP (Wildev, 1993) stack could be used.

The TCP/IP protocol is the protocol used on the Internet, and, as a consequence, the protocol used by WEB browsers and servers.

From the implementation point of view, network support will be automatically assured by the use of an operating system with native TCP/IP support.

In addiction to that, the real system will have some other network related considerations, which will be focused later on.

2.3. Interface with the industrial unit

In this proposal of implementation, the industrial unit will be based on 4 main modules:

- **Network Input/Output module:** This module is mostly supported by a WEB server. The management interface will be supplied to the operation and management system by this component. This module will be also the one who will receive the data supplied by the operator at the user's interface.
- **Configuration Database management:** This module handles the data archive responsible for the storage of the system's current configuration.
- **Actions generation module:** This module will be dependent of the specific industrial process. Its function will be the generation of actions and commands for controlling the configurable industrial process's components. It will get its input data from the database and, by interpreting the stored data, produces commands to control the physical components of the production unit.
- **Industrial process monitoring system:** This module will be responsible for feedback data transmission to the management and operation units. Depending on the specific system, there may be several types of feedback data produced by several systems' elements, such as video signals, alarms, reports, production ratios, etc.

Figure 4 shows the basic topology of the interaction system with the industrial unit:

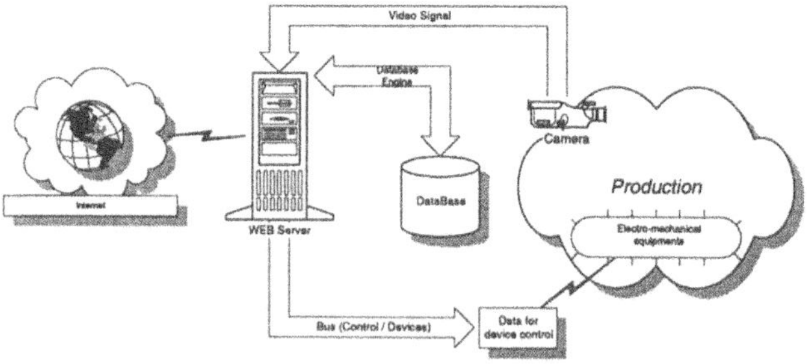

Figure 4 System's topology: Interaction with the industrial process.

2.4. The production process

The production process may be composed by a set of electro-mechanical elements controlled by one or several central control units (CCUs). The central control units must be able to receive the configuration data transmitted from the server for each of the process's elements, and keep track of its current states.
Figure 5 illustrates the concept of a process (a production line) controlled by a central control unit.

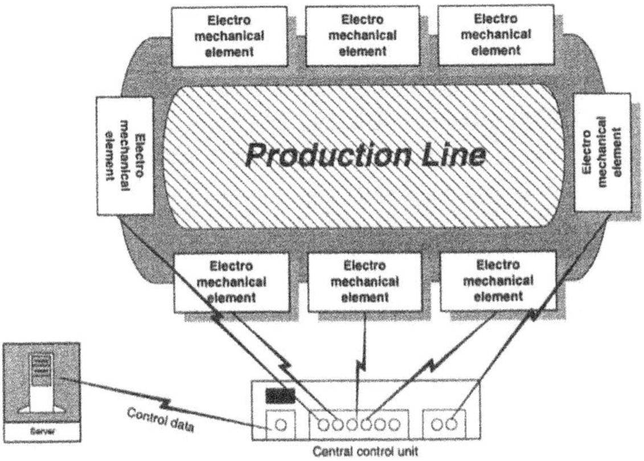

Figure 5 Diagram of a production line controlled by a central control unit.

3. THE WEB BASED CLIENT/SERVER SYSTEM

All the implementation issues related to the operation, control and management of the remote industrial process would be built using and HTTP client/server topology (Elbert/Martyana, 1994) (figure 6).

Using this approach there will be an HTTP client (a WEB browser) on the operator's side, and an HTTP server on the industrial process side.

From this point on, we'll refer to the system on the operator's side as the *Management and Operation Unit* and to the system on the production side as the *Production Unit*.

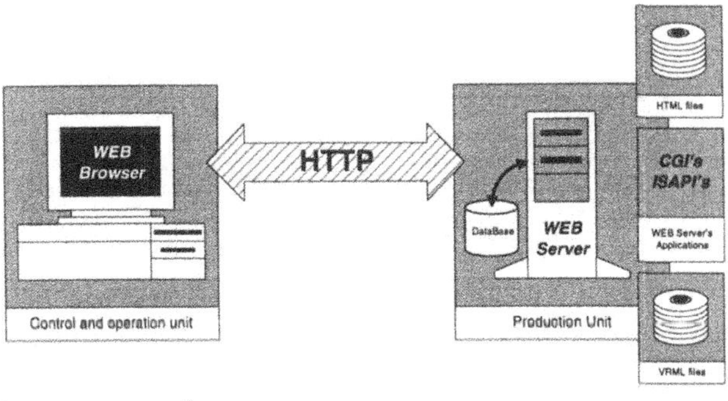

Figure 6 Client-server approach.

3.1. The management and operation unit: HTML[1] Interface

The implementation of the user's interface will be built over a WEB browser. This enables the use of any platform on which the chosen browser is supported. This is one of the greatest advantages of using this approach, since it allows this important system's element to be architectural independent. If we choose, for instance, the *Netscape Navigator*[2] as the browser to be used we can install management clients on any of it's supported platforms (Windows 3.1, 95 and NT, Macintosh, SunOS, HP-UX, Solaris, Linux, BSD and AIX).

In the proposed implementation, some browser add-ons or plug-ins will be used and, as a consequence, the browser should support them on each of the used platforms.

To minimize the effect of such a limitation and in order to be able to support less powerful platforms, the system will always offer, as a complementary option, standard HTML (December/Ginsburg, 1995) versions of the user interface.

The user approach to the management and operation interface will be built following a top-down perspective using this logical strategy:

1. As a first approach to the management and operation system, the operator will access a tridimensional model of the whole production facility, allowing him to "walk" inside it just by using a pointing device (usually the mouse). This is done using a plug-in for the browser, which enables it to render VRML (*Virtual Reality Modelling Language*) (Pesce, 1995) objects. If Netscape Navigator is 3.0 or Microsoft Internet Explorer[3] 3.0, the VRML is used, VRML extension is a standard. The use of this tool enables the representation of models of the reality just through interpreting simple text script files. Just navigating inside the 3D space of the process model, the operator easily reaches each one of its components (figure 7).

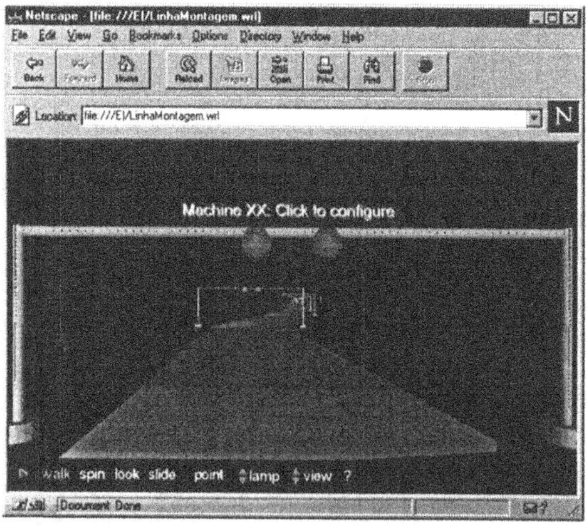

Figure 7 Example of layout of a production line implemented in VRML.

[1] HTML: Hypertext Markup Language.

[2] *Netscape Navigator* is a registered trademark of *Netscape Communicatins Inc.*

[3] *Internet Explorer* is a registered trademark of *Microsoft Corporation*.

2. By clicking on the representation of the process's components, an HTML page will be displayed on the browser, allowing the user to get information about the corresponding component; furthermore forms to input the data necessary to configure the component may be displayed (figure 8). This data will be stored on the server, located on the production unit, as will be discussed later.

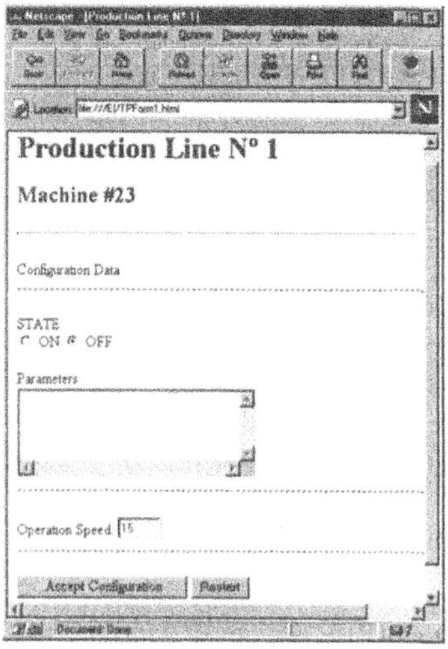

Figure 8 Example of an HTML form for data entry.

3. After the completion of the data acceptance and storing processes, the operator returns to VRML mode, enabling the configuration of other components (figure 9).

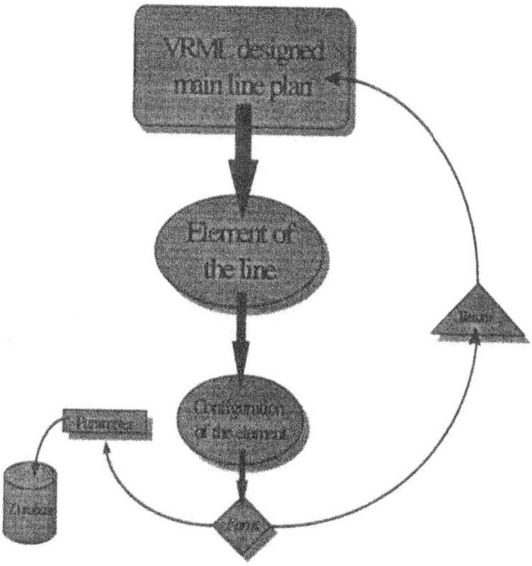

Figure 9 Top-Down diagram of the operator's actions.

3.1.1. The use of the Java language as a tool to enhance the user interface

The Java[4] (Thomas, 1996) language allows writing small applications (applets) which are executed locally by the browser (as far as the Java specifications are supported on it).

The main advantages of using Java are its object orientation, which makes it strongly modular, and its multi-platform characteristic: one same Java application can be executed on any machine independently of its platform, because is the browser that translates the pseudo-code files that arrive via HTTP to the machine's specific code.

In addiction, Java has an open API (*application programming interface*) (Yellin, 1996), which easily permits the enhancement of its basic possibilities.

Focusing on the system under discussion, the use of the Java language can be very interesting to solve issues related to the enhancement and increase of functionality of the user's interface.

As examples of this, Java will enable the implementation of an on-line help system embedded on the pages consisting on small mouse sensitive applets, or the integration of some multimedia gadgets.

3.1.2. Additional modules to the management and operation unit

Coexisting with the WEB browser, the operator needs some additional tools. On such a system two other modules are quite important: the alarm monitoring module and the videoconference client module

The function of the alarm monitoring module is to receive the alarm messages generated on the industrial unit's server whenever abnormal or error situations occur on the industrial process.

The videoconference client module will receive and display the video signals generated by the monitoring cameras, if installed on the industrial unit.

3.2. The production unit: A WEB Server based system

At the production unit side the interface to the Internet will be implemented through a server (figure 10). This server, that we'll call WEB server, will provide access to text files containing HTML and VRML scripts. Additionally the server will be used to store specific applications required to manage the processing of data received from the configuration forms, and other applications such as monitoring and communication tasks. Some of those applications are CGI[5] applications, which are launched by the HTTP service.

Coexisting with the services and associated applications, this machine will be the host for a database. The function of this database is the storage and management of the configuration data of the components of the production unit. On the proposed implementation this database will be based on a SQL Server (ANSI X3.135, 1989), for increased functionality and modularity of the system.

To enable the coexistence of all those services, the operating system to be used on this server must be a multitask operating system, with integrated network support.

The proposed solution is based on *Windows NT Server* (Solomon, 1996), which includes the entire network functionalities required embedded. Furthermore its version 4.0 includes also the IIS[6] which includes an HTTP server with support for CGIs, direct access to databases and

---

[4] Java is a registered trademark of *Sun Microsystems*

[5] CGI: Common Gateway Interface

[6] IIS: *Internet Information Server*, from *Microsoft Corporation*

an API called ISAPI[7], that allows to write applications for the WEB service, with some advantage over CGI applications as we'll discuss later.

Using the *Microsoft SQL Server* for *Windows NT* the connectivity and access to the database problem is also solved.

On this machine there will be also a service to process and transmit/receive the control data from/to the central control unit, and also a service to handle the feedback video signals, including its transmission to all the connected management and operation units.

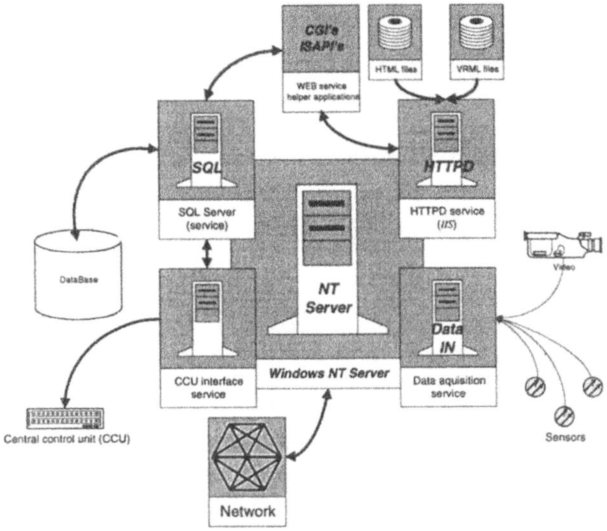

Figure 10 The server of the production unit.

3.2.1 The HTTP service on the WEB Server

This service has the responsibility to deliver text files to the clients, some of which containing HTML pages and data entry forms and others containing VRML scripts.

This service can also launch small applications at client's request. These applications are called CGI applications, since they follow the CGI norms. In the specific case of the adopted server (MS IIS), is also possible to have another type of applications: ISAPI's which are DLLs[8] loaded by the HTTP service itself, having some advantages over CGI applications, as we'll discuss later.

Figure 11 illustrates data flow on an HTTP server, emphasising its request/response orientation.

[7] ISAPI: Internet Server Application Programming Interface

[8] DLL: Dynamic Link Library

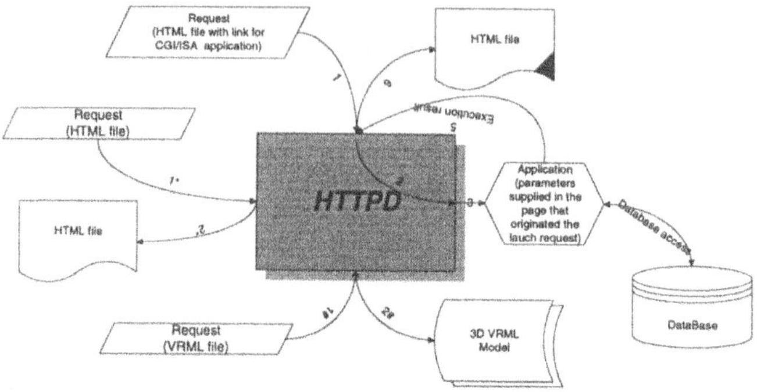

Figure 11 HTTP server functional diagram.

3.2.2 Data received from the management and operation units: acceptance and validation

The processes involved in data acceptance and validation are executed whenever a form containing configuration data is submitted to the HTTP server.

The main actions taken after data reception are the following:

1. **Testing data for validation:** When a data set is submitted to the server, sometimes it isn't considered as a valid set, because its originating form fields weren't correctly filled, or just because that data combination is not valid on the current system's state. (some actions over an element may be dependent of the state of other elements)

When such situation occurs, the server must generate an HTML page containing explicitly the error description and the situation which generated it.

This validation process is done by a CGI application (or an ISA) launched by the server after the form reception process. As result a response page will be generated and sent to the user.

2. **Database updating:** When a valid set of data is received on the server, the database will be updated with these data (on the table corresponding to the element which owns the data).

On a similar way, this is also done using a CGI or ISA, capable of receiving the data set as input parameters, and write it to the correct database table.

This application will interact with the database engine (the SQL Server on our implementation) using the ODBC (Open DataBase Connectivity) from *Microsoft* (figure 12).

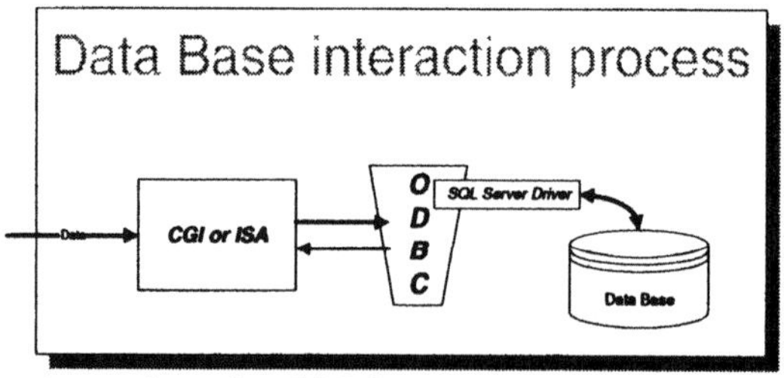

Figure 12 Data Base interaction using *ODBC*

Figure 13 illustrates the data validation and acceptance processes.

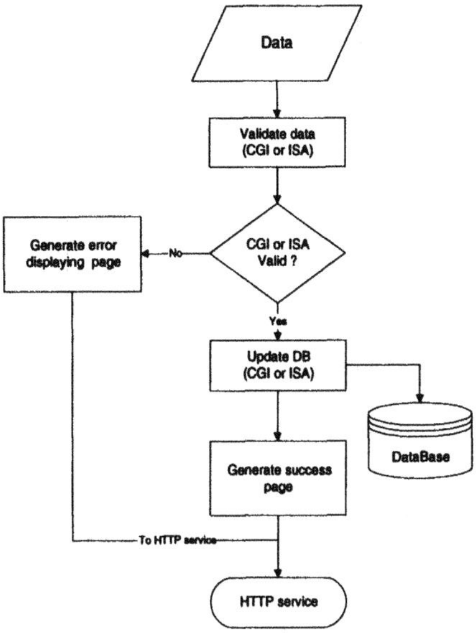

Figure 13 Flow diagram of the data validation and update process.

3.2.3 The Database

3.2.3.1 DATABASE ARCHITECTURE

The database will be used to store the actual status of all the configurable components of the industrial processes. It is composed by several tables each one representing one type of component, and include fields to represent all the element's possible states and properties.

The several tables should contain relationships between them, since there will be several cases where a change on a table's field implies the change of other fields from other tables.

3.2.3.2 THE SQL SERVER

The use of SQL Server provides to the system the power and efficiency of a database engine, with high performance and giving the possibility of distributing the stored data to its clients.

This functionality can bring great benefits to the system, essentially in terms of performance and flexibility. By enabling the independence between the database and the other services on the server, it will be possible to minimize the overhead associated with the database management, because the database may be implemented on a dedicated machine if the number of accesses per time unit is considered to be high (figure 14).

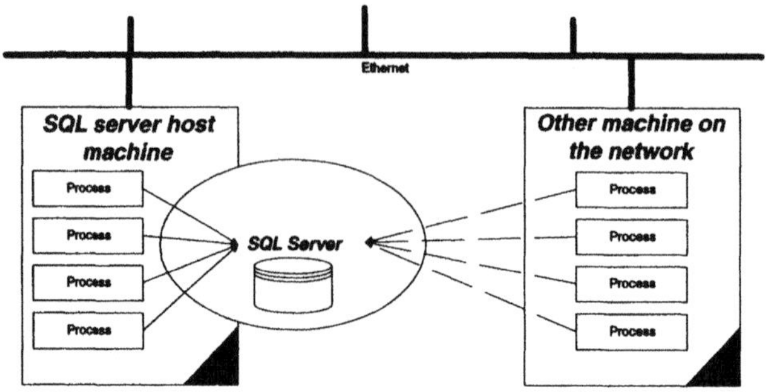

Figure 14 Data base distribution using the *SQL Server*

3.2.4. CGI Applications and ISAPIs

The WEB server to be used should be able to execute CGI applications and ISAPIs. In the described system will be used *Microsoft IIS* and the server applications will be ISAs[9]. The advantages of this type of WEB server applications over CGI applications are, basically, the following:

- Good language support for writing ISAs: it's possible to write an ISA using C++ language. The most recent C++ compiler - *Microsoft Visual C++ 4.2*, has even built in classes to easily use the ISAPI (figure 15).

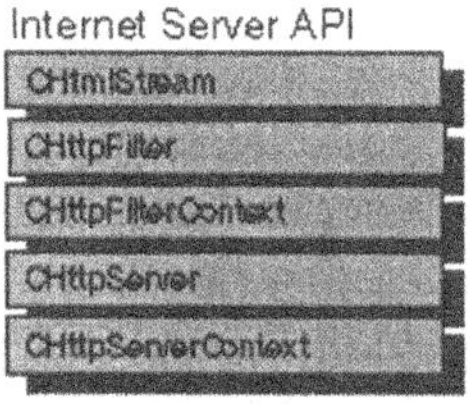

Figure 15 ISA related classes of the *Microsoft Foundation Classes*

- ISAs run on the same address space as the HTTP server, and they can access all its resources (what does not happen with CGI applications).
- ISAs have less overhead than CGI applications, since new processes don't need to be created, as with CGIs. This doesn't waste processing time necessary to do interprocess communication, since everything runs on the same process.

[9] ISA: Internet Server Application

A WEB client calls an ISA using the HTTP server the same way as with CGI applications. The following example shows how this invocation is performed:

```
CGI:
http://plant.corporation.com/CGIexample.exe?Param1
?Param2?.......?ParamN

ISA:
http://plant.corporation.com/ISAexample.dll?Param1
?Param2?.......?ParamN
```

3.2.5. Complementary services to HTTP Service

3.2.5.1 MONITORING DATA ACQUISITION SERVICE

The monitoring data acquisition service receives the data from the monitor devices placed over the industrial process and sends it to the connected management and operation units. This service can easily be divided into two sub-services: one responsible for the acquisition, treatment and transmission of video signals from the cameras placed on the line, and other service responsible for processing the alarms generated when an abnormal event occurs on the industrial process.

Analysing these two cases separately, we have:

Video acquisition service: This service receives the coded video signals from the hardware acquisition boards installed on the server (or on other machines placed over the LAN). and sends this data to the connected management and operation units.

Two different implementation approaches rise when it comes to the implementation strategy of this service:

On the first one (the most elegant one), the service, when requested by a client on the network, starts sending back the acquired video coded using a stream format. Under these conditions, whenever a client requests the visualisation of the data from one of the cameras, using its video-conferencing client style software module, the service will start the transmission of real time data to the client[10].

On the second case, the service stores data using stream format multimedia files. This format must allow the direct display of the video image using the client's browser, with real time download and visualisation from the network.

Using this strategy, one file will be created and used for each of the cameras installed on the industrial site, and with a cyclical re-write mechanism each of the files are substituted with more recent ones when their sizes reach a limit value. This process doesn't allow really real-time visualisation but, on the other side, it allows the inclusion of video frames within HTML pages, enabling the visualisation of a video window in the operator's browser client area.

As a tool to implement this solution, *Microsoft* already provides a native stream format and a SDK to develop applications for it. This format is called *Active Movie* and enables the encapsulation of most of the common multimedia formats in a single stream format[11]. This is a data type independent of its contents and allows the transport over several network physical topologies and logical protocols. It supports the video formats .AVI, QuickTime[12] and

---

[10] This process will be similar to several video-conferencing systems used on the Internet on a generalized way, as for instance, the system CU-See Me

[11] **.asf** format: *ActiveMovie Stream Format*

[12] *QuickTime* is a registered trademark of *Apple Computer*

MPEG, repackaging them to optimise its transport. With the .asf format, data reproduction will start immediately after the first packet of bits arrival to the client.

To correctly use this format it's necessary to install the correct plug-in for the browser in use on the management and operation units.

Alarm processing service: This service keeps listening to one of the serial ports of the server, where are supposed to arrive the alarm signals from the central control unit (figure 16).

When an alarm signal is received, a process is initialised that updates an alarm log file and also broadcasts the alarm message to all the connected clients.

The clients, which should have their own alarm monitors active anytime, will receive a message describing the error situation and the state of the element who produced it. This enables an immediate reaction to the alarm event.

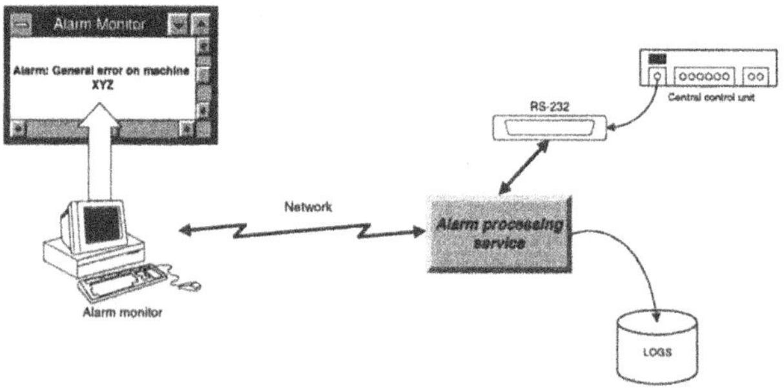

Figure 16 The alarm processing service

3.2.5.2. COMMAND GENERATION AND CENTRAL CONTROL UNIT (CCU) INTERFACE SERVICE

This process interacts with the CCU. It is a Windows NT service that runs on the server and will be always listening to changes on the configuration's database.

Whenever a database modification is performed by one of the WEB server applications (ISAs), a message will be sent to this service to notify that change. This message will have a specific format, containing a field with the element's identification affected by the modification. With this information, the service will request a database query to the SQL Server to read the corresponding element's information. Using that data, the system will construct a frame and will transmit it to the CCU, using a server's RS-232 communication port.

Figure 17 shows the processing flow of this service:

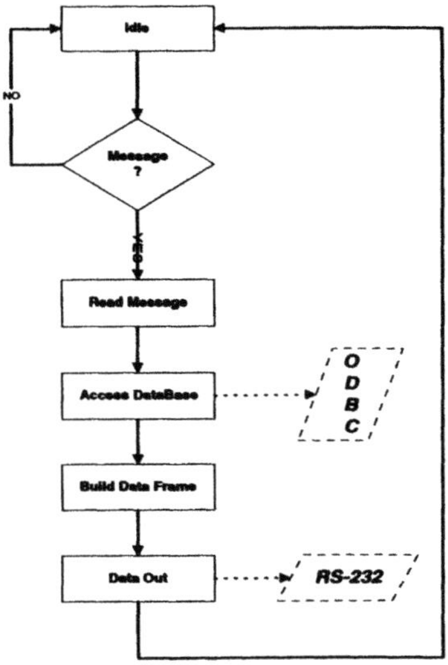

Figure 17 Flow diagram of the command generation and UCC communication service.

4. THE CENTRAL CONTROL UNIT AND THE ELECTRO-MECHANICAL ELEMENTS

The central control unit (CCU) will be a block based on a rack of communication ports (RS-232) and a message delivery module (figure 18).

On the proposed implementation two RS-232 ports will be present to communicate with the industrial unit's server, and a number of other ports to communicate with each of the controllable electro-mechanical elements of the production site (as many as the number of these elements).

The first two ports are used, respectively, to receive the control frames produced on the server and to transmit alarms generated by the industrial process to the server.

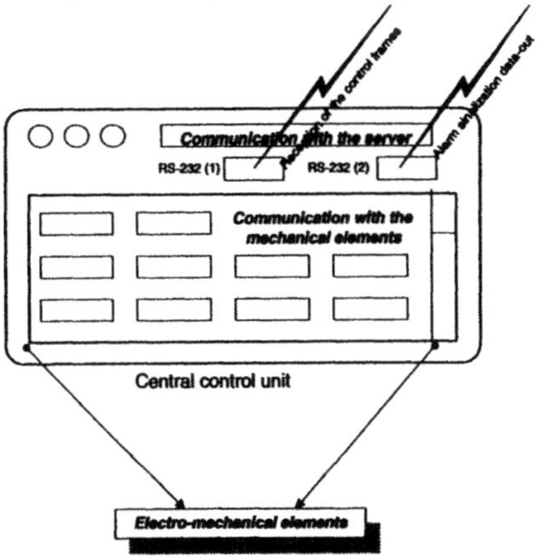

Figure 18 The central control unit

As a straight forward approach to the CCU's implementation it will be used an off-the-shelf PC equipped with a rack of serial ports, that will run a service dedicated to distribute the arriving control frames to the appropriate communication ports and transmitting all the received alarm signals to the alarms port. Such a system is easily implemented over a machine running the MS-DOS operating system.

A more elaborated solution will be the implementation of the CCU with dedicated hardware, by building a black box encapsulating all the unit's functionality.

5. NETWORK CONSIDERATIONS

As discussed above, a system like this uses the protocols and tools usually used on the Internet (browsers, WEB servers, the TCP/IP stack, etc). This means that such a system could be used in pure Internet environment, being possible to have several management and operation units distributed all over the world. Although this can be conceivable, this scenario presents some limitations: security (Karanjit, 1996) becomes a very important matter (new mechanisms should be taken in consideration), and performance, because bandwidth is limited and uncertain. This can cause us some new problems when it comes to real-time processing events on the system. The proposed implementation takes these constrain into account, and, because of that all the system features remain functional under Internet conditions, although system's responsiveness may be affected.

These disadvantages may be minimized if an Intranet topology is used: the support network is kept isolated from the rest of the Internet. In this scenario it will be easier to manage the available bandwidth, and restrict the security issues.

6. CONCLUSION

Current information and communications technologies enable the development of new applications that may contribute to substantial cost reductions: staff costs through the possibility to remotely control processes and investment costs through the capability to increase resource's sharing.

Teleproductics is a potential example of this, because it will enable remote access to industrial processes.

This paper discussed how current technologies could be used to implement a Teleproductics system based on the Internet. The system under development is focused to implement virtual laboratories for teaching mechanical engineering (Minoli, 1996).

7. REFERENCES

ANSI X3.135 (1989) Database Language SQL. American National Standards Institute, ANSI X3.135 (ISO 9075)

December, John and Ginsburg, Mark (1995) HTML and CGI unleashed. SAMS.net Publishing corp, Indianapolis.

Elbert, Bruce and Martyana, Bobby (1994) Client/Server computing: Architecture, Applications and Distributed Systems Management. Artech House, Boston-London.

Gesling, James; Yellin, Frank and the Java Team (1996) The JAVA application programming interface. Addison-Wesley.

Minoli, Daniel (1996) Distance Learning Technology and Applications. Artech House, Boston-London.

Pesce, Mark (1995) VRML: Browsing and building cyberspace. New Riders, Indianapolis.

Siyan, Karanjit and Hare, Chris (1996) Internet Firewalls and Network Security, 2nd Ed. New Riders Publishing.

Solomon (1996) Microsoft SQL Server 6.5 Unleashed, 2nd Ed. SAMS Publishing.

Thomas, Michael D. (1996) JAVA Programming for the Internet: A guide to creating dynamic, interactive Internet applications. Research Triangle Park, Ventana.

Wildev, Floyd (1993) A Guide to the TCP/IP protocol suite. Artech House, Boston-London.

8. BIOGRAPHY

João Paulo N. Firmeza, is currently working on his MCs degree, at the Electronics and Telecommunications Department, of the University of Aveiro. His work for the last two years has been related to telematics, with special focus on teleproductics, covering research, and development of a teleproductics prototype system. This work is being done with the collaboration of the telecommunications systems and services section of INESC and supported by JNICT's (*Junta Nacional de Investigação Científica e Tecnológica*) Praxis XXI program.
As an electronics and telecommunications engineering student on Aveiro's university, three years ago, he developed a Windows based ISDN videophone, and since than his work has been focused on computer based telematic systems.

Prof. Fernando Ramos, holds a PhD on Telecommunications from the University of Aveiro, Portugal, where he is currently Professor and head of the Department of Electronics and Telecommunications. He is also leading an R&D group at INESC in the area of telecommunications systems and service. His current research interests include open distance learning and telesurveillance systems based on ISDN, GSM and VSAT.

28

A Model of Methodology Need for AQAL Production System

Ziqiong Deng
Narvik Institute of Technology
Lodve Langes gt. 2 • 8501 Narvik • Norway
tel: (47) 76 96 62 15 • fax: (47) 76 96 68 10
e-mail: Ziqiong.Deng@hin.no

Abstracts

This paper firstly makes a survey on computer integrated manufacturing, lean production, "keiretsu", and agile manufacturing. Then, definitions for production system (product-alliance oriented corporation), four-external-factors and four-internal-factors of production system, and AQAL (Attractive, Qualitative, Agile, and Lean) production system are given. Finally, a model of methodology need for AQAL production system is given as well.

Keywords

Enterprise integration, lean production, agile manufacturing, production system, methodology

1 INTRODUCTION

From the end of the seventies, the market have evolved. The level of competitiveness increased enormously. For meeting the challenge, enterprises have made great efforts to improve their system performances. Aiming at this, many new concepts and philosophies have been widely discussed and developed for aiding the improvement. The concepts and philosophies of computer integrated manufacturing, lean production, agile manufacturing, the "keiretsu", etc., are typical examples.

Facing the rich and varied new concepts and philosophies, it is necessary to make a survey of them to see what they will benefit the enterprises and what should be done further for the enterprises.

For this purpose, firstly, there is need to analyze the basic external market requirements to the enterprises; secondly, there is need to study the basic internal performance factors that the enterprises have to work with internally for meeting the external market requirements; then, it is necessary to derive a model to explain the need of methodologies which might be used by enterprises to aid them for improving their internal performances. In what follows in this paper, above questions are discussed in details.

2 FROM COMPUTER INTEGRATED MANUFACTURING TO AGILE MANUFACTURING

From the end of the seventies, the level of market competitiveness increased with the appearance of new competitors coming from new geographical area. Then customers has become more and more exigent, searching cheapest products with appropriate quality, in shortest lead time.

At this time, on one hand, computer aided techniques for various functions in an enterprise such as Computer Aided Design (CAD), Computer Aided Planning (CAP), Computer Aided Manufacturing (CAM), Computer Aided Quality Control (CAQ), and Production Planning and Control (PP&C) were installed in advanced enterprises forming many automation islands in one enterprise. On the other hand, the development of information technology such as local-area network communication and database technology in industries enables the possibility of enterprise-wide integration of those automation islands. Therefore, people started to further strengthen the research and implementation of enterprise integration - Computer Integrated Manufacturing (CIM) for meeting the challenge of the increased level of competitiveness.

CIM involves in the internal integration within a single enterprise. A definition of CIM given by U. Rembold indicates that CIM combines the activities CAD, CAP, CAM, CAQ, and PP&C in one system (Rembold, 1993).

Parallel to the CIM research and implementation, almost at the same period, some groups had been researching to find the reason of why the Japanese can produce products so good with so cheap price. Some of them found that the Japanese "weapon" is the Just-In-Time (JIT) approach. It provides for the cost-effective production and delivery of only the necessary quantity of parts at the right quality, the right time and the right place, while using a minimum amount of facilities, equipment, materials and human resources. It is dependent on the balance between the supplier's flexibility and the user's flexibility (Voss). Here, people notice the necessity of integrating the host enterprise as an user with the supply enterprises to obtain the just-in-time production.

In 1990, J. P. Womack, et al published their book, "The Machine that Changed the World", in which they first time define the Japanese production philosophy with a terminology "lean production" which is now well known around the world. They affirmed that the Japanese "weapon" is a new set of ideas and a new type of production system - lean production system, not just a dedicated new approach for material flow. A comparison was made by the authors of that book indicating that the lean production is lean because it uses less of everything compared with conventional mass production - half the human effort in the factory, half the manufacturing space, half the investment tools, half the engineering hours, and half the time to develop new products. It implies that the *leanness* is the central factor of a lean production system. The elements of lean production were figured as (1) running the factory, (2) designing the car, (3) *coordinating the supply chain*, (4) *dealing with customers*, and (5) managing the lean enterprise (Womack, 1990). This book revealed that the lean production system is a system not only involving the production in a host enterprise, but also the production in the *supply chain*, i. e., a cluster of supply enterprises, and the activities in the "*dealing with customers*" companies, i. e., a cluster of selling agencies.

Since 1989, a Japanese word "keiretsu" migrated to America, Europe, and even Asia. The most common meaning of "keiretsu" is something close to the English words "link," "connect," or "series." The general subject is related to Japan's industrial "families" (the various kinds of keiretsu). A horizontal keiretsu is a group of very large companies with common ties to a powerful bank. A vertical or pyramid keiretsu is made up of one very large company and hundreds or thousands of small companies subservient to it. A good example would be a large manufacturer like Toyota with twin vertical keiretsu within its total "group": one pyramid producing goods and the other pyramid distributing and selling those goods (Miyashita, 1996).

In 1991, R. N. Nagel, et al wrote a famous two-volume report entitled "21st Century Manufacturing Enterprise Strategy" in which they first time raise the concept of agile manufacturing and virtual corporation. They emphasize the *agility* as the central factor for an agile manufacturing system just as that Womack emphasizes the *leanness* as the central factor in a lean production system. Nagel indicated that Factory America Net is envisioned as a utility that will enable inter-enterprise integration and the virtual corporation. Inter-enterprise integration will tie a corporation into its entire supply base and its customers. It will tie together relevant pieces in each company in pursuit of a specific market opportunity (Nagel, 1991).

From lean production to "keiretsu" and to agile manufacturing, from the integration point of view, people today all emphasize that the success in world market is dependent not only on the successful integration within one single enterprise, but also on the successful integration of the cluster of enterprises.

As mentioned above, the agility is emphasized in the report "21st Century Manufacturing Enterprise Strategy" as the central factor of an agile manufacturing system, and agile manufacturing is affirmed to be the 21st century manufacturing (Nagel, 1991). As well, in the book "The Machine that Changed the World", the leanness is emphasized as the central factor for a lean production system, and the lean production is affirmed also to be "the standard global production system of the twenty-first century" (Womack, 1990).

What will be the 21st century manufacturing? lean production, agile manufacturing, "keiretsu", or something else? From enterprises' standpoint, facing so many different new concepts and philosophies, how could enterprises absorb the advantages of those new concepts and philosophies to benefit their business? This is a matter which the author of this paper wants to discuss in the following sections.

3 PRODUCTION SYSTEM - A CLUSTER OF ENTERPRISES

As discussed above, we want to stand on the enterprises' standpoint to observe and absorb the advantages of new concepts and philosophies for benefiting the enterprises. From such a point of view, the market requirements are the main concern to the enterprises. None of enterprises would produce all components of a product, i.e., a product will be produced by a cluster of enterprises. Therefore, our discussion is focusing on a cluster of enterprises which are product-alliance oriented and we would *call such a cluster of enterprises a production system.*

Based on such an understanding to the production system, we would say that a production system is a chained enterprises based on a product alliance (or say a product corporation) which includes: (1) an *input subsystem* for inputs of raw materials and/or semi-finished products (supply enterprises), (2) an *output subsystem* for outputs of finished products for marketing (selling agencies), and (3) a *transform subsystem* for transforming the inputs into the outputs (host enterprise) (see Figure 1).

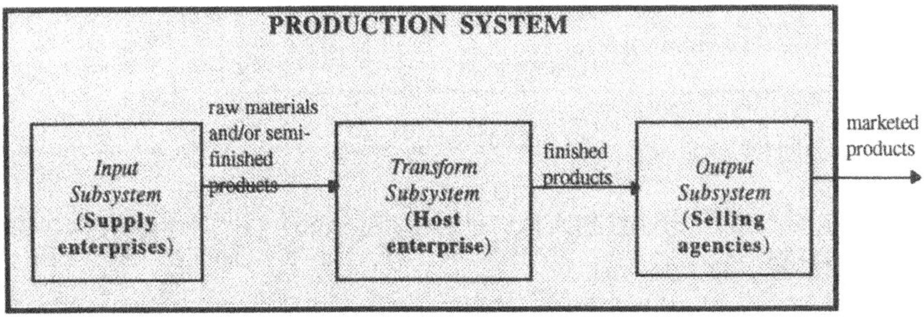

Figure 1 Concept of a production system

In Figure 1, the *host enterprise* plays a central role of producing products via transforming inputs into outputs. *A cluster of supply-enterprises* serves the host enterprise and playing a role for supplying host enterprise with raw materials and/or semi-finished products.

A cluster of selling agencies serves also the host enterprise for selling products and feedbacking the market requirement information to the host enterprise.

For example, a host enterprise can be a car manufacturing company having its final product - car (See Figure 2). One of the "raw materials or semi-finished product" supply-enterprises of the car company may be a car radio company, and one other "raw materials or semi-finished product" supply-enterprise of the car company may be a steel metallurgy company, ..., etc.. Those enterprises are composed to be a production system (or a corporation) which is shown in the shaded part in Figure 2. With the same principle, we may also say that the *car radio* itself is a final product, the car radio company itself is a host enterprise and possesses its own "raw materials and/or semi-finished products" supply-enterprises such as the LSI (large-scale integrated) chip supply enterprise, ..., etc., forming one other production system (See top-left part of Figure 2). Again we may also say that the *steel* is a final product and the steel metallurgy company is a host enterprise possessing its own "raw materials and/or semi-finished products" supply enterprises forming another production system (See bottom-left part of Figure 2). Furthermore, the LSI chip manufacturing company can also be a host enterprise possessing its own "raw materials and/or semi-finished products" supply enterprises forming its production system, ..., and so forth.

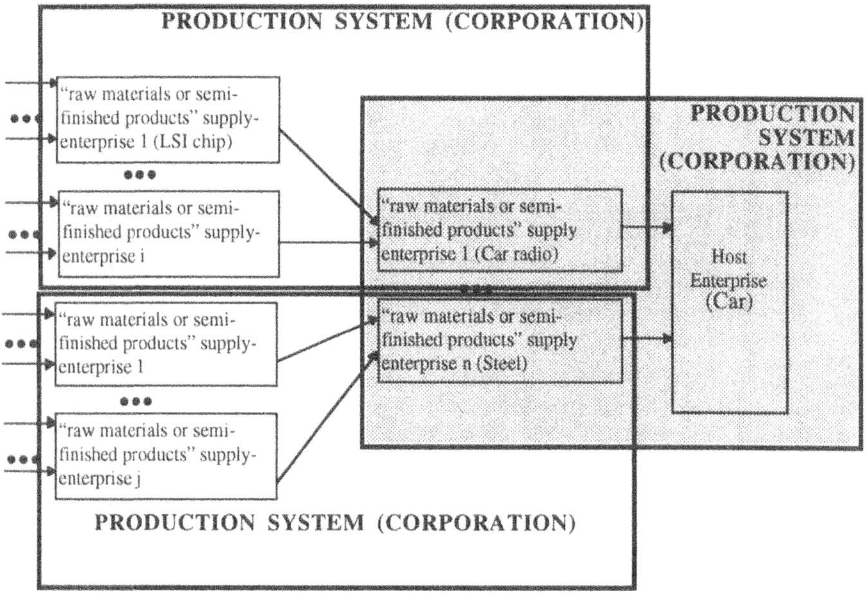

Figure 2 Hierarchy of corporations (production systems)

4 ATTRACTIVE, QUALITATIVE, AGILE, AND LEAN PRODUCTION SYSTEM

4.1 External market factors of a production system

The market concerns only products. Whoever can supply market with products which meet the customers' keenness and necessity with lowest price, satisfied quality, higher deliver speed of required quantity, who will take the competitive advantage. Therefore, we may say the external market factors which should be always considered by a production system are as follows:

- Keenness and necessity of customers (K),
- Quality required (Q),
- Speed and quantity (S),
- Price (P).

The market keeps changing all the time, but the market factors are invariant. Say it in details, the keenness and necessity of products from customers (K), the quality of products required from customers (Q), the deliver speed of products and the quantity required from customers (S), and the price (P) keep changing all the time, however, at any time, the four market requirement factors, K, Q, S, and P, which a production system has to analyze are invariant.

Those four external market factors are fundamental and indispensable to a production system (see Figure 3). A production system has to behave best on all the four external market factors but not only one, two, or three of them.

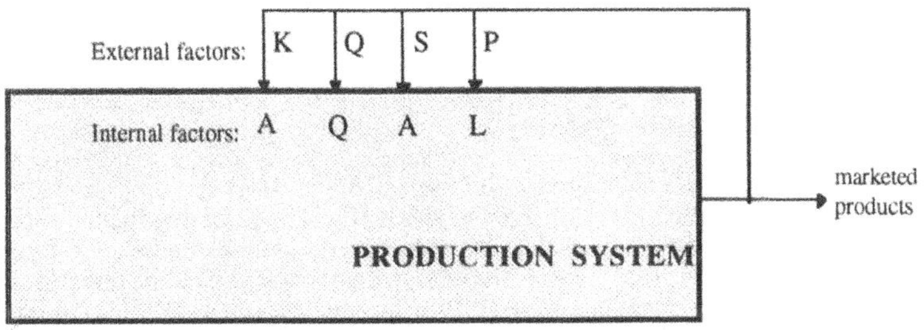

Figure 3 External and internal factors of a production system

4.2 Four internal working factors of a production system

As discussed above, a production system has to analyze and meet four external market factors. Obviously, inside a production system, all its internal functions and organizations should work together optimally and dynamically to improve its internal performances for meeting the changing external requirements.

What will be the common internal performances which the internal functions and organizations of a production system should work together to improve them for meeting the four external market requirements? It is easy to inference those performances as follows:

- *Attractiveness* of products (A),
- *Quality* of manufactured products (Q),
- *Agility* of production (A),
- *Leanness* of production (L).

The first external requirement, *keenness and necessity of customers*, keeps changing, a production system internally should often work to innovate its products making them as attractive (A) as possible to the customers.

The second external requirement, *quality required from customers*, keeps changing, a production system internally should often work to improve its quality performances (Q) such as quality design, quality manufacturing, quality control, and quality monitoring to meet quality requirement from customers.

The third external requirement, *deliver speed of products and quantity of products required from customers*, keeps changing, a production system has to improve its agility (A) to meet the changing external speed and quantity requirements.

The forth external factor, *price*, should be lowered again and again due to the market competitiveness, so a production system should internally try every best to decrease its production cost making its production as lean (L) as possible.

Obviously, there exists a one-to-one relationship between the four external KQSP factors and the four internal AQAL factors as shown in Figure 3 and Table 1.

Table 1 Relationship between the four-external-factors and the four-internal-factors

		External factors			
		Keenness and necessity of customers	Quality required	Speed and Quantity	Price
Internal factors	Attractiveness and necessity of products	x			
	Quality manufactured		x		
	Agility			x	
	Leanness				x

4.3 AQAL production system

To summarize above discussion, a production system has to meet all four external market requirements, but not only one, two, or three of them. Therefore, the production system has often to improve its all four internal performances, A (attractiveness), Q (quality of manufactured products), A (agility), and L (leanness), but not one, two, or three of them. In other words, the market requirements of KQSP factors are indispensable. Accordingly, the internal AQAL factors are also indispensable to a production system. Therefore, when we stand on the enterprises' standpoint, we say a production system could not only pursue to improve its agility, or leanness, or quality, or attractiveness, but all its AQAL performances. From this sense, we would like to say a production system should be an AQAL production system.

Of course, as an existing production system, sometime the product attractiveness is its main concern, so its internal functions and organizations would only concentrate their efforts to solve the product attractiveness problem. With the same principle, sometime the main concern of a production system may be the quality, or the agility, or the leanness, then their efforts may be concentrated on solving only one of those problems: the quality problem, the agility problem, or the leanness problem. However, when we try to plan and organize a new production system, then all AQAL factors should be taken into account. Therefore, in general looking, a production system should possess capabilities to solve all four indispensable AQAL problems.

5 METHODOLOGY NEED FOR AQAL PRODUCTION SYSTEM

We mentioned that the internal functions and organizations of a production system should cooperatively work together to improve their system to be AQAL. A further question is how they can make their system attractive, qualitative, agile, and lean. The answer is that they need methodology aid including tools. So in this section, we would first analyze the methodology need, then derive a model of methodology need to explain the scope of the need.

For this purpose, we need first to analyze the relationship and cooperative manner of the internal functions and organizations as follows.

5.1 Cooperative manner of Internal functions and organizations of AQAL production system

We have mentioned that a production system consists of a host enterprise, a cluster of supply-enterprises, and a cluster of selling-agencies (See Figure 1). If we take a bit detail looking into the production system, we may say each enterprise or selling-agency internally has following three parts of functions and organizations (See Figure 4):

- *Product-design (D) or Selling (Sel),*
- *Product-manufacturing (Mfg) or Service (Srv),*
- Supervisory control and *management (Mgt).*

From Figure 4, we see that, inside a host enterprise, three functions and organizations, Mgt, D, and Mfg, coordinate each other closely. Then functions in the host enterprise corporate also with the functions in the supply enterprises and selling agencies.

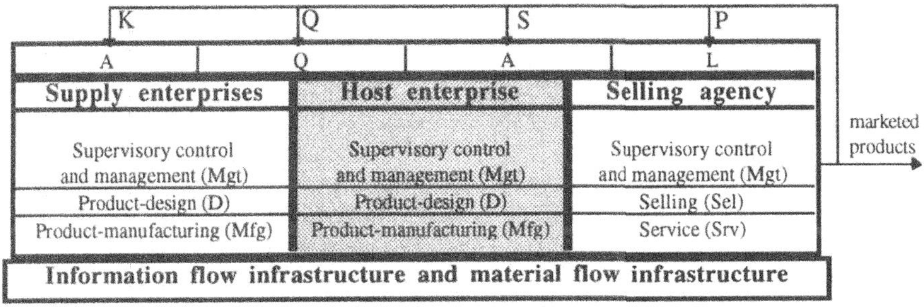

Figure 4 Relationship of internal functions of a AQAL production system

For example, if the system wants to improve their product attractiveness, then a work team may be organized. The team may be composed by persons from product-design organizations of host and supply enterprises, selling organization of selling agencies, manufacturing organization of host and supply enterprises, and management organizations of host and supply enterprises and selling agencies. The cooperative manners would be the team work manner with a concurrent engineering working principle. The working purpose is to improve the product attractiveness. With same principle, if the system wants to improve their leanness, or agility, or quality performance, or two, three, even all four of the AQAL performances, the work team can similarly be organized and working in team work manner with concurrent principle.

Incidentally, for enabling the team work and concurrent engineering, an efficient information flow infrastructure is needed making the team members distributed around a wide area being able to communicate each other of getting right information, at right time, to right location (See bottom of Figure 4). Also, for supporting agile, lean material (product) flow which starts from supply enterprises, to host enterprise, to selling agencies, and finally to customers, using for example JIT principle, an efficient material flow infrastructure is also needed to guarantee of getting right materials (products), at right time, with cheapest cost (See also bottom of Figure 4). In this paper, we will not further discuss the matters of infrastructures for brevity and will only discuss the methodology need for team work in a bit detail in what follows.

5.2 A model of methodology need for AQAL production system

From Figure 4, we mentioned that in a production system, no mater which function has the duty to work jointly for solving AQAL problems. So in Figure 5, we use one cube to represent host enterprise, another cube to represent supply enterprise. For brevity, we did not depict the selling agency cube in the Figure.

In Figure 5, we use three dimensions in a cube to represent three functions of an enterprise such as *product-design* function, *product-manufacturing* function, and supervisory control and *management* function. Each function has its duty to improve system performances of attractiveness, quality, agility, and leanness. Therefore, we mark every dimension with *attractiveness, quality, agility*, and *leanness*. In a sense, Figure 5 is a re-drawing of Figure 4. However, Figure 5 is better for expressing the methodology need than Figure 4.

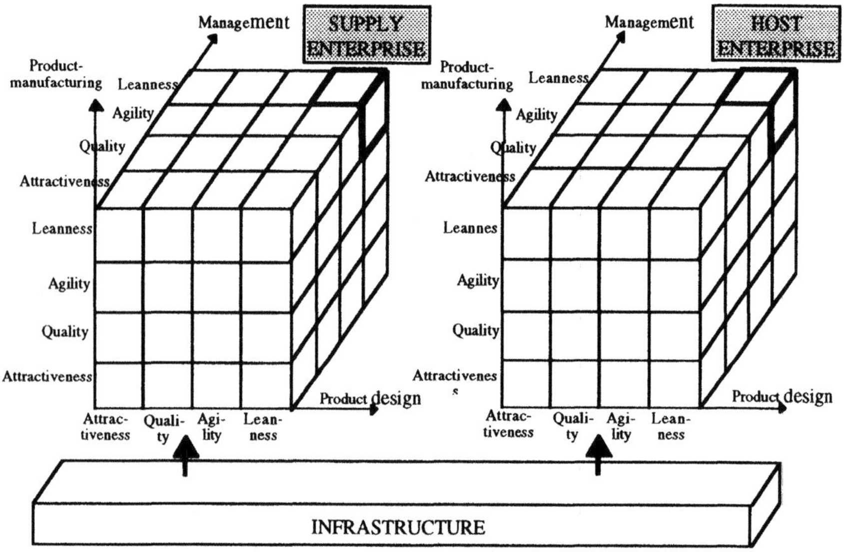

Figure 5 A model of methodology need for AQAL production

For example, if a production system wants to improve the leanness of the whole production system, then a work team should be organized which may be composed of partners coming from the management functions, product-design functions, and product-manufacturing functions in host enterprise, supply enterprises, and selling agencies (not shown in Figure 5). For the aids of fulfilling this task, a complete set of methodologies and tools are needed to help the team work for leanness improvement. What is the complete set of methodologies and tools for leanness improvement? they need to be further researched. We can first evaluate the existing available methodologies and tools. If they are powerful enough and fully match the working requirements, then we can fill them into the sub-cube which resides at the back-upper-right corners of both the host enterprise cube and the supply enterprise cube of Figure 5 where their border are depicted with heavy bold lines. If those methodologies and tools are not complete for our purpose, additional developments should be taken.

The bold-lined sub-cubes in host and supply enterprises' cubes means that persons from design, manufacturing, and management organization of host and supply enterprises jointly work together on leanness improvement. The ingredients of the sub-cubes (which are now not filled in Figure 5) will indicate the complete set of methodologies and tools which can be used for leanness improvement.

The same idea applies to other sub-cube. For example, if a team is organized by persons from D, Mfg, and Mgt organizations of both host and supply enterprises, then the front-lower-left corner' sub-cubes of host and supply enterprises are involved, in where we also need to fill in a complete set of methodologies and tools for aiding the attractiveness improvement.

Therefore, we use Figure 5 to define the scope of methodology need for the AQAL production.

6 CONCLUSION

In this paper, we mainly discusses what a production system and an AQAL production system are. Then, a model of methodology need is conceived which gives a scope of methodology need for AQAL production in where the need of how many complete sets of methodologies and tools is defined. However, there are a great bundle of further works needed for finding, editing, or developing the complete sets of methodologies and tools.

7 REFERENCES

Miyashita, K. and Russell, D. (1996) Keiretsu - Inside the Hidden Japanese Conglomerates, McGraw-Hill

Nagel, R. and Dove, R. (1992) Volume 1: 21st Century Manufacturing Enterprise Strategy - An Industry-led View; Volume 2: 21st Century Manufacturing Enterprise Strategy - Infrastructure, Iacocca Institute, Lehigh University

Rembold, U., Nnaji, B. O., and Storr, A. (1993) Computer Integrated Manufacturing and Engineering, Addison-Wesley

Voss, C. A., Just-In-Time Manufacture, Springer-Verlag

Womack, J. P., Jones, D. T., and Roos, D. (1990) The Machine that Changed the World, Harper Perennial

29

An interface integrating design and mfg in small size companies

MSc. C. Gimenez
UNICAMP - Faculdade de Engenharia Mecânica - DEMA
CEP 13.083-970 Campinas - SP - Brazil, telephone +55 192397966
fax +55 192393722, e-mail: claude@fem.unicamp.br

Dr. G. N. Telles
UNICAMP - Faculdade de Engenharia Mecânica - DEMA
CEP 13.083-970 Campinas - SP - Brazil, telephone +55 192397966
fax +55 192393722, e-mail: geraldo1@fem.unicamp.br

Abstract

Design tasks attempt to ensure that products meet market demands. Many programs have been developed to assist in the design task. Rarely do these programs permit to exchange of information with other departments without restrictions. This work deals with the development of an interface to extract information (e.g. the bills of materials), in a CAD (Computer Aided Design) environment. In addition, the materials are organised through MRP (Materials Requirements Planning) system.

Keywords

CAD/CAM system, interfaces (computer), materials management, process manufacturing, CIM systems

1 INTRODUCTION

Manufacturing automation allows high quality and low costs products, through the use of CAD, CAM (Computer Aided Manufacturing) and computer network. Product development in CAD environment works over all manufacturing variables (CAM environment). Therefore, some problems are brought forward to early production specification. Other important feature is the opportunity of obtaining a efficient production and breaking paradigms with the jobs made sequentially. In this environment, the jobs are made in parallel process.

The main objective of this work is to integrate CAD/CAM systems and reduce manual input of data. Therefore, the input of data will be quicker without errors. This work includes design (CAD) and manufacturing (CAM) technologies, essential to small sized companies which are competing in global market. This paper presents an interface development written in AutoLISP (Autodesk AutoCAD™). The interface extracts a bill of materials directly from the CAD environment. The materials are classified in standard, produced by another company, or internally manufactured. This classification is related to the competitive power of company. In addition, the materials are organised through MRP system developed using Microsoft Excel ™.

2 COMPUTER AIDED TECHNOLOGIES

Computer Aided Design (CAD)

CAD systems are used to help in creation, change, analysis, or optimisation of a project. Product quality and costs are determined by project decisions and performance planning. Little change in design phase is most precious to save resources that great change during the performance. Approximately 90% of product cost is defined during project phase so that, the planning activity is important to decrease errors (Berliner and Brimson, 1992).

Computer Aided Process Planning (CAPP)

Process Planning is the activity that determines the appropriate procedure to transform raw material (usually in some prespecified form) into a final product (Sluga et al., 1988).
The make-or-buy decisions are very important today, because the customers require high quality product and efficient delivery performance (Just-In-Time strategies). These decisions are strategic, because they allow fulfil market needs and have quick reaction in face to environmental, social and technological changes.

The historical development of CAPP was concentrated mainly in CAD, NC (Numerical Control) and other systems. In the past CAPP was known as the weak point in information integrated flow (Alting and Zhang, 1989; Tönshoff and Anders, 1990).
Software development to planning different process and different functions do not consider the hierarquical structure of process planning, operation planning and programming. Therefore, the efforts to develop independent systems produce the following problems (Eversheim et al. 1987, 1989):

- Different data structure;
- Storage of unnecessary data;
- Inconsistent database;
- Inadequate statement of technological data;
- Different understanding of features in CAD, CAPP and NC. In addition, the current efforts show protocols and standards to data interchange in definition phase.

The greatest challenge to CAD/CAPP integration is the translation of CAD language to CAPP language. As a consequence, a system which recognises the part features is necessary. This process is expensive and error prone (Madurai and Lin, 1992). Therefore, the CAD system do not perform well technological information inherent to the project because CAD is geometric. The use of new concepts, like features, reduce these difficulties.

Quality process planning is very dependent of planners (Emerson and Ham, 1982; Shunk, 1985). In many companies the process planning are multipliable because of the lack of information retrieving system and comparison techniques (e.g. two planners make different plans of a component) (Wolfe, 1985).

Computer Aided Manufacturing (CAM)

CAM is the activity that planning, management and control the manufacturing operations.

Materials Requirements Planning (MRP)

Materials Requirements Planning is usually made in a spreadsheet (MRP system) that defines materials quantities and materials delivery time to manufacturing. The spreadsheet can be made manually (in case of small quantities of materials). In case of large quantities of materials the spreadsheet should be made in computational systems. Some MRP systems have special features. The planning in MRP is based in Master Production Scheduling (MPS).

Small size agile manufacturing

Gillies, Nelder and Fan (1996) present four steps to small size companies getting agile manufacturing. Firstly, such companies must be helped to recognise the strategic value of their reactive ability and how to learn from their experience of changes. Secondly, the value of networking must be promoted and the networking skills of partner selection, trust, risk evaluation and relationship management must be understood and systematised. Then, Information Technology applications can be introduced to speed-up network transactions and to extend the boundaries and possibilities of the network. Finally, extended enterprise concepts can be applied leading to fully agile manufacturing in small size companies

3 CADSYSP INTERFACE DEVELOPMENT

Objective of the work

The principal aim of this work is develop an interface called CADSYSP (an interface between CAD environment and planning systems) to data extract. The project was made in AutoCAD program (release 13 for Windows).

The MRP spreadsheet, made in Excel program, only shows the data utilisation after the interface extraction. The objective was not to share software development in the market, but only to demonstrate the feasibility of the idea.

Structure of the program

The AutoCAD is a popular CAD system developed by Autodesk. The CADSYSP interface was developed in AutoCAD R.13 (for Windows environment). AutoLISP is a subset of LISP language. Autodesk maintains, eliminates, or adds commands in AutoLISP. AutoLISP is a specific AutoCAD language (Head, 1989).

Competitivity is the ability to get (and sustain) velocity, performance and flexibility in the company system. In a global market, a company needs to offer projects and products to the customers quickly. Competitivity arises in all parts of the organisation such as R&D, purchasing, supplier, and other areas that use CAD information, and not by the use of some AutoLISP routines.

Microsoft Excel 5.0 (for Windows) is a computer spreadsheet to save, calculate and analyse data. A MRP system was developed in Microsoft Excel. The objective of this system is only to carry out research.

The CADSYSP Interface

Figure 1 shows the connection between CAD and materials management across the CADSYSP interface.

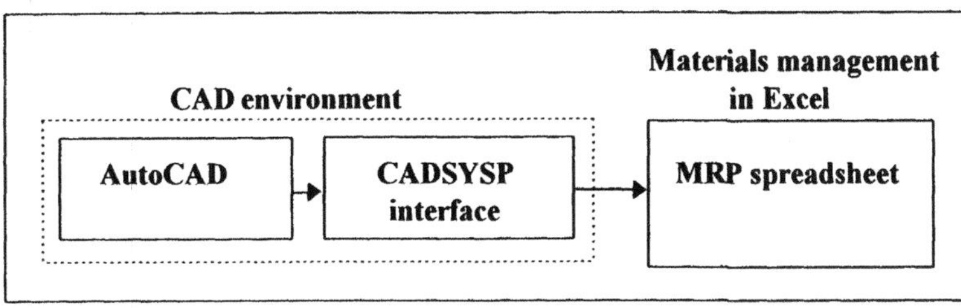

Figure 1 Connection between CAD and materials management across the CADSYSP interface.

The CADSYSP interface was written in AutoLISP. This interface extracts data concerning materials used in project. Both material and project are in the AutoCAD environment while the exported data are in ASCII standard. Therefore, the data should be used by a computer spreadsheet like Excel.

The AutoCAD program executes AutoLISP routines in CAD environment. The bill of materials is made in CAD environment. CADSYSP interface extracts data information, which contains materials data such as description, quantities and so on.

Figure 2 shows the cycle between project and planning since the company may buy any materials, these products, need to have high quality, low costs, and reliable delivery time.

The download of CADSYSP interface can be seen in figure 3. In this stage, the CADSYSP is transformed into a AutoCAD command. The mouse buttons select the information to copy to a ASCII file. This file is imported by Excel program. The source and the lead time are specified for each material. After this, the materials report was indexed, as the 'source' (see 'SC' in Table 1).

Table 1 shows the classified materials report. Materials classified as standardised or produced by other companies are sent to purchasing function. On the other hand, materials classified as internally manufactured are sent to Production Planning and Control (PPC) function.

As the MRP system was developed in an Excel spreadsheet, the cells were programmed by the user. The introduction of new products needs new spreadsheet program as well as design changes need spreadsheet changes.

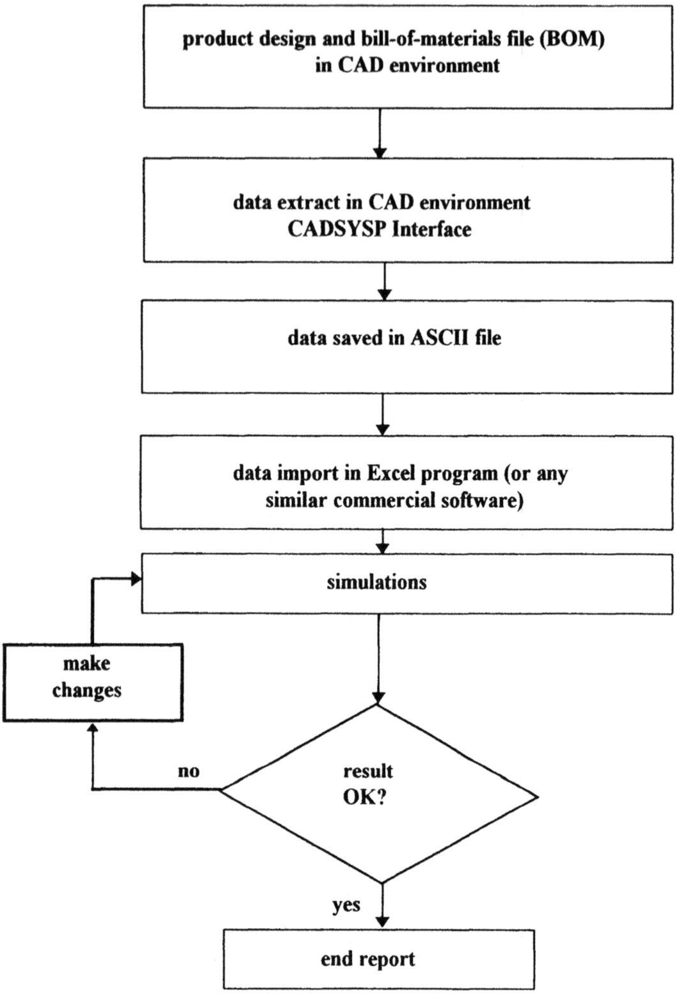

Figure 2 Cycle between project and planning.

The project may change in AutoCAD environment to attend market needs such as high quality, low cost and reliable delivery. Therefore, the company have an external material source to attend reliable delivery. The company may produce in house some materials that contain important know-how. This work does not considered product costs.

The project changes in CAD environment may decrease costs. At least 90% of products costs are defined in project phase.

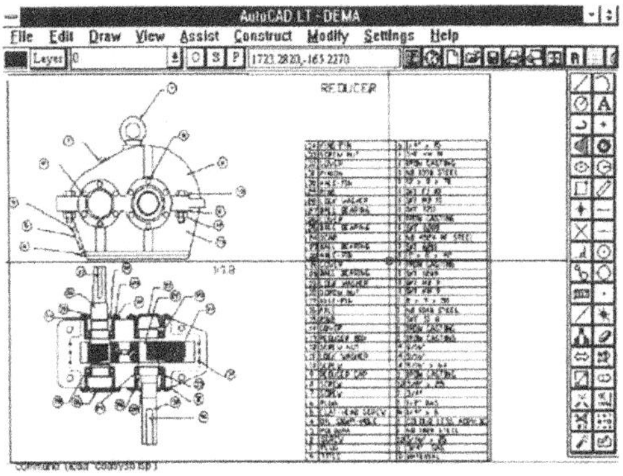

Figure 3 Download of CADSYSP interface - A typical assembly draft.

Table 1 Classified materials report

SQ	*title*	*QT*	*material*	*SC*	*LT*
1	plug	1	3/4" gas	1	0
4	oil sight-hole	1	colour less acrylic	1	0
5	flat-head screw	6	1/4" x 6	1	0
6	plug	1	1/2" gas	1	0
8	screw	12	3/8" x 25	1	0
10	screw	4	9/16" x 64	1	0
19	lock washer	1	SKF mb 9	1	0
20	ball bearing	1	SKF 1209	1	0
25	ball bearing	1	SKF 6209	1	0
33	screw nut	1	SKF km 10	1	0
9	reducer cap	1	iron casting	2	0
13	reducer box	1	iron casting	2	0
14	cover	1	iron casting	2	0
21	cover	1	iron casting	2	0
26	cover	1	iron casting	2	0
32	cover	1	iron casting	2	0
16	axle	1	SAE 1040 steel	3	8
22	axle-pin	1	12 x 8 x 42	3	8
24	gear	1	SAE 4524 af steel	3	8
31	pinion	1	SAE 1030 steel	3	8
34	king-pin	6	1/4" x 15	3	8

Source (SC): 1 = standard;
2 = produced by an other company;
3 = internally manufactured.

LT = Lead Time; SQ = Sequence; QT = Quantities

Note: Security inventory = 0; Minimum order = 1

Companies which have great product mix may execute MRP system only to high cost materials and low use (e.g. axle bed). Other materials (low cost and high use) like screws are controlled by replacement point. Data analysis should be executed in other programs like dBASE, Lotus 123, ACCESS and so forth.

Companies that use other CAD programs (except AutoCAD) need to change files to IGES (Initial Graphics Exchange Specification) standard to data extraction. The IGES file is imported by AutoCAD. The CADSYSP interface send materials data to MRP system. However, this work do not consider IGES standard to data extraction because AutoCAD has leadership in CAD program market. The principal advantage of AutoCAD is that it requires minimum hardware configuration in small sized companies.

5 CONCLUSIONS AND FUTURE RESEARCH

Manufacturing systems need to provide a wide range of products. Additionally, the life cycle of products has been drastically reduced. Companies need to integrated all functions through computational resources. Consequently, the interface development becomes an important element of computer integration.

This paper presented an interface development between CAD and materials management function. The main contribution of this work is the material database that may be used by other company's functions, or suppliers (e.g. materials standard or sourced by other companies). The interface extract data information. The output is an ASCII file. This file could be used by is DOS, Windows and Mainframe. Some commercial systems (like dBASE, Lotus 123, Excel, COPICS, and others) could use this file. Future research in this field will include cost analysis (e.g. ABC - Activity Based Cost) and sourcing strategies definition.

6 REFERENCES

Alting, L. and Zhang, H. (1989) Computer Aided Process Planning: The State-of-the-Art Survey. International Journal of Production Research, v.27, n.4, p.553-585.

Berliner, C. and Brimson, J. A. (1992) Gerenciamento de Custos em Indústrias Avançadas. T.A. Queiroz (editor), São Paulo.

Emerson, C. and Ham, I. (1982) An Automated Coding and Process Planning System Using a DEC PDP-10. Computers and Industrial Engineering, v.6, n.2, p.159-168.

Eversheim, W. et al. (1987) Changing Requirements for CAP Systems Lead to a New CAP Data Model. Annals of the CIRP, v.36, n.2, p. 321-326.

Eversheim, W. et al. (1989) Requirements on Interfaces and Data Models for NC Data Transfer in View of Computer-Integrated Manufacturing. Annals of the CIRP, v.38, n.1, p.443-446.

Gillies, J., Nelder, G. and Fan, I. (1996) The Applicability of Agile Manufacturing to Small and Medium-Sized Enterprises, Proceedings of the Twelfth International Conference on CAD/CAM Robotics and Factories of the Future, Middlesex University, London, England, 14-16 Aug, 1996, p.538-543.

Head, G.O. (1989) AutoLISP in Plain English, Ventana Press, Chapel Hill.

Shunk, D.L. (1985) Group Technology Provides Organised approach to Realizing Benefits of CIMS. Industrial Engineering, v.17, n.4, p.74-80.

Sluga, A. et al. (1988) An Attempt to Implement Expert System Techniques in CAPP. Robotics and Computer-Integrated Manufacturing, v.4, n.1/2, p.77-82.

Tönshoff, H.K. and Anders, N. (1990) Survey of Development and Trends in CAPP Research within CIRP. Annals of the CIRP, v.39, n.2, p.707-710.

Wolfe, P.M. (1985) Computer-Aided Process Planning is Link between CAD and CAM. Industrial Engineering, V.17, N.8, p.72-77.

7 BIOGRAPHY

MSc. Claudemir Gimenez graduated in production engineering at Methodist University of Piracicaba in 1992. In 1996 he received the MSc. in mechanical engineering at the State University of Campinas. He is currently working toward his PhD. in mechanical engineering at the State University of Campinas.

He joined the IBM Corporation in 1986 where he was a trainee in quality engineering. From 1987 to 1990, he was a process engineer at Texas Instruments Ltda. From 1991 to 1993, he was a product engineer at Robert Bosch Ltda. His current research interests include concurrent engineering, reengineering, and strategy development.

Dr. Geraldo Nonato Telles graduated in mechanical engineering at the State University of Campinas (UNICAMP), Brazil, in 1971. He then worked as a research engineer at the Technology Centre at UNICAMP (1972-1979), then at Nardini Machine Tools Company (1979-1982), and at the Centre for Computing Science of the Ministry of Science and Technology. Dr Telles obtained his PhD in 1990 and became a senior lecturer in the Mechanical Engineering Institute at UNICAMP. At present he is Associate Professor and his research activities involve CAD/CAM systems (interface development), CNC machines (simulators and automatic programming), industrial robots, FMS and CAPP (features strategies).

PART TWELVE

Systems Integration II

30

Contributions for Ethernet Based Networking in Manufacturing Systems

Sérgio E. Loureiro, Adriano S. Carvalho
Faculdade de Engenharia da Universidade do Porto
Instituto de Sistemas e Robótica - Grupo de Automação Industrial
R. dos Bragas 4099 Porto Codex - Portugal
Tel: 02 -2041619 email sel@tormentas.fe.up.pt, asc@garfield.fe.up.pt

Abstract

The work presented is this communication intends to evaluate the possibility of using Ethernet to perform integration in manufacturing systems. The principal obstacle is its random medium access time that is not suitable to real-time systems. It refers to some experimental tests based on a case study, which showed that Ethernet can be a correct choice in soft real-time systems. Furthermore, it is discussed the design and implementation issues behind the alteration of the medium access control of Ethernet to support real-time, without requiring any modifications of hardware.

Keywords

communication networks, ethernet, real-time communication, CNMA

1 INTRODUCTION

The integration between the different levels of a manufacturing system is a cumbersome task. This results from the diversity of equipments and different communication requirements among the several manufacturing levels.

Ethernet could be a useful alternative, due to its availability in most equipments and low price. However, this option has some associate problems, like the possible unlimited time response. This fact does not allow Ethernet utilisation at the lower levels of a manufacturing system, where hard real-time is usually required.

Nevertheless, in soft real-time systems, we can evaluate an ethernet based solution, realising that its medium access time is (probably) almost zero for light loads. This great advantage in comparison to another LANs has become more attractive with the availability of higher bandwidth Ethernets. On the other hand, this fact is not guaranteed, so the application must dynamically cope with possible late messages.

It is not possible to establish rules between load and access time, since there are factors that influence Ethernet medium access time, like traffic pattern and packet size (Molle, 1994).

So, if either this probabilistic approach does not cope with application requirements, or it is a hard real-time system, then Ethernet utilisation is not possible. The alternatives are fieldbuses, token-bus or Ethernet adapted to real-time. This last possibility must be evaluated in terms of cost, implementation and integration effectiveness.

2 SOFT REAL-TIME SYSTEMS

This section tries to clarify ethernet utilisation in soft real-time systems and to identify the associate problems.

Firstly, the medium access control is contention-based, that is, there is not any access restriction but the channel is idle. This implies the possibility of occurring collisions, that are resolved in a nondeterministic way. However, we must say that the time involved in resolving one collision is not the problem, because it is very low. The real problem appears for successive collisions with the same packet, which is unlikely to happen under light load.

Secondly, the service provided by Ethernet protocol at different stations is not fair (Boggs, 1988), and its performance is dependent on packet size.

Thirdly, there is the capture effect. This effect results from hardware evolution and affects the successful transmission probability of the packets that are involved in successive collisions, improving the conditions for their starvation (the retransmissions are aborted after 16 tries). This effect can happen at low utilisation, of about 45%.

Some experimental tests were conducted to clarify these points. These problems, namely the capture effect were reported in high performance platforms (workstations) and with traffic very distinct from the one found in manufacturing systems The main goal is to verify if the referred obstacles to Ethernet utilisation happened in lower performance platforms (PCs in our case), that are usually present in these systems and to try to evaluate their consequences. The test conditions were based on a case study described on (Loureiro, 1996) and (Praça, 1996) that concerns an integrated application constituted by planning, scheduling, simulation and monitoring modules developed to a textile industry.

A network constituted by three i486 based PCs running Linux v2.0.6 was used, with off-the-shelf Ethernet cards (SMC-Ultra and D-Link).

Firstly, we try to characterise the test platform. The results showed that just one PC can absorb the entire bandwidth, but only if it is dedicated to this job and with large dimension packets (1,5 Kbytes). With small dimension packets (64 bytes) the results found are very different; the maximum bandwidth occupied becomes about 15% with one PC.

This fact constituted a limitation of the tests because we cannot evaluate high utilisations with low dimension packets. In spite of using a workstation with the objective of absorbing bandwidth, we continued not to achieve high utilisations with small dimension packets.

After these characterisation platform tests, we have tested the Ethernet service time. We define service time as being the time necessary to transmit a request and receive the related acknowledgement. The sizes of packets tested were the small dimension ones because it is usually sufficient to transmit process variables, alarms, control commands and acknowledgements, and 1 Kbytes because it is the maximum size allowed by standard manufacturing protocols. The transfer of large pieces of data, for example the download of programs, is properly performed with large packets.

Nevertheless, we have registered packets dropped due to excessive collisions (and some packets with large service times), when about half of the entire bandwidth was occupied with large packets and we tried to transmit a few small dimension packets and related acknowledgements.

In cases with greater bandwidth occupied (inferior at 80%), but with identical dimension packets, we have not registered any dropped packets. This result reaffirms a guideline of (Boggs, 1988) that advises not to mix very different dimension packets in the same network segment if we want to achieve high performances.

In spite of the tests' limitations, namely the number of hosts, we believe that Ethernet utilisation can be a correct solution for soft real-time systems.

The main constraint is that it is not possible to limit network utilisation in a given moment, or to make a bandwidth reservation to a host. We have to pay particular attention to possible time slots occupied with large dimension packets.

3 HARD REAL-TIME SYSTEMS

If the application context is of hard real-time systems, we cannot use Ethernet. Nevertheless, we can make use of low cost Ethernet cards to develop a solution that supports real-time traffic.

The option of altering the Ethernet medium access control without requiring any modifications of existing hardware was made, not only for reasons of low cost, but also to allow a more generic and flexible implementation. The aim of this section is to discuss the design and implementation of a software based proposal.

After an analysis of the off-the-shelf Ethernet cards and network interface controllers, we realise that it is impossible to change the medium access method and the retransmission algorithm, since they have been implemented in firmware.

This fact does not allow to change the medium access to a method that resolves collisions in a deterministic way like CSMA/DCR or CSMA/PDCR (Turiel, 1996). When a packet is copied to the controller transmission buffer, we have no more control over it.

There are some network interface controllers that constitute an exception to the last two affirmations because they allow to disable retransmissions after the first collision of a packet. The use of any these proposals would restrict the final solution applicability.

What leave us the possibility of using methods that guarantee an exclusive medium access. These methods can always work in an exclusive manner or with a way of changing between two modes of operation: a controlled one and a normal one. The work presented in (Venkatramani, 1995) describes one protocol of the second type. However, this kind of protocol brings up an additional problem that is the transition between modes of operation. (Venkatramani, 1995) proposes a mechanism of broadcasting one message to perform it.

Another possibility is to define an interval where each host has exclusive medium access and can request a change of operating mode. The definition of this interval can be made by clock synchronisation or by an established order.

This interval of exclusive medium access can also serve to make a reservation of a time slot, or to establish priorities among hosts like in (Adelstein, 1994).

In applications where there are periodic sources, we can use the information about the size and periodicity of their transmissions to assign exclusive time slots to these sources by means of a distributed algorithm like the one described in (Yavatkar, 1992).

So, we can conclude that there are several methods to implement, and the pros and cons of each proposal result from a trade-off between efficiency, implementation and access priorities.

A drawback of this software based solution is that it does not guarantee deadlines if integrated into standard ethernet network segments, that is not implementing our medium access modifications.

4 IMPLEMENTATION

In the same tests platform, already described, we developed a generic software based solution that allows the implementation of any of these proposals.

There were developed two distinct modules: one that catches traffic information, and another that restricts the medium access.

The module that gets traffic information is done operating the network controller in promiscuous mode (present in the all controllers analysed). In this easy way, we can get all possible traffic information, that allows the design of other methods, namely predictive or closed loop methods. The only drawback of this solution is the extra processing time needed.

The module that restricts medium access, or better, card access is implemented at the device driver level, more specifically, in the device independent routines, which implies that our solution supports any network card, without requiring any modifications on the respective device driver. Also, it supports any protocol in spite of only having tested TCP/IP.

This module implied kernel modifications, so we created a channel between this kernel part and the user space, which allows an easier implementation of all the methods described.

This constitutes a generic platform, capable of performing the tests set discussed before and new tests involving other methods of control, in order to analyse solutions that can cope with the requirements of hard real-time systems.

5 CONCLUSIONS

The Ethernet protocol is an useful alternative to achieve manufacturing systems integration. In real time operating conditions, we must evaluate the application requirements and decide in conformance. If the communication requirements are not copped by this solution, it is necessary to implement appropriate medium access control. It is our intention to develop a method library that suits the most common application.

6 REFERENCES

Boggs, David R., Mogul, Jeffrey C. and Kent, Christopher A. (1988) Measured Capacity of an Ethernet: Myths and Reality. *Proceedings of the ACM SIGCOMM 88.*

Molle, Mart L. (1994) *A New Binary Logarithmic Arbitration Method for Ethernet.* University of Toronto.

Praça, Isabel C. (1996) *Balanceamento e Simulação de Linhas de Fabrico Manuais.* Faculdade de Engenharia da Universidade do Porto.

Loureiro, Sérgio E. (1996) *A rede Ethernet como elemento integrador dos sistemas de fabrico.* Faculdade de Engenharia da Universidade do Porto.

Turiel, Javier Pérez, Marinero, Juan C. Fraile and Gonzálex, Jose R. Perán (1996) CSMA/PDCR: A Random Access Protocol Without Priority Inversion. *IEEE Proceedings of the IECON.*

Venkatramani, Chitra and Chiueh, Tzi-cker (1995) Design, Implementation, and Evaluation of an Software-based Real-Time Ethernet Protocol. *Proceedings of the ACM SIGCOMM.*

Yavatkar, Rajendra, Pai, Prashant and Finkel, Raphael (1992) *A Reservation-based CSMA Protocol for Integrated Manufacturing Networks.* University of Kentucky.

31

Global AICIME - An integrated approach to support SMEs in the global electronic commerce environment

A. D. de Oliveira
M. F. de Oliveira
ALFAMICRO
Alameda da Guia 192 A - Quinta do Rosário
2750 Cascais - Portugal
Tel.: 351 1 4866784   Fax: 351 1 4866752
E-Mail: alfamicro@mail.telepac.pt

Abstract

This paper addresses the problems and solutions faced by SMEs in the context of the new business paradigms: globalisation and electronic commerce. Practical experience and advise is provided as a result of a number of projects which allowed the development and refinement of the Global AICIME (Advanced Innovation and Competitiveness in Manufacturing and Engineering) approach as well as its methodologies and tools including the development of products and integration strategies to tackle the SMEs needs on process re-engineering, re-organisation, management of change and integrated manufacturing management IT systems.

Keywords

AICIME, CIME, virtual enterprise, virtual process, globalisation, SME, electronic commerce

1 INTRODUCTION

Industrial and commercial organisations can be thought of as systems which are composed of operational processes structured and regulated by a set of functions that can become strategic; they are currently the object of intense environment pressures.

The mass production and rigid organisational hierarchy paradigms of the so called industrial revolution were replaced by quality and production control as the major areas for manufacturing improvement 10-15 years ago. In the late eighties the focus changed to strategy and business objectives. In the middle of the nineties the focus is changing to globalisation issues concerning access, collaboration and new markets. These are new areas of concern for SMEs which are not well equipped to answer these pressures. Never have these forces been so diverse: they are disrupting previous balances and call for rapid and coherent responses as shown in Figure 1.

Figure 1 Environmental pressures influencing trends and paradigms.

Major developments are happening in the supply chain management requiring tools and organisation to support the participation of the SMEs. The market globalisation creates new paradigms for the SMEs and multiple responses which imply co-ordination and integration in an approach with a clearly defined strategic character.

Only through the use of clear, simple, flexible and opportune strategy it is possible to overcome the extreme instability that is characterising the business environment.

The ability to use operational information, integrating it quickly, allowing the formulation of plans of action and supporting management decisions at operational and strategic level is critical to sustain the business competitiveness. Global AICIME has developed a methodology and tools which are able to identify the most effective business indicators in a specific organisation and obtain the relevant data models from structured data which is extracted from single or multiple databases.

The Global AICIME MDSS (manufacturing decision support system) integrates data from different sources: shop-floor, production planning and control and business system. This tool enables companies to formulate and achieve their objectives, allowing them to seize and to take advantage of opportunities as they arise, while at the same time, remaining in tune with their environment.

The changing environment of business globalisation is giving way to many new domains of electronic commerce where virtual manufacturing enterprises need to act seamlessly integrated in a network of companies which increasingly tends to be organised as a logistics chain combining effectively the physical flow of materials and goods as well as related information. The logistics chain enables companies to control the flow from downstream to upstream and to optimise, in terms of cost and level of service, the whole physical movement, pulled by demand as shown in Table 1.

Global AICIME approach is a far-reaching total approach which is both transversal and very ambitious. Its main role is to synchronise the overall physical flow, in permanent interactions with all of the functions of the company and its environment. The responses that companies formulate to respond to the global environment pressures imply combined and synergetic responses from complementary expertises and capabilities.

Table 1 Evolution of global logistics activities

	OPERATIONAL PROCESSES					
	Product Development	Materials and Parts Procurement	Production	Distribution	After Sales Service	Recycling
Distribution Logistics				✓	✓	
Integrated Logistics		✓	✓	✓	✓	
Integrated Logistics Support	✓	✓	✓	✓	✓	
Global Logistics	✓	✓	✓	✓	✓	✓

The required access and sharing of knowledge and information imposes new models for the organisation and IT tools to support them. The management of the simultaneous competition /co-operation paradigms is a critical factor in the SMEs networking at product development, materials procurement manufacturing, warehousing, distribution, after-sales service and the recycling processes.

The Global AICIME approach encompasses: strategy, methodologies and tools, which have been developed, tested and consolidated in a large number of projects addressing the needs of the Portuguese, Greek, German and Italian companies. Global AICIME has developed and acquired experience in the following domains:

- Effective knowledge transfer.
- Technology transfer between European regions.
- Virtual team work with experts from different European countries.
- Virtual software development at remote locations.
- Integration of different products.
- Distributed design and manufacturing.
- Virtual product development.
- Advanced integrated logistics.
- Personal mobility: cultural knowledge and adaptability.

The functional architecture of a company must continually adapt to the existing trends originated by external and internal pressures, as new trends are originated and new business paradigms are created as a response. These paradigms although broadly recognised by the industry are slower to be adopted by the SMEs due to the difficulty of providing reference criteria and evaluation methods necessary to achieve "tangible" evidence of their advantages.

The market needs are today the driving force of the industry requiring fast responses and leading to the emergence of new trends. In the immediate future a new trend is emerging and gaining acceptance as different technologies converge to provide a novel global market in the format of the global electronic commerce. The Global AICIME approach intends to provide answers for the changes faced by the SMEs in this new environment.

2 AICIME APPROACH

ALFAMICRO in co-operation with other European partners has conceptualised, developed and applied a well defined model represented in Figure 2, that enables industrial companies to embrace the new business paradigms minimising the existing barriers in their adoption, but fully benefiting from their advantages.

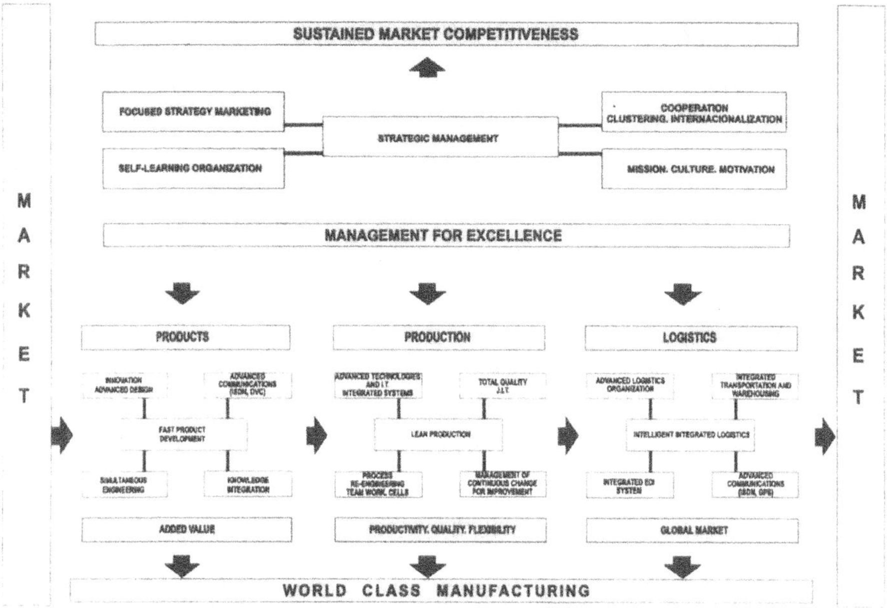

Figure 2 AICIME intervention model.

The organisational and technological approach evolves from a comprehensive understanding of the business strategy and processes leading to the design and application of their re-engineering through the management of organisational changes, process optimisation and sustained self learning processes supported by a dynamic approach to innovation. This methodology is supported by different tools and IT products which support the design of the AICIME solutions built around the functional IT model shown in Figure 3.

One of the major advantages of AICIME approach is the integration of already existing information islands in a company. This is a strategy which can have a major impact in many other SMEs which hesitate to progress for more sophisticated access to the advantages of information technology due to the barriers created by isolated PC based systems which are not synchronised and many times do not follow a common data structure.

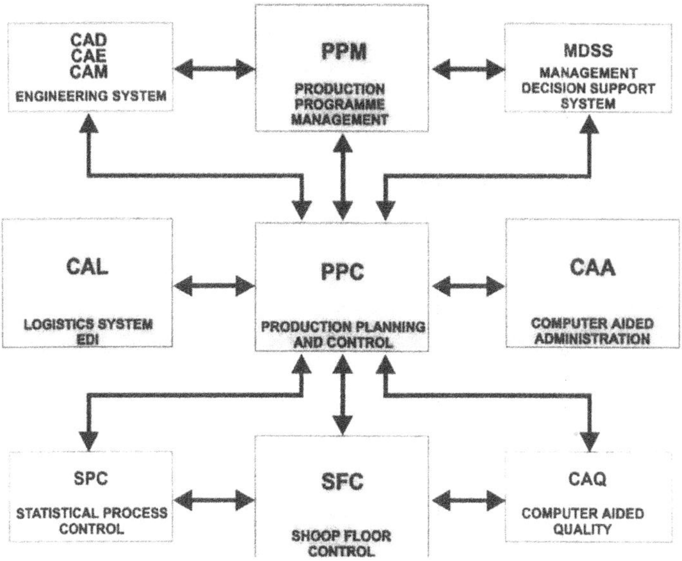

Figure 3 AICIME functional IT model.

3 GLOBAL AICIME APPROACH

The AICIME approach presented in the previous section is the result of a dynamic process which is evolving continuously, receiving direct feedback from the experience gained in several applied demonstrator companies. The business globalisation and the electronic commerce environments led to an evolution of the strategies, methodologies and tools which are presented in the Global AICIME IT model shown in Figure 4. Global AICIME architecture comprises three main levels:

- Strategic.
- Tactical .
- Operational.

The SFC system at the operational level controls the production on the shop floor and collects the necessary data to manage local decisions and supply the PPC level the MS-maintenance and CAQ-quality systems. They include preventive and corrective maintenance, SPC and quality.

The tactical level includes the PPC, the CAPD ad the IBS:

- The CAPD is the integration of advanced IT to support virtual processes bringing forth a more productive approach to product design (ISDN videoconference).
- The IBS is the integration of the logistic system with the business system.

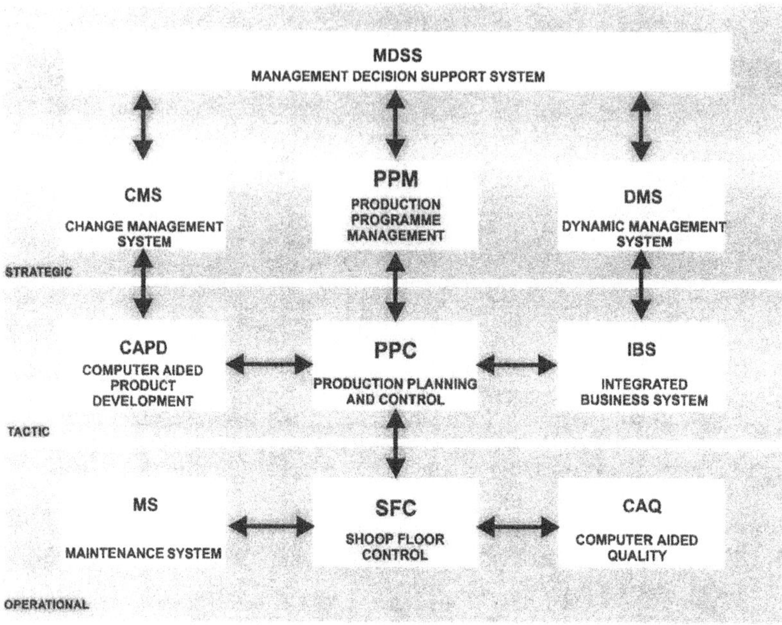

Figure 4 Global AICIME IT model.

The PPC receives and sends data allowing to plan and control all the manufacturing processes. The innovative aspects of Global AICIME towards the AICIME approach is most evident at the strategic level where the MDSS supports the management decisions. The MDSS processes the knowledge relative to the entire productive process, receiving data from the PPC through the PPM. This knowledge is completed with the input from the process simulation and forecasting tools CMS and DMS:

- The CMS handles the management of changes providing a dynamic modelling tool for new processes and optimisation of existing ones. The same system provides local and remote learning to support the skills required by the new organisation.
- The DMS takes in consideration historical and present data together with seasonal and environment trends to build a reliable forecast tool which supports the management of an extended enterprise in the global market.

4 GLOBAL AICIME EXPERIENCE

The extensive experience of ALFAMICRO contributing to the overall Global AICIME architecture is the result of a coherent and determined strategy aiming at gaining the relevant experience of different domains through the practical implementation of a variety of projects as shown in Table 2.

Table 2 Global AICIME Projects

	Product Development	Materials and Parts Procurement	Production	Distribution	After Sales Service	Recycling
AICIME I & II		✓	✓	✓		
LOGSME		✓	✓	✓	✓	
MDSS	✓	✓	✓	✓	✓	✓
PCNET	✓					
ATHENA	✓		✓			
ICIMS-NOE	✓	✓	✓	✓	✓	✓
IiM-B	✓	✓	✓	✓	✓	✓

The fact that most of these projects were or are being developed in LFR demonstrates that it is feasible to provide SMEs localised in these regions with the right tools and approaches to achieve a fast and efficient cohesion with the more advanced environments. The elimination of the competitiveness gap imposes technical, organisational and cultural changes which need to be addressed in an integrated way by a multi-skilled team of experts probably assembled and managed as a virtual team originated from different sources and gathered to achieve a common objective in a specific project. This is one of the main strengths gained by the virtual team managed by ALFAMICRO. This experience has been acquired in Portugal and also applied elsewhere in other regions of the European Union. The Global AICIME approach is characterised by:

- Market focus
- Market globalisation
- Electronic commerce
- Virtual enterprising
- Process re-engineering. Management of changes
- Advanced technologies
- Low cost solution
- Pro-active dissemination of results
- Use of reference models: demonstration and benchmarking
- Effective management of changes
- Multi-disciplinary teams operating as a virtual enterprise
- Reusable experience strategy

This approach has been completed with the implementation of a number of products which support all the major processes of the extended enterprise. ALFAMICRO has gained experience in developing, implementing and integrating these products. Summarised in Table 3 is the range of products which are part of this experience and knowledge.

Table 3 Global AICIME Products

Global AICIME Products	Product Development	Materials and Parts Procurement	Production	Distribution	After Sales Service	Recycling
PCNET	✓					
PPC pro			✓			
TRITON	✓	✓	✓	✓	✓	✓
PROFIT			✓			
LOG FORECAST	✓	✓	✓	✓	✓	✓
EDI MANAGER		✓		✓		
MDSS/DMS /CMS	✓	✓	✓	✓	✓	✓

5 FUTURE WORK. CONCLUSIONS

The continuous and dynamic changes characterising the market globalisation and the ever increasing competitive pressures are taken by ALFAMICRO as a challenge to be fully explored, to the advantage of the SMEs. The benefits of the modern communications such as ISDN and Internet and multimedia applications are the strategic low cost tools to achieve sustained competitiveness.

These tools allow simple technical business networking where individual competencies and skills can be brought together as a virtual enterprise linked at different bonding levels.

The experience of AICIME demonstrates that it is possible to achieve advanced levels of consultancy, engineering and supporting information technology using virtual teams with the appropriate profile.

The successful AICIME strategy provides solutions for the SMEs problems, taking in consideration their present resources and culture.

The evolution for virtual enterprising in the proposed GLOBAL AICIME approach allows the SMEs to handle to their advantage the new paradigms of market globalisation and electronic commerce including the far reaching capabilities of virtual work environments for product development and dynamic management decision support tools.

PART THIRTEEN

Vision and Sensors

32

A Measurement Software for Vision Machine Applications

L. C. R. Carpinetti
Escola de Engenharia de São Carlos -USP
C.P. 359, 13560-970 São Carlos SP Brazil, FAX: +55 16 271 9241, email: carpinet@tigre-prod.prod.eesc.sc.usp.br
R. Spragg
Cranfield University - Bedford, MK 43 OAL, England, FAX: +44 234 750875

Abstract

This paper presents the development of a software for measurement of linear and angular dimensions and geometric deviations of two-dimensional views of prismatic and cylindrical components. This software was developed in cooperation with Integral Vision, a Software House based in Bedford, England, for application in vision systems. The software organization and its different modules are presented. A final discussion is made in which the merits and drawbacks of such a measurement system is considered.

Keywords

Machine Vision, Metrology Software

1 INTRODUCTION

Machine Vision systems are playing an increasingly important role in automated inspection. This technology is being used for such diverse applications as looking for defects in parts, ensuring correct configuration of parts in assembly lines and dimensional inspection. They bring special advantages to the automated factory such as high speed of measuring and inspection, high flexibility, in process, on-line inspection and low cost. In addition to this, the contactless type of inspection and the ability to inspect small or restricted areas make such a measurement system appropriate in many inspection situations.

Applications of machine vision in automated inspection cells are well demonstrated in the automotive industry. A number of automotive plants, including General Motors of Canada, Chrysler Motors Newark (DE) (Pastorius, 1989) and Austin Rover (UK) (Mullins, 1987) have implemented vision systems for 100 percent dimensional process control of car body assembly. These systems provide measurement capabilities not previously available to the car body assembly plants.

At Austin Rover (UK), for example, a vision inspection cell inspects car body dimensions to an accuracy of 0.1 mm. Operating on a two minute cycle time, 62 laser cameras make 88 checks on the body shell (Mullins, 1987).Moreover, recent developments in the field of image analysis are increasing the resolution and accuracy of vision systems.

In the field of Image and Vision Computing, the approximation of digitized curves into segments and arcs is an image analysis technique presently available (Rosin e West, 1989). Roche (1989) offers new evidence of applying this technique for automatic segmentation of digitized profiles of objects. This article reports the development and implementation of a vision machine measurement software (Carpinetti, 1990) that further refine the results of a procedure for automatic segmentation of a contour of an object, (Roche, 1989), defined by a set of points extracted by using a video camera and a vision system for edge detection.

2 MACHINE VISION SYSTEMS IN INSPECTION AND DIMENSIONAL MEASUREMENT

Machine Vision systems basically employ digital representation of the scene under investigation (Nash and Webster, 1984). In a typical arrangement, the camera viewing area is divided into a matrix of picture elements called pixel. The larger is the number of pixels the higher is the resolution. Pixels are typically structured in arrays of 256.000 (512 x 512 matrix); the array size depends upon the hardware utilized.

Once the image of a part is spatially sampled by each pixel, its signal is sent to the processor and treated by image analysis techniques implemented in software. Digital data produced by vision systems may be either binary or gray scale. Binary systems translate each pixel into either black or white. Gray scale systems use a certain number of bytes of information for each pixel to record different levels of light intensity. Because of the amount of information to be processed, gray scale systems are slower than binary systems and require more memory to manipulate the data.

The cameras used are basically of two types: Vidicon and solid state cameras. Vidicons are similar to closed circuit television cameras. They produce an analog signal that must be converted to the digital format. Solid state charge couple device (CCD) and charge injection device (CID) cameras are based on integrated circuit technology and provide digital information with no need for translation. As a result, they provide more accurate information than a Vidicon camera does because of the distortion that analog to digital conversion produces. By changing the focus of the camera, or by using different arrangements of lenses, users can get higher resolution over a smaller area, or can look at a larger area with less resolution.

Vision systems generally rely on one or two forms of illumination: backlighting or frontlighting. Backlighting produces a high contrast image of the part's silhouette. An advantage of backlighting is speed: it produces less information than other sources of illumination and so requires less time for analysis. Binary systems usually use backlighting illumination because of the need for high contrast. But backlighting is impractical in many industrial situations or in cases requiring inspection of features that do not have silhouettes. Frontlighting depends on reflected light from the part being inspected. Unlike backlighting, frontlighting can provide information about the surface of the part. Features to be inspected must be distinguished by enough light intensity to neutralize any light variation coming from the environment. Common sources of light include incandescent and fluorescent lamps, lasers, strobe lamps, radiant energy from a heated part and arc lamps.

The operations performed by a machine vision system in digitizing the profile of an object for posterior measurement may be divided as follows:

- image acquisition;
- image binarization;
- extraction of the pixel coordinates of the profile;
- conversion of pixel coordinates into real world coordinates.

The accuracy of machine vision systems is dependent on the resolution at which the image of the object is digitized and the algorithms for extraction of the pixel coordinates of the profile.

3 IMPLEMENTATION OF SOFTWARE FOR METROLOGY

For any particular measurement, the software implementation comprises the following steps:

1. mathematical modelling: Mathematical modelling, for metrology, may be defined as the construction of a system of mathematical relationships or equations that connect the measurements made in the test procedure to the various assessment parameters which describe the geometric properties of the workpiece. For example, the British Standard Institution (BSI, 1989) recommends that a line in a specified plane should be defined by either one point on the line and information about the orientation of the line or by two points on the line. Another example, the departure of any point from a straight line is given in terms of the "euclidian" distance between the point and the line. The mathematical modelling should have the property that small changes in the geometric element result in correspondingly small changes in the equation parameter values.
2. numerical analysis: numerical calculations are employed to find the position and orientation of the geometric element that best fits the measured points. The fit is then represented by the value of the parameters of its mathematical model. For example, a set of points taken around one section of a cylinder will define the centre coordinates and the radius of a circle, which are the parameters of its equation. Erroneous computation of these parameters will result in poor estimation of the geometric references and will strongly affect the software accuracy [Cox, 1985]. Different criteria for specifying the best fit are possible. In general, the criterion should be to make some combination of the residuals (or departure of each point from the reference) as small as possible. The British Standard Institution (BSI, 1989) and the National Physical Laboratory (Forbes, 1989) recommend the least-squares technique for computing the best fit geometric element to data. Establishing a least-square best-fit geometric element to the data is based on the minimization of the summation of the square of departures of each point from reference, hence the term least-squares fit. This technique is quite used and extensive literature regarding it is available.
3. design of algorithm: Any particular software for metrology will entail several separate tasks: eg. collecting data, checking data, setting up equations, computing the assessment parameters and their statistical uncertainties. It is very desirable that the software be well-structured, or modular, in the sense that such natural divisions be reflected in it by keeping each in a separate module (subroutine). Such modules, where appropriate, can themselves be subdivided into smaller modules (Anthony and Cox, 1984].

4 MEASUREMENT SOFTWARE SPECIFICATIONS

The Measurement Software was specified for two dimensional applications only. This is quite adequate for many industrial applications, since many situations involve a two dimensional view. The input is a set of measured points previously obtained by a machine vision. The output is the result of the assessment. The process is schematically shown in figure 1.

The procedure for the automatic segmentation of profiles was incorporated into the measurement system as it was originally proposed (Roche, 1989). This procedure calculates a first approximation of the original set of points into geometric entities, each entity being either an arc or a segment. The steps for this procedure are as follows:

- Open the file containing the set of measured points.
- Proceed the vectorisation and segmentation.
- Save the results in a text file.

These operations are carried out for one contour at a time. If, however, an object having many contours is to be dealt with, these operations may be applied for each contour and then combined into the same file. The procedure to obtain a segmented contour is based on approximation, which depends on the parameter-of-vectorisation (for reducing the original curve in a contour of vectors) and the parameter-of-segmentation (for approximating the contour of vectors for extraction of boundaries and curvature calculations). The right choice of parameters is a key requirement in achieving the correct representation of the contour in terms of arcs and lines.

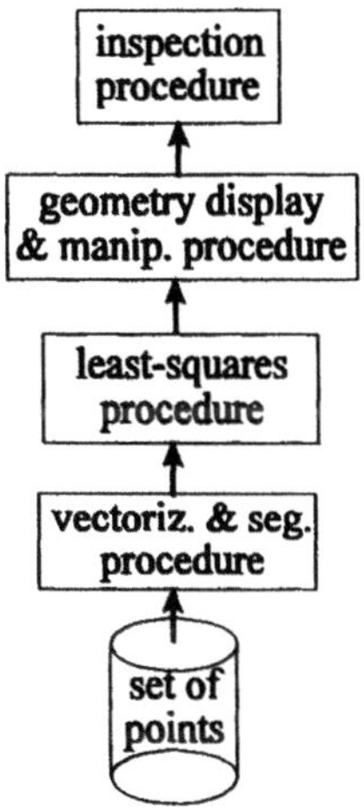

Figure 1 Overview of the process.

In order to investigate dimensions and geometric shape of a profile, this approximation method can be further refined by applying a curve fitting procedure which can allow a more accurate approximation of the centre and radius of arcs as well as the slope of lines. Therefore a least squares curve fitting procedure is proposed. The least-squares fitting is proceeded by following the steps bellow:

- Read the file containing the set of measured points.
- Read the file containing the results of the vectorisation and segmentation process.

- Proceed the least-squares fitting by taking in to account the set of original points between the two boundaries of each entity.
- Keep the results in a array of geometric elements for posterior use or save in a file.

After this, the user selects the features to be inspected. The functions for inspection are presented to the user through the User Interface in a menu selection mode.

The set of points describing any profile of a workpiece is assumed to be contained in a binary file. The points are defined in terms of cartesian coordinates. This file is used as input to the vectorisation and segmentation procedure, as well as to perform the Least-Squares Fitting procedure.

The Software structure is schematically shown in figure 2. Internally, the software is basically organized in three different layers, which correspond to different levels of action as well as different functions.

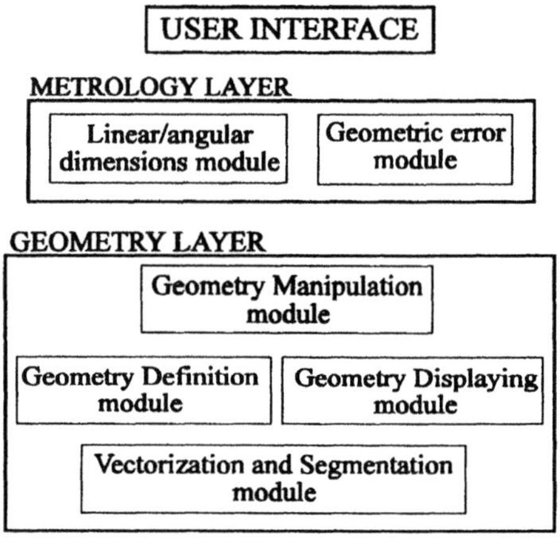

Figure 2 Software organization.

The Geometry Layer is the bottom level of the software, concerned with the definition, manipulation and displaying of the geometric elements used as references at the metrology level.

The Geometry layer is divided in four modules:

1. Vectorisation & Segmentation Module: containing the procedure for automatic segmentation of profiles as proposed by Roche (1989). The results of the Vectorisation and Segmentation procedure are stored in an array of geometric elements. The information for

each geometry entity includes its type (either a line or an arc) and the indexes of the extremity points in the array of (x, y) coordinates.

2. Geometry Definition Module: encompassing the mathematical functions for:
   - computing the Least-Squares best fit geometric elements to cartesian data points. The algorithms adopted employ stable parametrization of geometric elements and are based in numerical algorithms proposed by Forbes (1989). The specific geometries considered in this work are lines, circles and arcs in a specific plane. The information about the extremities points of each geometric entity returned by the vectorisation and segmentation procedure is used to select the points pertaining to each entity in the array of data points, and the entity type is used in applying the apropriate least squares algorithm (either line or arc). In the case that the entity type is an arc and the first and last points coincide, then a least squares circle algorithm is applyed.
   - redefine the intersection points between the entities redefined by the least squares procedure.
3. Geometry Manipulation Module: containing the functions for manipulation of the geometric entities, such as opening and closing the file of geometric entities, selecting, retrieving and putting them back into the file.
4. Geometry Display Module: containing the hardware dependent functions for displaying of the geometric references.

The Metrology Layer is the middle level layer, concerned with the evaluation of position, size and deviation from nominal dimensions of the geometric elements previously defined in the geometry layer. It is in turn divided in two modules:

1. functions for measurement of linear/angular dimensions: the mathematical modelling of these functions are based on planar analytic geometry. The basic functions involved in performing measurement of distances/angles are for:
   - measurement of distance from point to line: the normal distance from a point to a line. The point may be the centre of a circle, the peak of an arc, or the intersection between any two geometries.
   - measurement of distance from point to point: the distance may be the x projection, the y projection or the actual distance. The points may be the centre of a circle, the peak of an arc or the intersection between two geometries
   - measurement of distance from line to segment: by taking the average of the normal distance of the extremity points from the line.
   - measurement of distance between two segments: by taking the average of the normal distance of the extremity points of each segment from the opposite one;
   - measurement of angles: the angle between two lines.

2. functions for evaluation of geometric errors: the mathematical modelling for the calculation of the errors are based on definitions of geometric errors stated in national standards (BSI, 1990, BSI, 1989) and on planar analytic geometry. The inspection of geometric errors is divided in:
   - measurement of form error: straigthness and roundness;
   - measurement of attitude error: parallelism, squareness and angularity;
   - measurement of location error: concentricity and symmetry;

The User Interface is the top level layer, through which the users and computer interact. The User Interface adopted is of a menu selection type. Menu selection is regarded as a mode of operation well suited to human-computer interaction, as it can eliminate training and memorization of complex command sequences (Sutchliffe, 1988). The menu structure is organized in layers, converging on specific actions. The first layer is concerned with the presentation of the possible types of measurement and a list of options of features which may be inspected (figure 3). The terminology used is based on definitions given in international standards (BSI, 1990). The second layer is mainly concerned with data acquisition.

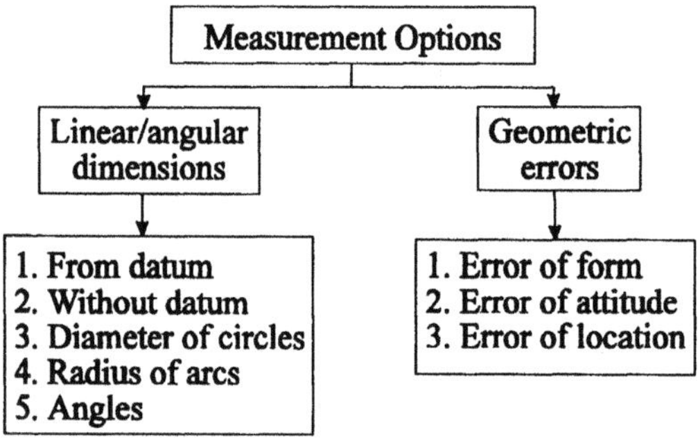

Figure 3 Measurement options.

5 MEASUREMENT SOFTWARE IMPLEMENTATION AND TESTS

C language was adopted for implementing this software,since it is being used more and more in industry, as it provides several advantages, viz: powerful structural features; standard dynamic memory management; low and high level functions and modularity.

The software was implemented in cooperation with Integral Vision Ltd, Bedford, England, a software-house in the field of vision technology. Therefore, some of its in-house facilities and product definition software tools were adopted for reasons of efficiency and existing software compatibility. These are as follows:

- The menu windows structure used for interfacing the system with the user was created by using the company's software libraries for windows set up and display (Integral Vision, 1990).
- The functions for graphic display were based on the company's low level software library (Integral Vision, 1989) and computer hardware facilities.

The computer used was an IBM PC AT compatible. A remote monitor was used due to the impossibility of working with the computer monitor in graphical mode (software limitation). The monitor was a HITACHI VM 900 monochrome, 500 lines, 23 centimetres. It was linked to the computer by a Visionetics VFG 512 grey level framestore of 512 by 512 pixels.

A file containing the set of measured points was generated by using a two-dimensional machine vision system (Auto-Vision®, by Integral Vision Ltd). This system is based on a microcomputer IBM AT compatible, with a Visionetics VFG framestore, 512 by 512 pixels in 256 grey levels. The camera was a Pulnix monochrome CCD, 510 by 580 pixels.

The output of the image aquisition operation is a matrix of pixels, each pixel having a value varying from 0 to 255 defining the light intensity of the associated portion of the scene.

The light intensity of each pixel is reduced to either black or white (image binarization), depending on whether the light intensity exceeds a given threshold level.

The extraction of the pixel coordinates of the profile is achieved by using an edge detection algorithm called "sniff". A line is defined crossing the contour. This line is then scanned, by using the sniff algorithm, until a transition from a white pixel to a black pixel or vice versa is founded. This transition is followed all around the shape of the object and the position of every pixel belonging to the transition is saved in dynamic memory.

The conversion of pixel coordinates into real world coordinates is accomplished by multiplying the x and y pixel coordinates by a transformation matrix relating the size of one picture element to the size of the object contained in a picture element.

After this, the set of data points describing a particular profile is available for measurement. The results of the measurement procedures are saved in a file for posterior analysis. The results are saved in sequence, following the sequence that the measurements are performed. The information saved, for each measurement performed is as follows:

- for linear/angular dimensions: nominal dimension, lower and upper tolerances and actual dimension;
- for geometric shape: geometric tolerance and actual geometric error

Tests for correctness were carried out testing each function inside a module separately and then testing it in its particular context. The functions for measurement of linear/angular dimensions and geometric errors were tested for correctness and the results were verified to be satisfactory, although the results were affected by the accuracy of the machine vision in the edge detection and calibration processes.

6 DISCUSSION AND CONCLUSION

The low level functions used for setting up and displaying the menus imposed setting the computer monitor in text mode and the use of a remote monitor for graphic display. This led to a less standard hardware configuration. This disadvantage can be solved by using more sophisticated user interface software, running in graphical mode and with pop-up/pull-down menus.

This measurement system was implemented in manual mode. In other words, the user has to go through the menus each time a measurement is to be performed. However, this can be automated for example by implementing a function chart for sequential processes. In this case, the user goes through the menus only during the set up phase, and then the defined sequence of options are saved for posterior automatic execution.

The execution time for the automatic segmentation, least squares curve fitting and measuring functions were of the order of tenths of a second in the worst cases. The overall execution time was not possible to measure due the fact that the set of measured points taken by the machine vision were first saved in a file for subsequent use. However, this step can be avoided by interfacing the machine vision with the measurement system and keeping the results in dynamic memory.

Although the accuracy of measurement of machine vision systems is unsatisfactory for high precision measurements, it is very much adequate in cases requiring low to medium precision, therefore making vision technology an alternative in many dimensional inspection situations.

The inclusion of the automatic segmentation technique in the measurement software does improve the apropriateness of vision systems in automated inspection activities.

The implementation of the procedure for application of the Least-Squares Curve Fitting Method on the results of the automatic segmentation procedure resulted in a better approximation of the geometric elements and therefore improved the accuracy of the measurement system as a whole.

The limitation of this Software to measurement of prismatic and cylindrical workpieces does not invalidate the usefulness of the system as these encompass the majority of the sort of shapes of manufactured components.

Finally, the Measurement Software, as it was implemented is machine independent, in the sense that several "measuring machines" can be used in the data acquisition task and integrated under the same measurement software.

7 REFERENCES

Anthony, G. T. and Cox, M. G. (984) *The design and validation of software for metrology*, National Physical Laboratory, Londres, NPL Report DITC 50/84.

BSI (1990) *Engineering drawing practice. Part 3: Recommendations for geometric tolerances*, London: British Standard Institution. BS 308: Part 3.

BSI (1989) *British standard guide to assessment of position, size and departure from nominal form of geometrical features*, British Standard Institution, Londres, BS 7172.

Carpinetti, L. C. R. (1990) *Development of a Machine Independent Measurement Software.* Msc thesis, Cranfield Institute of Technology.

Cox, M. G. (1985) Software for computer-aided measurement. *Laboratory Practice*, June, 59 - 63.

Forbes, A. B.(1989) *Least squares best fit geometric elements.* National Physical Laboratory, Londres, NPL Report DITC 140/89.

Integral Vision (1989) *Auto Vision library*, (program). Beta release. Bedford, UK: Integral Vision,Ltd.

Integral Vision (1990) *Miscell library*, (program). Beta release. Bedford, UK: Integral Vision,Ltd,1990.

Mullins, P.(1987) European quality in production *Production*, **99**(12), 50 - 53.

Nash, W. L. and Webster, F. M. (1984) Machine vision systems today and tomorow. *Quality progress*, **17** (6), 13 - 17.

Pastorius, W. J.(1989) Vision measures up in automated assembly. *Manufacturing Engineering*,April, 81 - 83.

Roche, H. (1989) *Automatic segmentation of profiles resulting in a perfect geometrically defined, continuous and bounded object.* Msc thesis, Cranfield Institute of Technology.

Rosin, P. L. and West, G.(1989) Segmentation of edges into lines and arcs. *Image and Vision Computing*, **7** (2), 109 - 114.

Sutcliffe, A.(1988) *Human-computer interface design.* Macmillan Ed. Ltd, 156 - 180.

8 BIOGRAPHY

Dr. L. C. R. Carpinetti is an assistant professor of quality control at the University of São Paulo where he teaches statistical quality design and process control. He received a M. Sc. in metrology and quality assurance from University of Cranfield, UK, and a PhD in engineering from University of Warwick, UK.

Prof. R. Spragg has recently retired as professor of metrology and quality assurance at the University of Cranfield, UK. He also worked for several years as director of Rank Taylor Hobson Limited, a British manufacturer of precision systems.

33

Design and uncertainty of an opto-mechanical device for checking CMM touch trigger probes

P.A. Cauchick Miguel
CT - UNIMEP, Rodovia Santa Bárbara-Iracemápolis, Km 1, 13450-000 Sta Bárbara d'Oeste, Brazil - Tel. +55 19 463 2311, Fax +55 19 455 1361

Tim G. King
Man. & Mech. Eng., The University of Birmingham, Birmingham B15 2TT, UK - Tel. +44 121 414 4266, Fax + 44 121 414 6800 (e.mail t.king@bham.ac.uk)

Abstract

This paper presents aspects of design of a test apparatus for checking touch trigger probes. It describes the probe test rig whose main objective is to check the probe repeatability and pre-travel variation (lobing effects). The results indicate that such a verification has been achieved with the additional advantage of testing the probe independent of the CMM error sources. The paper details the design characteristics of such a system for which the combined uncertainty is computed. Finally, it concludes that its accuracy is good enough for CMM probe performance verification.

Keywords

CMM probe, probe performance verification, probe test rig, touch trigger probe

1 INTRODUCTION

Undoubtedly the probe system is one of the key elements of CMMs (Coordinate Measuring Machines), especially touch trigger probes. These are the most common probe-type applied (approximately 90% of the market). In addition to the appropriateness of using touch trigger probes for CMM measurements, there are applications other than those of coordinate metrology. For example, a touch trigger probe can be incorporated into a machine tool for in-process control. One of such application has been to fit a probe system to a machining centre to establish the centre line on a crank seal bore, as employed in the Jaguar plant at Radford, near Coventry, UK. However, the probe performance is a major factor contributing to CMM measurement uncertainty. Inaccuracies in the probe mechanism will affect CMM measurement results. Therefore, the nature of probe-related errors and uncertainty must be quantified (e.g. see Butler, 1990; Miguel, 1996). Many tests have been devised to established the magnitude of possible error contributed by the probe. None of them, however, consider a test independent of the machine so that errors in the probe are usually superimposed on those of the CMM. This paper covers the performance verification of the probe system, particularly touch trigger probes, applying the concept of a new probe test apparatus, including analysis of the main error sources of such a system.

2 TOUCH TRIGGER PROBE VERIFICATION USING A TEST RIG

Traditional probe testing methods involving artefacts, whilst useful, are not always indicative of the probe performance as a completely independent functional unit. Hence, a probe test rig has been developed for checking kinematic-resistive probes, presented in the following sections.

Probe Test Rig Design

The required hardware of the probe test rig consists of a motorised traversing table coupled with a laser interferometer which enables the laser to be employed as the table measuring scale. A gauge block, is then fixed on the traversing table providing a reference surface for triggering the probe. Figure 1 illustrates the schematic arrangement of the probe test rig including all components.

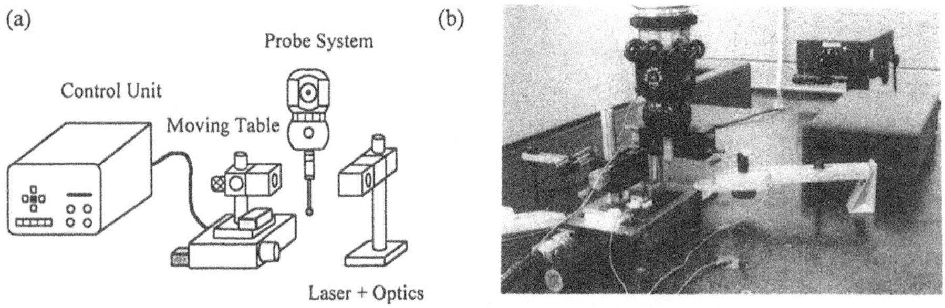

Figure 1 Probe Test Rig Arrangement (a) and Actual CMM Set up (b).

The components of the proposed test rig can be basically divided into; a translational stage with controller, a laser interferometer, and a probe interface. A commercial motorised stage with a travel range of 50 mm was used as the moving element for triggering the probe. The design of the carriage utilises linear ball bearings with twin track bearings disposed laterally on each side on the carriage slides. It has a dc servo motor to propel the carriage along the table base. The table is interfaced to a controller providing forward and reverse traversing movement, with automatic origin search.

The test procedure consisted of measuring the distance from the table origin to the point where the probe was triggered. The table was moved forward until the gauge block touched and triggered the probe. The laser display reading was taken and the table was moved back to its home position. The probe head was then rotated through a 7.5° interval and the stage moved forward to trigger the probe again. This cycle was repeated until the probe head has been rotated by 360° of indexing angle (± 180°) making a total of 48 points. The collected data were used to calculate both the standard deviation (repeatability) and the pre-travel variation over 360° of the head rotation.

The laser interferometer is linked to the probe system to provide a signal for reading the laser display, as illustrated in Figure 1. When the surface of the gauge block touches the probe ball tip, the signal from the probe is sent to the laser computer via a PCMCIA card. The system employed the laser 'TPin' facility. It allows data to be captured by the laser system upon receipt of a trigger signal initiated from the machine under test. This is provided from a relay within the CMM controller, via the probe interface module. Data capture may be triggered either asynchronously or synchronously. In both modes, the trigger signal must be a clean debounced TTL, CMOS, or SSR signal. The mode used was TPin synchronous data capture which provides a higher speed hardware trigger facility greatly reducing the delay between the leading edge of the TPin pulse and the instant that the laser reading is recorded.

The motorised probe head is used for rotating the TP2 probe to successive angular positions. In order to retain the function of the head while the probe is triggered, it is necessary to link the probe interface unit to the probe head controller (housed in the CMM cabinet). The interface unit used is capable of automatic recognition and interfacing the probe system. Figure 2 shows all necessary links between the laser interferometer, probe head controller, and probe interface unit. The main advantage of this configuration is the good compatibility between all components since they are produced by the same manufacturer.

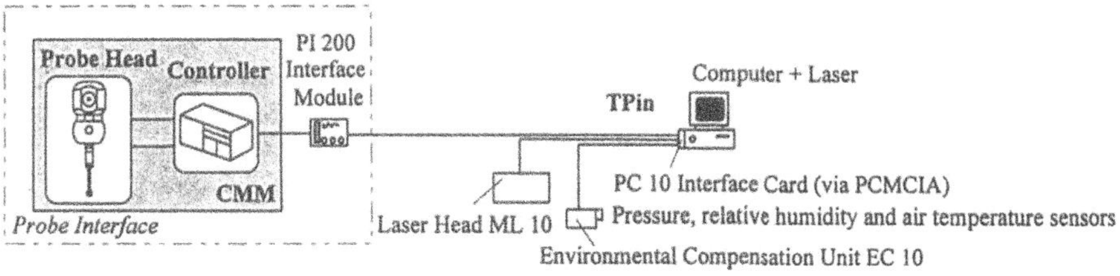

Figure 2 Interface between the CMM and Laser System.

Experimental Set up of the Test Rig on the CMM

The set up of the experiment comprises the alignment of the stage, retroreflector, laser head, and the gauge block (see Figure 1b). First, the table was set up on the CMM table by aligning it parallel to the X-axis using two reference points measured by the CMM. Second, the reflector was also aligned with the table and consequently with the X-axis direction of travel. Third, the laser head was properly adjusted in relation to the table carriage (to which the retroreflector was fixed). Then, the linear interferometer was placed in a fixed position between the laser head and the moving retroreflector. After fine adjustments, a maximum laser strength signal was obtained. The translational table could then be moved incrementally over the whole traverse range while readings were taken by the laser. Although the experiment has been set up in a controlled environment (temp.: 20°C $\pm$ 1°C and humidity: 50% $\pm$ 5%), air and material temperature sensors were placed at the test area. Pressure and relative humidity were also taken from the environment unit in order to automatically compensate variation in the refractive index of the air. Finally, the gauge block was aligned, using the CMM, perpendicular to the table's traverse movement. In all cases, the alignment was considered good enough when the CMM coordinates of the reference points were the same.

3 ERROR SOURCES IN THE PROBE TEST RIG

There are several sources of errors associated with the proposed test rig. Some of them are related to the motorised table itself and others to the laser system. These error sources have been described, in detail, in previous publications (Renishaw, 1989; Steinmetz, 1990; Miguel, 1996) and are summarised in the following sections.

Table Repeatability and Angular Errors

It is clear that, irrespective of the motion mechanism employed, the proposed system depends upon adequate repeatability and low rotational errors to perform effectively. It is therefore necessary that these errors are measured to determine whether the translational table is suitable for this particular application. Such verification has been presented and discussed elsewhere (Miguel, 1996). The angular motions associated with the moving stage are pitch and yaw angular errors. In terms of yaw, it can be assumed that the centre of rotation is in the lead-screw. If the gauge block (in fact, the spatial position of the centre of the probe ball tip) is maintained aligned with the screw centre line, the yaw errors will be null, or negligible. Investigation on the characteristics through the traverse range revealed a region of constant rate of change for pitch errors (a large interval between 15 and 30 mm from the datum). This interval could be the best for using the table to check the probe system. However, small intervals around 20, 25 and 35 mm seem to have a smaller spread. Assuming the centre of pitch rotation in the centre line of the screw, it is possible to calculate the variation in the table traversing movement for those intervals. The minimum spread of pitch angular error was 1.9 arc-seconds (position at 20 mm). Since the distance from the screw centre line to the probe ball centre is measured as 26.85 mm, the variation in table traversing position due to the errors in the pitch direction will be 0.25 μm (in the position 25 and 35, the error will be o.35 μm and 0.45 μm, respectively). Therefore, the position of 20 mm was considered the best one for positioning the table when triggering the probe. In fact, the maximum standard deviation for three runs was 0.58 arc-seconds which yields a variation in the traversing table position of just 0.08 μm. The previous analysis has been made to check the influence of pitch and yaw on the gauge block position when touching the probe ball tip. Actually, pitch error is relevant to the test while yaw is not (see next section). Furthermore, there is another parameter which is affected by these angular motions. This is the Abbé offset (or Abbé error), described below.

Abbé Error

Abbé error occurs when the measuring point of interest (in this case the centre of the stylus tip) is displaced from the actual measuring scale location (the position of the laser beam, which is a line through the apex of the cube-corner reflector and parallel to the beam direction). For this error to be avoided, the measurement system must be placed co-axially with the line in which the displacement is to be measured. However, this is not possible on the test rig because the probe system must be shifted from the laser beam path. Abbé error can be represented as $d \times \tan\beta$, where d is the vertical or horizontal offset distance between the laser beam and the centre of the ball tip. The angle β is the variation caused by angular motion of the carriage of the table. It should be noted that yaw affects horizontal Abbé error while pitch affects vertical Abbé error. There is no Abbé sensitivity to roll angular motion. The pitch and yaw errors were found to be 1.9 and 1.4 seconds of arc for pitch and yaw at the chosen table test position (i.e. the best interval, identified in the last section, to be used for testing the probe). The horizontal Abbé offset was not taken into account; it can be considered negligible due to the fact that the gauge block is positioned on the screw centre line. The Abbé vertical offset (due to pitch) was reduced by mounting the reflector as low as possible, i.e. as close as possible to the gauge block top surface. The Abbé vertical distance was measured as being 20 mm. The variation in angular movement for the vertical plane resulted in an error of 0.18 μm.

Atmospheric Conditions

As relatively well-known that the interferometric measurement accuracy is directly determined by how accurately the ambient conditions are known and their stability during the test. The laser wavelength is a function of the refractive index of the ambient atmosphere in which it operates. During linear measurements variations in the refractive index of the air can lead to significant measurement error. The refractive index is dependent upon the temperature, pressure, humidity and chemical composition of the ambient atmosphere through which the beams travels. Therefore, in order to obtain proper results, the laser readings must be corrected for any change in the wavelength of the measurement beam. Errors in the laser wavelength can be reduced by applying either manual or automatic compensation using Edlén's equation (Edlén, 1966) for the refractive index of air under standard conditions of temperature and pressure. This equation is accurate to approximately 0.1 ppm (ANSI/ASME, 1995). This form of equation also assumes atmospheric air with the normal mixture of gases. Atmospheres that deviate significantly, particularly in regard to CO_2, can lead to measurable errors. If this situation is suspected, appropriate corrections should be applied. However, previous study conducted by Mainsah (1994) indicated that there was little reason to suppose that the concentration of CO_2 was high enough to significantly affect the accuracy to which the measurements were made. These errors can be minimised to less that ±1.1 ppm by using the laser's environmental compensation unit in automatic mode, providing the environment remains stable during the course of measurement. The Centre for Metrology the CMM resides in a room with controlled conditions of temperature and relative humidity. However, the temperature variation measured over a period of 2 hours was 0.31°C; and over a longer period (7 hours) was 0.79°C.

Laser Wavelength

The laser source of any interferometric system has some type of frequency stabilisation to maintain wavelength accuracy and repeatability. The laser system used in the experiments was a frequency-stabilised He-Ne laser operating at a nominal wavelength of 633 nm. The wavelength stability gives a theoretical accuracy of linear measurement of ± 0.1 μm per metre measured, or ± 0.025 μm with compensation for refractive index changes in air *Downs, 1991).

Deadpath Error

This is an error associated with changes in the environmental conditions during the course of a measurement. The deadpath error (ε_{DP}) is related to the distance between the two optical elements, when the system is datumed (Figure 3). This occurs due to an uncompensated length of the laser beam and arises when atmospheric conditions surrounding the laser beam change while a measurement is being taken, thereby causing alteration in the refractive index of the air and hence in the wavelength of the laser beam. Deadpath error is not a function of the measurement length but of the distance DP (see Figure 3). It is a systematic error which is constant for a given system set up.

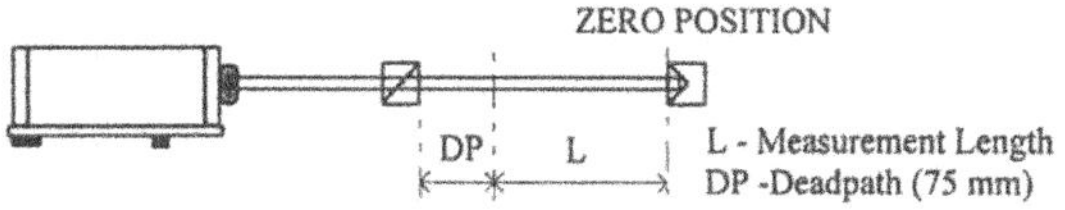

Figure 3 Optical Configuration with Deadpath.

The deadpath error can be expressed by the dead path distance multiplied by the variation in the environmental factor (change in the wavelength compensation) during the measurement time (Steinmetz, 1991). The environmental factor (EF) is the inverse of the refractive index (or λ_A/λ_V, where λ_A is the wavelength in air and λ_V is specified as the wavelength in vacuum). In order to minimise this, the stationary and moving optics were mounted as close as possible following recommendations provided by Downs (1991). He also recommended that whilst this error is inevitable if proper alignment is carried out, its magnitude is usually small enough that detailed quantification is unnecessary. Nevertheless this error was quantified for the probe tests and it was found to be 0.05 μm, when the EF was recorded a few times during 1.5 hours of test.

Material Thermal Expansion

Another factor which could affect the probe test results is the thermal expansion of the table carriage and other CMM components, e.g. the motorised probe head, through the heating generated by their motors. Temperature changes introduce further error sources which can be relevant since solids expand when heated. In order to verify the effects of temperature, sensors were placed at different locations in the traversing table, gauge block, reflector, and CMM surface table. The temperature variation is particularly unwanted in the gauge block as it could expand during the tests. Considering a uni-directional thermal expansion in the gauge block (towards the probe ball tip), the temperature gradient can lead to an increase in its length up to 0.42μm for a 50.8 mm gauge block (rate of about 1.4°C/h). It was also possible to observe a temperature gradient even in the reflector fixed to the mounting post (0.33°C/h). Therefore, precautions had to be taken in order to minimise this effect. Two materials were tested to isolating the base plate (where the gauge block and the reflector were mounted) from the table carriage. The first material was a layer of polystyrene, well compressed by the clamping system. The second one was a layer of Tuccite™, which is a material used in the assembly of lathes, which is considered good in isolating temperature effects from the motors to the guideways. The temperature increases due to heat transfer from the drive motor can be seen in Figure 4. The results show a great deal of thermal isolation was achieved with both materials when compared with the un-isolated case, since the rate of heating is much reduced. However, the layer of polystyrene achieved a better result compared with the Tuccite™ material (rate of 0.58°C/h and 1.13°C/h, respectively). Therefore, the expansion of the gauge block, using polystyrene, would be reduced to 0.17 μm. However, it was possible to reduce the length of the gauge block to 20 mm yielding an expansion of 0.06 μm.

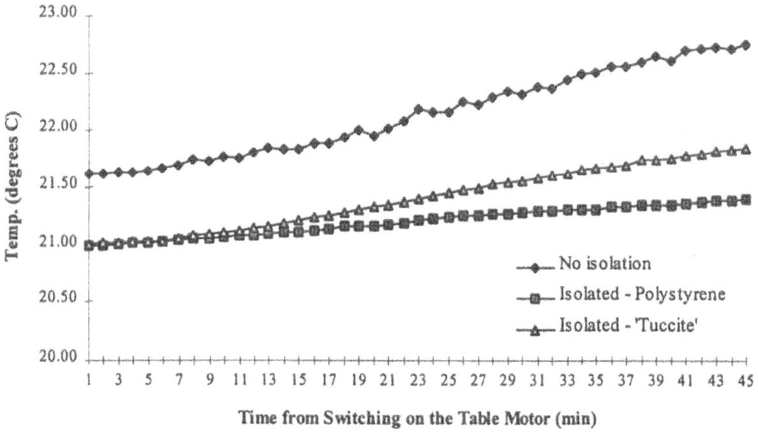

Figure 4 Temperature Variation in the Gauge Block after Isolation.

Cosine Error

Misalignment of the measuring axis (laser beam) with the axis of movement (traversing table) results in a difference between the measured distance and the actual distance travelled, called cosine error (ε_c). It is called cosine error because its magnitude is proportional to the cosine of the angle of misalignment. The resulting measurement error is then a function of the distance measured by the interferometer. Therefore, the cosine error can be represented as indicated in Figure 5.

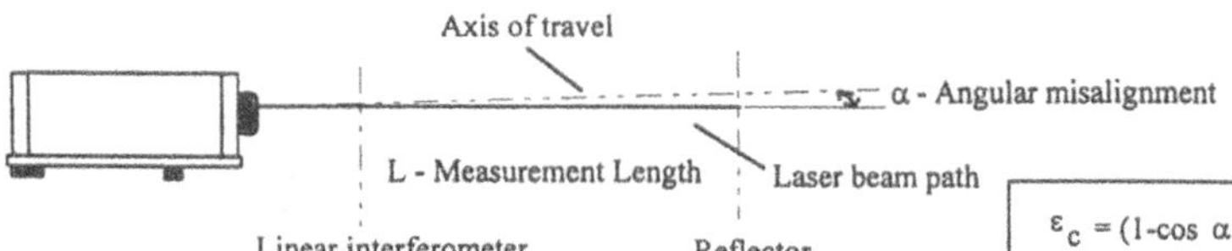

$$\varepsilon_c = (1-\cos\alpha) \times 10^6 \text{ (ppm)}$$

Figure 5 Cosine Error.

The cosine error can be eliminated by orienting the laser beam parallel to the actual axis of travel. However this is not entirely possible and there always will be some misalignment. By following proper alignment procedures, cosine error can be minimised. This was achieved in this experiment by taking some precautions with regard to such a laser beam alignment. Firstly, the reflector was aligned with the axis of travel by probing two reference points in its lateral surface using the CMM until the coordinate (in Y direction) showed the same values for both reference points. Secondly, due care was taken when aligning the laser beam making sure that the reflected beam was in the centre of the laser output aperture (using the provided target) over the full range of the table. Finally, the signal level on the laser display was also used to reduced this error. This was done by fine tuning the alignment to achieve a constant and maximum strength reading over the whole table traversing range resulting in an effective beam alignment. Downs (1991) stated that, with the aid of a position sensitive quadrant detector system, it is possible to align the optics such that the cosine error can be reduced to 0.0015 μm/150 mm. However, such a system was not available for this work. The cosine error has been investigated by Mainsah (1994) and it has been shown that it is practicable under normal laboratory conditions to reduce the error to well under 0.1 μm/50 mm.

Other Error Sources

Additional error sources are the electronics error and optics non-linearity. The electronics error stems from the method used to extend basic optional measurement resolution in an interferometer system. The basic resolution of an interferometer is 1/2 λ (where λ is the laser wavelength) when using a cube-corner reflector and can be electronic or optically extended beyond 1/2 λ. The interpolation factor in the laser interferometer used is 256 which, combined with the double path of normal linear displacement optics, gives a resolution of λ/512 which is 1.24 nm. The electronics error equals measurement resolution. The optics non linearity is the ability of an interferometer optical element to perfectly separate the two laser beam components in vertical and horizontal polarisations (Steinmetz, 1991). The non-linearity affects all interferometer systems whether they are single or two frequency. The specifications of these error sources were not available from the laser manufacturer. However, Hewlett Packard provides them for its laser systems. Owing the fact that there are no significant differences between the optics for both systems (Steinmetz, 1991), the values recommended by HP were used for calculating the combined uncertainty.

4 PROBE TEST RIG ERROR ANALYSIS

The results of the analysis of the total error budget associated with the probe test rig (the combined uncertainty). The combined uncertainty (U_C) can be thought of as representing the measuring uncertainty resulting from combining all known sources of uncertainty in a root sum of squares (RSS) manner (by assigning them a variance). It resulted in an $U_C = 0.25$ μm. This approach is based on the CIPM method, according to the ISO Guide to the Expression of Uncertainty in Measurement (1993). Further improvement in the uncertainty model will be done as a future work.

5 CONCLUSIONS

The achievements using the probe test apparatus described in this paper have shown that it serves the purpose of checking the main error sources in the probe system within an accuracy which can be considered satisfactory for this stage of development. Nevertheless, the limitations to the accuracy attainable to the test rig have been quantified including a number sources of uncertainty for this particular application. The error source which contributes most to its uncertainty was the Abbé offset, but this occurred due to physical constraints of the system. The U_C value has been calculated for the worst possible situation but it would be expected to be smaller than the 0.25 μm. The full results from probe tests will be reported in a future paper. It is believed that the main advantage of the probe test rig is to support CMM users when inspecting their probes either in their acceptance or during their useful life as an important aspect of industrial production.

6 ACKNOWLEDGEMENTS

The authors would like to thank to Prof. Ken Stout, Head of the Centre for Metrology for providing CMM facilities, to Mr Bob Aston and Mr Paul Saunders for their technical support as well as Dr Álvaro Abackerli (*Centro de Tecnologia, UNIMEP*) for discussions on probe performance verification. Thanks are due to the CNPq, Brazil, for sponsoring one of the authors.

7 REFERENCES

ANSI/ASME B89.4.1 (1995). Methods for Performance Evaluation of Coordinate Measuring Machines. The American Society for Mechanical Engineering, New York, December Draft.

Butler, C. (1991) An Investigation into the Performance of Probes on Coordinate Measuring Machines. Industrial Metrology, **2(1)**, 59-70.

Downs, M.J. (1991) Optical Metrology: The Precision Measurement of Displacement Using Optical Interferometry, in *From Instrumentation to Nanotechnology* (ed. by J.W. Gardner and H.T. Hingle), Gordon and Breach Science Publishers, 213-226, USA.

Edlén, B. (1966) The Refractive Index of Air. Metrologia, **2(1)**, 71-80.

Mainhah, E. (1994) Investigation of Pre-characterisation Aspects of Three-dimensional Micro-topography. PhD Thesis, The University of Birmingham, UK.

Miguel, P.A.C. (1996) CMM Performance Verification, Considerations for a Virtual Artefact Approach. PhD Thesis, The University of Birmingham, UK.

Renishaw Interferometer System User Handbook (1989). Renishaw Transducer Systems Ltd.

Steinmetz, C.R. (1990) Performance Evaluation of Laser Displacement Interferometry on a Precision Coordinate Measuring Machine. Industrial Metrology, **1(3)**, 165-191.

8 BIOGRAPHY

Paulo A. Cauchick Miguel

Paulo Miguel graduated in Industrial Engineering at the Universidade Metodista de Piracicaba (UNIMEP) in 1986. He then worked as a manufacturing engineer for Allied Automotive and Varga companies until 1990 when he became a lecturer in the Centro de Tecnologia at the UNIMEP. In October 1993 he moved to the University of Birmingham, UK, to join the Mechatronics group in the School of Man. & Mech. Eng. His research activities include quality and metrology developments, and engineering education. Paulo Miguel completed his PhD in December 1996.

Tim G. King

Professor Tim King BSc (Eng), MDes *RCA*, PhD, FRSA is Professor of Mechanical Engineering at the University of Birmingham where he heads the Mechatronics Research Group in the School of Man. & Mech. Eng. His current research includes development of ultra-high speed piezoelectric actuators, piezoelectric motors, calibration techniques for surface metrology and CMM applications, and microprocessor and DSP based systems for computer-visual inspection and controls of deformable patterned materials. He is author of more than 90 journal and conference papers.

PART FOURTEEN

Virtual Reality and Education

34

Virtual reality full immersion techniques for enhancing workers performance

CASAS, L. A. A.
Professor DAIE/FIPS/UNSA. Arequipa - Perú, M. Eng. Produção, Doutorando PPGEP/UFSC, Brazil
casas@mbox1.ufsc.br

BRIDI, V. L.
Medicine, Mestranda PPGEP/UFSC, Brazil
bridi@cce.ufsc.br

FIALHO, F. A. P.
Professor PPGEP/UFSC, Dr. Eng. Produção
Departamento de Expressão Gráfica -CCE/EGR, UFSC - Florianópolis-SC-CEP:88040-900, Brazil
ffialho@matrix.com.br

Abstract

The aim of this paper is to analyze the potential use of computers as educational tools for enriching organizations educational programs with modern learning environments. The theoretical basis of the article relies in Human Cognition, mainly taking its biological support into consideration. An investigation was performed, pointing out properties and advantages of Virtual Reality Immersion and its contribution for knowledge construction. The idea is that continuous education activities are needed and that modern technologies can support it. Finally, we establish our conclusions and suggestions for future developments in the field.

Keywords

Virtual Enterprise, Computer Assisted Learning Environments, Virtual Reality, Knowledge Construction by Immersion in Virtual Reality.

1. INTRODUCTION

The emergence of new technologies, the fast changing economical and juridical variables, caused by countries integration in multinational organizations, together with changes occurring in the social tissue, happening nowadays, make obsolete the knowledge acquired in the school or university, turning then insufficient to face new situations occurring in work or daily life. (Fernandez, Lopez, Kumpel, Villa, 1992, p. 333). The natural obsolescence, due to this fast changing technology added to the forgiveness of acquired knowledge, due to automation, implies in a reduction of the overall company workforce efficiency. The educational system, responsible for individual formation, is not prepared to follow the rhythm of these mutations occurring both in technology as in industrial processes.

The new organization is leading to a new social order, where only qualified workers will have a place. The instruction or training performed as it was done in mass production systems are not sufficient for companies working under a Lean Production System paradigm. At the same time, the increasing diversification of the work force, requires new capacity approaches, where the new paradigm is learn while working (Borges, Borges, Baranauskas, 1995, p. 154). The same argumentation is valid for Learning to Learn Organizations.

The educational segment is growing faster than any other, inside industries, as an answer to these challenges, in order to make the work force follow this flow (Reinhardt, 1995). It is an organization need to prepare their employees for a more active participation in the decision making process.

New approaches for employees training that results in something interesting, stimulating, able to supply the answers asked under this new enterprise point of view, are requested. These new approaches must be designed to expand employees vision, beyond its routine work limits. In order to succeed in attaining a global participation in the work, questions like how to interpret and to produce an overall optimization of the processes, are a challenge whose answer can be found in training techniques, mediated by the use of computers.

2. COMPUTERS AS AN EDUCATION AID

Literature on Computer Aided Instruction (CAI) presents three generations of educational philosophies, prior to the advent of virtual reality, the fourth. The first generation was based on the behavioral approach (Winn, 1993, p. 5), that is:

- Student behavior can be foreseen if one has sufficient knowledge about the intended instruction results, the employed methods and the training conditions (Reigeluth, 1983);
- Students knowledge and skills that are beyond the teachers understanding, can be standardized and reduced, using appropriated analytical techniques, toward atomic components, in which domain one can work it out to obtain a desired behavior (Landa, 1983);
- The prescriptive instructional theory is sufficiently reliable to affirm that the design procedures guarantee that the instruction, developed by such a systematic application, will be effective, without the need of professors or designers (Gagne, Dick, 1983).

These argumentations were strongly refuted by Streibel (1991), and Winn (1990, 1993). In spite of this, the idea of a computer assisted instruction developed in accordance with this philosophy is still followed.

The second CAI generation states that design must focus on how to present the information to the students and not on the content of the presentation itself (Flemming & Levie, apud Winn, 1993, p.5). The emphasis on the processing aspects of instructional messages, resulted from psychologists perception that the behaviorist theory is incomplete. Cognitive theories about human learning are more satisfactory for models 'design orientated' (Bonner (1988); Campagne, Klopfer, Gunstone (1982); DiVesta, Rieber (1987); Tennyson, Rasch (1988), Winn (1990)).

The emergence of the second generation implied in a significant impulse on CAI systems. Since Gardner (1983, 1993) considerations that two students are not similar in its psychological conditions, and Cronbach, Snow (1977) considerations that these individual differences are important enough, that designers are faced with the prescription of developing different instructional methods considering not only students aptitudes but also, as signaled by Tobias (1976, 1989), skills.

The third generation of CAI systems comes from the belief that, although men machine iteration nature is a determinant of the learning process, with an importance equal or greater than the content itself, the way the information is presented is also meaningful. This orientation is strongly based on cognitive sciences. Truly, cognitive theories, such as ACT (Anderson apud Guine, 1991, p. 253), are being used as a support for developing intelligent tutors, highly interactive (Wenger, 1987). One of the recent and strongest approaches is given by Merrill (1991, 1993), "*Instructional Transaction Theory*", based on the idea that all learning results from transactions between students and software.

An approach for CAI that bets on the understanding on how the student interacts with learning resources is just a first step for the fourth generation in which the central point is the belief that knowledge is built by the students themselves and not supplied by a third person. Bartlett (1932) was the first to propose, based on Jean Piaget, that learning occurs when an individual build "schemes" representing some particular aspect of the world he lives in. Neisser (1976) suggests that schemes are active structures that drives the way people search for information in their surroundings. Recent theories about knowledge construction, like the work of Spiro (1991) "*Cognitive Complexity Theory*", and Bransford (1990) "*Anchored Instruction*", are based in cognitive theories.

Ideas such as the metaphor linking 'human mind working' and computers, and that cognition consists of symbolic manipulation (Boden, 1988; Jackendoff, 1987; Johnson-Laird, 1988; and Pylyshym, 1984) are common in Cognitive Sciences. Marr (1982), in his seminal work on vision, suggests that the mind is too complex to be understood and that the only possible approach consists in explain cognition as computations based on mathematical models related to a chosen cognitive model.

The 'human mind' - computer metaphor, results in the belief that cognition is structure and symbolic dependent. Computer interfaces limit students experiments to "third person ones". The corollary is that first person psychological activity, non reflexive, non symbolic, that occurs when the person interacts directly with real or virtual worlds, have no place in Cognitive Sciences Theories (Winn, 1993, p.7). This is a fatal omission, as pointed out by constructivists and critics of cognitive sciences (Dreyfus (1972); Edelman (1992), Searle (1992)).

Constructivism describes the ways of knowledge, the process of cognitive structures and functions construction, as well as the related learning processes. While these considerations do not form an unified theory, three hypothesis about knowledge construction give support for learning experiments and Virtual Reality applications.

- Maturana and Varela (1972) work, together with Thompson and Rosch (1991) developments, play a particular influence on some constructivists like Cunningham (1993). Maturana and Varela propose that living beings, including humans, do not take information from the outside but react to perturbations occurring in the environment through adaptations of existing internal structures. Interactions with the environment do not add directly to the organism physical structures. Interaction promotes indirect quantitative and qualitative exchanges in symbols and mental structures. The skill of detecting perturbations and the class of structural change they produce are determined by the species evolution experience (Maturana, Varela, 1992, p. 69) and the individual adaptations, or ontogenesys (Maturana, Varela, 1992, p. 67). The interpretation and representation (symbolization) of the world depends on structural adaptations due to the interaction of a symbolic and real world with these perturbations. In accordance with Maturana (1970), learning is not a process of accumulating media representations. It is a continuous behavior transformation process.
- Construction in each human being is unique. Nagel (1974) pointed out that even being able of building a description of some other experiences, talking in third person, we will never be able to know how the other feels. This means that the experience of one entity with the world of anybody else can only be a third person world description experience (Winn, 1993, p. 7).
- These two previous suppositions guides contructivism toward a solipsism. If we are closed in terms of information (Maturana, Varela, 1992, p. 109) and if the objective world is not a pattern, so we are condemned to the impossibility of communicating with others.

Nevertheless, communication turns possible by what Maturana and Varela (1992, p. 49) call "structured coupling". Organisms of similar species basically share similar apparatus for detecting and adapting to perturbations. Moreover they inhabit similar environments where there is a trend of finding similar perturbations. As a result, the story of their structural adaptations, can be similar. Their structures are coupled. That is why we can communicate with other human beings.

In accordance with Vigotsky (1978), to make communication possible, we have to have an approximation about the meaning of the symbols given by others. Negotiations between group members about meanings can only result in temporary agreements. Due to this fact, in accordance with McMahom & O'Neil (1993), the constructivist praxis frequently insist in presenting learning opportunities that demands the students to work in groups searching for a general agreement about meanings.

The environment where the students can build knowledge must be created. Proposals of environmental learning construction recommends instruction (Brown, Collins & Duguid, (1989); Brown & Duguid (1993); Lave & Wenger (1991)) and reflexive praxis (Schon (1987)) as the best method to allow students knowledge construction departing from "authentic" activities. Kozma (1991) states that technology is reaching a state of the art where it is possible to create artificial learning environments that could not have being created following traditional strategies and that this quality is the one that allows the development of superior pedagogical methods.

The coming "fourth generation" of ICAI (I for intelligence) is based on constructivist learning theories.

Each individual learning mode is a combination of how he perceives, organizes, and processes information. Although it is possible to give different names for the several learning styles, one can, in a simple way, to split then in two main learning categories (Guillon, Mirshauka, 1995, p. 20).

- Modality, defined as the way we understand the information.
- Cerebral dominance, defined as how we organize and process information.

Among learning modalities we quote (Guillon, Mirshauka, 1995, p. 21) visual, audio, and Kynestesics. In situations where we know the visual, audio, and kynestesics characteristics of a cognoscente individual, it is more simple to "tune" the learning process to the modality more adequate. This process could be realized through the use of Information Technologies, using resources like multimedia, hypertext, hypermedia, virtual reality, and telematic, that can give teaching flexibility, personalization, interactivety and quality.

✎ Multimedia. The use of texts, graphs, sounds, images, animation and simulation, interactively combined to provoke a previously determined effect.

✎ Hypertext use, allowing to situate different topics that are inter related one with the others in several levels, turns possible to personalize the learning process letting the student find his rhythm, level and style.

✎ Hypermedia: A good hypertext combined with multimedia, offers what Nielsen calls "hypermedia".

✎ Virtual Reality is defined by (Rios, 1994, p. 1), as "the environment and men sensorial mechanisms computer simulation, that supply users with the sensation of immersion, providing also the ability of interacting with artificial environments".

3. VIRTUAL REALITY

Virtual reality is one of the newest computer sciences branches. There are several terms related to the same concept, like "Synthetic Reality" and "Cyberspace".

There are several definitions for Virtual Reality. One can take Rios (1994, p. 1) definition, others can say that "Virtual Reality is just a mathematical model that describes some 3D space". Virtual Reality can be defined also as the way humans can visualize, manipulate and interact with computers and extremely complex data. For some people, Virtual Reality is any interactive simulation, for others virtual reality is only obtained when you are in a network and several people share their realities between themselves, like in virtual communities (BBS) and schemes MUD (*Multi User Dungeon*). There are still researchers that limit the term virtual reality to the use of_sophisticated equipments, like Head Mount Devices, that is, 3D synthetic realities. In accordance with other investigators, through virtual reality it's possible to experience tangible models of places and things, where tangible means that the model can be directly perceived by the sensorial apparatus and not through language or mathematical models.

Virtual Reality is a step beyond computer simulation, dealing with interactions occurring in real time, a dynamic simulation.

In accordance with Rios (994, p. 2), the characteristics of a virtual reality system, that discriminate it from other information systems, are:

- Immersion: property that provides the user with the sensation of being inside a 3D world.
- Existence of a observation point or reference that allows the determination of the user position and situation inside the artificial or virtual world.
- Navigation: property that allows the user to change his observation point.
- Manipulation: characteristic that makes possible interaction and transformation of the virtual environment.

The elements present in any Virtual Reality system are:

- Interaction: allowing the explorer to control the system. Absence of interaction reduces the system to a movie or video. To allow interaction several interfaces, from keyboards, data gloves, and special dresses, can be used.
- Perception: comes to be the most important factor. Some systems are addressed directly to the senses (visual, tactile, ear); other systems try to reach the mind directly, avoiding external sensorial interfaces, and still others, more humble, recurs to the imagination strength to allow human beings to live a virtual reality experience.
- Simulation: simulated worlds do not have, necessarily to adapt themselves to natural physical laws. This is the characteristic that makes virtual reality applicable to any human activity. We accept that some applications are more proper than others. Virtual Reality is more than a simple simulation, offering interaction with the model, allowing users presence inside the virtual world. Through this capability it is possible to realize tasks inside a remote world, or a computer created one, or a combination of both.

In accordance with W. Robinnett the models to be used in Virtual Reality systems can be grouped in the following categories:

- Models captured by scanners, digitizer apparatus, transferring elements from the real world; telepresence systems using video camera (one for each eye), to explore the real world from a remote place, sound records for each ear registering a distant world audio model.
- Calculated models, that are mathematically computed, allowing a posterior building-visualizing-manipulation, which are generally used in complex or abstract models, like air flux models in a air turbine or the lava flow of a volcano in a valley.
- Models build for artists, that are generally generated in CAD systems, created with complete coordinated structures. These models can be based on fictitious or real spaces. A cuisine of the future or a spatial landscape, for example.
- Models edited from a combination of contents captured by scanners, calculated or created by artists.

In accordance with its elements, a Virtual Reality can be classified in the following categories:

- *Desktop* systems of Virtual Reality: Comprehends all kind of applications that shows 2D or 3D images in the computer video, instead of projecting then in a *Head Mounted Device* (HMD). Due to the fact that they represent 3D worlds, the explorers can navigate in any direction inside these worlds. Characteristic examples of these environments are computer flying simulators. Shortly, *Desktrop* virtual reality shows 3D worlds through 2D screens. Some of then incorporate complex interfaces like data gloves, control commands, customized cabins, but all of then will present the above mentioned characteristic (3D in 2D).

- Immersion Systems: Those are the ones that submerge or introduce the explorer in a close relation with the virtual environment they are dealing, through the use of visual systems like HMD, gestures and movements following devices, like sound elements processing systems, giving the participant the sensation of a very close relation with the artificial environment and, at the same time, isolating then from the real one.

Computer user interface elimination is a needed condition for virtual reality immersion. The participant "dresses the computer". He is part of the data. As a result, applicants can interact with the virtual world, which can include some simulation or aspects of the real world, some abstraction instance that could be only accessible as numerical data, or the creation of some fantasy from imagination, like we do in our daydreamings. This is a great advantage over natural interactions. (Winn, 1993 : p. 2).

These 3D immersion worlds can be obtained through the creation and sending almost similar images to the eyes, which provides depth, perspective and dimension sensation. What each player sees and experiences must be computed for each eye, for each detected movement, in order to present a new adequate sound or image position.

Immersion virtual reality systems allow the navigator to go to any place inside de assembled structure, to cross walls, float, levitate toward the sky, penetrate the earth core, etc.

- Second person VR: The immersion difference arising from being in a second person systems (or *unencumbered systems*) involves real time perception and answers to actions of others participating in the experience, which can be performed by the use of gloves, HMD's, conducting wires or any other interface.

In spite of this, in second person systems, the navigator knows that he is inside the virtual world because he can see himself inside the scene, that is, he is part of it. To allow this feature his image is added to the environment image (*chroma-keyed*) in order to build the complete image the explorer have access to. Through a contour detecting software it is possible to realize manipulations inside the scene, which can be visualized in the video monitor. More then imitating real world sensations, a second person system changes the rules, based in the old notion that "we need to see in order to believe", inducing a being present sensation.

- Telepresence systems. Immersion systems simulate real world perception. The travelers coming from the real world knows that they are inside because the sounds of the virtual world, and head movements, are answered in a similar way it would be done in a real environment.

4. LEARNING BY KNOWLEDGE CONSTRUCTION

The use of technology to improve the quality of constructivist learning environments, nowadays, is centered in the creation of computational tools and virtual representations that students can manipulate (Dede, 1995, p. 1). Perkins constructivist approach (1991), includes: Data banks, construction kits, and administration tasks.

Interpreting apprentices experience on refining their mental models by the use of these tools, we conclude that computational tools, designed for aiding students memory and intelligence associated skills, are available to teachers. Parallel to these applications, transitional objects (similar to the Logo turtles) are used to facilitate the transference of abstract symbols obtained from personal experience (Papert (1988); Fosnot (1992)).

Constructivist knowledge, improved by technology, focus, currently, in how representations and applications can mediate the interactions between apprentices and natural or social phenomena.

The key for the compatibility between Virtual Reality and constructivism arises from the immersion notion. First hand experience considers the processing of our world activities and the learning resulting from these processing. First hand experience occurs when our interaction with the worlds do not enroll symbol use or conscious reflection.

In accordance with constructivist theory, knowledge building emerges exactly from these first person experiments, the ones that can never be completely shared. Immersion experiences allow this kind of experiences by the elimination of user computer interfaces (Winn, 1993, p. 15). In other words, immersion in a virtual world allows us a knowledge building derived from direct experimentation and not through a description of another person experience.

Any kind of learning is mediated by a system of symbols: a text, some spoken language or computer. It is a reflection on others experiments. Any specification we use in a symbolic system to communicate about the world we build for another person will never allow this other person to know this world as much as we do. Constructivism theory describes how the worlds of the first person come to be internalized, arguing that any imposition of symbolic representations requires negotiation (Flores, Winograd, 1989, p. 79).

Virtual Reality Immersion provides a class of interactions that is similar to our natural interaction with real world objects. If cognition is non symbolic, and learning is associated with action, through interaction with the virtual world, knowledge can be build. Immersion, or the sensation of being present, of being inside and surrounded by the virtual environment, is analyzed in terms of individual differences: motivations, environment visual characteristics, interactivety, and the role of the several sensations in the assembling of a Gestalt. (Psotka, 1996, P. 1).

Papert (1991) and his colleagues use the world "constructionism" to describe knowledge building arising through physical interaction with world objects. Immersion provides these physical and perceptual interactions.

Due to the fact that the virtual environment is computed from collected data, three classes of learning constructions experiences that are not possible in real world but that show a high interest for education, are possible: These concepts are called: "*Dimension*", "*Transduction*", and "*Reification*" (Winn, 1993, p. 16).

- In real world an object seems bigger when I get close and diminishes when I move apart. In spite of this, there are limits. There is a point where I can not get closer and this point signals the maximum apparent dimension of an object. In a similar way there is a point where the object disappears. In a virtual world I can get as close as I wish or be so far away and still see the object. Before colliding with a virtual wall, for example, I can get approximations in such a way that details, smaller and smaller of the wall material, can be unveiled. I can appreciate the cellular structure of a wood panel and, equally, travel through the atoms that constitute its molecules. In the other extreme, I can go from the wall to the house, and then to the city, the planet, and so on, without violating the four immersion conditions. (Like in *Cosmic Zomm*, produced by the *National Film Board of Canada*, that follows these ideas, going beyond what we are describing here).

The advantages of providing these dimensions changing properties for education are meaningful.

It is possible, for the student, to penetrate an atom, inspect it, and change the orbits of their electrons, altering atoms valence and their combination capacity to form other molecules. It is possible also, for the students, to get the notion of the relative distances between the planets, just flying from one to another.

- Transducers are similar to *earphones* and are used in Virtual Reality hardware to present information and to convert participants behavior in commands to be interpreted by a software. Transducers are devices that transform information not accessible to our senses in other forms.
- Changes in dimensions and transduction provide a first person experience impossible to be obtained in other learning environments. Some of these experiments emerge from the simulation of events or aspects of real objects. Others come from the representation of abstractions, that can only be perceived as non physical objects or events, like algebraic equations, or the dynamic opening of a flower. Reification is the process of creating these forms. Reification contrasts with simulation. In simulation, the virtual world contain *facsimiles* of real objects with their behavior.

5. PROBLEMS RELATED TO AN OPEN VIRTUAL IMMERSION

Immersion characteristics were analyzed in a *U.S. Army Research Institute* research, through questionnaires designed to access individual susceptibilities for immersion and how deep was this experience to the participants.

In the research reported by Psotka (1996), the questions were carefully build to cover the cognitive factors mentioned in the literature as meaningful for virtual reality environments. A five point scale of categorical answers was used. Two psychological factors considered as dominants for predicting immersion deepness were: the imagination needed for accepting another reality (a living imaginary turns the participation integral and satisfactory) and the concentration, attention, and autocontrol needed to exclude real world distraction effects.

The extension of the visual field in a HMD, the precision of the egocentric localization, audio precise information (synchronized with visual changes like rotations, affirmation head movements, or accelerations), are determinant factors of an in depth immersion. A "*cognitive tracking*" paradigm was used to get visual-vestibular interactions. The precise synchronization of the head spatial movements and the changes in visual perspective result in a deeper immersion, even with an imprecise coupling between head movement and changes in the visual presentation (Psotka, 1996: p. 1).

Cognitive factors were grouped in two categories: Immersion susceptibility and quality.

- Susceptibility depends on imagination (dreams, a gift for replacing old beliefs), a living imaginary (dreaming, previous expectations about virtual reality environments), concentration and attention (attention filtering, cognitive conflict by two recursive immersions, spatial navigation), and self control (active participation and catharsis).
- Quality depends on the environmental resources provided by the virtual reality immersion (object persistence, sensorial perfection, interactivety, environment realism, delays, visual field, precise localization of the ego center or corporal image, pleasure and satisfaction with the new experience), distractions due to the real environment (noise, tactile presence, tiredness, irritation with the equipment (size or restrictions), similarity between real and virtual world), psychological effects (simulator disturbances, disorientation after immersion), and other effects (preference for a lonely immersion, surprise when the HMD is removed).

Movements in the virtual environment, at the present stage of technology, can originate several errors (an extremely slow answer, failures). The main causes of movement errors are: imprecise scenery for the inter ocular distance, absence of convergence and accommodation signals, lack of good texture gradients for deepness and improperly designed models. They are easily perceived and remains a disturbance factor in several virtual reality applications.

Postka (1996: p. 3) reports that with a exhibition vision greater than 60° (*field of view,* FOV), several individuals relate some level of discomfort although not mentioning nausea. In accordance with Wertheim (1993) and Wolpert (1990) the relationship between behavior and the study about self oriented emotions are strongly dependent on FOV. The most influent requirement for immersion is a solid coupling between head movement and visual exhibition. Data input delays in visual exhibition using HMD immersion results in a brisk interruption of spatial orientation and a poor immersion experience (Psotka, 1996)

Immersion can be understood as a dual phenomena; from one side it depends on subconscious skills and, from the other side it depend on our voluntary attention abilities that depends on self control, self consciousness, will, expectations, etc. Those two factors (implicit *versus* conscious immersion control) are captured by the found correlation that immersion is more complete in some one that is able to have colored dreams (Psotka, 1996a).As those two components interact is a mystery. The implicit and conscious components seem to perform different things without affecting, one the other, directly. These factors appear so briskly that they make themselves visible in all three sets of correlation: in the susceptibility factors, in the immersion factors, and in their intercorrelations. Implicit factors present a kind of dominance. If the delay between the iteration and the visual *feedback* of a movement perceived by a hand, in Virtual Reality, is to big, in such a way that there is no image filtering technique sufficient to reduce this delay; disturbance occurs. When the visual system indicates that the egocentre or corporal image is in a certain location but the kynestesic centers are in some near place, there is no way to integrate these two positions. It can be necessary a long time for the learning and adaptation processes to change the cognitive machinery (Psotka, 1996a).

Instead of supplying access to a greater cyberspace, the equipment used nowadays is still causing sensations like claustrophobic, nausea and confusion after the experiment (Psotka, 1996a)

6. CONCLUSIONS

When a baby is born, as stated by most psychologies (including psychogenetics), he is nothing more then a set of parts: hairs, skin, ... There is nothing one can call Ego, that is, nothing showing consciousness, that thinks about himself as "I". If there is not an 'I', we can not talk also about "Others". There is no interface between interior and exterior. In other words, there, in the origin, we have our so searched 'non interface', pure absence of differentiation, Jacques Lacan concept of "Real".

The first logical question is: How and when this "I-candidate" assumes his place as an individual? In psychoanalysis what we are discussing here is how we get the "I" form an image - Moi - that represents the individual for the Others - Je -. In other words, we are talking about an identification process that allows the subject to function as a being able to exchange information with mother, father, or simply, the Others.

"There is a magic moment in this evolutive process that occurs when the baby looks at his own face in the mirror, recognizing himself as someone apart from the Real World. *A symbolic interface that will mediate our external or internal interactions is then established*". The baby sees his image because her mother eyes (personification all the Others) sustains this knowledge. The baby sees himself through the eyes of her mother. His mother provokes the first Learning. A Learning about himself

The important thing is not the mother's eyes, but her strong desire, a "symbolic matrix" offered to the infans (Lacan, 1949: 87). The baby holds himself to this image to be the object of her mother desire. We have here a three personage representation. The reflected image, the subject, and the acting of a third person. Teaching is always a representation involving three persons at least: the student, the image this student has about himself, and the teacher.

What is there, reflected in the mirror, represents the individual for the others and even for the individual himself. But this image is not the individual. Jacques Lacan says that the I (moi) is constituted by the sacrifice of the Je (no translation at my possibilities). Anyway that is a natural tension between the 'Moi' and the 'Je' that *provokes learning searching*. In other words, we are condemned to lose our equilibrium and search for a new equilibrium state through learning.

The materialization of the perceptual and cognitive components of immersion seems to be a long and arduous task but, for sure, the results will compensate all the effort. In spite of this it is particularly satisfactory, nowadays, for training and educational proposals, because we already know that most of the cognitive representations assume the form of mental models, that are used for understanding complex systems. Virtual Reality promises to turn education and training in tools much more direct and effective than nowadays (Psotka, 1996b).

It is primordial to insist in a continuous education, open, flexible, personalized, allowing the individual to update and adequate his knowledge throughout all his professional life. The creation of distributed environments for a constructivist learning is a challenge. Research in this field is needed for the development of cooperative learning tools able to facilitate and motivate learning.

The development of intelligent didactic systems is complex demanding the support of knowledge coming from different fields.

A multidisciplinar team, including pedagogues, educational psychologists, specialists in the knowledge domain, technicians in computer graphics, programming, multimedia, virtual reality, project management, and others, are needed.

A sustainable industrial production will not be possible in organizations designed under some mechanical metaphor. Organizations are complex systems that can be modeled as higher order autopoietic living beings. Survival in the fast changing environment we are living in demands for a continuous adaptation. Immersion techniques seems to be the tool we need to enhance the chances of organizations survival, training and educating people to face the challenges of concurrence in a global economy.

Virtual Immersion introduces new resources for inter individual communications that deals with reality interpretation. In this paper we tried to outline the beginning of a long journey on how imagination power can be used for the collective creation of worlds able to improve learning.

7. BIBLIOGRAPHICAL REFERENCES

BORGES, E. L., BORGES, M. A. F., BARANAUSKAS, M. C. C. (1995). Da simulação à criação de modelos - um contexto para a aprendizagem an empresa. *Simpósio brasileiro de informática na educação*, VI, Florianópolis - SC - Brasil, p. 154-165.

DEDE, C. (1995). *The Evolution of Constructivist Environments: Immersion in Distributed, Virtual World.* Http://www.virtual.amu.edu/pdf/constr.pdf

GUIN, M. D. (1991). *Nécessité d'une spécification didactique des environnements informatiques d'apprentissage.* Cachan. Les Editions de l'Ecole Normale Superieure de Cachan. 262 p.

GUILLON, A. B. B., MIRSHAWKA, V. (1994). *Reeducação. Qualidade, produtividade e criatividade: caminho para a escola exelente do século XXI.* Makron Books do Brasil. São Paulo.

FERNADEZ, M. KUMPEL, D. LOPEZ, A. De la Rica. VILLA, A. de la. (1992). *Multimedia y pedagogía.* Un binomio Actual. Congreso Computadoras Educación y Sociedad, República Dominicana.

FLORES, F., TERRY, W. (1989). *Hacia la comprensión de la informática y la cognición.* Barcelona. Editorial Hispano Europea. 266 p.

MATURANA, R. H., VARELA, G. F. (1992). *El árbol del conocimiento.* Santiago de Chile. Octava edición. Editorial Universitaria. 172 p.

MAZZONE, J. (1993). *O sistema "enxuto" e a educação no Brasil. Em computadores e conhecimento - Repensando a educação.* J.A. Valente (ed), Gráfica Central da Unicamp. Campinas.

PSOTKA, J. (1996). *Immersive training systems: immesion in virtual reality and training.*: http://205.130.63.7/vrTraining.html

PSOTKA, J., DAVISON, S. (1996). *Cognitive factors associated with immersion in virtual environments.* http://205.130.63.7/vrfopub.htm

PSOTKA, J. (1996), *Immersive tutoring system: virtual reality and education and training.* http://205.130.63/its.html

REINHARDT, A. (1995). As novas formas de aprender. *Byte*, Março.

RIOS, H. F. (1994). *Potencial de la Realidad Virtual*: http: //lania.xalapa/ spanisch /publications /newleters /fall947index.html

WINN, W. (1993) *A conceptual basis for educational aplications of virtual reality.* http: //www.hitl.washington.edu/ projects/ education/ winn/winn-R-93-9.txt

WHITNEY-SMITH, E. (1996) *Conversation, Education, Constructivism and Cybernetics.* http: //www.well.com/ user/ elin/ edhom.htm

8. BIOGRAPHY

Francisco A. P. Fialho graduated in 1973 in Electronic Engineering at Pontifícia Universidade Católica do Rio de Janeiro, Brazil, where, in 1980, he specializes in Propagation of Electromagnetic Waves. He got his M. Eng. in Knowledge Engineering and his Dr. degree in Production Engineering (1994) at Santa Catarina Federal University. His special areas of interest are Virtual Immersion, Cognitive Ergonomics, Artificial Intelligence, Piaget, and Lacan. Dr. Fialho authored three books and has more than one hundred articles published in technical magazines and Congress Annals.

35

Mechatronics education: examples in the UK and a proposal in Brazil

G.N. Telles
DEMA - FEM UNICAMP, Cidade Universitária "Zeferino Vaz", CP 6122, CEP 13081-970 Campinas, SP, Brazil, Tel. +55 19 239-7966, Fax + 55 19 239-3722 (e.mail: geraldo1@fem.unicamp.br)

P.A. Cauchick Miguel
CT - UNIMEP, Rodovia Santa Bárbara-Iracemápolis, Km 1, 13450-000 Sta Bárbara d'Oeste, Brazil - Tel. +55 19 463 2311, Fax +55 19 455 1361

T.G. King
Man. & Mech. Eng., The University of Birmingham, Birmingham B15 2TT, UK - Tel. +44 121 414 4266, Fax + 44 121 414 6800 (e.mail t.king@bham.ac.uk)

Abstract

This paper describes the structure and contents of a mechatronics engineering course in the State University of Campinas (UNICAMP) in Brazil. Firstly, the paper outlines the UK experience along with two examples of the existing courses there. The second part highlights the reasons for proposing a mechatronics engineering course in Brazil including the required disciplines demanded by the Ministry of Education. In its last part, the paper describes the overall structure of the curriculum, which is a multi-disciplinary effort jointly undertaken by the faculties of Mechanical and Electrical Engineering. Finally, it concludes that the existence of mechatronics engineering courses is an essential step towards the development of a more complete understanding of the challenges of this growing field of engineering.

1 INTRODUCTION

The word Mechatronics was first used by the Japanese in the late 1970s to describe the philosophy they adopted in design of electro mechanical products to achieve optimum systems performance (Day, 1995). Although there are definitions of Mechatronics there is still considerable debate about what it means (Hewit and King, 1996). It has been formally defined by the European Community as *"The synergistic integration of mechanical engineering with electronics and intelligent computer control in the design and manufacture of products and process"*. A broader definition of mechatronics in the context of machine and product design is (King, 1995): *"Mechatronics is the design and manufacture of products and systems possessing both a mechanical functionality and an integrated algorithmic control"*. Anyway, mechatronics has gained currency in recent years by providing a multi-disciplinary approach to engineering design in which a symbiosis of mechanical, electrical, electronic, computer and software engineering is used to create new design solutions to engineering problems. Such innovations vary from development of machine tools, portable video cameras, CD players, and others but it is also applied to process control and measurement technology (Doray and Bradley, 1994; Milne, 1995).

The adoption of the mechatronics approach through education and training should produce engineers with a wide range of, and perhaps new, skills and attitudes. These professionals should be able to work across the boundaries of engineering disciplines in a true multi-disciplinary approach. Therefore the educational institutions must respond to these challenges, offering mechatronics engineering programmes at undergraduate and postgraduate levels. In recent years, mechatronics has been gaining an increasingly prominent place in higher education all over the world, as evidenced by a number of courses implemented so far. This paper principally details the development of a mechatronics engineering degree in Brazil. It also outlines some experiences in the UK providing a list of existing course as well as some detail of such programmes. The paper concludes with a recognition that mechatronics education is vital to face the new challenges of this growing field of engineering.

2 MECHATRONICS COURSE - EXPERIENCE IN THE UK

In recent years, mechatronics engineering has gained a distinguish place in higher education all over the world. This can be seen from the rapidly increasing number of courses offered by higher education institutions in many countries such as Australia (Evans et al., 1995), China (Ye et al., 1995), England (Seneviratne, 1996), Finland (Airila, 1995), Germany (Janocha, 1993), Hong Kong (Venuvinod and Rao, 1994), Japan (Emura, 1995), Russia (Vodovozov, 1995), Scotland Fraser and Milne, 1993), and USA (Rizoni, 1995). More specifically, mechatronics is gaining a prominent place in UK higher education as evidenced by a number of undergraduate and postgraduate taught courses being provided (Hewit and King, 1996). At the present it is estimated that there are around 34 universities and colleges offering degrees and Master's in mechatronics. Table 1, expanded from the work of Acar (1996), summarises courses offered by various institutions in the UK.

The mechatronics engineering programme at the University of Hull serves as a good example of a successful mechatronics curriculum. The department has been active in the field of robots and mechatronics for over 15 years in research as well as education. There are also a wide range of research projects at the Department of Electronics including lasers for assembly of electronic components, virtual reality robot, computer controlled variable-ratio gearbox, and a distributed controller for a six-robot gantry system. The use of novel materials such as magneto-and electro-rheological fluids, ferro fluids, magnetostrictive ceramics and artificial muscles based on chemically actuated polymers are also currently actively investigated. In 1988 the School of Engineering and Computing identified the need for an undergraduate programme to address the topic of mechatronics. A multi-disciplinary approach was created, from which the combined characteristics of computing, electronic, mechanic and human interaction disciplines are developed effectively through a project-based course with additional support of conventional lectures and tutorials. The School is well-equipped with undergraduate laboratory and computer facilities appropriate for mechatronics. There are a number of projects set for the students at different levels, e.g. a mechanical-based project, followed by another which considers software, control, interface and computer control, then a project which is directly related to a industrial problem, and a final year individual project. Recently, the department has been awarded a prize in an undergraduate student robotics project at the Robotix '96 Exposition, in Glasgow (see Figure 1).

Table 1 Mechatronics Undergraduate and Postgraduate Courses in the UK

University	Undergraduate	Postgraduate (MSc)
Bedford College	•	
Blackpool and Fylde College	•	
De Monfort University	•	•
City University		•
Crawley College	•	
Falkirk College	•	
Glamorgan University	•	
Glasgow University	•	
John Moores University	•	
King's College London	•	•
Lancaster University	•	•
Leeds Metropolitan University	•	
Loughborough University		•
Manchester Metropolitan Univ.	•	
Middlesex University	•	
Nottingham Trent University	•	
Teesside University	•	
University College Warrington	•	
University of Abertay Dundee	•	•
University of Brighton	•	
University of Central Lancashire	•	
University of Dundee		•
University of Hull	•	•
University of Leeds	•	
University of Surrey	•	
University of the West England, Bristol	•	
University of Walles College	•	
University of Westminster	•	
Salford University	•	
Sheffield Halam University	•	
Staffordshire University	•	
Stoke on Trent College	•	
Sussex University	•	
Swansea Institute of HE	•	

• = Existing Programmes.

Figure 1 Mechatronics Project Awarded a Prize at Robotix'96.

Another example of a mechatronics degree in the UK is provided by Manchester Metropolitan University. In fact, two courses are available, a BEng Integrated Engineering Systems and a BEng Mechatronics. These courses were proposed to serve the needs of industry for graduates with wide ranging competence, especially in computer technology and modern techniques applied to manufacturing and process industries. Each of them has different emphasis. The Integrated Engineering Systems degree, as its name suggests, is aimed at systems design and implementation and looks globally at the characteristics of equipment and processes rather than being concerned with detailed design. The mechatronics degree is also concerned with systems, but in particular for automation in the manufacturing industries. There is a great emphasis in this degree on detailed engineering design particularly in areas of robotics and computerised control. Both courses appear to offer good job prospects and meet the general objectives of a broad based engineering degree.

There are not only established but also emerging mechatronics engineering programmes in the UK. For example, Sheffield Hallam University will launch its mechatronics engineering course in the academic year of 96-97. The proposed course has been developed on the basis of a design oriented programme with some specialisation in the final year to allow student to focus on a theme of their choice.

3 MECHATRONICS INITIATIVES IN BRAZIL

One of the first mechatronics engineering courses to appear in Brazil in 1988 was in the Polytechnic School (POLI) at the University of São Paulo (USP). The course emphasises systems automation, as the main objective is the integration of mechanical and electronic engineering with computing science. The course was primarily concerned with industrial automation due to a request from the Federation of Industries of the State of São Paulo (FIESP) but now involves aspects of design theory, and bio-engineering (Miyagi, 1996).

The implementation of a mechatronics course described in this paper is being carried out at the Mechanical Engineering Institute (FEM) at the State University of Campinas (UNICAMP). UNICAMP was founded on October 1966, but even within the reality of the Brazilian university setting, where the oldest university has been operating for less than 60 years, UNICAMP may be considered a young institution. Nevertheless, it has already attained a strong tradition in education and technological research. Today, UNICAMP has some 8627 undergraduates in 46 courses, and 6129 enrolled in its 144 graduate level courses (Master's and PhD levels). The Mechanical Engineering Institute is one of the faculties at the UNICAMP. It is divided into seven departments: Computational Mechanics, Energy, Manufacturing Engineering, Materials Engineering, Mechanical Design, Oil Engineering, and Thermal Energy and Fluids.

The basic engineering curriculum of the Mechanical Engineering Institute at the UNICAMP consists of a five year programme. Each year is subdivided into two semesters. The first two years of study are devoted to a basic scientific education; mainly mathematics, physics, chemistry, material engineering, engineering design, and some laboratory activities. However, even considering the success of its traditional mechanical engineering course, the necessity of development of a multi-disciplinary approach to engineering has started to gain academic and industrial acceptance. Consequently, it has been decided to propose a mechatronics engineering programme.

Looking into the activities performed by a professional in the field of mechatronics two relevant areas were be highlighted (Altemani et al., 1996): the conception, design, and operation of automated production systems, and the design and manufacturing of automated systems involving software and hardware engineering. Considering these areas, in 1994, the Brazilian Ministry of Education defined the disciplines required for a mechatronics engineering programme, namely:

Process Control, Industrial Systems, Instrumentation, Discrete Mathematics for Automation, Industrial Computing Science, Manufacturing Management, Systems Integration and Assessment.

It is quite clear that these disciplines require a strong integration between mechanical, electrical engineering and computing science in order to propose a undergraduate course in mechatronics. In the beginning of 1995, a committee was formed to elaborate a proposal for a mechatronics engineering degree, described below.

3.1 Structure of Mechatronics Engineering at FEM-UNICAMP

There are various factors which justify the implementation of a mechatronics engineering course at FEM-UNIC(Altemani et al., 1996). After extensive discussion, the committee has identified some needs and requirements for a mechatronics engineering curriculum. The curriculum structure should involve, of course, the disciplines required by the Ministry of Education as well as a large number of subjects incorporating laboratory activities, i.e. 'hands-on' experience (correspondent to 20% of the total number of hours). It is worth noting that the laboratory activities were scheduled after the corresponding taught disciplines in order to have a theoretical foundation for them. Furthermore, a balance between the disciplines of mechanical, electrical engineering and computing science should be achieved considering that the basis of this programme should be in mechanical engineering. Another interesting aspect is the effort to allocate technical disciplines in the early semester to enthuse the students right from the beginning of the course. The course structure along with the proposed disciplines can be seen in Table 2 at the end of this paper. These disciplines have, on average, three hours per week.

Figure 2 shows the distribution of the disciplines by area. Note that some of the areas shown in Figure 2 (Biology, Chemistry, Energy, etc.) are demanded by the Ministry of Education for any engineering course in Brazil. It can be seen that the course has a strong orientation towards design, computing science, electronics engineering, and manufacturing processes. The proposed programme is to be launched at the beginning of the 1998 academic year, with an initial intake of 50 students per annum.

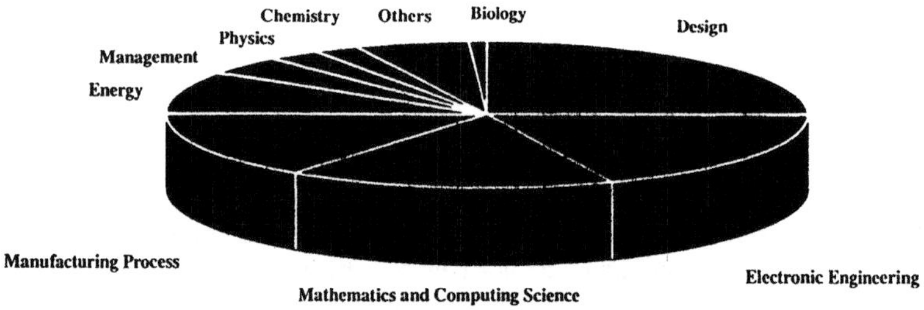

Figure 2 Distribution of Areas in the Mechatronics Programme.

4 CONCLUSIONS

In recent years there are changes in industry to more integrated production especially with the introduction of advanced manufacturing technologies (e.g. FMS and CIM) in addition to new design philosophies to obtain the best possible solution to products or processes. Design of such products or process must, therefore, consider multi-disciplinary approaches. Undoubtedly mechatronics engineering solutions are the key factor to achieve such goals. However, the present driving force for further development of mechatronics must embrace the need of an engineer working in a hybrid environment involving mechanics, electronics and computing. In response to these challenges the educational system must adapt to this new reality. Education establishment have been developing courses aimed at producing the technologically aware engineers for the next century. It is believed that these programmes must consider:

- Activities based on team working and case studies.
- Balance between taught courses and project based learning.
- Design based projects and hands on experience in problems solving.
- Ability to plan and apply a mechatronic approach to a technological problem.

It has been shown that the aspects of a mechatronics approach to education must involve the previous statements. The results of such an approach should bring many benefits for both industry and academia. This paper has highlighted the importance of mechatronics education. It listed mechatronics programmes currently applied in the UK and described the philosophy, structure and contents of a mechatronics engineering degree in Brazil. The aim of such a development is to build a more complete understanding of this growing field of engineering.

5 ACKNOWLEDGEMENTS

The authors would like to express their appreciation to the committee members of the mechatronics engineering course, Dr C. Altemani, Dr A. Diniz, Dr E. Nóbrega, and Dr S. Button of the Mechanical Eng. Institute (FEM), and to Dr M. Andrade Neto of the School of Electrical and Computing Eng. (FEE), all of them from the State University of Campinas (UNICAMP). Thanks are also due to Dr K. Selke, of the University of Hull, and Mr A. Verwer from Manchester Metropolitan University, both from the UK, who helped in offering an informative visit to their department. The authors would like to thank Dr K. Dutton, from Sheffield Hallam University, UK, and to Dr P. Miyagi from USP, Brazil, for providing useful information about their mechatronics courses. Additional thanks to Prof. R. Parkin of Loughborough University for helpful discussion concerning mechatronics education.

6 REFERENCES

Acar, M. How Does the Higher Education World Respond to the Mechatronics Challenge (1996). Colloquium on Mechatronics in Education, Institute of Electrical Eng., Manchester.

Airila, M. Ten Years of Mechatronics Education at Helsinki University of Technology (1995). Int. Conference on Recent Advances in Mechatronics, 266-271, Istanbul, Turkey.

Altemani, C.A.C., Diniz, A.E., Nobrega, E.G.O., Telles, G.N., Andrade Neto, M., Button, S.T. (1995) Undergraduate Course on Mechatronic Eng., FEM, UNICAMP.

Day, C.C. The Mechatronic Manufacturing Solution - Myth or Panacea (1995). International Conference on Recent Advances in Machatronics, Istanbul, Turkey, 259-265.

Dorey, A.P. and Bradley, D.A. (1994) Measurement Science and Technology - Essential Fundamentals of Mechatronics. Measurement and Science Technology, **5**, 1415-1428, 1994.

Emura, T. Mechatronic Education and Research at Tohoku University (1995). International Conference on Recent Advances in Mechatronics, 255-265, Istanbul, Turkey.

Evans, B.J., Shahri, A.M., Naghdy, F. and Cook, C.D. (1995) Developing a Mechatronics Laboratory at the University of Wollongong. Second International Conference on Mechatronics and Machine Vision in Practice, 315-319, Hong Kong.

Fraser, C.J., Milne, J.S. and Logan, G.M. (1993) An Education Perspective on Applied Mechatronics. Mechatronics, **3(1)**, 49-57.
Hewit, J.R. and King, T.G. (1996) Mechatronics Design for Product Enhancement. IEEE/ASME Transaction on Mechatronics, **1(2)**, 111-119.
Janocha, H. (1993) Mechatronics from the Point of View of Geman Universities. Mechatronics, **3(5)**, 543-558.
King, T.G. (1995) Millrights to Mechatronics: The Merits of Multi-disciplinary Engineering. Mechatronics, **5(2/3)**, 95-115.
Milne, J. (1995) Mechatronics: An Integrative Approach. Manufacturing Engineer, 159-162.
Miyagi, P. (1996) Mechatronics Engineering at the POLI-USP. Private Communication.
Rizzoni, G. (1995) Development of a Mechatronics Curriculum at the Ohio State University. Proc ASME International Mechanical Engineering Congress & Exposition, San Francisco, 1-4.
Seneviratne, L. (1996) Mechatronics at King's College London. Colloquium on Mechatronics in Education, Institute of Electrical Engineering, Manchester.
Venuvinod, P.K. and Rao, K.P. (1994) A Mechatronic Engineering Degree Course to Meet the Needs of Hong Kong. Proc. of Mechatronics and Machine Vision in Practice Conf., 52-57, Toowoomba, Australia.
Vodovozov, V.M. (1995) The Educational Resources of Mechatronics. Mechatronics, 5(1), 15-24.
Ye, Z.Q., Xun, L.B. and Tse, P.W. (1995) Development of Mechatronic Engineering in the Special Economic Zone, Shenzhen, China. Proc. of the 2nd Int. Conference on Mechatronics and Machine Vision in Practice, 331-332, Hong Kong.

7 BIOGRAPHY

Geraldo N. Telles

Dr Geraldo Telles graduated in Mechanical Engineering at the State University of Campinas (UNICAMP), Brazil, in 1971. He then worked as a research engineer at the Technology Centre at UNICAMP, Nardini Machine Tools Company, and at the Centre for Computing Science of the Ministry of Science and Technology. Dr Telles obtained his PhD in 1990 and became a senior lecturer in the Mechanical Engineering Institute at UNICAMP. At present he is Associate Professor and his research activities involve CAD/CAM systems (interface development), CNC machines (simulators and automatic programming), industrial robots, FMS, and CAPP (features strategies).

Paulo A. Cauchick Miguel

Dr Paulo Miguel graduated in Industrial Engineering at the Methodist University of Piracicaba (UNIMEP), Brazil, in 1986. He then worked as a manufacturing engineer for Allied Automotive and Varga companies until 1990 when he became a lecturer at the UNIMEP. In October 1993 he moved to the University of Birmingham, UK, to join the Mechatronics group in the School of Manufacturing and Mechanical Engineering. His research activities include quality and metrology developments, and engineering education. Dr Paulo Miguel completed his PhD in December 1996.

Tim G. King

Professor Tim King BSc (Eng), MDes *RCA*, PhD, FRSA is Professor of Mechanical Engineering at the University of Birmingham where he heads the Mechatronics Research Group in the School of Man. and Mech. Eng. His current research includes development of ultra-high speed piezoelectric actuators, piezoelectric motors, calibration techniques for surface metrology and CMM applications, and microprocessor and DSP based systems for computer-visual inspection and controls of deformable patterned materials. He is author of over 90 journal and conference papers.

Table 2 Structure of the mechatronics engineering course at the Mechanical Engineering Institute (FEM) at UNICAMP in Campinas, Brazil.

Year	*Semester*				*Module Organisation & Curriculum Structure*			
1	1	Calculus 1 Analytical geom. Tech Chemistry			Mechatronics introduction	Workshop lab	Computing lab CAD 1 Algorithms	
	2	Calculus 2 Physics 1 Physics lab		Engineering materials 1 (properties & structure)	Electrical circuits (ac, dc)	Statistical methods	Numerical methods Numerical methods lab	
2	3	Calculus 3 Static		Continuos mechanics Materials test lab	Electronics introduction Electronics circuit lab	Introduction to Man. Eng.	Prog. by objects Prog. by objects lab	
	4	Calculus 4	Heat & fluid flow 1	Dynamics Strengh of Materials	Logic circuits Electronic lab			
3	5	Basic for discrete mathematics	Heat & fluid flow 2	Vibration Mech. of machines Mech. systems control 1	Microprocessors Electro-magnetic theory Logic circuits lab	Machining process		
	6		Thermal systems Heat & fluid flow lab	Mechanical systems Digital systems	Microprocessors lab Electrical machines Instrumentation	Manufacturing automation	Programming in real time Intro. to FE analysis	
4	7		Fluid-mech. systems Thermal machines Thermal systems lab	Control of mechanical systems 2 Dynamic tests lab	Introduction to robotics Electro machine devices Electric machines lab	Industrial installation	CAM/CAPP Prod. Planning & control (by computer) Software engineering	
	8		Thermal machines & fluid mech. systems	Advanced control		Quality engineering Industrial autom.	Local Networks CAD 2 Artificial Intelligence	
5	9				Mechatronic design Topics in mechatronics 1	Quality systems Industrial environment control	Simultaneous engineering	Organization & management Law & legislation Undergraduate project 1
	10				Topics in mechatronics 2			Undergraduate project 2 Technical training

PART FIFTEEN

Manufacturing Management Strategies

36

Cost-Oriented Strategic Manufacturing and Quality Planning for BPR

I. Mezgár
Computer and Automation Research Institute,
1111 Budapest, Kende u. 13-17. Hungary,
Phone:36-1-1811143, Fax:36-1-1667503, E-mail: mezgar@sztaki.hu

C. Basnet and L. R. Foulds
The Univ. of Waikato, Private Bag3105, Hamilton, New Zealand,
Phone:64-7-8384562, Fax:64-7-8384270,
E-mail:chuda@waikato.ac.nz

Abstract

The different manufacturing strategies can influence production to a significant extent and, through this, manufacturing process and product quality. A stochastic simulation-based approach has been developed and carried out for estimating product quality on strategic level. The methodology is based on the Taguchi method, and consists of a simulation of the manufacturing system with a view to identifying manufacturing strategies that are relatively insensitive to process fluctuations and are cost effective. The experimental analysis is carried out with the strategies as controllable factors, and process fluctuations as noise factors. The response variable of the experiment is the unit cost of the finished product. While the Taguchi method is usually applied to the tactical design of production systems, we have shown how it can be applied to the strategic design and analysis of manufacturing systems. This approach can help the Business Process Reengineering (BPR) as well.

Keywords

Cost reduction, manufacturing management strategies, product quality, simulation, Taguchi method.

1. INTRODUCTION

Traditional manufacturing has focused on a low cost strategy to capture the market, and consequently, to improve the bottom line. The techniques of flow design, work design, and automation have been the traditional mainstay in accomplishing this strategy. Recently, competitive pressures have pushed quality to the centre stage of manufacturing management. Adoption of a quality strategy has entailed worker involvement, statistical process control, just in time, automation etc. The process-oriented, higher level description of manufacturing-, quality-, and management activities can be integrated in the business processes. The reengineering (flexible adaptation to the actual market environment) of these business processes has a key role in the market competition. In designing manufacturing systems with the goal of high quality, we need to adopt quality strategies which can cope with the vagaries of manufacturing such as process variabilities. This does not imply that cost is no longer a consideration in manufacturing. These types of manufacturing systems has to be designed and operated in a way that yield consistently high quality and, which are, at the same time, cost effective.

There are many factors that influence product quality during the product life-cycle. The technical related variables during the design and manufacturing phase can be called as primary factors (e.g. tolerances), but the management-related factors have significant influence on product quality as well. The realisation of both the quality- and manufacturing management strategies need a longer introductory time, and the real results of these technologies will appear only after a longer period.

So, the thorough analysis and selection of the different quality- and manufacturing management strategies have a very important role. The stochastic simulation is a proper tool to do this analysis and evaluation before introducing any quality, or manufacturing management strategies in a manufacturing system. An additional advantage of the preliminary analysis can be that a robust strategy can be selected by combining the Taguchi methodology with the design of experiment approach. By doing such analysis industrial firms can make more reliable decisions concerning their planned quality assurance and manufacturing strategies without disturbing the actual continuous production. The Taguchi methodology can be applied also in the parallel technical design of products and manufacturing systems as described in Mezgar (1995).

Taguchi has proposed a method of parameter design which provides a means of selecting process parameters which are least sensitive to process variations, i.e., are robust. This process selection is at the tactical level - detailed day to day operation of the manufacturing plant. In this paper we extend the Taguchi method to the strategic design and analysis of a manufacturing system and present a methodology for selecting quality strategies which are robust and cost effective. In evaluating quality strategies, we consider not only the cost of such strategies, but also the benefits (improvement in process capability). This provides a quantitative basis for the selection of quality strategies. The manufacturing strategies of SPC, JIT, and automation are tested in a simulated manufacturing scenario, and their effects on quality costs are investigated. From these, the strategies are evaluated on their robustness from the viewpoint of quality.

2. QUALITY- AND MANUFACTURING MANAGEMENT STRATEGIES

The quality assurance and manufacturing management strategies have a great importance in the market competition especially in economies that are in transient period from any aspect (like in Central Europe as introduced by Foulds and Berka (1992)), or that intends to raise their competitiveness radically (Hyde et.al., 1993). New forms of manufacturing architectures, structures have been developed as well that can give more appropriate answers for the market demands. These strategies are part of the business processes reengineering.

In order to keep their market positions manufacturing enterprises have a strong motivation to move from large, hierarchical organisations to small, decentralised, partly autonomous and co-operative manufacturing units, which can respond quickly to the demands of the customer-driven market. Parallel with these organisational changes the type of production is also changing from mass production techniques to small lots manufacturing. The autonomous production units geographically could be found both inside the enterprise, but outside as well, physically in a long distance. In all of the new forms of the manufacturing organisations different optimal quality- and manufacturing management strategies can be applied.

2.1 Quality assurance strategies

The quality of the product is an aggregate of the quality of individual features (geometrical characteristics) and properties (e.g. material). The main difference between traditional and new quality philosophies is that traditional quality approaches focus on correcting mistakes after they have been made while the new philosophies concentrate on preventing failures. For measuring the quality (based on tolerances) the process capability index and the quality loss function can be used. Quality should be designed into the product and into the processes involved when designing, fabricating and maintaining it through its life cycle. Total quality management (TQM) means the continuous satisfaction of user requirements at lowest cost by minimum effort of the company. Taguchi is noted chiefly for his work on designing products and processes with a view to making the quality of the product robust (insensitive) to process or product variations (Taguchi et.al., 1989). His approach have found widespread dissemination and application (e.g. Ross, 1988).

The main reasons why product quality characteristics can deviate from the predetermined values during manufacturing are the inconsistency of material, tool wear, unstable manufacturing process, operator errors, low level system management. In on-line quality control those methods are applied that help to raise the efficiency and stability of the production process (e.g. SPC, diagnosis). In Drucker (1990) the author highlights the role of statistical quality control (SQC) in manufacturing. He sees SQC as improving not only the quality and productivity of an organisation, but also the dignity of labour.

2.2 Manufacturing management strategies

The automation which is based on computer technology can give a manufacturing company a big competitive advantage. The different forms and levels (e.g. high-level integrated automation in Computer Integrated Manufacturing Systems vs. balanced automation) of automation needs different manufacturing strategies.

The human resource management strategy influences all important competitive factors. The education level of the workers, employees can increase/decrease the quality, the time-to-market as well.

The inventory management strategies has a big influence on cost and on throughput time. The introduction of Just-In-Time (JIT), or other strategies need a careful analysis because there is a big financial risk which one to choose.

3. THE TAGUCHI METHOD

The methodology for quality control developed by Taguchi et.al.(1989) covers the quality assurance involving both product design and process design stages. Taguchi follows an experimental procedure to set parameters of processes. In Taguchi's method of parameter design, the factors affecting the outcome of a process are separated into controllable factors and noise factors. Controllable factors are the parameters whose value we seek to determine. Noise factors are factors which cause variation in the response variable, but are normally not amenable to control (but they are controlled for the sake of the experiment). Treatment combinations for controllable factors are set by an experimental design called the inner array, treatment combinations for noise factors are called the outer array. Usually, to save cost and time, a fractional design is used for these arrays, where there are three controllable factors and three noise factors. The controllable factors are set in a full factorial design, but the noise factors are set in an L4 array. Usually, an analysis of variance is not the object of Taguchi-style experiment. The object is to identify the set of parameter values that is most robust to the variations in the noise factors. Taguchi proposes a measure called Signal to Noise (S/N) Ratio. When the response variable is such that a lower value is desirable, one measure of S/N ratio is:

$$\text{S/ N Ratio} = -10\log_{10}\left(\frac{1}{n}\right)$$

where y_i is the response variable, and there are n observations in a run. A high value of S/N ratio indicates that the corresponding combination of controllable factors is robust.

4. STRATEGIC DESIGN AND ANALYSIS OF MANUFACTURING SYSTEM

With the development of modern simulation languages, simulation has established its place in the design of manufacturing systems. Simulation permits modelling of manufacturing systems at any desired level of detail. Controlled experiments can be carried out on a simulated environment at a low cost.

We suggest the use of this environment for the selection of manufacturing strategies following the Taguchi method for manufacturing systems both under design and in operation. The effect of these strategies on the quality levels can be modelled, and experimented upon. For example, the continuous improvement aspect of JIT can be modelled by using a learning curve model. The Taguchi method permits us to select robust strategies. An application of this simulation methodology is presented next.

4.1 Selecting performance measure and define the target

The goal of the experiment was to identify levels of the manufacturing strategies which were most robust to quality problems. We used cost per unit of good finished product as the surrogate for the cost effective quality level achieved by the manufacturing system. Consequently, this cost was the response variable. The analysis gave an answer that on what levels of the noise factors were the production cost on minimum while the selected management strategies (factors) are the most insensitive on the noise in the given manufacturing system.

4.2 Identify factors

The controllable factors are the manufacturing strategies that we wish to investigate. The three strategies and their levels are (a full factorial design is used for these factors):
1. Statistical process control (Yes /No)
2. Automation (Yes /No)
3. JIT learning (Yes/No)
The noise factors that cause variation in the quality of the product and their levels are:
1. Nominal process capability of the processes (Higher/Lower)
2. Complexity - number of serial processes (Low / High)
3. Process shift (Low/High)
All the factors are at two levels. An L4 design is used for the noise factors.

4.3 Conduct experiment

A simulated environment of a manufacturing system was created using the simulation language VAX/VMS SIMSCRIPT II.5. (introduced e.g. in User's Manual, 1989). The following assumptions apply to the system. It comprises a number of serial processes. All the processes work to their nominal process capability. The processes are liable to a process shift which causes the process capability to decrease. The distribution of the interval (number of parts) between process shifts is Poisson. However, if nothing is done to the shifted process, it will reset to the nominal capability after another interval, which has the same distribution. There is a nominal cost of processing for each part. After all the operations (serial) are carried out, the quality of the part is checked against the upper (USL) and the lower (LSL) specification limits for all the processes, and if any measurement is out of specification, the product is scrapped.

Three quality strategies are considered: statistical process control (SPC), just-in-time (JIT), and automation. When SPC strategy is adopted, samples of size n are taken at an interval of f parts. Upper and lower control limits (UCL and LCL) are calculated on the basis of 3 sigma, and if the process is found out of control, the process is reset to the nominal process capability. There is an additional cost (Cspc) of manufacturing, attributable to the adoption of SPC.

As pointed out earlier, many benefits of automation strategy have been identified in the literature. As far as the quality of the product is concerned, the adoption of automation strategy results in an increase in the process capability. There is an additional cost (Cauto) per part associated with this strategy.

Similarly, the advocates of JIT have claimed many advantages for this strategy. It is agreed that JIT is both a philosophy and a discipline which results in a continuous improvement of the processes of a manufacturing organisation. In our simulated environment, the adoption of JIT causes the process capability to increase over time, following a learning curve . Again, there is an additional cost (Cjit) incurred for each part. Some of the parameters and their values which were applied during the investigation of the manufacturing strategies are the following:
Nominal process capability: 0.9 (high) and 0.8 (low)
Amount of process shift: 0.4 (high) and 0.3 (low). When the process shifts, the process capability decreases by this amount.
SPC cost Cspc (additional to the nominal cost of $1): $0.15
Increase in the process capability due to automation: 0.2
Automation cost Cauto (additional to other costs): $0.20
Learning curve under JIT (percentage): 99.5%. Doubling of the number of products causes 99.5% increase in process capability. JIT cost Cjit(additional to other costs): 0.1

4.4 Analyse Results and Select Manufacturing Strategy

The Taguchi method and inside this the S/N ratio formula for the smaller-is-better has been used since the goal of the experiment was to minimise cost. The result of the simulation experiment is shown in Table 1.

Table 1. Results of the Experiment

			Nominal C_p	2	2	1	1		
			Process-shift	2	1	2	1		
			Complexity	1	2	2	1		
Run	*SPC*	*JIT*	*Auto*	*Average cost/unit for 3 replications*				*Average*	*S/N*
1	1	1	1	1.43	1.27	2.08	1.47	1.56	-4.04
2	1	1	2	1.45	1.35	1.84	1.47	1.52	-3.73
3	1	2	1	1.37	1.27	1.85	1.42	1.48	-3.50
4	1	2	2	1.45	1.39	1.72	1.47	1.51	-3.59
5	2	1	1	1.59	1.45	2.24	1.66	1.73	-4.91
6	2	1	2	1.60	1.59	1.97	1.63	1.68	-4.55
7	2	2	1	1.54	1.43	2.01	1.59	1.64	-4.38
8	2	2	2	1.61	1.55	1.89	1.64	1.67	-4.49

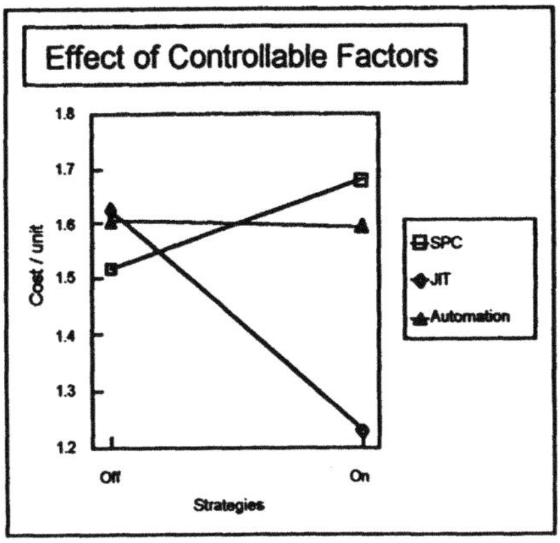

Figure 1. Main Effects of Controllable Factors.

From this table, it can be seen that the most robust strategy for this scenario is the use of JIT, without the adoption of SPC and automation strategies. This gives the lowest average cost per unit of $1.48, and the highest S/N Ratio of -3.50. The main effects of the controllable factors are shown graphically in Figure 1. Obviously, JIT has a profoundly desirable effect on quality measured as cost per good units. The effect of automation is to decrease the cost only slightly, while the use of SPC strategy increases costs.

5. CONCLUSIONS

We have presented a methodology for the selection of manufacturing strategies. This methodology is based on the Taguchi's method of parameter design and takes into account the current process capability, possible process shifts, the complexity of the existing processes, and the costs and the benefits of the proposed strategies. The methodology was demonstrated on an example hypothetical manufacturing system.

The Taguchi method is normally applied to the tactical design of the production system. Our methodology presents an application of the method to the strategic design or analysis of manufacturing systems. This methodology is extendable to include other manufacturing strategies such as total quality control and supplier certification. The methodology also can be used to analyse the different subprocesses of business processes during reengineering. As a result of using the methodology the firms (mainly little firms - SMEs) can adapt themselves easier, quicker and with less risk to the fast changing market demands by changing their production strategies and process & product quality levels.

6. ACKNOWLEDGEMENTS

The first author is grateful for the Academic Research Visitors Grant that he received from the School Research Committee of the University of Waikato, which made it possible for him to carry out the work.

7. REFERENCES

Drucker, P.F. (1990) The Emerging Theory of Manufacturing, *Harvard Business Review*, May-June, 1990, 94-102.

Foulds, L.R. and Berka, P.(1992) The achievement of World Class manufacturing in Central Europe, In *Proc. of the IFORS SPC-2 Conference on "Transition to Advanced Market Economies", June 22-25, 1992, Warsaw*, Eds.: Owsinski, J.W., Stefanski, J., Strasyak, A., pp139-145.

Hyde, A., Basnet, C., Foulds, L.R. (1993) Achievement of world Class Manufacturing in new Zealand: Current Status and Future Prospects, In *Proc. of the Conference on "NZ Strategic Management Educators", Hamilton, New Zealand*, NZ Strategic Management Society Inc., pp.168-175.

Mezgar, I.(1995) Parallel quality control in manufacturing system design, *in the Proc. of Int. Conf. of Industrial Engineering and Production Management, April 4-7., 1995. Marrakech, Morocco*, Eds. IEPM-FUCAM, pp. 116-125.

Ross, P.J. (1988) Taguchi Techniques for Quality Engineering, McGraw-Hill Book Comp., New York-Auckland.

Taguchi, G., Elsayed, E., Hsiang, T.C. (1989) Quality Engineering In Production Systems, McGraw-Hill Book Company, New York.

VAX/VMS SIMSCRIPT II.5 User's Manual, (1989) Release 5.1, CACI Products Company, February 1989.

8. BIOGRAPHY

István Mezgár [Diploma in m.e.: 1977, special Diploma in automation engineering: 1988, dr.Univ in m.e.: 1990, all from the *Technical University of Budapest (TU)*, C.Sc. in m.e.: 1995, from the *Hungarian Academy of Sciences*] is a senior researcher at the *CIM Research Laboratory, CARI, HAS.* Visiting researcher/professor at the *University of Trento*, at the *University of Genoa* (Italy, in 1989-91), at the *University of Waikato* (New Zealand, in 1995), and at the *CAD/CAM Research Laboratory, Korea Institute of Science and Technology* (South-Korea, in 1995-96). IPC member of several international *IFAC, IFIP* conferences. More than 70 scientific publications. His current interests focus on manufacturing automation, integration of product and enterprise models, agile manufacturing systems, design methodologies (e.g. concurrent engineering) and complex quality control.

37

Experiences in information systems development for maintenance management: a techno-organisational view

Paula A. Silva[1], António L. Soares[1,3], José M. Mendonça[2,3]
[1]INESC - Manufacturing Systems Engineering Unit, R. José Falcão, 110, 4050 Porto, Portugal. Email: {psilva,asoares}@inescn.pt
[2]Agência de Inovação, R. de Sagres 11, 4150, Porto, Portugal. Email: jmm@adi.pt
[3]FEUP/DEEC Univ. of Porto, R. dos Bragas 4099, Porto Codex, Portugal.

Abstract

The need of a functional integration in the manufacturing area (Production, Quality, Maintenance), outlooking for a new manufacturing agility, causes the emergence of concerns with reliability, maintenance and security of the manufacturing equipments at the shop-floor. Several models and methodologies, encompassing technical, human, social and organisational aspects such as Total Preventive Maintenance (TPM) or Continuous Improvement (Kaizen), are today available to help in implementing new paradigms in this area. In this paper, we briefly relate our experiences in the development of a maintenance management information system in a shoes manufacturing company, highlighting the socio-organisational context influence on this development. This project was part of the ESPRIT project Real-I-CIM, aimed at providing low cost shop floor advanced management tools, inside an open and distributed architecture.

Keywords

Maintenance management, shop-floor management, techno-organisational development.

1. INTRODUCTION

The openings to new markets more and more competitive and the rapid development of technologies, globally available, compel industrial companies in searching improvements, both technological and organisational. Several models and methodologies, encompassing technical, human, social and organisational aspects are today available to help in implementing new paradigms such as Total Preventive Maintenance (TPM) or Continuous Improvement (Kaizen). In those paradigms, it is very clear the need of a functional integration in the manufacturing area

(Production, Quality, Maintenance), outlooking for a new manufacturing agility. From this need emerges in particular the concern with reliability, maintenance and security of the manufacturing equipment at the shop-floor. Therefore, one can say that the Maintenance function, embedded in an adequate organisational model, comes out to a key position side by side with the Production and Quality functions.

2. ORGANISATIONAL MODELS OF MAINTENANCE AND INFORMATION SYSTEMS SUPPORT

2.1 Centralisation vs. decentralisation

In organisational terms, the Maintenance function can be classified as centralised or decentralised, both at the physical and the decision-making levels. Traditionally, a hierarchical centralisation should allow easy cost management, process standardisation and communication, the homogeneous attendance of equipments and its breakdowns and a good maintenance staff management. A centralised Maintenance is indeed characterised by an isolated department where are concentrated all important resources (tools, materials and manpower) to the Maintenance function and where all the decisions concerning planning, intervention, reaction, purchasing, etc. are made. This type of traditional Maintenance management stills diffused in Portugal, and presents only some adequacy to companies with few production equipments and simple design, as well as breakdowns with a small weight in the company budget.

In a decentralised organisation, people not belonging to the maintenance department is actively involved in this function. The formal existence of a maintenance department, holding specific knowledge and historical data, provides (in principle) the capability to undertake all the required maintenance assigned tasks such as reacting in face of breakdowns, planning preventive interventions, repair spare parts, continuously improve the equipment, security areas and work environment. The distribution of some of these tasks for other manufacturing related departments, has the potential to simplify the maintenance tasks and involve a significant number of people, enabling a better planning, repairing costs and time reduction, production stops due to breakdown, etc.

2.2 Total Productive Maintenance and Continuous Improvement

In Takahashi and Osada, 1990, it is advocated a decentralized organisation, with a total involvement of everyone in the company, adopting an equipment oriented management from which depends the reliability, safety, maintenance and operational characteristics of the shop-floor. The main goal of this concept is the way to "zero defects" and "zero breakdowns". The implementation of a Total Productive Maintenance (TPM) philosophy implies a decentralisation in decision making regarding maintenance, requiring the serious concerns about pervasive aspects such as the workers professional qualification, incentives creation, vocational training, professional careers, responsibility assignment, subcontracting quality, inter-departmental co-operation, etc. Correctly considered these aspects, where the key element is the human factor (actually more important than technological or organisational impositions), the result is the reliability improvement in the shop-floor and as consequence the improvement of the internal competitiveness factors: productivity, quality, delivery on-time and costs (Takahashi and Osada, 1990; Ishikawa, 1982). The fulfilment of these goals in a scope of a decentralised philosophy raises, as referred, communication and interaction needs between the different company departments, excluding some isolating perspectives in functions such as Quality Control, Industrial Engineering, Production Control, Production Maintenance and Cost Control.

Continuous Improvement philosophy (Kaizen), foresees an involvement of every manager and worker in continuously improving the company competitiveness, particularly through adequate Quality and Maintenance systems. Once more, it is fundamental the use of technical systems supporting a decentralized organisation through integration and communication mechanisms (Mizuno, 1988).

The mentioned organisation philosophies (TPM and Kaizen) develop naturally in a decentralised organisation. Currently, in the portuguese scenario, we can find some difficulties in their implementation due to deeply rooted Tayloristic concepts in the organisation, together with an average low skilled work force, both in technical and social competencies. However, survival lead those companies to (at least) assume the importance of decentralising decision making particularly in Maintenance management.

2.3 Information systems support

In this context it is clear that tools supporting the implementation and exploitation of TPM and Kaizen can play a role of utmost importance. However, IS's developed under the requirements of a tayloristic and hierarchical organisational structure can greatly hinder the adoption of TPM and Kaizen, since frequently result in tools supporting individually the shop-floor management functions, satisfying departmental confined requirements. One counter-example is the concept of Process Condition Monitoring - PCM (Ribeiro et. al., 1996) emerging from the need of an integrated view of the operations at the manufacturing execution level: production, quality and maintenance management. This view requires technical solutions that integrate information usually managed and used independently, reflecting the vision of a decentralised organisational structure. In the particular aspect of maintenance, tools based on this concept enable an improved maintenance policy and a better maintenance personnel performance.

In the following sections, we briefly relate our experiences in the development of a maintenance management information system in a shoes manufacturing company. This development was part of the ESPRIT project Real-I-CIM, aimed at providing low cost shop floor advanced management tools, inside an open and distributed architecture.

3. THE SOCIO-ORGANISATIONAL CONTEXT

The company considered in this case study is a shoes manufacturer that, as many other companies in this industrial sector in Portugal, relied its competitiveness on the manpower low costs. However, the increase of those costs in Portugal, when compared with other countries of the Asian south-west, triggered changes in the strategy i.e., the increasing of the manpower qualification in order to produce with more quality and competitive prices, and to fulfil delivery times. This change in strategy called for the use of advanced technologies of production, maintenance and quality, and profound changes to the actual organisation.

3.1 Social characteristics

The human resources characteristics such as qualification levels, age levels, seniority in the company and schooling levels, have a marked influence on the people attitudes towards a techno-organisational change. Participation and change acceptance or resistance are always framed by the company's social context. In the initial phases of the prototypes testing, a study was conducted regarding the analysis of the relevant company's social characteristics using data from the last 3 years (Moniz and Soares, 1995). The main conclusions of this study clearly identified factors capable of positively influencing a process of technical and organisational change: an on-going personnel re-qualification process, a strong job stability, an age structure very young, a significant increase in the schooling levels, an increase in the technical skills and a mix of a large experience with solid technical knowledge. Nevertheless, some negative factors were also identified: an increasing absenteeism rate, very few vocational training actions, a difficulty to recruit personnel with specific technical skills.

A subsequent field-study further extended our understanding of the socio-organisational context by collecting the subjective views of the operators and managers influenced by the information system, concerning technological and organisational innovation within the company (Santos et al., 1996).

3.2 Attitudes towards new organisational forms

The company was facing a restructuring process of its production organisation, although the traditional model was prevailing. Work organisation forms based on individual enriched jobs (more complex tasks, fulfilled with initiative and responsibility) and team work were being adopted, looking for an increased manufacturing flexibility. There was a general positive feeling, from both managers and operators, towards the adoption of team working as the way to reach better results in productive and organisational terms. Managers and operators perceptions and opinions didn't differ to much and are resumed in the following attitudes:

- strong adherence to working groups organisational form,
- preference for a multi-skilled kind of work,
- acceptance of an increased intervention and participation of operators, particularly in preventive maintenance and quality control tasks,
- foreseeing the need to have competencies, at the middle management level, that associate a strong technical competence to a capacity to disseminate information and co-operate with the operators,
- foreseeing the need to change the middle management role from line supervisors to group leaders,
- foreseeing the need to change the role of operators in direction to an increased responsibility sense, initiative and change capability,
- willingness to see implemented broader forms of participation.

However, some limiting socio-organisational factors could restrain the change process. From the same group of people, the following negative attitudes were registered:

- some resistance to change the traditional power rationality restraining the participation of the operators in decision making related with planning and scheduling,
- limited management vision of team work being sometimes considered only as combining individual (simple and repetitive) tasks and doesn't implying necessarily an enlargement or enrichment of tasks,
- more concern in obtaining flexibility and multi-skilling than in increasing the level of operators decision/intervention in managing and organising working teams,
- feeling of a reward system not adjusted to the new competencies and organisation, raising some conflicting situations.

3.3 Attitudes towards information technology

The attitudes towards new technologies, in particular information technology, is another important indicator in the techno-organisational change process. The positive attitude found in the company concerning the introduction of new technologies can be illustrated in the statement of expected benefits both from operators and middle-managers.

In the perspective of the operators, the main advantages of new equipment are mainly the possibility of working in better conditions, an increase in productivity levels and higher quality performance. From the point of view of middle-managers, the perceived advantages are the possibility to increase technical skills and productivity levels, higher quality performance, easiness in the work execution and keeping the pace with technological development in order to stay competitive.

Summarising, there is generalised opinion that the introduction of new technologies at the manufacturing level would enable to manufacture with better quality and an increase in the skills of the organisation members majority. Also, new technologies would contribute to the company modernisation, coping better with the market and making the company more competitive.

3.4 The maintenance organisational context

In regard to the maintenance department, the implemented policy involves a mix between corrective and preventive actions, with more emphasis in the corrective ones. The organisation is centralised i.e., only maintenance staff carry out maintenance tasks. In terms of organisational chart, the maintenance manager reports to the production manager, meaning that all maintenance decisions reflect heavily the production manager views. This has obvious effects not only in the efficiency of the maintenance tasks, but also in the autonomy and responsibility of the organisational unit members.

3.5 The Maintenance Management System features

In functional terms, before the Real-I-CIM installation, the maintenance operators undertook corrective actions when an intervention request was made directly by the production operator to the maintenance department. The breakdown could be immediately solved or transferred for another day if the equipment could be replaced. There was not any time control and consequently time analysis (manpower, equipments, etc.) was not achieved. In a broader scope, the performance evaluation of the department was never accomplished due to the lack of relevant indicators. Other maintenance management tasks such as stocks management, equipment statistical analysis and predictive maintenance were never realised and were not part of the requirements for this system due to the very nature of the shoes manufacturing process and equipments.

The Real-I-CIM tool-box provides a module in the Industrial Maintenance area with the goal of supporting the maintenance tasks (Silva, 1995). The generic features of this module are the following:

- modelling of the physical, informational and functional aspects of resources, orders and manufacturing processes through the use of a powerful set of building blocks;
- easy integration with existing enterprise IT systems through access to manufacturing programmes issued by an upstream MRP system, machine data, etc., by using standard interfaces;
- coding, grouping and characterisation of the enterprise equipments;
- data acquisition and shop floor interaction through appropriate shop floor communication networks;
- continual information updating to the historic generation and important resources management for the maintenance intervention;
- Maintenance intervention planning according production planning;
- Production and Maintenance times optimisation provides the communication between departments to ensure the best planning;
- different levels of user interaction in an integrated, automatic and efficient way;
- time analysis for performance evaluation and improvement support;
- cost analysis and Maintenance policy evaluation;
- functional behaviour of equipments during its different life periods;
- extraction of indicators to estimate the industrial processes performance and/or the management policy;
- shop floor alarm notification.

The module specific features, together with the system envisaged architecture were designed aiming at an effective adaptation to the organisational requirements of each company. Particularly in this company, the system features where directed to following broad functional requirements: equipment inventory management, equipment dossiers, corrective and preventive maintenance management, cost analysis, record and control of maintenance times and reports and extraction of performance indicators.

4. THE PROJECT PROGRESS

From the software engineering viewpoint, the system development life cycle followed a mix between the traditional "waterfall" model and an evolving system prototyping approach. Detailed requirements gathering and analysis followed by the system specification were undertaken with several iterations. The prototypes were developed and installed in different phases, being tested and validated at the end of each one. In this section we analyse the IS development from the point of view of the workers participation and the organisational constraints influence.

4.1 Workers participation

During the process, the well known difficulties of an waterfall approach to systems development revealed, being somewhat compensated later in the process during the phases of prototyping. Particularly during requirements analysis and system specification, several technical and organisational hindrances prevented a smooth running of the project:

- users found difficult to state their requirements for a technology that they did not understand,
- to foresee technical and organisational opportunities for improvement motivated by a new information system was even more difficult,
- technical questions conducing to further details in the system requirements were also found hard to answer by the users,
- different visions for a matching technical and organisational system made complex an agreement about the requirements for the system supported procedures and interactions;

These difficulties were somehow increased by some maintenance operators inhibition in stating their requirements, once this task was seen by them as a production manager responsibility. The formal organisational structure was, in a great extent, the main cause of this behaviour because all the "important" decisions (where the ones concerning the design of technology and work are included) were pulled to the production manager. Moreover, an inculcated tradition of autocratic management did not induce an effective relevant participation in the requirements analysis and system specification.

4.2 Organisational constraints

Regardless the restructuring process and the people feelings about technical and organisational innovation, the company's management imposed an unchanged manufacturing organisation as a major constraint for the new information system. The maintenance centralised organisation, the hierarchy, the use of the already implemented information systems (such as the data collection system), were to suffer minimal changes as to adapt to the new system.

However, in order to promote a best maintenance management service i.e., to be possible the time and maintenance costs analysis and to make the intervention more fast and efficient, the an integration between the Maintenance system and the data collection system was accorded. This fact allowed the direct communication between the shop floor and the maintenance department and changed the behaviour either on production staff (when they ask repairing after a breakdown) or on maintenance staff (to solve a breakdown). In the first prototype the Corrective Maintenance management and the Preventive Maintenance management modules were implemented. The earlier results pointed for a good system adaptation, either in the technical aspects or organisational aspects, and its enormous usefulness in the selection of new shop floor performance indicators.

5. ORGANISATIONAL CHANGE SUPPORT

Despite the requirements imposed by the company management, we believe that a profound change in the maintenance department organisation could be undertook with the support of the technical system described. In such comprehensive changes — possibly undertook within a Business Process Reengineering (BPR) — all the strategy must originate on the top management. Nevertheless, tactical and operational issues of change (both in the functional and social dimensions) strongly rely on everyone within the organisation. Management (in broad) and external consultants establish the vision, strategy and general tactical directions and the rest of the organisation implements it. Thus, management must create conditions for the effective participation of the people involved in those processes. Usually there are several options when it comes to restructure a process or part of a process as established by the BPR plan. Redesigning those parts should take into account socio-organisational and human-computer interaction best-practices. This is the case of the manufacturing reorganisation of this company, and particularly the maintenance services optimisation. In figure one are summarised the main factors influencing the techno-organisational change process.

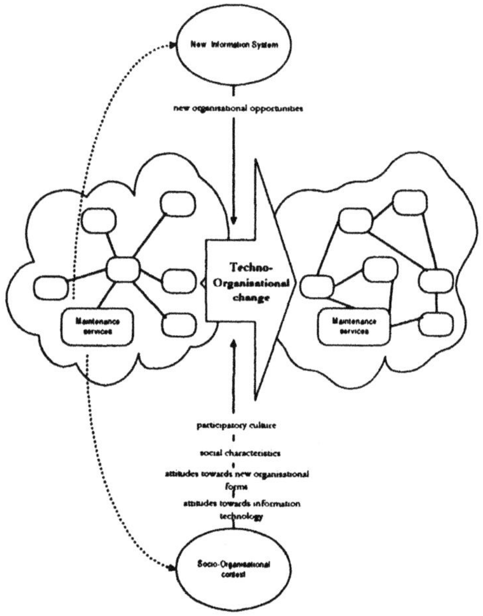

Figure 1 - Some factors influencing techno-organisational change

In the first place, the maintenance department autonomy, reaching a peer position with the production department, would in principle increase the reliability, security, quality and other operational characteristics in the shop-floor, if a decentralisation strategy is pursued. In the second place, this decentralisation would increase the flexibility of the maintenance tasks as well as the motivation potential of the maintenance related jobs. However, this was seen very sceptically by the company management that links this change to high investments without a clear return.

On the other hand the maintenance department had no way to prove, both qualitatively and quantitatively the advantages of the proposed change. It was thus, very important to make use of tools that could help to ascertain the current performance and enable in foreseeing future results.

The maintenance management system, developed within the Real-I-CIM project, can be used to support a decentralised organisation offering features such as:

- cost indicators collection, enabling the evaluation of the costs per maintenance resources and per maintenance intervention, the maintenance policy comparison i.e., corrective vs. preventive, and the quality of the equipment maintenance actions,
- temporal indicators collection, for the evaluation of immobilising times, maintenance times, etc.
- statistical indicators calculation concerning the availability, reliability, confidence levels, etc.

With this basic set of features and information, the maintenance responsible is able to evaluate, foresee and if necessary prove if the actual maintenance policy and maintenance organisation are the best for the company, and this way negotiate with the management the best future strategy concerning maintenance.

6. DISCUSSION

From the experiences depicted in this paper we intended to highlight three main points. Firstly, the technical and organisational requirements for a maintenance management information system. TPM and Kaizen are seen as the main inspiration sources for the maintenance organisation and generate the generic requirements for the information system support.

Secondly, the importance of the socio-organisational context in the system requirements definition and development and, more generally, in a more comprehensive change process such as BPR. It is essential the full participation of the company's stakeholders particularly the end-users, at least in the requirements definition phase, in order to obtain an effective and usable system. Aiming at this goal, the company's social characteristics (qualification, job stability, training, age structure, etc.), and the people's attitudes towards techno-organisational change must be assessed, providing the information on what to expect concerning the type and quality of participation. In this case study, we found good conditions in the company for an involvement of the end-users in designing the new system. However, this involvement was hindered by some organisational constraints and lack of willingness from the management side.

Thirdly, the influence of a maintenance management information system in the organisational change towards a decentralised organisation and in the implementation of best-practices in maintenance management. This is a most difficult topic because it involves: the *system developers* in orienting the system implemented features and the system adaptability towards a range of organisational needs; the *system end-users* in generating the requirements for a system in a maintenance organisation with redesigned tasks and structures; and finally the *company's management* in providing the vision and commitment as well as resources for the undertaking of the project. In the paper we identified some basic features in this system that could support an alternative maintenance organisation within this company.

7. REFERENCES

Ishikawa, K., 1982, "Guide to Quality Control", Quality Resources, White Plains NY.

Mendonça, J.M., Schulte, J.B., Távora, J., "Production Planning and Control Systems and their Factory Implementation ", Sharing CIM Solutions, K.H. Knudsen et al. (Eds.), IOS Press, 1994.

Mizuno, S., "Management for Quality Improvement: The 7 New QC Tools", Productivity Press, Inc., Cambridge MA, 1988.

Moniz, A., Soares, A., 1995, Techno-Organisational Analysis for Systems Implementation: Social Characteristics Analysis, EP 8865 Real-I-CIM deliverable D1.2.2.2.

Nayatani, Y., T. Eiga, R. Futami, H. Miyagawa, "The Seven New QC Tools: Pratical Applications for Managers", Quality Resources, White Plains NY, 1994.

Ribeiro, B., Silva, P., Mendonça, J., 1996, "Process Condition Monitoring - a Novel Concept for Manufacturing Management Tool Integration"; presentation document, Março.

Ribeiro,B., Silva, P., Távora, J., Mendonça, J., Romão, N., "Requirements Specification for the Growela Pilot Site", Real-I-CIM deliverable, 1994.

Santos, M., Urze, P., Moniz, A., Soares, A., 1996, Techno-Organisational Analysis for Systems Implementation: Real-I-CIM Support to Organisational Change, EP 8865 Real-I-CIM deliverable D1.2.2.2.

Silva, P., "A Relevância da Função Manutenção no Contexto dos Sistemas Integrados de Apoio à Gestão do Fabrico", in Actas do 2º Encontro Nacional do Colégio de Engenharia Electrotécnica, Lisboa, Portugal, 14-15 Dez 95.

Silva, P., "Manutenção Condicionada Baseada na Análise de Vibrações", Dissertação de Mestrado FEUP-DEEC, 1993.

Takahashi, Y., Osada, T., 1990, "TPM", Asian Productivity Organisation, Japan.

8. BIOGRAPHY

Paula Alexandra Silva, holds a MSc. in Industrial Informatics from the University of Porto. She is currently a researcher at INESC-UESP and her research interests are maintenance management, information systems for maintenance management.

António Lucas Soares is an Electrical Engineer teaching at the Faculty of Engineering of the University of Porto and doing research work at INESC-UESP. He holds a MSc. in Industrial Automation and is currently undertaking its PhD work in the area of collaboration models for techno-organisational development. The main research interests are the interdisciplinary collaboration in the development of integrated manufacturing systems and socio-technical systems.

José M. Mendonça is an Electrical Engineer and received the PhD in Electrical Engineering in 1985 and is currently Associate Professor at the University of Porto as well as one of the Innovation Agency directors. His interest areas are: manufacturing management, enterprise modelling, shop floor control and management of innovation.

38

A Methodology on Work-flow Control System in Large-scaled job shop line

T. Kaihara[1] *and H. Imai*[2]
[1]*University of Marketing and Distribution Sciences*
3-1, Gakuen-nishi, Nishi, Kobe 651-21, JAPAN
Phone: +81-78-796-4402, Fax: +81-78-794-3054
E-mail: kaihara@umds.ac.jp
[2]*Mitsubishi Electric Corporation*
Manufacturing Engineering Center
8-1, Tsukaguchi-Honmachi, Amagasaki, 661, JAPAN

Abstract

Work flow of very large-scaled job-shop type manufacturing systems is complicated due to a large number of processes and various types of work flow patterns. Sophisticated work flow management policy is required for efficient manufacturing operation, since the investment into such a large scaled manufacturing factory is considerably huge. A work flow control methodology which is based on future WIP(Work In Process) location estimation with virtual manufacturing model is proposed. The Architecture is simple and PC-based, and that facilitates the integration with conventional manufacturing system. The methodology has been proved to be efficient enough to apply into practical fields. Finally our concept on fully automated real time control system is presented.

Keywords

Schedulling, Simulation, Object-oriented, Manufacturing management system

1 INTRODUCTION

Work flow of very large-scaled job shop type manufacturing systems, such as semiconductor factories, is complicated due to a large number of processes and various types of work flow patterns. Sophisticated work flow management(or control) policy is required to attain highly performed manufacturing systems (Uzsoy, 1992)(Fuyuki, 1992). Conventionally work flow management is based on the knowledge of some skilled operators in manual handling manufacturing lines, although it is not proved optimum dispatching strategy. The manual operation based on the experience is required frequently even in the fully automated lines, since FA(Factory Automation) control system can't handle unexpected irregular occasions, such as long time machine faults. Additionally it is necessary in FA control system to reduce the manufacturing POP(Point of Production) data size for the response in time, despite the real time control generally collects the huge amount of the data. Since the investment cost into such a complicated manufacturing line is increasing gradually, it is essential to develop the work flow management system which manages to operate the manufacturing line efficiently (Yuki, 1996).

Virtual manufacturing model built in computers enables the previous performance evaluation and the decision makings about the factory planning, development, and management. The evaluation by the virtual model has been mainly used for planning phase or static evaluation of the factory, although the concept, that the model is applicable to dynamic factory operation, has been suggested by some research groups(Alting, 1988)(Wu, 1989).

The flexibility is one of the most important features in general software development. Hard-coded manufacturing control program, which requires much efforts and time for the modification, is not useful in practical use. Object-oriented programming paradigm facilitates efficient manufacturing system modelling(Kaihara, 1993a)(Kaihara, 1993b). Object-oriented programming is useful especially to human interface development in work flow control system, because that consists of many similar graphical windows. Prototype method leads to quicker development of the software.

2 OBJECTIVES

There are two types of manufacturing line in large-scaled job shop factory in terms of factory control automation level, fully automated and semi automated. Process machines and material handling facilities are controlled automatically in the former type, whereas the work flow control (schedulling and dispatching) are manually decided in the latter. Our research is mainly focused on the semi automated work flow control in this stage, since most factories adopt this type of control policy. The moderate cooperation between human experience and control logic is important and practical, because it is quite difficult to acquire expert's logic or instinct.

The major target of our research is the performance improvement of large-scaled job shop type manufacturing systems. Valuable information should be presented timely to operational staffs for higher decision makings by collecting, calculating, and filtering the factory data. Only present WIP(work in process) data is available in the conventional work flow control policy. The information of future WIP location helps greatly even experienced staffs to plan the appropriate schedulling or dispatching sequence into each resource. The manufacturing system performance is expected to be kept totally high, because the whole manufacturing system events should be included in the estimation. If something unexpected happens, such as machine failures or the increase of defect products, the higher decision makings with the virtual manufacturing model facilitate quick recovery.

Secondly the line operation efficiency on work flow management is expected. It is effective especially in semi automated operation factory, since it sometimes takes long time to negotiate to decide the WIP input order and such decision making is useful only temporarily. The WIP input sequence has been proved previously effective in the virtual manufacturing model, and few negotiations are required at the shop floor level.

3 CONCEPT AND SOFTWARE ARCHITECTURE

3.1 Concept

The concept of our methodology is the interface agent based on the future WIP estimation by virtual manufacturing model with discrete-event simulation engine. Factory staffs can previously grasp the operational strategy to control manufacturing line appropriately, the amount of product output, lead time, and machine availability, and check if they meet the original production plan.

The future WIP location data is required at decision making occasions, perhaps several times in a day, in practical operation. Once the decision is fixed by the virtual manufacturing model, the work flow control sequence during the simulation scenario is obtained as the work tracking log data in the simulation results. In case the estimation results differ from actual factory performance, the comparison between simulation log and actual work flow enables to reveal the overestimated operational factors of the factory, because theoretically they are identical under correct factory parameters Proposed methodology is regarded as a leverage to exterminate hidden defects of the factory. Highly performed effective manufacturing system is attainable gradually through that feed back procedures.

3.2 Structure

Our prototype system based on the proposed methodology consists of four software modules, Work flow management module, Factory model module, Data translation module, and Automatic calculation module, shown in figure 1. The work flow management module handles the work flow data and controls user I/F. It is developed with object-oriented programming paradigm in Visual Works language(Goldberg, 1986), so that it is easily applicable to other systems.

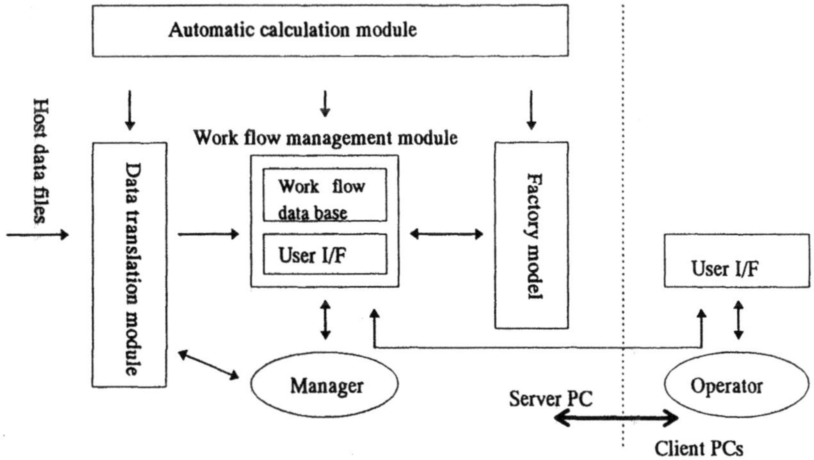

Figure 1 Software structure

General purpose virtual manufacturing modelling environment with discrete-event simulation engine is newly developed and applied into the factory model module in order to estimate the WIP location precisely. Template-based modelling method in which factory model is classified into two types, skeleton and specific models, is proposed and helps to facilitate efficient modelling. It is written in Microsoft Visual C++(Microsoft, 1996). This environment is applicable to factory planning and development as well as daily operation.

Manufacturing data files fetched from the upper main computer(Host computer) are translated and stored into the work flow data base automatically via the data translation module.

The automatic calculation module enables default decision making without the operator in actual use.

The major features of the system are as follows:

- Work flow management system with precise virtual manufacturing model
- EUC(End User Computing) environment based on MS Windows graphical user I/F
- OOP paradigm for software reusability

Figure 2. shows the system structure, which consists of Server PC(Windows NT) in management staff room and client PCs(Windows 3.1) in manufacturing shop floor.

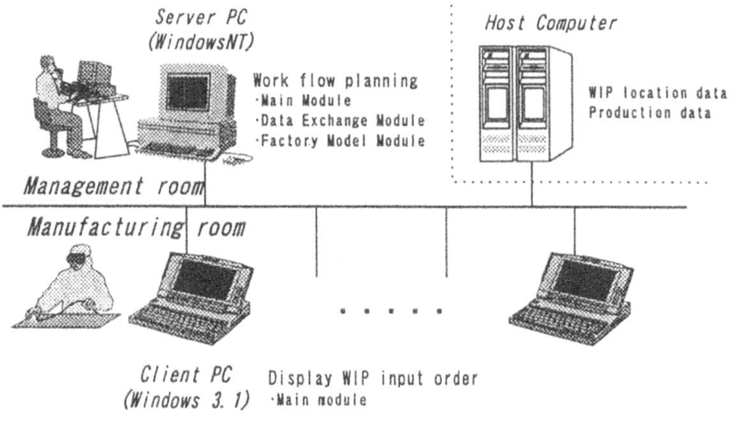

Figure 2 System structure

The host computer collects and maintains the manufacturing standard data and real-time information, such as process data, process flow, current WIP location and so forth. The server computer fetches several host data files for the WIP location and simulation via the data translation module. The work dispatching lists of each machines are created automatically from the work tracking log data in the simulation result, shown in figure 3. Some sophisticate dispatching rules which have been endorsed by static simulation are installed in the main module. In case some unexpected event happens, managers can try some simulation scenarios under new dispatching policy, compare the results with the original plan, and finally decide the appropriate strategy. The fixed dispatching order is transferred into each client PC to display the dispatching list to shop floor operators. The communication between the server PC and the client PCs is established by standard OOP protocol, CORBA. Object images produced in the server can be handled in each client transparently. After a product set into the process, operators are supposed to put a mark at the product on the display. The current manufacturing progress compared with the original plan is shown on the screen, then the manager can realize it anytime at a glance.

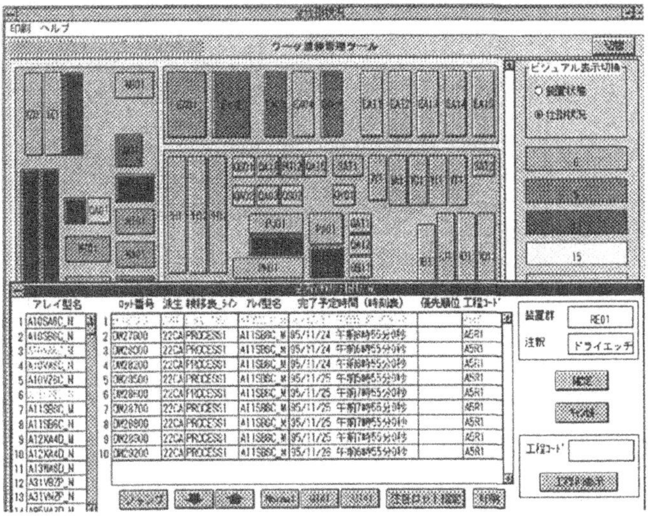

Figure 3 A Display example of User I/F(WIP layout map & dispatching plan)

4 CASE STUDY

The prototype system has been installed into practical large-scaled job shop lines including about 60 process machines. Several satisfactory results were obtained:

1. Decrease in production lead time by 10%
2. Reduction of the amount of shop floor operation by 10,000 hours/year
3. Easy installation and modification of the system
4. Significant reduction of the total investment cost into the PC-based work flow control system

5 CONCLUSIONS

Proposed work flow control methodology with virtual manufacturing model is quite simple and has been proved to be practical with human-computer interaction.

This methodology is also applicable in fully automated FA factories. Our opinion is that real time work flow control is not always suitable for large-scaled job shop manufacturing system due to complicated control algorithm and huge amount of data size to handle. The efficiency is not so obvious in spite of large investment. According to our methodology, the work flow control standard table is created automatically at each decision making point, perhaps several times in a day, and FA controller could just follow the control table for the time period. If the difference would happens between the location of virtual and actual works, the information could be feed backed and reset at the next decision making time. New research project is now in progress focused on FA factories.

6 REFERENCES

Alting, L.(1988) Extended applications of simulation in manufacturing system; *Annals of the CIRP*, Vol.37, pp.417-420

Fuyuki, M. and Inoue I.(1992) LISP-A Simulation System for Large-Scale-Production Planning Activity Support - Model and Simulator; *Journal of the Japan Society for Simulation Technology*; Vol.11, No.4, pp73-83

Goldgerg, A. and Robinsons D.(1986) *Smalltalk80: The Language and its Implementation*, Addison-Wesley

Kaihara, T. and Besant C.B.(1993a) Object-oriented flexible and integrated manufacturing system modelling; *1993 CompEuro Proceedings*, pp.312-319

Kaihara, T. and Besant C.B.(1993b) Manufacturing System Modelling Structure based on Object Oriented Paradigm; *The Proceedings of the 1993 Summer Simulation Conference*, pp.831-836

Microsoft Corp.(1996) *Microsoft Visual C++ Version 4.0 Development System for Windows 95 and Windows NT*

Uzsoy R., Lee C. and Martin-Vega L.A.(1992) A Review of production planning and scheduling in the semiconductor industry; *IIE Transactions*, Vol.24, No.4, pp.47-60

Wu, S.D. and Wysk R.A.(1989) An application of discrete-event simulation to on-line control and scheduling in flexible manufacturing; *Int. J. Production Research*, Vol.27, No.9, pp.1603-1623

Yuki, H., Saito H., Imai H., and Kaihara T.(1996) Work-flow management system with virtual manufacturing model; *Advances in Production Management Systems*, pp55-58

7 BIOGRAPHY

Toshiya Kaihara received the B.E. and M.E. degrees in precision engineering from Kyoto University, Kyoto, Japan, in 1983 and 1985, respectively. He joined Mitsubishi Electric Corp. in 1985, where he worked at manufacturing system development. He received the Ph.D. degree in mechanical engineering from Imperial College, University of London, London, UK, in 1994. He joined University of Marketing and Distribution Sciences in 1996.
He is a member of IEEE and Japan Society for Precision Engineering.

Hiroshi Imai received the B.E. degree in mechanical engineering from Tokyo University, Tokyo, Japan, in 1976. He joined Mitsubishi Electric Corp. in 1976, where he works at manufacturing system development. His current interests are PC-based FA control system using precise manufacturing simulator and real time DB, and decision support system for shop floor operator.

PART SIXTEEN

CAD/CAM

39

Enhanced pocket milling strategy for improved quality

R. M. D. Mesquita
Associate Professor, Instituto Superior Técnico, Mechanical Engineering Department
Av. Rovisco Pais, Lisboa, Portugal
F. Ferreira Lima
Research Assistant, Instituto Tecnológico para a Europa Comunitária, Advanced Manufacturing Technology Center

Abstract

Computer aided part programming applications, available within state-of-the-art CAD/CAM systems, fail to generate high quality part programmes, in most cases due to a poor understanding of shopfloor operation and constraints. Usually, there is lack of manufacturing culture and expertise in these systems. The development of applications that implement sound machining practices and provide manufacturing expertise to part programmers is a key issue, if robust part programmes, requiring no editing at the shopfloor level and enabling the use of high material removal rates, are to be achieved. This paper presents an enhanced rectangular pocket machining strategy, based on the data derived from an experimental work on the effect of machining strategies and parameters on the quality of machined parts.

Keywords

CAD/CAM, computer aided part programming, milling, pocket machining

1 INTRODUCTION

High productivity and high quality milling, as measured through the machining time, set-up time, part programme editing time, rework cycles, scrap rates and achievable dimensions and tolerances, is determined mostly during the part programming phase, either manual or computer assisted.

One of the most common operations in NC milling, is pocketing, both of rectangular or complex shaped contours, with or without islands. Pocketing operations are carried out with end mill/drills and require axial feeding and both, full immersion and non-full immersion cutting passes. This procedure is repeated several times until the full depth is attained. The number of passes is dependent on the allowable axial depth of cut. It follows that, together with tool geometry and material, cutting speed and feedrate, axial and radial depths of cut are key parameters that have to be defined before NC programme generation. While cutting speed and feedrate can be optimised / tuned by machine-tool operator, axial and radial depths of cut determines the actual structure of NC programme. The resultant toolpath is highly difficult to be edited at the shopfloor. Editing of an improperly generated part programme can be a time consuming operation, prone to errors and inconsistent part quality.

The work by Tarng (1993) has shown that during pocketing operations, radial depth of cut changes with cutting direction, traveling paths and position of path. To take that fact into account, and produce dynamically acceptable toolpaths in order to avoid chatter, conservative machining parameters are selected and the material removal rate reduced. A system was developed where feedrate is automatically adjusted to compensate for variable radial depth of cut. It was showed that a 20% decrease in pocket machining time could be achieved with this feed optimisation procedure. A chatter-free toolpath planning strategy to be used in CAD/CAM systems was also proposed by Weck (1994). Stability charts which contain chatter-free axial and radial depths of cut at practical cutting speed range are used to implement dynamically correct toolpath for pocketing.

Tlusty (1990) showed that internal corners change (increase) radial depth of cut, which has a strong effect on stability, reducing the quality of the part. Consequently, it is argued that pocketing strategies with internal corners should be avoided.

It can be concluded that, together with the above mentioned machining parameters, the machining strategy, as defined through the toolpath or distribution of machining passes required to clear out the material, in order to produce even a simple rectangular pocket, is of paramount importance.

Kline (1982) have shown that in end milling process, the cutting forces during machining produce deflections of cutter and workpiece which result in dimensional inaccuracies or surface errors on the finished component.

Milling mode (up or down-milling) is also an important parameter and has a key effect on part quality. Budak (1994) stated that in up-milling mode, cutting tool is pushed into the part side wall, increasing pocket width / length due to the end mill deflection, while in down-milling, tool deflection shifts the tool axis away from the pocket wall, which relieves the cut and decreases pocket width / length. However, during finishing operations the effect of the ploughing forces cannot be omitted. Consequently, even in the case of up-milling, when the radial depth of cut is very light, tool can deflect away from the pocket surface. One concludes that milling mode can affect part quality, particularly in finishing operations.

Commercial pocket cycles, available within state-of-the-art Numerical Controls and Computer Assisted NC programming systems can be classified into four broad categories: spiral-out, spiral-in, zig-zag, linear. In all these cycles, axial and radial depths of cut are fixed and cannot be modified during the elemental passes required to machine each pocket. Cutting speed and feedrate cannot be modified either, to accommodate for changes in radial depth of cut which arise at the end of each elemental pass required to enlarge the pocket. Toolpaths are distributed according to particular algorithms that consider only geometry-related constraints, such as, pocket geometry, tool diameter and specified stepover. They are considered as geometry-oriented pocketing cycles being not process-oriented and not technologically optimised. It was this understanding that induced the work of Tarng (1993), Weck (1994), and Tlustly (1990) above mentioned.

Our research work had a twofold objective: primarily to identify the effect of several machining parameters and pocketing strategies on part quality; secondarily to propose an enhanced and technology-oriented pocketing cycle to be implemented in a commercial CAD/CAM package.

2 EXPERIMENTAL WORK

The objectives of the experimental work were, the assessment of the effect of machining parameters and pocketing strategies on part quality and machining time. The standard *Mastercam* pocketing strategies were used in order to determine:

1. Influence of roughing milling strategy on machining time and dimensional accuracy.
2. Influence of finishing milling method on machining time and dimensional accuracy.

The machining tests were carried out on a CNC Hermle milling machine with a 12 mm diameter, 40 degrees helix angle, 3 fluted high-speed steel end mill. The work material was a 7079-T651 aluminium alloy. Machining parameters were determined from tool manufacturer machinability databases.

The pocketing strategy efficiency assessment is based on the comparison of pocket machining times and achieved dimensional accuracy. This experiment included the machining of a series of rectangular pockets, carried out with identical machining parameters, using the following *Mastercam* roughing strategies: one-way conventional (Owcv), spiral-in (Spi), spiral-out (Spo) and zig-zag (Zz). The sequence of operations to machine these pockets and the used geometrical machining parameters for each operation are presented in Table 1.

Table 1 *Mastercam* pocket strategies and elemental operations for pocketing

Machining parameters for each operation

Elemental Operations	no.of passes	depth (mm)	radial depth (mm)
pocket roughing	4	4.875	8
pocket bottom finishing	1	0.5	8
pocket side finishing	1	20	0,5
pocket dry passes	1	20	

Elemental operations for pocketing

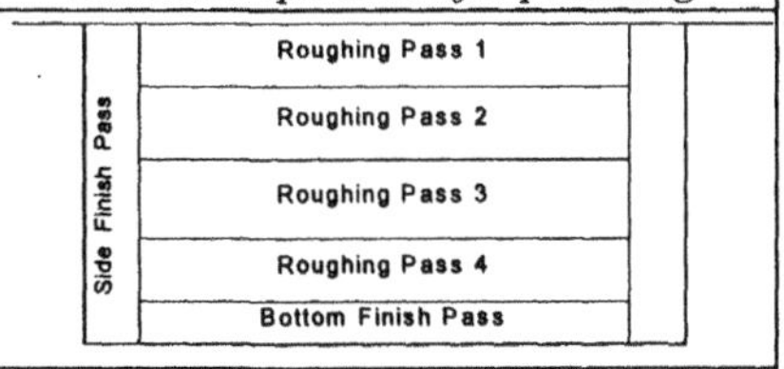

Dimensional analysis was carried out on a DEA coordinate measuring machine (CMM) in order to reduce measuring time and to achieve a suitable measuring accuracy. Dimensions were taken both in the straight sections of the pockets and at the corners. The results are presented in this section.

Roughing milling strategy analysis

Figure 1 presents the effect of pocketing strategies on machining time. As expected, zig-zag strategy is the fastest strategy, being the lowest material removal rate achieved with the one way strategy.

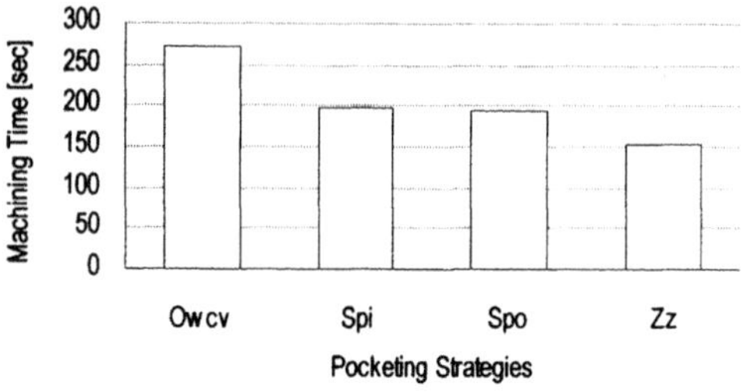

Figure 1 Influence of pocketing strategies on machining times (60x40x20 mm pockets).

Figure 2, displays the simulated *Mastercam* pocketing cycles, presents the geometries generated by roughing passes as calculated by the routines, and quantifies both the maximum stock values left for finishing passes and the dimensional deviations that were found in the corners of the pockets, for each one of the pocketing strategies.

Figure 2.a presents the stock geometry left for side finishing passes with the one way strategy. Maximum stock values of 6.07 mm (which greatly overlaps the programmed value of 0.5 mm) occur at the top right corner (TRC) and bottom right corner (BRC). These corners correspond to the ones where the tool retracts at the end of the extreme end straight cuts. The algorithm used by this *Mastercam* pocketing cycle seems not to be able to handle properly the situation. A suitable compensation is not carried out since the routine offsetted the pocket x side in the y direction with no change the endpoint x coordinate for the first and last moves. Similar error occurs in the zig-zag strategy, but since tool is interpolated along the BRC and top left corner (TLC), the material left by the improperly offsetted function, is removed. Identical limitation was found with the Heideinhain TNC 425 controller zig-zag strategy.

As a result, this error is propagated during the finishing operation. One can observe in figure 2.e and 2.f, that the highest dimensional deviations, with values ascending to 350 to 400 microns, can be found in TRC and BRC for the one way strategy, and BLC and TRC for the zig-zag strategy. It is shown that the highest dimensional deviations occur in the corners where the biggest stocks arise. These deviations can be explained as follows:

- The cutting forces have a direct influence on cutter and workpiece deflection and the resulting surface error, and are directly associated to the radial immersion. It is expected that higher radial immersion ratios will produce higher cutting forces and cutter deflections, and result in larger dimensional deviations. Therefore, the largest is the material stock left for the side finishing cutting pass, the largest will be the cutting forces, together with the tool deflection. It is believed that, to a certain extend, the geometrical deviations that were found in the machined pocket corner profiles can be associated to the stock geometry left for the side finishing pass by the roughing passes. In fact, in the pockets produced by zig-zag or one way strategies, the stock geometry left for the finish pass is not regular with peak stock values exceeding 10 times the defined value.

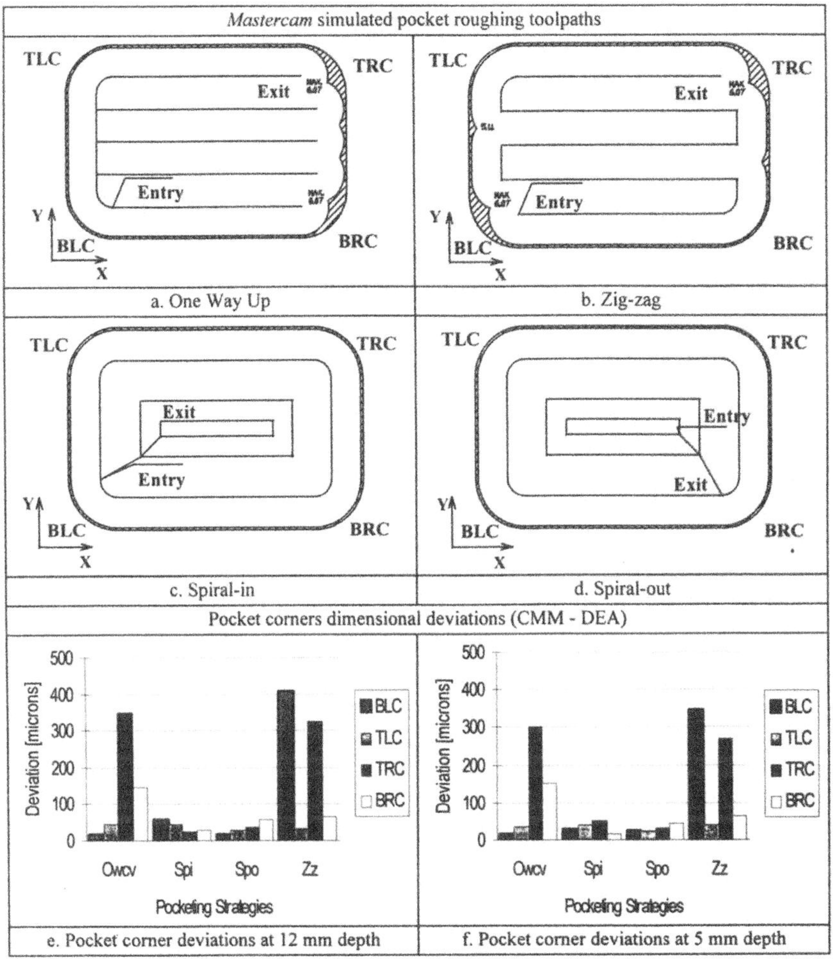

Figure 2 *Mastercam* pocketing roughing strategies and pocket accuracy analysis.

A good correlation can be established between the stock geometry left for the side finishing pass, the maximum stock values locations and the highest dimensional deviations observed in the pockets after the finishing operation.

For both spiral strategies, the stock left for the finishing passes is very homogeneous following the desired pocket contour. Consequently, the dimensional deviations after the finishing pass are the lowest among all pocketing strategies.

The comparison of the *Mastercam* roughing pocketing strategies, considering the criteria machining time and dimensional accuracy, allows the extraction of the following information:

1. The zig-zag strategy. presents, as expected, the lowest machining time of all strategies.
2. The spiral strategies, can produce the pockets with higher dimensional accuracy, due to improper tool path compensation in the corners when the zig-zag strategy is used.
3. The maximum deviations observed in the areas close to the pocket corners are critical in the dimensional accuracy comparison, once these values are generally 8 to 10 times higher than those observed in the pocket basic dimensions.
4. A good correlation can be established between the finishing stock geometry left for the finishing pass and the final machined surface accuracy. The maximum deviations occur where the highest stock values left for finishing stock arise.

Finishing milling method analysis

The objective of this set of machining tests is the comparison of the finishing pass methods: up-milling method, and; down-milling method, and its influence on the final pocket dimensional accuracy and pocket surface quality. This experiment included the machining of a series of rectangular pockets, carried out with identical cutting machining parameters, except the milling method used in the finishing pass.

Figure 3 presents the influence of milling method on dimensional deviation. One can observe that lower dimensional deviations can be achieved if up-milling is used in the finishing pass.

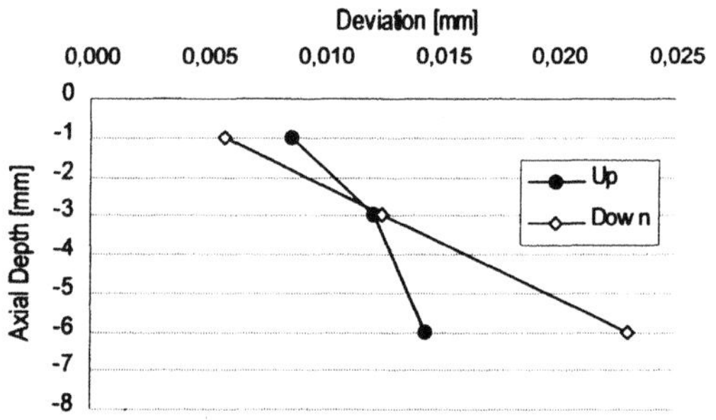

Figure 3 Influence of milling method on dimensional deviation.

The explanation for this behavior can be found once again on the effect of the cutting forces. During down-milling operations, the cutting forces produce cutter and workpiece deflection that shifts the cutter away from the nominal surfaces. Theoretically, these deflections result in pockets with smaller dimensions than desired, with the larger dimensional deviations being at higher axial depths. Usually, for roughing cuts, the opposite is equally true for up-milling.

For the down-milling experiments, the obtained surface profile error, along the axial depth, is as predicted. The pocket is smaller than desired and the larger errors occur at higher axial depths. However, in the up-milling method experiments, the observed surface error profiles have not the expected slope. This can be explained by the combined effects of the cutting forces and ploughing forces on the resulting cutter deflection:

- The interference between the tool flank and the machined surface results in the generation of ploughing forces acting in the radial and tangential directions, pushing the tool away from the surface. In down-milling, the effects of these ploughing forces are added to the ones derived from the cutting forces, deflecting the cutter and producing a surface with a high slope. In up-milling, the forces are applied in opposite directions. For light radial immersions, the ploughing forces effects are probably stronger than the cutting forces effects, which results in inverted cutter deflections and justifies the surface error profile found. The magnitudes of the dimensional errors in down-milling are larger than in the case of up-milling due to: the above mentioned combined effects of the ploughing and cutting forces; the magnitudes of the radial cutting force which are larger in down-milling; and finally, because in up-milling the tool is pushed into the material, which resists the deflection, while there is no resisting contact stiffness in down-milling, since the tool deflects away from the workpiece toward the air.

The analysis of the pocket dimensions that were obtained with the finishing pass methods, up-milling and down-milling, gave the following indications:

1. Higher dimensional accuracy is achieved if up-milling is used in the finishing operation.
2. For the range of feedrates considered (0.11 - 0.173 mm/tooth), the milling method had no influence on the pocket walls surface quality.
3. As expected, the milling method (up/down) have shown no influence on the pocket machining time.

As a conclusion of the experimental work, one can derive the following rules to be applied:

1. Spiral-in and spiral-out strategies will produce the “best” roughing quality with a marginal increase in machining time, when compared with zig-zag pocketing.
2. Spiral-in and spiral-out strategies create the best conditions for a smooth finishing operation and enable the control of roughing milling mode (up/down).
3. Up-milling mode should be used for finishing the pocket side walls.

However, if we take into consideration the work by Tlusty (1990), the spiral-out strategy should be discarded, since systematic internal cornering is present. This fact determines that even if we select, for the cuts preceding the corner, one half-immersion pass, there will be a time during the corner where the tool is slotting. Since radial immersion has a strong effect on stability, one should avoid systematic changes from non-full immersion to full immersion conditions (Budak,1994).

The spiral-in strategy seems to be the best strategy for rectangular pocketing. In its very basic structure, spiral-in includes a frame cut followed by several offsetted contouring cuts (towards pocket center). The first frame cut is carried out in full immersion mode (slotting) and the remaining contouring cuts in non-full immersion as determined by the specified and controllable stepover. This structure of the spiral-in routine, as implemented in CAD/CAM systems, entails a strong disadvantage. A unique axial depth of cut and feedrate has to be programmed for both full immersions and non-full immersion cuts. It follows that, to avoid chatter, conservative parameters are used in pocket machining. The improvement of both, quality and productivity pocket milling can be achieved if we split the cycle in two different routines, one for slotting (the outside frame) and the other for spiral-in in non-full immersion cutting, each one with a different set of machining parameters, which are dependent of tool diameter and workpiece material.

3 ENHANCED POCKET MILLING CYCLE

As a result of the limitations of the existing strategies, that were found in the experimental work, a new Enhanced Pocket Cycle was developed for *Mastercam* software environment, enabling the manufacture of the pocket using the following sequence of machining operations: 1. Pocket Framing; 2. Pocket Roughing; 3. Pocket Finishing; 4. Pocket Additional Finishing (optional), and; 5. Pocket Bottom Finishing.

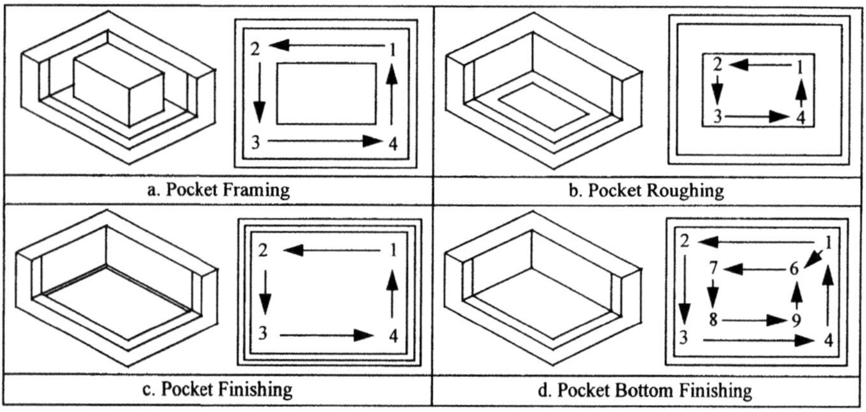

Figure 4 Enhanced sequence of operations for pocketing (passes at the final depth).

The new enhanced strategy is independent of tool diameter and workpiece material.

Pocket framing operation consists in slotting the periphery of the pocket in as many passes as required. The tool passes through the positions marked 1, 2, 3, 4 and 1 along the pocket periphery, as shown in Figure 4.a. The suitable cutting parameters for this specific operation (the most restrictive case) are used.

Considering that full immersion cuts are carried out, axial depth of cut can be reduced, only for this section of the pocket. This operation is followed by pocket roughing, which consists in machining the central area in as many passes as required, using a spiral-in strategy. If up-milling is the suitable method, clockwise direction is required and the tool passes through the positions marked 1, 2, 3, 4 and 1, as shown in Figure 4.b. Otherwise, the cutter follows these positions in the reversed order. The pocket dimensions determine the number of cutter motions required at each depth. The optimised cutting parameters for this operation can be used. Radial immersion can be half of the tool diameter in order to avoid the simultaneous use of up and down-milling methods. Axial depth of cut can be increased (comparing with the one used in the preceding frame cutting), decreasing the required number of passes.

The pocket finishing operation aims the finishing of the pocket side walls. The suitable milling method determines the direction (CW/CCW) in which the finishing pass is performed. The passes along the pocket boundaries follow the toolpath as shown in Figure 4.c. If up-milling is the suitable method, clockwise direction is required and the tool passes in order through the positions marked 1, 2, 3, 4 and 1. Otherwise, the cutter follows the positions marked in the reverse order. The optimised cutting parameters for this specific operation can be used also.

Additional finishing operation (dry passes) can be used for further improvement of the pocket side walls accuracy. The direction (CW/CCW) in which the dry passes are performed is similar to the finishing passes direction. This machining operation is optional, and the suitable number of dry passes (0, 1 or 2) are performed, as determined by shopfloor expertise built-in the system.

Finally, pocket bottom finishing operation is carried out in order to finish the pocket bottom, using the spiral-in strategy and the optimised machining parameters (Figure 4.d).

The enhanced rectangular pocketing cycle was implemented in the *Mastercam* milling package. The new cycle included both, the new operations sequence, partially driven by CAD data and complemented by on-line user input, and an optimised spiral-in toolpath. With this enhanced cycle, CNC programmes can be generated with "local" optimisation of machining parameters for several elemental machining passes which are required to complete the full cycle.

The developed application, although implementing a complexer toolpath, through the splitting of the initial spiral-in cycle, still guarantees a reduced effort in the user input. Machining expertise is also provided to the user, since it is built-into the system. A programmer with a lower process knowledge can use the system and still providing improved quality part programmes.

4 CONCLUSION

An enhanced rectangular pocket cycle was proposed. The improvements achieved with the new cycle can be summarised as follows:

1. Use of only one routine to apply an optimised sequence of operations to produce a pocket (framing, spiral-in roughing, finishing, additional finishing and spiral-in bottom finishing).
2. Use of a complete optimised set of machining parameters for each one the elemental operations, automatically determined by the system.
3. Use of an optimised spiral-in machining strategy for pocket roughing operation and pocket bottom finishing operation.

4. The reduction of pocket machining time can be achieved together with the control of chatter.

5 REFERENCES

Budak, E. and Altintas, Y. (1994) Peripheral milling conditions for improved dimensional accuracy. Int. J. Mach. Tools Manufact., **34**, 879 - 891.

Kline, W. A., DeVor, R. E. and Shareef, I. A. (1982) The prediction of surface accuracy in end milling. ASME J. Engng Ind., **104**, 272 - 278.

Tarng, Y. S. and Shyur, Y. Y. (1993) Identification of radial depth of cut in numerical control pocketing routines. Int. J. Mach. Tools Manufact., **33**, 1 - 11.

Tlusty, J., Smith, S. and Zamudio, C. (1990) New routines for quality in milling. Annals of the CIRP, **39/1**, 517 - 521.

Weck, M., Altintas, Y. and Beer, C., (1994) CAD assisted chatter-free NC tool path generation in milling. Int. J. Mach. Tools Manufact., **34**, 879 - 891.

6 BIOGRAPHY

Professor, Ph.D. Ruy Manuel Dias Mesquita
received his Ph.D. in Mechanical Engineering at the Technical University of Lisbon in 1988. He his associate professor in Mechanical Technology at IST since 1991. His research activity has been developed in the area of Production Engineering. At present, he is the director of the Institute of Materials and Production Technologies at INETI-National Institute of Engineering and Industrial Technology.

Engineer, Pg.D. Francisco Ferreira Lima
is graduated in Mechanical Engineering at the Polytechnic of Setúbal and received his Pg.D. in Computer Aided Engineering at the School of Engineering of the Staffordhire University, UK. He his a research assistant in the Advanced Manufacturing Technologies Centre of ITEC, being involved in several projects in the areas of CAD/CAM. Currently, he develops a M.Sc. thesis on Computer Numerical Controlled End Milling.

40
PROBACOM
Computer Aided Project Based on Features

João Rafael Galvão
Instituto Politécnico Leiria - Escola Superior Tecnologia e Gestão - Morro do Lena - Alto do Vieiro - 2410 Leiria - Portugal - tel. 044-32168; fax. 044-814254; e-mail: estglei@mail.telepac.pt

Abstract

In this work the wide concept of feature will be analyzed in the area of CAD/CAM (Computer Aided Design/ Computer Aided Manufacturing). An application program on a group of interdependent features was developed, with the support of a commercially CAD program available. One of the objectives is the expansion of the capacity of the CAD package and the satisfaction of the needs of the users, which simultaneously are the designers and schemers and should be able to have a new group of options, in the sense of increase the power of the chosen graphic system and to improve the timing of the design of a product called mould.

Keywords

Feature, geometric feature, integration, modeling, CAD

1 INTRODUCTION

These called computer aided design/project based on features, introduced at the beginning of the 80s, is the development of the computer aided design for the conception of the product drawing through the relationship among entities, which is also supported by a model of geometric representation. Our practice shows that it is more intuitive to manufacture a model in terms of certain particularities or features, such as holes or cavities, applying stored information associated to each one of those particularities, than to think out parts of the geometric model and to find the identification, Anderson & Chang (1990).

Throughout a number of years the notion of feature in drawing was restricted to processes planning or parts to be manufactured in spite of some forms classifications.

Nowadays some different classifications have been emerging in a higher multiplicity of applications areas such as in numerical programming and in drawing by geometric modeling.

The work of Pratt & Wilson (1988) pioneer in the cataloguing of features, applying them to the geometric models conception.

The conception of feature

By **feature** we mean the relationship between components or parts of a geometry which may be used as the origin of other forms initially dependent on this.

The features occur as a possible solution for the **connection and integration between CAD/CAM** emerging a multiplicity of definitions. To mention the definition of a standard feature becomes a difficult task, depending mostly on the perspective we visualize the geometry of a certain object, as well as the perception and knowledge of the person that manipulates it. For this reason, a certain speculation arises among the authors of the field on which best definition to adopt.

In a wide context of features definition, we soon conclude that conception and modeling features are also of product and technology, along the assert for that a feature is a set of geometric or technologic information.

Others definitions may also occur, as being any form or entity whose presence or dimensions are relevant for one or more manufacture functions integrated by computer and whose use for the designers is a primitive in the drawing process. In such definition the feature is described as a set of properties able to be codified, deriving also from a classification of forms to schematize particular classifications or specific geometric configurations in a surface or edge, this allow the modification of the extend shape, so that an intended function would be performed as a part of a drawing description such as, for example, a hole, a cavity, a slot, a channel or any other detail.

The design by features has begun in geometric models and the emergence of algorithms for the recognition of the drawing or manufacture details, having a purpose for representation parts of separate pieces or models, appeared to the designers as an accessible and of easy to use tool.

A geometric feature of forms has various types of manufacturing processes: it is common to specify the nature of the geometry or technology used, which has to do with the dichotomy CAD/CAM. To sum up, the design by features allows a high level of abstractions in the draw of graphic entities geometry.

Examples will be given mainly in the areas of **design, process planning, computer numerical control programming and testing/control.**

As results from those experiments we give evidence of the fact that different areas relevant to a sequence of operation, which provoke the emergence of **specific particularities** or, in a general way, **features, which may become data to use in a latter phase of the CIM** (Computer Integrated Manufacturing) **chain, when the objectives is to achieve better quality of any product**.

On conclusion the perception of a certain object is influenced by the knowledge that the user has of this object, this way **different types of features** will provoke a greater **connection of the areas of CAD/CAM** and expand the concept of CIM.

Objectives of the work

The author's professional experience achieved for two years and a half in the moulds industry, allows him to assert that a product's drawing component is of crucial importance from the moment that it is conceived throughout all the phase of manufacturing. That experience was crucial in the increase of interest in the potentialities of the design aided by computer based on features.

In specifying the practical objective, it was taken into account the especifity of the mould plastics industry *, situated mainly in the coast area of the country (Leiria, Marinha Grande, Oliveira de Azeméis), being mainly constituted by companies or groups of small companies whose activity is based on the design and manufacture of moulds for external market.

At the beginning of the 90's, the design systems aided by computer were present in most of the productive process, because the potentialities of the PC's personal computers and their low price which were the main factors in its acceptance in this type of companies

The main objective of the current work was the mixing of various created functions, as a solution for different problems met by the users in the area of design of moulds for plastic, this is done within a fast repetitive tasks hypotheses in the act of conception of the mould, creating dependence relations or features among graphic entities, for a better efficiency in the conception of the article (design/drawing) and its transference for the machining process.

The creation of those relations between graphic objects directed for the areas of design, process planning, testing and control, were achieved by mean of programming the development module of the chosen system, using the instruction of the ADS-AutoCAD Development System AutoDESK(1994), which is similar to C++.

In order to attain the given objective, programming analyses from CAD to PC were put into practice. The PC programs more common in Portugal for the industry of moulds for plastic are: AutoCAD, Cadkey, Microstation PC, these systems having possibilities of transferring files into graphic stations, like SUN, DIGITAL or HP. Yet they have different geometric modelers and some of them using mixed modeling systems like B-Rep and CSG, Rembolt (1993).

The AutoCAD was chosen in a PC environment with processor 486, 100 MHz, on which features functions were implemented about the two-dimensional entities.

Adding to AutoCAD existing menus, new options were created to provide the expansion and to create others solutions that apply to the dependence relations or geometric features among graphic entities.

A moved production cycle that needs the drawing details of the specified article consist on: **Design/Design consulting room**, **Planning/Work preparation** (studies and methods), **Milling** (remove the exceed steel material by CNC-computer numerical control), **Lathe** (conventional and CNC), **Erosion** by CNC, **Grinding** (plain and cylindrical), **Workbench** (assembly and improvement), **Control and Testing.**

We emphasize that the information about the parts currently being manufacture in that area of production is conveyed, in this industry, in a two-dimensional way.

---

* We emphasize that the meaning of some words in the productive sector have a very specific meaning: **mould** - tool in steel to produce a product by the means of an injection of plastic stuff; **article** - product to be produced.

2 ARCHITECTURE AND FUNCTIONS

For an easy usage of the commands of the present applications and to provide a agreeable man/machine interface, we had to create a menu with new options or geometric features. The menu is stored in the file "acad_1.mnu", which is a transcription of the main menu of the AutoCAD named "acad.mnu", with some changes, namely making use of the menu POP10, Omura (1991), that until then was not used and is shown in the top right corner of the main screen in Figure 2.

In order to make clear the potentialities and objectives of these applications, which we intend to be an extension of the capacities claimed by the AutoCAD main program, some explanatory pictures are present as a way of involving the capacities of the option, in the effective resolution of some needs presented by the base program.

We apply these new geometric features as a tool in the achievement of fast solutions to the designer work with this type of graphic computing environment.

The explanatory sequence of the new options is about a mould design conceived for industry, following various stages of conception, respectively the **preliminary design** and the **final design,** in order to create a design/design of easy comprehension and reading and respecting the manufacture rules.

The first option of the menu that comes up to the user after the installed of the PROBACOM is the option of loading the various constituting files using the option [LOAD]. Some brief information related with the name of the application also exist, that is the option [Sobre] (About). Only then the areas of the general options follow, grouped within a main menu with various possibilities **[PESQUISA, ALTERAÇÃO, CÓPIA, GERAÇÃO]** (SEARCH, CHANGE, COPY, GENERATION) as seen in Figure 1, these options allow the user to access a number of related operations at the level of the two-dimension drawing being automated, namely having access to some properties information like area entity co-ordinates and radius. Then in a fast and interactive way, the graphic entities positioned in the screen which have the dependence relations among them, can be edited by using search, change, copy, identification and calculus of the number of objects, with the same properties present in the drawing.

The set of options [PESQUISA] (SEARCH) as seen in Figure 2, apply to entity (circle or arc) and the options [Proc. Circ. Raio], [Proc. Arco Raio] are used for fast search of entities with the same radius being the options [Proc. Circ. Dam.] and [Proc. Arco Dam.], similar to the previous ones, but the search of the entity is liked to the diameter. At the level of the code of language ADS the function *ads_rtos ()* is used to convert real values in sets of strings, that may be used as textual information. We use this function to convert the radius or diameter value (numerical values) into string of values in a way that it can be handled as text. The option [INFORMAÇÃO] gives to the programmer details of the graphic entity to be used in the language code ADS, at the level of implementation.

All the functions, in a direct or indirect way, manipulate elements of a list of the project entities being that list created by *ads_buildlist ()* and is built for the text typed entity, in which various fields are defined and allow the text identification control, to be introduced in the drawing through the manipulation of the existing variables in the list.

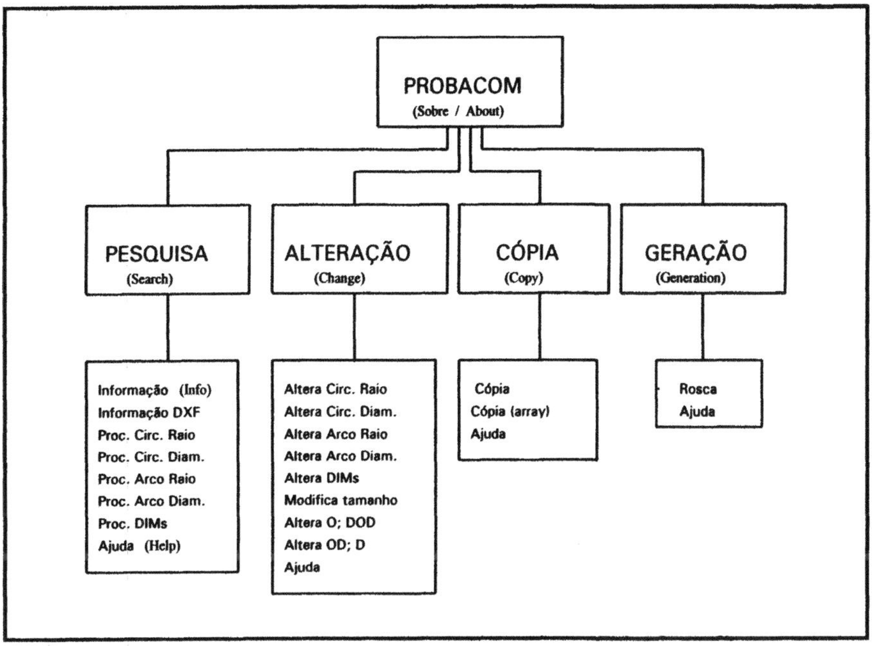

Figure 1 Diagram of the areas and options developed.

The option [Procura DIMs] allows the user to know how many entities of type DIMs co-ordinates exist and which are their particularities and their value.

The area of function [ALTERAÇÃO] (CHANGE) has various options with different characteristics, but the option [Altera OD; D] as shown in Figure 3, (O original entity; D descendent entity) being this strictly related with the option [CÓPIA] (COPY), because it is a command that allows to modify, simultaneously, the original picture and all copies (created by the option [CÓPIA]), due to the **relations of mutual dependence** generated when the copies were created. If the options [Altera OD; D] is used in an entity that is a copy and not the original picture, the changed pictures will be only those that are copies of this same entity.

When the selected entity is a circle or an arc the user is asked to introduce a new value of the entity radius. After the introduction, this value is positioned in the corresponding field and then an entity is changed using the function code *ads_entmod()*. After the original picture is changed, the program searches among all the entities of the drawing those which are copies of the original entity and changes them in the same way. For polygons, and after the introduction of the new co-ordinates, the function updates the vertex position through the code function *ads_entmod()*, being the copies of the original picture also changed.

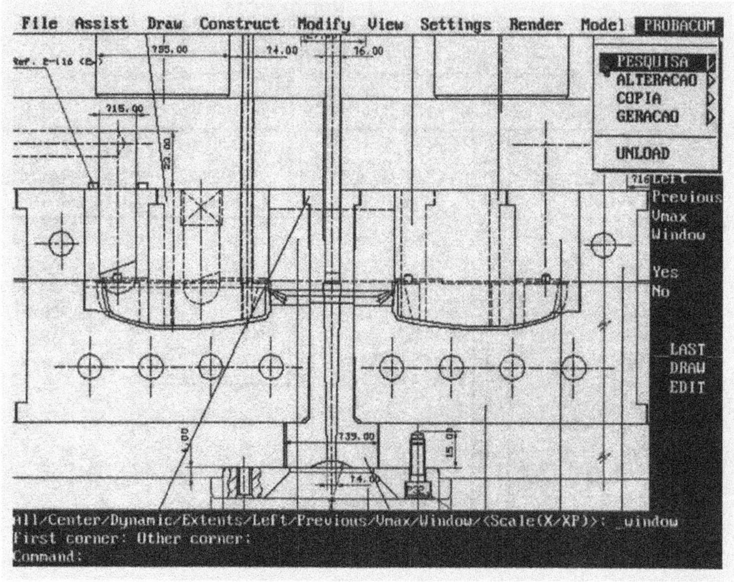

Figure 2 The access to information about the entities is one option [PESQUISA]

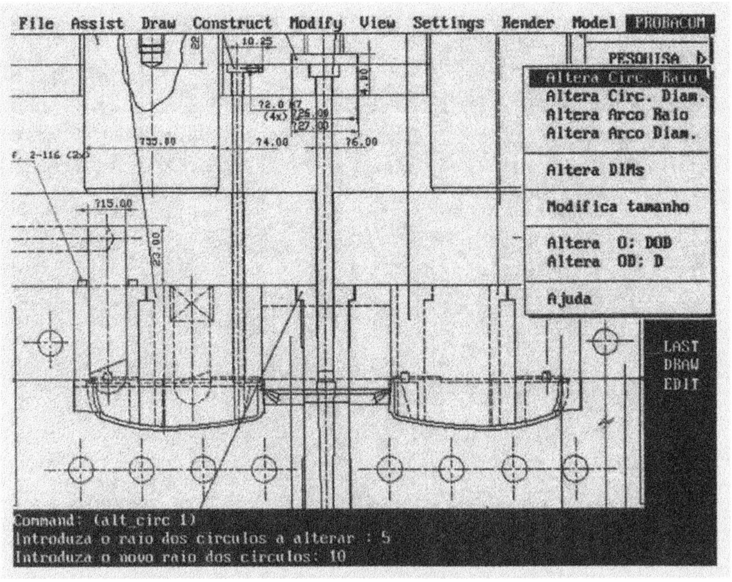

Figure 3 Change the value radius 5 to 10 in circles

The options [Altera Circ. Raio] and [Altera Arco Raio] as shown in Figure 4, are used to change the circle and arc entities with the same radius, being the options [Altera Circ. Diam.] and [Altera Arco Diam.] to do a similar change, but by the entity diameter.

We give evidence to the option [Altera DIMs] (Figure 5), that changes the co-ordinate entities of a given project drawing by a given scale factor that the user introduces through a dialogue box among inches, millimeters and meters.

In the option [Modifica tamanho], the user may modify the point by stages or percentages.

After the drawing entity selection by the code functions *ads_entsel()* and *ads_entget()* this latter is put in the respective buffer and its type is verified. When there is a change of the entity's buffer there is the drawing updating through *ads_entmod()*

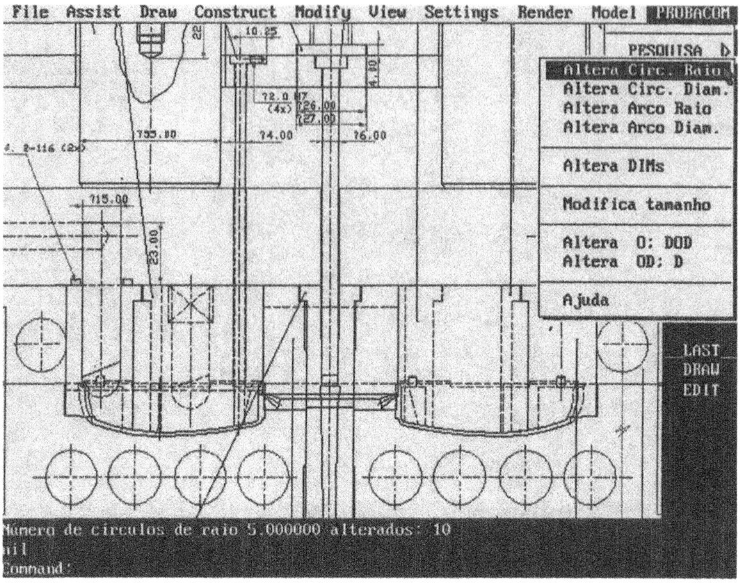

Figure 4 After the simultaneous change of the circles they have now radius 10

The option [Altera O; DOD] distinguishes itself from the others, because each of the entity copy, suffers a change including the original picture, every time a descending image is changed.

Within the area [CÓPIA] reference to an internal function *registar()* must be done that is only found in this option and that is addressed to perform in an ordered way the usage of the function *ads_tbsearch()* that records the name of the option [CÓPIA] in the table APPID, AutoDESK(1994), which is necessary for the correct functioning of the function.

The option [CÓPIA] allows to copy every entity, circles, arcs and polygons, generating **dependence relations** among copies and their original pictures, in a way that, when any change is made in the original drawing, using the option [Altera OD, D] every copy of this drawing will be simultaneously changed.

After the selection of the type of entity to create, circle or arc the program asks the user to input the center of the new entity and puts this value in the field 10 codes DXF, Omura(1991), creating then a new entity using the code function *ads_entmake()*.

In case the entity is a polygon, the copy is more complex because we are facing a group of entities, that are vertexes of that polygon. The option [CÓPIA(array)] is a copy of entities version, but it has the particularity of performing it in a certain direction (horizontal, vertical or leaned) and specify the request number of copies of the entity as well as the distance between them. The functions used are basically the option ones [CÓPIA], maintaining the features of the geometric dependence among entities.

The area [GERAÇÃO] mediates the option [ROSCA] that is an application particularization and allows drawing various entities (screws) producing a dependence relation between the circle and the arc constituting the screw thread. Thus when the option [Altera OD; D] is performed over the circle of the screw thread, the arc will suffer a change in order to maintain the correct proportions between its radius, using the functions *ads_entmake()*.

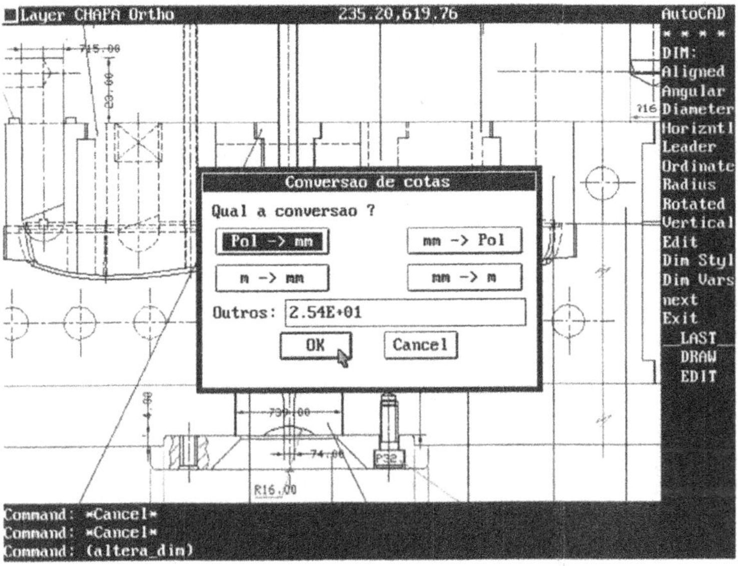

Figure 5 Dialogue box to access to options to convert the co-ordinates

In the option [GERAÇÃO] we intend to present a **pre-defined technological feature** [ROSCA], having geometric relations. We may compare with the type of features, where the designer will access a components library which has different sizes, but in this case of the screw thread the designer will define the size that better suits him.

The designer still has different information tables, that inform him when he intends to evoke any unknown or complex entity, as an auxiliary and fast help of the design accomplishment. This way, **the preliminary design** of a mould gathers the study of the object, the implementation of the object in the moulding position, the dimensions definition of the different components of the object namely, the cavity definition, the core, the study of the moulds mechanics and the special elements, such as movements and unscrew.

In the phase of **the mould final design** an implementation of the different components is put into practice, such as the guides, coupling-box, screws and extractor.

A definition of the final dimensions about the previous constituting parts is given and this is followed by the study of the plastic stuff injection, as the definitions of the moulding areas, cooling circuits (bores, tampons, o-rings) and injection circuit (nozzle of injection, runner system, manifolds), being necessary to improve every process stage.

To conclude this preliminary explanations about the mould building we must speak about object extraction stage, which involves the moulding areas definition and the maximization of the extractors position, cooling circuits and mould supports.

In the representation and design of all these components through a graphic computer program we must use a layer technique of the building stages, associating to them some properties such as type, line thickness and color for an easy reading and interpretation of the drawing phases, creating different layers, such as **moulding core, cooling core, support, screws, core dimensions, cavity dimensions and material lists.**

The application areas of these features function and grouped in [PESQUISA] (SEARCH) (Figure 1) are mainly for the design and processes planning, the ones grouped in [ALTERAÇÃO] (CHANGE) are for design, being necessary to emphasize the geometric features implementation in a one direction and two direction interdependence relations. Within this group a sub-option [ALTERA DIMs] appears as a feature belonging to the testing and controlling area. The option [CÓPIA] (COPY) entails two ways of performing the entities reproduction having relationships between them and it is used in the areas of design.

The option [GERAÇÃO] is a technological feature that applies a mathematical formula that represent the entities type and its consequent reproduction in different sizes and it is inserted in an area of design and processes planning.

3 CONCLUSION

These new options on features in application PROBACOM, maximize the conception of the design stages, but these are dependent on the user skills in the manipulations of the base system menus for a fast design building. The current application becomes useful in the designs and design rooms environment that have computer aided drawing systems for the drawing of two-dimension entities.

In this PROBACOM application the dependencies are mainly built upon single pictures such as circles, arcs and two-dimensional, polygons, taking the real advantages in the manipulation of those entities by saving edition timing by mainly when handling dozens of those entities, which is common in the case of the moulds drawings.

To obtain the AutoCAD base commands associated to these new functions commands is a development of the program itself, because the user may avoid in different circumstances the repetition of actions to achieve a picture change on a group of descending pictures of their copies.

The feature concept is going to evolve within each area according to the geometric entities to being develop. Motivated by one or another detail in the geometry of solids being models other feature type will emerge, with different ends and specific particularities to obtain the geometric pictures of the wished entities.

The approached concepts were focused starting with a bibliographical research and followed experiments in the industry area in order to create homogeneity of the studied topic and connecting it with the theoretical framework that is available as support of these system. This is illustrated with real cases in a mould manufacturing, where the application of new features options are used.

It was our intention to create an application that could be used in the daily life of the productive area, in a sense of improving the interaction between the application and the users, including their work performance.

Comparing this program to other graphic PC programs and by means of common language in engineering as the "C language" and the ADS´s functions, we obtained a better performance of the basic options of the AutoCAD, being difficult to obtain the same results with other programs, such as Cadkey or Intergraph PC, due to their peculiar developed languages, which are less well known and requires more learning time and consequently increasing costs..

In the current program the restrictions or interdependencies are only generated among simple pictures, being one aspect to reformulate by the creation of more complex with those properties.

Another aspect to be improved is the building of three-dimension pictures, where the API module, AutoDESK(1994), should be used so that its dependence relations do not create insatiability between faces or vertexes in order to avoid distortions.

The association between pictures would be a way to complete the performed work as well as the mechanisms of interaction between the user through a set of feature windows to be applied to the pictures. The difficulties found throughout the development of the PROBACOM application are mainly concerned with the complex nature of the data structure of the AutoCAD graphics program.

The PROBACOM application was developed mainly using the version 12 of the AutoCAD but are also compatible with the version 13.

4 REFERENCES

Anderson, D.C. & Chang, T.C. (1990), *Geometric reasoning in future based design and process planing.* Computers Graphics vol. 14, Nov. 2, pp. 225-235.

AutoDESK Incorporated (1994), *Reference manual AutoCAD v. 12 and v.13, Manual ADS's* AutoCAD - U.S.A.

Omura, G. (1991), *Mastering AutoCAD,* Sybex, 4ª edition.

Pratt, M.J. & Wilson, P.R. (1988), *A taxonomy of features for solid modeling.* Geometric modeling for CAD applications, Elsevier Publishers-IFIP, Holland.

Rembold, U. & Nnaji, B. & Storr, A. (1993), *Computer Integrated Manufacturing and Engineering.* Addison Wesley Publisher´s Company.

5 BIOGRAPHY

The author is a research and teacher in Instituto Politécnico de Leiria. He received a MSc. in electrical and computer engineering by Instituto Superior Técnico, Lisbon Technical University - Portugal, actually is PhD. student in same area. His current research interests include multimedia, computer graphics, neural networks and virtual reality.

41

Manufacturing Process Modeling Method for CAD/CAM and Flexible Manufacturing Systems

LÁSZLÓ HORVÁTH professor
Department of Manufacturing Engineering, Bánki Donát Polytechnic
Népszínház u. 8. Budapest H-1081 Hungary
phone: (36-1)-210-1449, fax: (36-1)-133-6761
e-mail: lhorvath@zeus.banki.hu

IMRE J. RUDAS professor
Department of Information Technology Bánki Donát Polytechnic
Népszínház u. 8, Budapest H-1081 Hungary
phone: (36-1)-133-4513, fax: (36-1)-133-9183
e-mail: rudas@zeus.banki.hu

Abstract

Knowledge based automation of the manufacturing process planning is an evergreen problem area of integrated design and planning (CAD/CAM) systems as well as flexible manufacturing systems (FMS). The authors propose a Petri net and rule based modeling method for manufacturing process planning. The generic nature of this model makes it possible to handle all of the process variants. The model is then evaluated by process planning and production scheduling procedures. Models are created on the basis of data of form features oriented part model and manufacturing environment. Petri net objects are created using manufacturing process objects and process structure information. The paper is organized as follows: First the aims of the research are outlined and the characteristics of manufacturing process models of this type are summarized. Next, the method of modeling, the structure of the process model and the process model entities are detailed. Following this, a generation procedure of the process model is presented. Finally, an example of process model is explained.

Keywords
Modeling of Manufacturing Processes, Petri Net Model, Knowledge Based Generation of Model Entities, Generation of Petri Nets.

1. INTRODUCTION

Flexible generation of manufacturing process plans is a main concern in both CAD/CAM and flexible manufacturing systems (FMSs). Integrated product modeling concepts stimulate research in this field. Important result of product modeling efforts in the last decade is the STEP (Standard for Exchange of Product Model Data, ISO 10303). In this work the product model approach of the STEP has been taken into consideration. It is well-known that decision intensive nature of manufacturing process planning calls for knowledge based methods. A product model consists of partial models. One of the problematic partial model within product models is the manufacturing process model (Eversheim, W. and Marczinski, G. and Cremer, R., 1991). Methods for creation of model entities are in the center of development work at present day computer aided design and planning systems. Future systems demand knowledge based procedures for creation and evaluation of manufacturing process models. Conventional knowledge based methods for modeling applied at computer aided manufacturing process planning have not enough to fulfill the requirements of present day CAD/CAM and flexible manufacturing systems.

The paper is organized as follows. First the aims of the research are outlined and the characteristics of manufacturing process models of this type are summarized. Next, the method of modeling, the structure of the process model and the process model entities are detailed. Following this, a generation procedure of the process model is presented. Finally, an example of process model is explained. The work presented in this paper has been restricted to modeling of manufacturing processes of mechanical parts.

2. MODELING OF MECHANICAL PARTS

Comprehensive range of methods and tools for shape modeling, form feature modeling, finite element analysis, tool path generation and other tasks are available in these days for engineers. Manufacturing process modeling has been placed into other engineering modeling activities and the correct functional and data communication connections have been revealed. This was the basis of the process modeling research that made by the authors. They made proposal for manufacturing process model structure, Petri net representation for process model entities and application of knowledge based paradigms to process modeling (Horváth, L. and Rudas, I. J., 1994, 1995). The effects of human-computer procedures on design and planning work have been analyzed (Horváth, L. and Rudas, I. J., 1994/2).

Figure 1 shows product modeling activities that are in relation with manufacturing process modeling. Main classes of these activities are modeling by geometric, engineering (H. Grabowski and R. Anderi and M. J. Pratt, 1991), and manufacturing objects of the part. Communication between computer programs and humans greatly affects these modeling activities. Manufacturing of mechanical parts is characterized by less or more number of suitable process variants taking into account a given manufacturing capacity. The process model must contain information on all available process variants. This gives flexibility for production scheduling. Entities that are involved in the proposed manufacturing process model carry information for all suitable variants of a process or a process element for a cluster of manufacturing tasks. At the same time model entities carry knowledge to select the most suitable variant. Because both conditions in the manufacturing environment and the production schedule are subjected to changes in the every day industrial practice, the manufacturing process must be modified frequently. This can be done by re-evaluation of the proposed generic process model. Application of advanced form feature based shape modeling techniques are assumed at the product model.

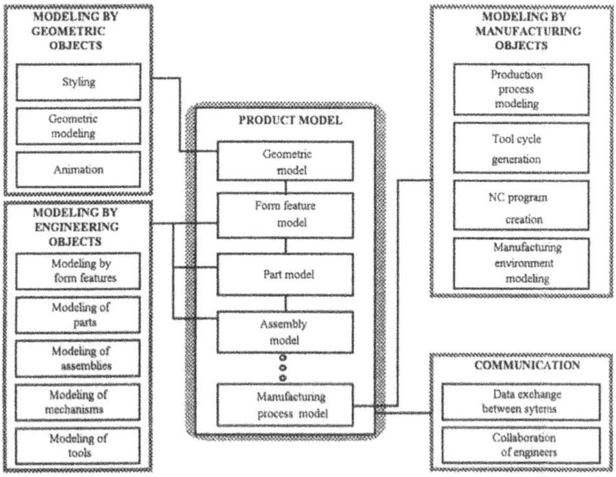

Figure 1 Modeling of products.

3. FOUR LEVELED MANUFACTURING PROCESS MODEL

Activities for development of advanced modeling methods in these days involve implementation of the concept of application oriented features. Models are built by features or features are recognized in models. The authors made an attempt to elaborate a feature based method for manufacturing process modeling (Horváth, L. and Rudas, I. J., 1994, 1995). Manufacturing process feature is any element of a manufacturing process that helps to understand a manufacturing related to a part, a form feature, a machine tool, or a cutting tool. Model entities are instances of manufacturing process feature classes. The feature instances have generic characteristics. The feature class is a global term whereas feature instance is for a given machining task and manufacturing environment. Feature libraries may contain features and their instances as usual in object oriented computer aided design and planning systems.

The authors have proposed a four leveled model in which model features are placed at the process (L1), setups (L2), operations (L3) and numerical control (NC) cycles (L4) level as it can be seen at Figure 2. Model features are processes, net of setups, net of operations and sequence of cycles. Nets of operations are for setups and manufacturing of form features. On the level 1 process variants are mapped for manufacturing of the part. Level 2 involves process features that describe all possible sets of setups. A setup is part of the manufacturing process that is done using a single machine tool with the same clamping position. Level 3 contains possible sets of operations. Operation is a manufacturing that is realized using the same cutting tool. Level 4 is for tool cycle sequences.

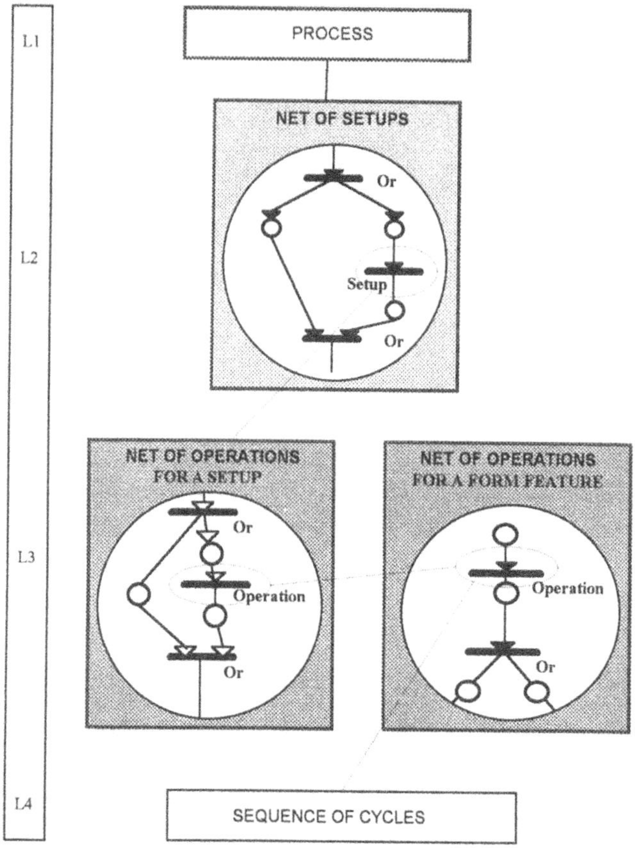

Figure 2 Levels and features in the process model

There are process features that have a net structure. These are net of setups and net of operations process features. Elements of the structure are setup or operation manufacturing process objects and have attributes. The actual state of the process feature represents a sequence of setups or operations. The actual state is generated in course of evaluation of the net using built-in knowledge. The actual state is changed by repeated evaluation process taking into account modified initial conditions. Repeated evaluation is initiated by an appropriate procedure or an engineer on the basis of changes in product data, production environment data or preferences for production scheduling. In common sense the process entity offers a manufacturing process or subprocess that is changeable according to the changing requirements. That is an important characteristic of the proposed process model that makes it suitable for application at flexible manufacturing systems.

After an assessment on the nature of modeling task and the needed computing power, implementation of ordinary marked Petri net formalism with some extensions had been decided. These Petri nets consist of places and transitions (Figure 3.). Transitions represent setup or operation process object. Description of process object by their attributes is attached to the transition. Special transitions represent AND and OR splits and joins similarly as proposed in (Kruth, J. P. and Detand, J, 1992). These transitions are used for creating branches. Place carries marking and identification of rules for selection the subsequent transition. Rules are placed in the knowledge base. Transition can have *in-process* or *out-process* status. In-process status of a transition represents process object that is in the actual process. Process objects that are represented by out-process transitions are not in the actual process but those transitions may gain in-process status after a new evaluation procedure.

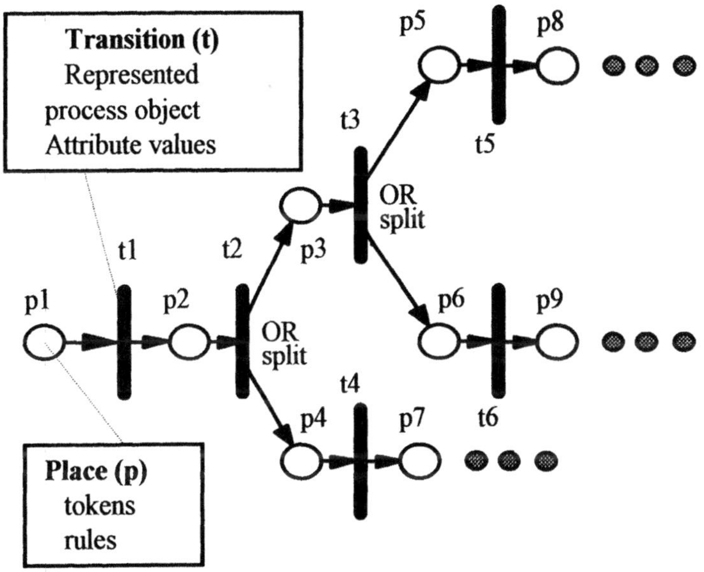

Figure 3 Petri net representation

Marking of the Petri net is configured on the basis of suitability of process objects represented by transitions for the part manufacturing task. Evaluation starts on the first level of the model and advances level by level. Selection of process objects on the present level determines process features to be evaluated on the succeeding level. The evaluation procedure always is going on a given Petri net representation. The evaluation of a process model feature basically consists of two stages, namely checking the process objects involved and execution of the Petri net. There is an initial marking of the Petri net. Previous to firing a transition the availability and suitability of the process object are checked and values its attributes are calculated using rules and equations. If a process object is not suitable, the marking is changed to make the transition disabled for firing. OR splits are used for making branches that represent variants. A branch is inactive if its first transition is disabled. If the execution of the net can not be continued because of the lack of enabled transition, the decision on the branch is revised. If checking of a process object leads to halt of the execution of the net, the process feature is not suitable for the manufacturing task. The process objects represented by the fired transitions are the actual ones. The model is evaluated from the point of view of both process planning and process scheduling. The extra features of the Petri net used here are:

- Special transitions for creating branch (AND, OR).
- Attaching rules to a place and a transition to make a validation of process objects possible previous to firing of the transition.
- Active and passive tokens. Token at a place becomes passive when a process object represented by the following transition is not suitable for the manufacturing task. Passive tokens are not taken into consideration at execution of the Petri net.
- Making it possible to use the Petri net model during both process planning and production scheduling.

If a transition is enabled, the token is moved in the usual way to the place that follows the fired transition. Value of status of the transition is changed from the *out-process* default value to *in-process*. If a transition is not enabled, the value of its status is *out-process*. At production scheduling, repeated evaluation of the process model so creating new process plan variant can be initiated. AND splits take it possible to fire all subsequent places. In this case all branches must be valid.

Repeated evaluation may be needed after any changes of manufacturing task description. Task description for the process planning involves data for

- class of part and form features,
- attribute values of part and form features,
- capabilities of manufacturing processes,
- available equipment and tooling and
- preliminary decisions on manufacturing process.

Petri net representations are interconnected within the generic process model but they are not synthesized into a complex net. Consequently, not only evaluation but also change of process features may take place. This characteristic gives flexibility for the modeling.

Other important and keenly researched application of Petri net is modeling and analysis of manufacturing systems (A. A. Desrochers and R. Y. Al-Jaar, 1995). These models can be interconnected with manufacturing process models in the future.

4. GENERATION OF PROCESS MODEL ENTITIES

The authors aimed to develop procedures for creating process model entities in a generative way using feature structure description and rule sets stored in a data base. This knowledge contains description of process feature classes, knowledge for selection of feature classes and knowledge for evaluation of process model entities. Generation of Petri net model representations can be followed on Figure 4.

Accuracy of description of manufacturing task is very important. On the basis of task description manufacturing process classes are selected. Classification tree gives class for given values of task parameters. Individual rules are applied where trees do not give solution. The next step is creation of basic process feature structure. Structure description is used and attributes are attached to manufacturing process objects that are involved in the structure.

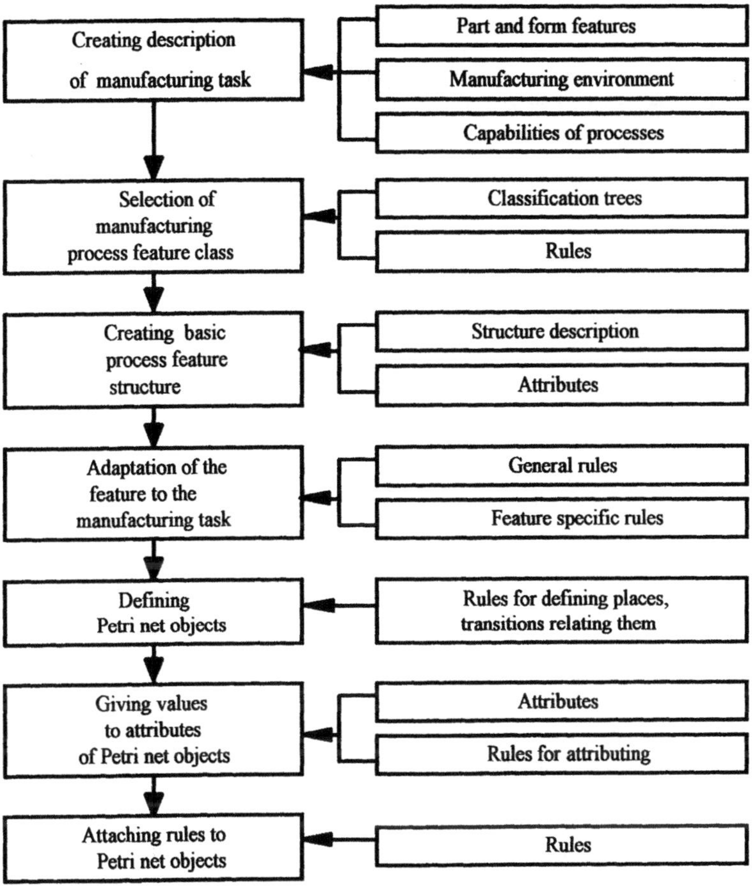

Figure 4 Generation of Petri net model representations

Figure 5(a) shows process structure with nine manufacturing process objects (Mo1-Mo9). As an example, in this description Mo1-Mo9 process objects can be for center drilling (Mo1), drilling with two stepped twist drill (Mo2), drilling (Mo3), Boring (Mo4), drilling with two stepped precise drilling tool (Mo5), reaming (Mo6 and Mo7), counterboring (Mo8) and deburring (Mo9). A possible form of structure description:

START

Mo1 *net starts with Mo1 process object*

Mo2 or Mo4 or Mo3 *net splits, Mo2, Mo4 or Mo3 can be selected*
IF Mo2 THEN Mo7
IF Mo3 THEN Mo5 subs Mo6 *on the Mo3 branch Mo6 follows Mo5*
IF Mo7 OR Mo4 or Mo6 THEN Mo8 *Mo7, Mo4 and Mo6 is joined to Mo8*
Mo9 *the last process object is Mo9*
END

The feature is adapted according to the description of manufacturing task. Figure 5(b)) shows a net in which Mo7, Mo4, Mo6 and Mo8 are unnecessary (on the basis of analysis of the manufacturing task and the manufacturing environment description) and so they have been deleted. For example, reaming and counterboring are not necessary for the given manufacturing and presently there is not available tool for the Mo5.

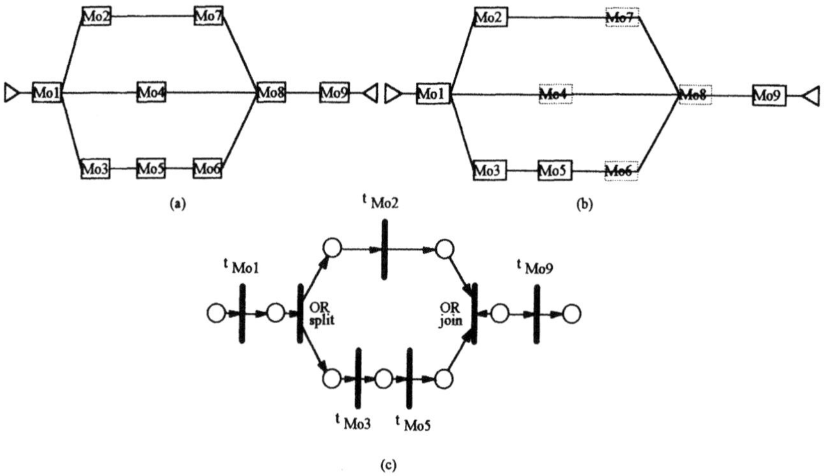

Figure 5 Creating process feature structure

Using the modified basic net structure the Petri net model of the process feature is generated (Figure 5(c)). First transitions are generated for each process object. Then splits and joins are generated. Following this, place objects are generated and put before the transitions and initial marking is created. If rules are available for a transition, their identifiers are placed in the appropriate attribute of the process object represented by a transition. Rules for selection of a given transition are attached to the place that precedes it. These are simple IF ... THEN ... ELSE rules. They contain attribute values for manufacturing tasks, process objects and manufacturing environment objects. In this manner structured and unstructured information for the given manufacturing process feature is organized into a Petri net structure. At this point the Petri net representation of a generic process feature is ready for evaluation.

Knowledge and dedicated procedures serve calculation of attribute values of the manufacturing process objects represented by transitions. Changes in initial conditions of the process planning may require changes not only in sequence of process elements in the process plan but changes in values of their attributes too. Elements of a Petri net model can be interrelated using special purpose rules. Process model features and Petri net elements are defined as objects. Petri net objects are net, place, transition, arc and token. Instances of these objects are generated as manufacturing process model entities.

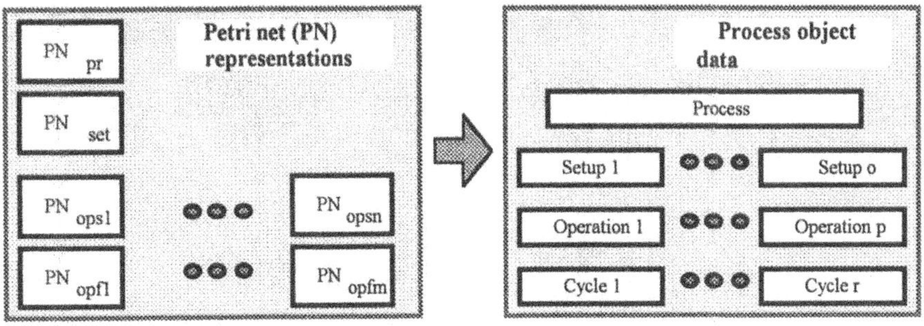

Figure 6 The process model for a part

The process model for a manufacturing task is outlined in Figure 6. Petri net representations are related through transitions. There is a Petri net for the process (PN_{pr}) and another for the net of setups (PN_{set}). Two or more nets of setups can be defined for a process as it can be seen in the example later in this paper. Two sets of Petri nets of operations are stored for setups (PN_{ops1}-PN_{opsn}) and for manufacturing of form features (PN_{opf1}-PN_{opfm}). Transition has pointer to the process object data. Manufacturing process generation procedures search rules in a structured central rule base where IF-THEN rules can be placed by manufacturing and systems engineers. Rules have been introduced by the authors in (Horváth, L. and Rudas, I. J., 1995).

5. PETRI NET FEATURE REPRESENTATION EXAMPLES

A typical example of manufacturing process model for a prismatic part machined in a flexible manufacturing system was generated. In the course of the run experiments the effect of typical initial conditions on the manufacturing process was simulated.

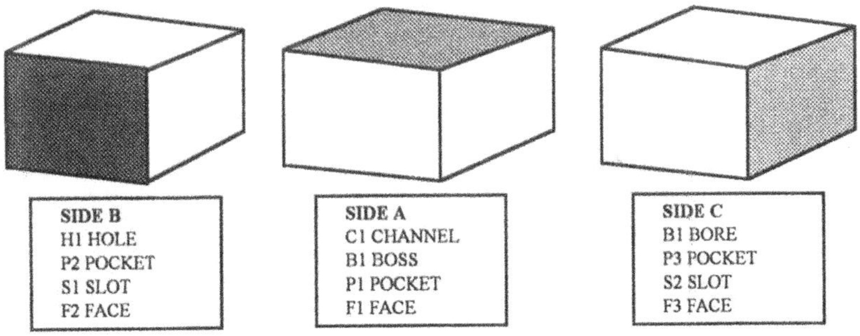

Figure 7 Form features on the test part

Figure 7 shows the form features that were defined on the prismatic test part. Typically form features are arranged on the six sides of a prismatic part. At this example only three sides contain features to be machined. Manufacturing process can be realized by two different nets of setups (Figure 8). These alternatives are represented by the t_{nset1} and t_{nset2} transitions in the Petri net.

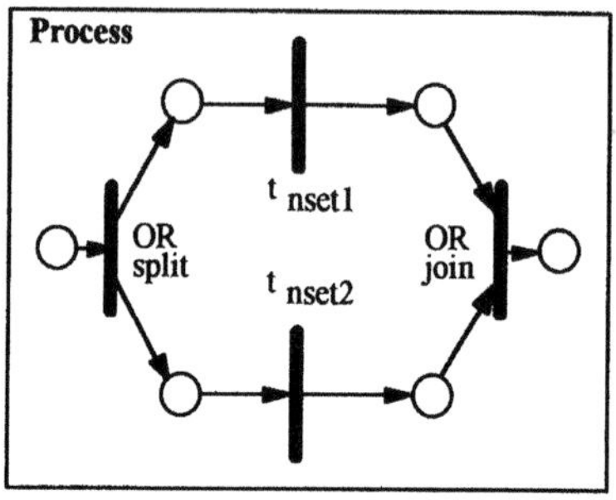

Figure 8 Process feature

The net of setup nset1 is represented by a Petri net that contains transitions for setups applied at manufacturing of the sides of the part (Figure 9). The part manufacturing starts with a setup for machining datum surface and ends with an end inspection. Two alternatives are machining sides in two (set2 and set3) or three (set4, set5 and set6) setups.

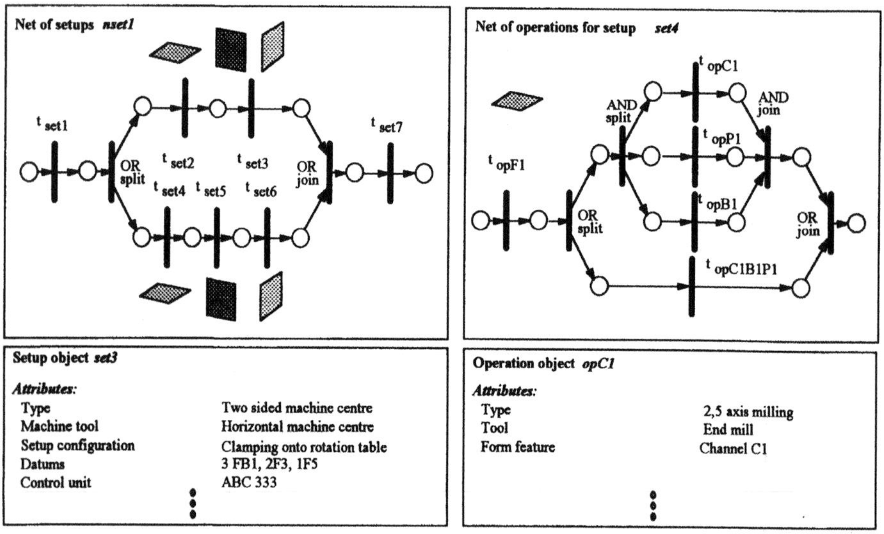

Figure 9 Net of setup and net of operations features

Attributes of the setup object describe the type of setup, the machine tool, the datum surfaces, etc. Figure 9 shows the net of operation feature for the set4 setup. This setup involves operations that are needed for machining of form features on the upper side of the part. After machining the face F1 completed, the channel C1, the boss B1 and the pocket P1 are machined in separated or common operations. The latter is the case of using common tool for manufacturing of these form features. The structure of the Petri net was created to represent the above process variants.

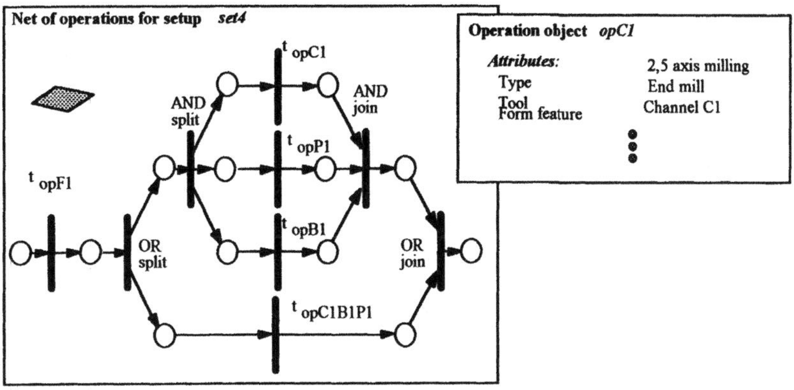

Figure 10 Net of operations feature

The focus in this paper is limited to the generation of the process model. Data base that contains information about the required knowledge is assumed. Evaluation procedures are not discussed. In practice, procedures for manufacturing process planning are attached to software packages included in CAD/CAM systems, production planning and scheduling systems and flexible manufacturing systems. The integration of manufacturing process model generation methods into these environments requires additional investigation. An another problem is related to interactions between the generation procedures and the human planner (Horváth, L. and Rudas, I. J., 1994/2). The cooperation of well-known problem solving, process planning and Petri net evaluation procedures with the process modeling environment and the related interfaces are topics of future research.

6. CONCLUSIONS

A method for modeling of manufacturing processes has been proposed by the authors. Demands of CAD/CAM and flexible manufacturing systems have been taken into consideration. Feature based modeling approach and Petri net representation for model entities were applied. The main contributions of this paper are:

- A product model approach to manufacturing process modeling was shown.
- Petri net representations of generic manufacturing process features were introduced. These features were placed into a four leveled model structure.
- Generation of process model entity using classification trees, structural description and unstructured knowledge was explained.

The proposed method has proved to be suitable for modeling real world processes. Models can be evaluated by executing the Petri nets taking into account requirements of both process planning and production scheduling. The results of experiments are promising. Integration or interconnection manufacturing process models with manufacturing system models (Kovács, G. and Mezgár, I.. and Kopácsi S. ,1994) is an exciting task for future research.

7 ACKNOWLEDGMENTS

This paper was written within the frame of Hungarian-Portuguese inter-governmental scientific and technological cooperation, as a result of the research cooperation supported by the OMFB and its Portuguese partner the Junta Nacional de Investigacao Cientifica e Technologica (JNICT).

8 REFERENCES

Eversheim, W. and Marczinski, G. and Cremer, R. (1991) Structured Modeling of Manufacturing Processes as NC-Data Preparation, Annals of the CIRP, 1991/1.. pp. 429 - 432

Horváth, L. and Rudas, I. J. (1994) Manufacturing Process Planning Using Object Oriented Petri Nets Supported by Entropy Based Fuzzy Reasoning. Integrated Systems Engineering, PERGAMON, Oxford, 1994, pp. 335-340.

Horváth, L. and Rudas, I. J. (1995) Modeling of Manufacturing Processes Using Object Oriented Extended Petri Nets and Advanced Knowledge Representations, Proceedings of 1995 IEEE International Conference on Systems, Man, and Cybernetics, Intelligent Systems for the 21st Century, Volume 3, pp. 2576-2581

H. Grabowski and R. Anderi and M. J. Pratt (1991) Advanced Modelling for CAD/CAM systems, Springer Verlag, 1991

Horváth, L. and Rudas, I. J. (1994/2) Human - Computer Interactions at Decision Making and Knowledge Acquisition in Computer Aided Process Planning Systems. Proceedings of the 1994 IEEE International Conference on Systems, Man and Cybernetics, IEEE Systems, Man and Cybernetics Society, San Antonio, 1994, pp. 1414-1419

Desrochers, A. A. and Al-Jaar, R. Y. (1995) Applications of Petri Nets in Manufacturing Systems, IEEE PRESS, New York, 1995

Kruth, J. P. and Detand, J (1992) A CAPP System for Non-linear Process Plans, Annals of the CIRP, 1992/1. pp. 489-492

Kovács, G. and Mezgár, I.. and Kopácsi S. (1994) A Hybrid Simulation-Scheduler-Quality Control System, The 10th ISPE/IFAC International Conference on CAD/CAM, Robotics and Factories of the Future, Ottawa, Conference Proceedings, p: 584-589.

9 BIOGRAPHY

László Horváth received the MSc degree in Production Engineering from the Technical University of Budapest in 1971, received the Dr. Techn. degree in computer aided process planning systems from the Technical University of Budapest in 1984 and received the Ph.D. degree in knowledge based process planning systems from the Hungarian Academy of Sciences in 1992. He is active as professor at the Bánki Donát Polytechnic, Budapest, Hungary where currently teaches courses on CAD/CAM systems engineering and knowledge engineering. He is member of the New York Academy of Sciences, Senior Member of IEEE Industrial Electronics Society, IEEE Systems, Man and Cybernetics Society and IEEE Human-Computer Interaction Technical Committee. His research interests include modeling of manufacturing processes, application of artificial intelligence in manufacturing process planning, integrating design and planning systems, concurrent and collaborative engineering, product modeling and human-computer interaction.

Imre J. Rudas graduated from Bánki Donát Polytechnic, Budapest in 1971 and received the Master Degree in Mathematics from the Eoetvoes Loránd University, Budapest while the Ph.D. in Robotics from the Hungarian Academy of Sciences. He is active as professor and Head of Department: Information Technology at Bánki Donát Polytechnic, Budapest. He is an active member of the New York Academy of Sciences, Senior Member of IEEE, and registered expert of the United Nations Industrial Development Organization. He is also a member of the editorial board of Engineering Application of Artificial Intelligence and Control Engineering Practice, member of various national and international scientific committees. His present areas of research activity are: Robot Control, Soft Computing, Computed Aided Process Planning, Fuzzy Control and Fuzzy Sets. He has published more than 150 papers in various journals and international conference proceedings.

PART SEVENTEEN

Performance Indicators

42

Performance Indicators and Benchmarking in Manufacturing Planning and Control Systems

A.K. Kochhar, A.J. Davies and M.P. Kennerley
UMIST
Manufacturing Division, Department of Mechanical Engineering, Manchester, M60 1QD, United Kingdom
Telephone +44 161 200 3801, Fax +44 161 200 3803,
E-mail A.J. Davies@umist.ac.uk, M.P. Kennerley@umist.ac.uk,

Abstract

The recent, much publicised, interest in Benchmarking and Performance Measurement has not resulted in many of the anticipated benefits. Much of the benchmarking carried out in manufacturing organisations relates to top level issues. For benchmarking to be of real value, the best practices associated with lower levels of an organisation, which contribute to its overall performance, also need to be considered. It is particularly important to take account of manufacturing planning and control systems which interface with almost all of the activities carried out by a manufacturing organisation. This paper describes the findings of detailed fieldwork undertaken to investigate the use of benchmarking and performance indicators in manufacturing planning and control systems. It is concluded that a considerable amount of research work is needed to develop a detailed understanding of the relationships between performance measures, the contributing variables and the associated best practices.

Keywords

Manufacturing planning and control systems, performance measurement, benchmarking, best practice.

1. INTRODUCTION

In recent years, a considerable amount of new thinking has emerged in the area of manufacturing. Traditional financial measures of performance are no longer regarded as adequate for the new style of highly competitive, world class manufacturing systems. Consumers are becoming ever more powerful and only companies that are sensitive to their changing requirements and are able to give them what they want will survive and prosper in the marketplace. As market requirements change, the strategies employed to maintain or achieve a higher share of the market must also change. These changes must be reflected in the measures used to evaluate the success of the strategies.

The actual manufacture and assembly of products represents the most significant and complex part of any manufacturing organisation. It is within the design and execution of manufacturing that the greatest impact on the competitive parameters of quality, cost, delivery and flexibility is made. The largest payback in any operation occurs at the point of manufacture(1). In order to be flexible, decisions need to be made at the lowest possible level to allow rapid action. The quality of these decisions is directly related to the information available. Hence accurate information is required at the point of use, at the right time, and in a form that clearly indicates the required action. A good, well-planned performance measurement system should provide a framework that will give people lower down the organisation the performance data they need to make decisions. This can move decision-making down to the operational level which should allow improved performance in each of the parameters of quality, cost, delivery and flexibility.

Measurement must be followed by actions which lead to improved performance. Such actions may be based on existing well-known best practices. Alternatively, these actions may be based on the application of benchmarking techniques. Benchmarking involves looking at best practices, external to the company, to adapt them, for use in the company, in order to achieve superior performance. Anderson Consulting(2) regard benchmarking as the most powerful tool for assessing industrial competitiveness and for triggering the change process in companies striving for world class performance. After a benchmarking study one can no longer hide behind excuses or put the clock back - one has to face the facts, however unpleasant, and act(2). Benchmarking is a very realistic way to assess where a company is in relation to its competitors. It also provides realistic achievable targets and establishes how much a company needs to do to reach world class levels of functional and cross-functional performance(3). Miller et al(4) classify benchmarking into four types, namely strategic, product, functional and best practice, and recommend that all four types should be used so that every member of the organisation is involved.

The recent high level of interest in the two related areas of Performance Measurement and Benchmarking, has resulted in actual benchmarking being undertaken mainly at top levels of manufacturing organisations. In contrast, real sustainable improvement of manufacturing activities requires an assessment of performance at lower levels of an organisation and the use or adaptation of best practices. For example, manufacturing planning and control systems make a very significant impact on the achievement of the competitive priorities of quality, cost, delivery and flexibility, and all of the detailed activities undertaken in a manufacturing organisation. Very little performance measurement and benchmarking has been undertaken in this area. A large number of manufacturing planning and control performance indicators can be and should be used. Typical performance indicators include master production schedule adherence, purchase order issue timeliners, supplier due data compliance, shop floor schedule adherence, work in progress stockturns, and so on.

It is not the purpose of this paper to discuss the well rehearsed details of typical manufacturing planning and control systems. Many publications, for example texts by Vollman et al[5] and Green[6] provide comprehensive details of the functionality and operation of typical manufacturing planning and control systems.

As part of ongoing research in performance measurement and benchmarking[7], detailed field studies were carried out to investigate the extent of performance measurement and benchmarking at the manufacturing planning and control system level, and the use of associated processes, performances measures and practices.

2. FIELD STUDIES

The field studies were designed to achieve the following related objectives:

- identify the measures of performance used to assess the efficiency and effectiveness of the manufacturing planning and control system;
- assess the priority associated with individual measures of performance;
- ascertain the actual values for specified measures of performance;
- assess the critical issues in the performance of manufacturing planning and control systems;
- assess the understanding of causal relationships;
- assess the cost and difficulties of collecting data relating to measures of performance;
- assess the effect of manipulating variables on performance indicators;
- assess actions which help to achieve higher levels of performance;
- confirm the practices and benchmarks which are considered to be 'best'.

An overall framework was created to structure the data collection and analysis process. The overall field study involved the following processes.

1. Preliminary interview with the senior management to obtain background information about the company. Questions at this introductory high level related to company products, strategy, organisational structure, major financial and production data, and the company environment.
2. Definition of the manufacturing planning and control systems used by the company, procedures for assessing performance and identification of the 'owners' of the particular modules and hence the interviewees for the main field study questions.
3. Identification of any organised benchmarking activities undertaken by the company.
4. Identification of critical success factors for the organisation.

5. For each of the major modules of the manufacturing planning and control system used by the company, the following information was collected.

   (a) The process of the module in terms of:
      - objectives;
      - inputs;
      - outputs;
      - boundary conditions;
      - tools and techniques used;
      - activities undertaken.

   (b) Measures of performance (MOP's) used to assess the performance of the module.
      - formal measures for assessing:
         - effectiveness of the module;
         - efficiency of the module.
      - informal measures.
      - detailed information relating to the data collection, analysis, reporting, ownership, costs, controllability, visibility, and actions for each measure of performance.
      - targets, their setting and review process.
      - best known benchmark for each measure of performance.

   (c) Best practices used to improve the values of individual measures of performance.

   (d) Awareness of other best practices and why they are not used.

   (e) Overall benchmarking, as well as manufacturing planning and control system benchmarking, carried out by the organisation.

   (f) Relationships between manufacturing planning and control system MOP's and the overall performance of the organisation.

   (g) Planned developments in relation to measures of performance and best practices.

Four companies were visited and studied in very considerable depth. Although four companies constitute a small sample, their characteristics are such that they are representative of a significant proportion of manufacturing industry engaged in the batch manufacture of engineering products.

Each field study involved a considerable amount of preparation by the staff of the company involved as well as the interviewers. Actual data was collected during face to face interviews with a number of people who 'owned' and/or operated particular modules of the overall manufacturing planning and control system. The data was collected by two researchers over a period of one week in each company. This was followed by further visits to obtain clarifications and missing data. A very detailed report was prepared for each field study company and sent to the company management. Subsequently, a verbal presentation was done during which the main findings were presented, followed by an assessment of the findings. The presentations took the form of an interactive workshop which resulted in the development of action plans to be used within the company. The detailed reports of all field studies have been analysed to bring together the main findings as discussed in the next section.

3. MAIN FINDINGS

The main findings of the field studies are as follows:

- There was little measurement of the effectiveness and efficiency of manufacturing planning and control (MPC) systems, particularly in relation to the strategic objectives. Performance measurement concentrated on the execution areas of the MPC system rather than the planning areas. There was no consistent approach to performance measurement at any of the sites and there was little formalised or structured understanding of cause and effect relationships.
- Very few formal structured benchmarking projects have been carried out in the area of MPC systems. Company visits were undertaken and provided a useful insight into improvement opportunities but a more formal benchmarking process is required to maximise the benefit of such activities. There was no formal benchmarking strategy at any of the sites leading to confusion as to what benchmarking involves. As a result there tended to be a preoccupation with metrics rather than the practices behind superior performance.
- Overall production/business planning was not explicitly carried out on any of the sites. It was largely a budgeting activity. There was integration with the master scheduling process and shop floor scheduling as the same personnel seemed to be involved in each process although feedback tended to be informal. There was a relatively little formal feedback to the production plan to revise financial and capacity plans.
- There was little formal performance measurement undertaken in the sales order processing module. There tended to be monitoring of progress against budgets for individual products/projects but there was no measurement across the company as a whole. Such measurement would indicate the areas in which performance should be improved. There tended to be good communication between the master scheduler, shop floor and sales in terms of agreeing due dates.
- There was no measurement of the accuracy of demand forecasting. The accuracy of demand forecasts is important for all types of companies in order to improve resource planning. The lack of consideration of forecast accuracy at the planning stages pushed uncertainty down to the execution areas where it was more difficult to cope with.
- The performance of the master production scheduling module was measured at the shop floor level. Schedule adherence was the main measure although this measured adherence to the plan rather than how accurate or achievable the plan was or how well it satisfied customer requirements.
- In all companies, rough cut resource planning was an informal and simple process. No consideration was given to parts availability within rough cut resource planning and there was no formal measurement of the availability of capacity or the accuracy of this process.
- In the companies visited, the MRP systems tended to deal with procurement rather than the issuing of shop orders. There was no regular measurement of the accuracy of the information in the system used for the planning of material requirements at any of the sites.

- There was no measurement of BOM accuracy at any of the sites and there was little feedback from shop floor to identify performance problems.
- There were formal procedures and clear ownership for the engineering change control process in place at all sites. There was little formal performance measurement with preference being given to informal reviews. Where measures were used, or planned, there was a tendency to make them over complex in this area.
- Purchasing was the area with the most developed strategically based performance measures and where most significant effort had been made to implement practices to improve these measures. Some supplier development had been undertaken at all sites.
- Stock accuracy was a major issue in all companies and was measured at each of the sites. There was a need for more stock control in relation to semi-dormant and obsolete items and more regular reviews of current and active stock.
- No formal detailed capacity planning was undertaken at any of the sites.
- The shop loading and scheduling process tended to be clear and simple and was mainly concerned with the allocation of labour. Due date appeared to be the main priority rule.
- Shop floor monitoring and control was one of the areas in which most measurement of performance was undertaken. However, there was little prioritisation of measures and measurement was not always strategically based. Ownership of measures of performance was delegated well at two of the sites.
- Little finished goods inventory was held by any of the sites, hence finished goods inventory control was informal and simple. Simplicity and visibility were important for this process in each of the companies.
- The areas that had the most developed measures of performance and practices tended to be the areas that historically have been the focus of attention, mostly due to financial measurement and control. For example, purchasing because it tended to be the main area of expense within the organisation, and the shop floor as it dealt with output and work in progress. Also, these areas are the easiest to measure. Although the execution areas were the most measured areas, it was often the planning areas which caused poor performance and these areas were not measured. This problem is exacerbated by the fact that there was inadequate feedback from the execution areas back into the top level plans in order to create more realistic plans.
- The implementation of good practices within manufacturing planning and control systems appeared to be strongest in the area of purchasing, particularly in relation to supplier development. This was due to the vast amount of literature that has paid attention to this area in recent years.
- Ownership and definition of modules within the manufacturing planning and control systems was often unclear. This caused problems in the ownership of measures and practices.
- High product quality appeared to be implicit throughout the four organisations, with quality departments and a large number of measures supporting this. However, the other competitive priorities of delivery and flexibility were not as well communicated, or made the focus of attention at all levels of the organisation.

- Many companies advocated due date compliance as one of their critical success factors, although there was a lack of communication and emphasis on this. In many cases, it was not measured until the end product was finished. There was a need in all companies for more focus to be placed on this measure at all levels of the manufacturing planning and control system.
- There was a lack of structured framework or strategy for measuring performance within organisations. As a result, there was little evidence of strategically based measures being in place. There was also little education on the need and use of performance measurement.
- On the whole, benchmarking was very limited and there were many misconceptions about what benchmarking involves. Many managers had a preoccupation with metrics, to the exclusion of practices.
- Cause and effect relationships between objectives, measures, variables and practices were not clear and thus were difficult to identify within the organisations studied.

4. CONCLUSION

Much performance measurement and benchmarking is carried out at business level. While this is of value, to be truly effective, benchmarking and the associated measurement of performance need to be aimed at other, contributory levels within a business. As a result, the measurement of, and targets for, strategic objectives can be developed throughout the organisation and focused on the areas that can satisfy them. This has particular importance within manufacturing planning and control systems as they are designed to co-ordinate the execution of manufacturing activities to support customer requirements and strategic objectives.

The findings of the fieldwork, outlined in this paper, clearly indicate lack of effective measurement of performance and benchmarking in relation to manufacturing planning and control activities. These findings support the belief that, in many organisations, computer-based manufacturing planning and control systems are simply no more than transaction processing systems which contribute relatively little to effective decision-making and continuous improvement of manufacturing systems. While there is a reasonable knowledge of best practices, especially in relation to execution modules, there is relatively little understanding of cause and effect relationships. A considerable amount of further research is necessary to develop this knowledge and improve the decision-making at lower levels, in manufacturing organisations, which contribute to the overall performance of an integrated and sustainable industrial system.

5. ACKNOWLEDGEMENTS

The authors are grateful to the Control, Design and Production Group of the British Engineering and Physical Sciences Research Council (EPSRC) for the award of a research grant to undertake the research described in this paper. They are also grateful to the companies in which field studies were undertaken.

6. REFERENCES

(1) Martin, P.G. (1993) *Dynamic performance measurement: The path to World Class Manufacturing*, Van Nostrant Reinbold.
(2) Anderson Consulting (1993) *The Lean Enterprise Benchmarking Project, Arthur Anderson & Co.*
(3) Walleck, A.S., O'Halloran, J.D. and Leader, C.A. (1991) Benchmarking World Class Performance, *McKinsey Quarterly*, Number 1, p3-24.
(4) Miller, J.G., Meyer, A. and Nakare, J. (1992) *Benchmarking Global Manufacturing*, Irwin.
(5) Vollmann, T.E., Berry, W.L. and Whybark, C. (1992) *Manufacturing Planning and Control Systems*, Irwin, Third Edition.
(6) Green, J. H. (1987) *Production and Inventory Control Handbook*, APICS, McGraw-Hill, Second Edition.
(7) Kennerley, M.P., Davies, A.J. and Kochhar, A.K. (1996) Manufacturing Strategy, Performance and Best Practices - the contribution of the Manufacturing Planning and Control Systems, *Proceedings of 3rd International Conference of European Operators Management Association*, London 2-4 June, 363–368.

7. BIOGRAPHY

Professor A K Kochhar is the Lucas Professor of Manufacturing Systems Engineering and Head of the Manufacturing Division of the Department of Mechanical Engineering at UMIST. His main research interests are Computer-Aided Production Management Systems; and the Design, Planning and Control of Manufacturing Systems.

Amanda Davies is a research assistant in the Manufacturing Division of the Department of Mechanical Engineering at the University of Manchester Institute of Science and Technology (UMIST). Her main research interest is Benchmarking and Best Practices within Manufacturing Planning and Control Systems.

Mike Kennerley is a research assistant in the Manufacturing Division of the Department of Mechanical Engineering at the University of Manchester Institute of Science and Technology (UMIST). His main research interest is Performance Measurement of Manufacturing Planning and Control Systems.

43

Dynamic Tableau-de-Bord: concept, tools and applications

*J. M. Neta, P. A. Silva, B. Ribeiro, J. M. Mendonça**

INESC - Instituto de Engenharia de Sistemas e Computadores

Rua José Falcão, 110 - 4000 PORTO - Portugal

T: +351 2 2094314/7   F: +351 2 2008487

e-mail: jneta@inescn.pt

Abstract

In order to more accurately evaluate the performance of a manufacturing enterprise, new concepts are needed for measurement and evaluation. The evolution of concepts in this area clearly indicates that non-financial indicators are the relevant information at all stages in manufacturing.

These indicators must be always available, up-to-date, relevant, easy to understand, simple to calculate and anyone executing productive tasks should be committed to monitor and react to them.

However, measuring all the needed features is a very complex and time-consuming task. Inevitably, it must be supported by new software tools.

Since Manufacturing Execution Systems are becoming a common solution in the majority of Portuguese companies, an adequate architectural system approach must be found in order to sustain new developments as these trends become more and more clear.

Within the scenario above, work is being undertaken in Real-I-CIM, an ESPRIT project which aims to develop a tool-set for building customer-driven manufacturing management solutions. Within the scope of the project, a novel concept has been introduced - the Process Conditioning Monitoring, emphasizing that continuous process monitoring is essential to build a constantly improving performance measurement system.

The Dynamic Tableau-de-Bord application, a key element in this context, will implement a software platform for continuous and reactive measurement of performance related parameters.

* Faculdade de Engenharia da Universidade do Porto - Departamento de Engª Electrotécnica e de Computadores - Rua dos Bragas - 4050 PORTO - Portugal

T: +351 2 6103359/60 - Fax: +351 2 6103361 - e-mail: jmm@adi.pt

1. INTRODUCTION

1.1 Competitive production and performance measurement

The competitive enterprise of the future has to strive for continuously ensuring the improvement, in terms of value, of their products and services. This process of continuous improvement, frequently linked with novel organizational paradigms and manufacturing strategies (cells, cooperative work, Kaizen, TQM, etc.) can and should be quantitatively assessed through adequate performance measurement.

Performance measurement is now recognized of paramount importance regarding both more traditional aspects, such as cost, quality and time, and crucial emerging factors, such as precision of delivery, reactivity, flexibility and time-to-market. The acknowledgement of these emerging factors brought in new requirements, that cannot be met by present systems and solutions for manufacturing management, and that played a significant role in the motivation for the Real-I-CIM* project.

1.2 Key Performance Indicators

1.2.1 Measurement of Quality Performance

There are several definitions of quality. The most popular are [2][4]:

- "conformance to requirements",
  meaning, conformance to engineering specifications;
- "fitness to use",
  meaning, conformance to user requirements;
- "customer satisfaction",
  giving a final emphasis in customer demands fulfillment.

Quality assurance means continuous improvement aiming at the systematic elimination of rejection causes. To measure quality, several indicators related with different areas of the company are used. The most widely used are:

- Vendor quality;
  establishing special relationships with the suppliers in order to increase and control the quality of incoming raw materials is one of the major guidelines; incoming quality calculations can be performed by counting the number of rejections (in lots or units) for each supplier and product and the percentage over the total quantity;
- production quality;
  it can be assessed through rejection rates analysis and non-conformities analysis; rejections mirror the percentage of the production classified as defective; non-conformities are the products quality characteristics that contributed to the product rejection;
- data accuracy;

---

* Real-I-CIM is the ESPRIT 8865 project undertaken by INESC (Porto-Portugal), TECNOTRON (Lisboa-Portugal), Fhg-IPA (Stuttgart-Germany), AESOP (Stuttgart-Germany) and DICONSULT (Munchen-Germany). The consortium has two pilot-sites: GROWELA (Maia-Portugal) and BMW (Lohhoff-Germany).

data accuracy is ensured through inventory accuracy checks, BOM and routing checks and forecast accuracy checks;

- effectiveness of preventive maintenance;

  applying a preventive maintenance program can improve quality assurance by increasing machine reliability; the most common indicators are the down-time over a time range, the number of machine unscheduled interventions and the average time between breakdowns;

- quality costs;

  quality costs are an essential measure of the efficiency of the company Quality System; usually, they are divided in: non-quality costs which are split up into internal costs (scrap, rework,...) and external costs (warranty charges, client devolution, etc....); appraisal costs, incurred to discover the conditions of the product (inspection costs, evaluation of stocks, ...); and prevention costs incurred to prevent non conformities, i.e., minimizing the two previous categories (training, auditing, etc.).

1.2.2 Measurement of Maintenance Performance

Maintenance can be defined as "a set of actions or operators that maintain or reestablish a system state or ensure a given task" [1].

To ensure such functions in a maintenance management system, a number of performance indicators is needed:

- Time analysis.

  Equipment availability and productivity are of prime importance in manufacturing companies, and simple and powerful steady-state equipment indicators can be obtained through MTBF (Mean time between failures) and MTTR (Mean time to repair). From these, it is usual to derive the reliability rate (λ), the repair rate (μ) and the equipment availability (also called the up-time ratio).

  The equipment availability is obtained through a simple calculation, in percentage format:

  $$\lambda = \frac{1}{\text{MTBF}}, \ \mu = \frac{1}{\text{MTTR}}, \ \text{Availability} = \frac{\text{MTBF}}{\text{MTBF} + \text{MTTR}}$$

  Maintenance interventions and manpower resources can be analyzed with specific time indicators. These enable splitting maintenance time into operators time, equipment time, testing time, corrective maintenance and/or preventive maintenance times. These values are necessary to plan periodic preventive maintenance actions, to schedule operators allocation according to their availability and performance, to continuously improve maintenance procedures and to calculate maintenance costs.

  All indicators referred to above are analyzed on a equipment-to-equipment basis, and can be consolidated to equipment families, work centers or the entire plant.

- Maintenance costs.

  The maintenance policy is usually closely linked to cost analysis. The maintenance policy should always evaluate the current costs of maintaining the equipment in a preventive maintenance program, repairing, purchasing surrogate equipment, etc.. One good reason to continue using some financial related costing is that the maintenance department often strongly impacts on the company's global costs. Precise budget and deviations analysis are thus very important.

These costs indicators can be used for budget forecasting, budget execution, to set the preventive maintenance level, to decide on subcontracts and acquisition of external manpower, to decide whether to repair or to replace equipment, etc.

- Global performance indicators.

  Some special ratios are important for evaluating the maintenance group performance.

  The number of reported breakdowns is important to measure the disruption frequency versus production. These numbers can be extended to determine the weight of corrective maintenance actions over preventive ones.

  The costs ratios between intervention costs and total breakdown costs are also important and the occupation time of the maintenance team over their capacity is frequently used.

  All such ratios are analyzed over pre-defined time intervals.

1.2.3 Other Important Indicators

Other measures may complement those presented above, covering for other aspects in manufacturing. A selection of those found more important is given below.

- Delivery Performance and Customer Service.

  The key areas of interest are vendor delivery performance, production schedule adherence, order and schedule changes, customer service level and lost sales analysis.

- Measurement of Process Time.

  It includes the break-down of production times into the various components of the cycle time. Cycle time is the time required for manufacturing a product. It includes the time spent in all the required equipment, queuing, transportation, inspection or other activities.

  $Ct = pt + mt + wt + it + st$, where

  Ct - cycle time (total shop-floor manufacturing time)

  pt - processing time (time required to actually perform the operations)

  mt - move time (time spent in movements)

  wt - wait time (time spent in queues or buffers)

  it - inspection time (time required for quality inspection)

  st - setup time (time required for preparation of production equipment)

  The goal in measuring cycle time is to increase the importance of the value-added activities versus the non value-added ones.

  Other commonly used indicators are the D:P ratios (D stands for the lead time offered to customers and P stands for the production lead time, which is the total time required to do all the required sub-tasks including purchasing raw-materials), the material availability, the distance covered by material movements during production and the customer service time.

  One should therefore, add to the Ct the time spent in purchasing raw-materials, or in ensuring in any way the material availability and the customer delivery time.

- Measurement of Production Flexibility.

  It aims to measure the degree of adjustment to new customer order specifications. It can only be measured controlling indirect indicators such as the number of different parts in the BOM, the percentage of standard, common and unique parts within the BOM, the number of different production processes, the position within the production processes where the products are differentiated, the number of levels in the manufacturing BOM, the introduction of new products, the cross-training of production personnel and the comparison of production output and production capacity.

- Measuring Social Issues.

  Here one tries to control some of the most subjective and hard to measure indicators. These can be found through the analysis of the morale of the work force and teamwork, the involvement of employees, leadership ability of managers and the company's commitment to safety and environmental procedures.

1.3 The Real-I-CIM project

The novel concepts behind REAL-I-CIM build on the innovative integration of a few strong, i.e. proven, widely accepted and simple, ideas which are laid out below.

- it aims at providing decision-making for factory management at different levels supporting:
  - improved productivity (costs, time, etc.),
  - improved quality,
  - high flexibility and reactivity to internal and external perturbations;
- it allows user-driven customized solutions to be built using:
  - an integrated tool set based on market standards, both business and IT standards,
  - problem-oriented analysis and implementation methodologies enabling organizational changes and stressing the management of change;
- integration with the existing environment is achieved through an open architecture and infrastructure and a core information model which allow an high degree of interoperability, and therefore adaptable and extendible solutions can be built and interfaced.

The results of REAL-I-CIM can be looked at as "a tool-set and an implementation methodology for building advanced shop-floor management systems".

All the applications developed within the ESPRIT 8865 (RIC) project will be bundled in PROFIT* (Production Optimization through Flexible Integrated Tools) as full commercial products.

1.4 Process Condition Monitoring

1.4.1 The PCM Roots

The RIC Process Condition Monitoring (PCM) concept builds on more classical concepts, specifically those of Machine Conditioning Monitoring, Process Monitoring through SCADA (Supervisory Control and Data Acquisition) Systems and Continuous Improvement (Kaizen engineering) [1].

---

* PROFIT is a consortium among R&D partners and software-houses. These are INESC (Porto) and Fhg-IPA (Stuttgart), for the R&D solution providers, TECNOTRON (Lisboa) and AESOP(Stuttgart) as the marketing partners.

1.4.2 The PCM Control Loops

The RIC PCM concept integrates real-time monitoring with advanced control at a lower (on-line/shop floor) level, and integrated performance monitoring and evaluation with intelligent data analysis and knowledge extraction at an higher (off-line/engineering) level.

For monitoring and controlling purposes the PCM tool set includes modules performing shop-floor data collection, data logging and simple data analysis and interpretation, within the scope of the classical quality, maintenance and WIP/order tracking functionalities. These modules collect and process relevant information and offer specific decision support and controlling functions within the scope of the specific users environment.

These facilities will be available in a readily configurable manner as to build PCM loops with integrated functionality. These local real-time information and decision control loops will allow higher level integrated PCM client applications to source information from the database server providing aggregate information and decision support.

The modules/functionalities provided are now presented in more detail, grouped in three main streams: quality, maintenance and dynamic tableau-de-bord.

1.4.3 Quality Tool-set

One of the goals of the RIC Quality Tool-set is to help manufacturing companies in the implementation and maintenance of their Quality Systems. The RIC Quality Tool-set specification was based in the ISO 9000 and FORD Q101 standards (see **Table 1-1**), the Quality standards adopted in most industrial sectors.

Another goal of this tool-set is to provide an adequate infrastructure embedded in all relevant areas of the quality function, i.e. allowing quality information gathering and availability at the different areas of the company.

Finally, this tool-set should also be easily customized to the organization needs since the procedures of a Quality System are highly dependent on the specific organization.

These goals also comply with the encompassing PCM concept, as the RIC Quality Tool-set provides a set of highly flexible and easily customized Quality tools.

Table 1-1: ISO9000 requirements versus RIC requirements

ISO 9000 Requirements	RIC
Management responsibility	
Quality system principles	
Internal quality audits	
Quality Costs	✓
Contract review	
Design control	
Document Control	✓
Purchasing	✓
Purchaser supplied product	✓
Product Identification and Traceability	✓
Process Control	✓

Inspection and testing	✓
Inspection, measuring and test equipment	✓
Inspection and Test Status	✓
Control of non-conforming product	✓
Corrective Action	✓
Handling, storage, packing and delivery of materials	
Quality Records	✓
Training	✓
Servicing	
Statistical Techniques	✓

1.4.4 Maintenance Tool-set

The Maintenance application includes flexible and customizable maintenance management functions capable of full integration in a real-time interactive production management system. The following features are provided (see **Table 1-2**):

- easy integration with company legacy information systems;
- grouping, coding and description of the company manufacturing equipment;
- data collection and shop-floor interaction using appropriate terminals;
- fast and simple answers to the Maintenance department requests and, as a consequence, to the Maintenance operators;
- continuous information updates for historical purposes and better management of Maintenance interventions;
- maintenance actions planning in agreement with the Production Master schedule;
- improved planning through the optimization of both Maintenance and Production times;
- inter-department time cross-analysis to improve the performance of their actions and to support improvement procedures;
- maintenance costs analysis and evaluation of the Maintenance policy (adequate balance between Preventive and Corrective Maintenance is searched for);
- study of the functional behavior of the equipment in their different life periods;
- extraction of indicators for the performance evaluation of industrial processes and management procedures;
- exception handling at the shop-floor level.

Table 1-2: Available versus specified functions in RIC maintenance application

Module Name	**RIC**
Corrective Maintenance	✓
Preventive Maintenance	✓
Predictive Maintenance	
Time Analysis	✓
Maintenance Costs	✓
Maintenance Statistical Analysis	✓

Document Control	✓
Training	✓

1.4.5 Dynamic Tableau-de-Bord

The dynamic Tableau-de-Bord is an application for handling performance indicators and for defining a control board in a management context, using a very graphical user interface for data exploring. It can thus be used as a decision support tool.

Indicators are grouped in decision boards, which are called *Tableau-de-Bord*, in fact, sets of indicators in a management context which enable a synthetic view over a system. The indicators in a Tableau-de-Bord can be correlated, consolidating information and deriving more elaborate indicators [5]. They must be animated using powerful graphical facilities giving them an enormous expressive power.

The big difference to conventional reporting tools is the dynamic presentation facility. In this sense, indicators are dynamically updated as the system information is.

1.4.6 Other Modules

There is also an extra PCM level [1], that includes knowledge-based (KB) tools. Advantage will be taken of the historical recordings with information from lower levels, to extract, organize and improve the knowledge on the process, allowing reference values, corrective actions and controlling strategies to be derived and deployed.

One of the basic functions in RIC, yet one of the most important ones, is the Alarms system. This application enables the registration and acknowledge of exceptions during the operation with RIC applications.

1.5 Manufacturing Execution Systems

1.5.1 Information and Control Model

All information handling problems within a company are structured in accordance to a hierarchy. In order to support the new manufacturing techniques or novel forms of organization, there must be a very well integrated solution addressing all levels of information needs.

Roughly, problems can be addressed in the following categories [7][11]:

- Planning level, which includes functions like forecasting, budgeting, logistics, order management and MRP (materials requirements planning); more advanced systems already include MRP II facilities (manufacturing requirements planning).
- Control level, that includes functions such as process, machine and equipment control. It encompasses equipment interfacing, communications and data collection/acquisition systems.

For these two levels, the market presents a wide variety of solutions. Some are general purpose, some very process optimized. However, the gap between those information levels is immense. Production dispatch is only accomplished through manual records or theoretical forecasting. Actual plant information has to follow tricky paths to go from shop-floor to the planning department. There is actually no real-time reactive ability in the system, since information from the production line does not reach in time the planning area.

Manufacturing Execution Systems (MES) appear to fill in the space between planning/business systems and their related process control systems [9].

1.5.2 The MES Layer

A MES is the solution for building a consistent real-time information infrastructure. A good system in this layer enables the company to implement continuous improvement, JIT, customer oriented production strategies and other modern production concepts. The core functionalities provided by RIC are at the MES level.

The major guidelines for the positioning of a MES action are [8]:

- plant management, including functions to manage resources, capacity scheduling, maintenance and product distribution;
- plant engineering, enabling company-wide document management and process optimization; it manages routing plans, schedules and implementations, keeping written records at all stages;
- quality management, implementing the inspection and analysis procedures for Total Quality Management; it includes Statistical Process Control (product or process quality reporting, supply inspection and rating, certification reporting, failure analysis, identification and traceability, Pareto analysis and SPC charts), quality costing and inspection, measurement and test equipment management;
- supervisory control and monitoring, including process management and shop-floor data collection.

1.5.3 MES Performance Measurement Functions

Specific high-level indicator functionalities are needed within a MES environment. Indicators are useful at any level but specially at this one, for three different reasons:

- all real-time control operations are held at this level;
- this is the closest point to both planning information and shop-floor data collection;
- indicators should be available to any operator.

One of the most important RIC goals is the construction of an application which embodies these features. That application is called Dynamic Tableau-de-Bord (DTdB) and will be now presented.

2. THE DYNAMIC TABLEAU-DE-BORD APPLICATION

2.1 System Integration

In order to implement the DTdB application, indicators must have some physical meaning within the RIC system. Indicators are delivered by system entities called *data servers* (see **Figure 2-1**). These entities provide high level *services* that implement the indicators, among other features [2].

The main goals of data servers are to provide a fully integrated and unified information model, suitable to every application in the system, and to provide integration with other systems (the so called legacy systems, such as MRP, LIM and managerial type applications).

Therefore, a data server is defined as a source of structured information concerning a certain domain (Quality Control, Maintenance Management, Reactive Scheduling, Simulation or Shop-Floor Monitoring).

2.1.1 Data Servers Conceptual Architecture

RIC applications have a typical Client/Server architecture which means that each one of them has a data server and, at least, one client. Typically, the data server component will be responsible for all major data processing features and the client side will deal with the user interaction.

The following requirements concerning data access were identified for RIC applications:

- data should be encapsulated in data servers, according to its information domain;
- data access has different levels of abstraction, from plain *database access* through to the complex *business rules*;
- moreover, these business rules usually remain unchanged, even when the user interface and/or the data source structure or format are changed;
- the format of the data source must be transparent to the applications;
- the data servers should rely on a common distributed infrastructure.

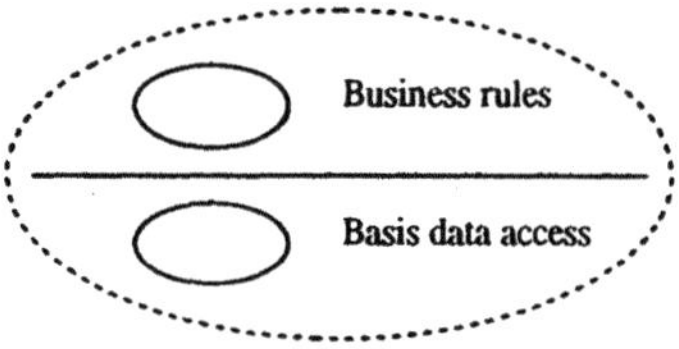

Figure 2-1 - Functional levels of a data server.

A data server has thus two levels of functional abstraction: business rules, which implement core services related with its information domain, and basic data access functions (see **Figure 2-1**). It supplies a set of services encapsulating the information model to the applications. Since this model usually changes in each system installation, the proposed architecture will promote reuse.

A data server should provide data access to their clients, services access over data by encapsulating procedures in one common structure, and a mechanism to trigger shared events (a signal sent to client applications upon occurrence of certain conditions) notifying client applications.

2.1.2 RIC Architectural View

The concept of Client/Server computing has been evolving and different implementation models have been proposed.

In RIC, information was split into several domains which exist in a decoupled way. Therefore, it is easier to develop and maintain applications which are built in different platforms, by different partners and for different end users.

For these reasons, the integration problem has been divided into the following categories:

- the information domain problem, which is dealt with by defining as many data servers (logical servers, not necessarily physical) as different domains in the specification for a specific company;

- the client/server architecture, which was developed as the natural upgrade from two-tiered systems into three-tiered ones [10].

According to this description, RIC applications may have the following type of connections [3] (see **Figure 2-2**):

- SQL connections to their data source, since SQL has been the well accepted standard for database interfacing and relational databases are, most of the time, the widely used data store;
- connections to the RIC system servers, which are message oriented applications and, for that reason, they rely on a common message transmission system.

Besides database servers, RIC applications may have the need to retrieve data from foreign systems (MRP's, Data Collection Systems, etc.). In each case, there is a distinct interface to those systems.

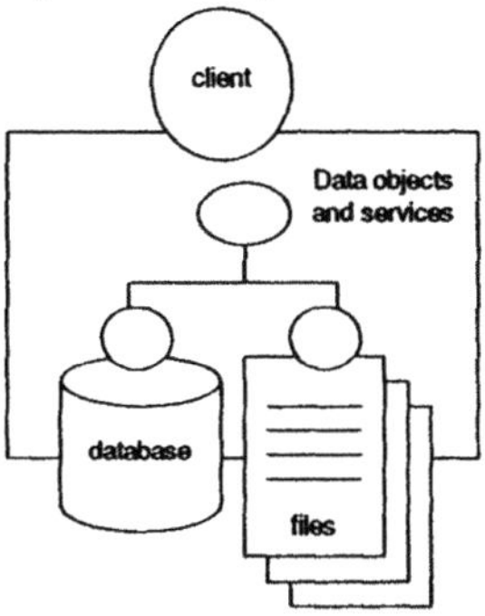

Figure 2-2 - Logical and physical view of RIC data servers.

The system thus evolved according to the following needs of encapsulation:

- data access, so that different kinds of data sources and structures could be hidden from the users and applications point of view;
- business rules, so that the procedural description of value added services could be hidden from the application itself.

A major goal was to provide service orientation instead of data orientation. Since not all RIC applications could easily cope with this goal, the Dynamic Tableau-de-Bord application was used as the test bench for these novel concepts.

Furthermore, the RIC architecture was designed according to an object-oriented paradigm. Each object comprises the following components [2][3]:

- the client;
- the ORB - the object request broker;
- the object implementation.

The ORB layer is responsible for all the mechanisms required to find an object's implementation for the request, to prepare the object implementation to receive the request, and to communicate the data accomplishing it.

Clients have no knowledge of the implementation of the object, its location, or the exact ORB that manages that object.

An object implementation provides the semantics of the object, usually by defining data for the object instance and code for object's methods. Often, the implementation will use other objects or additional software to implement the behavior of the object. In some cases, the primary function of the object is to have side effects on other components that are not objects.

A variety of objects implementations can be supported, including separate servers, libraries, a program per method, an encapsulated application, an object oriented database, etc.

The many tiered architectures have advantages over classical two-tiered systems:

- scalability, that is, the size of the problem can always be overcome with physical distribution of processing resources, maintaining all the major interface features;
- performance, since it has the system load more balanced and distributed;
- reusability, since objects encapsulate details that only concern to a certain context in the hierarchical structure; in this way features can be more easily added or changed without major changes in the whole system;
- specialization, that is, tasks are better divided between client and server; clients only deal with user interface problems.

Classical architectures are almost always dependent on SQL engines, and all the processing is bundled within them. The database server process is a bottleneck, although some engines already have distributed features. Nevertheless, this implies that large investments must be made in hardware and software.

When an application wants to access a service provided by a server that is not a SQL engine, an interface must be provided at the client level. With many tiered architectures, this problem is solved at a lower level.

Some disadvantages may arise. When an application needs to manage very large amounts of data which do not require really heavy processing, with low complexity, many-tiered systems may be redundant, and overheads may affect the overall performance. This happened with all major functions for Quality and Maintenance tool-sets, since they were very data oriented. For this reason, those two applications were kept on a so-called classical architecture.

2.1.3 The DTdB Implementation Process

The prime goal of the implementation process was to test all these innovative concepts. So, the new directions were:

- to test and validate the advantages of three-tier architectures over two-tiered ones;
- to present concrete solutions to the architecture transition problem;
- to study new development tools which are more suited for deployment according to the new concepts;
- to study new solutions for data storage and comparing its performance and ease of use with present technologies.

Therefore, the DTdB application used:

- object-oriented methodologies for design and implementation;
- dedicated data servers which encapsulate application specific indicators;
- conventional databases (although one of the main features of this approach is the possibility of connecting with different types of data storage, such as state-of-the-art object-oriented databases);

- an object request broker for object distribution;
- a very graphical user interface tool over a windows-based system (X-Windows or MS Windows).

The entire DTdB application already has a prototype available. Its deployment was devised in three stages:

- Non-configurable programming interface.

  The programming interface is static and only available for the application programmer. The effort is put in complying with the user interface and system requirements. Therefore, data servers provide performance indicator services, the DTdB is able to connect to them and the results are displayed according to the user interface requirements.
- Configurable programming interface.

  The user can redefine a Tableau-de-Bord in a WYSIWYG (what you see is what you get) manner. There are no further major improvements, at this stage.
- Deriving new indicators.

  A programming rule-based language is provided in order to put some programmable processing facilities at the client side. New indicators can thus be derived without re-programming the system data servers.

2.2 User interface

As it was already stated, a very graphical user interface is a must. The indicators must be presented to the user in the following manners:

- table or grid formats;
- graphical charts.

Charts have several types, according to the final goal:

- line or bar charts, that are more suited to see trends, indicator evolution across time or any other category, etc.
- pie or scattered bar charts, which are more suited for comparisons among several categories;
- three-dimensional charts, when analyzing indicators with three or more categories.

For specific reasons, combinations of these techniques can be used to produce special type of graphs such as ABC or Pareto-type graphs and Gantt charts.

The system must be capable of presenting results:

- On the screen.

  The dynamic requirement imposes some animation which will only be relevant on screen presentations.
- On paper, through a printer.

  That is to say, it must have conventional reporting facilities.

The definition of a Tableau-de-Bord must be simple and easy to understand. The indicators set structure will evolve with time, so a new Tableau-de-Bord will have to replace older versions.

The connection to the system infrastructure should be intuitive, since indicators are set upon system services. The programming skills of the system users should not be a relevant pre-requisite to define new Tableau-de-Bord.

3. CASE STUDY: THE GROWELA PILOT SITE

Growela has been developing, along with an industrial consulting company (Roland Berger), a set of indicators that should help them to monitor several internal processes performance, to define intervention areas and corrective actions, to set goals, to take strategic decisions and to monitor improvement actions.

3.1 The Indicators System

From the set of indicators, Growela is expecting to obtain:

- improvement measures;
- extra involvement from the staff in obtaining the aimed results;
- identification of procedures that are not producing the required results;
- ease of implementation and adjustment of new management procedures.

Each indicator is described through the following information:

- a description,
- a measurement unit,
- a calculation rule,
- a calculation frequency,
- a level of detail,
- a responsible,
- a data source,
- a target group,
- a current value,
- previous values, and
- a final goal.

These indicators were defined according to a top-down approach, defining levels at which detail and measuring frequency needed to be increased.

The definition of the indicators set was undertaken by a working group specially put together with that purpose.

3.2 Performance Indicators

Growela analyzed their logistic processes, from an high level point of view, and identified their major fields of interest (see **Figure 3-1**):

- commercial department;
- technical design;
- planning;
- purchase department;
- production management;

- quality assurance;
- maintenance control;
- delivery.

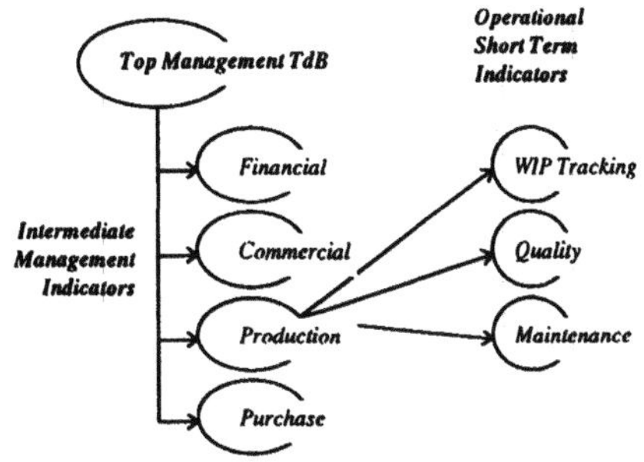

Figure 3-1 - Levels of abstraction in GROWELA indicators set. As the level decreases, the calculation frequency increases.

As an example, for quality assurance, the following indicators were chosen:

- inspection reject rates, per product, per section or operation, per supplier dissecting for each raw material, and per product dissecting for each section;
- non conformities, per product, per section, per disposition for a specific product, per non conformity cause, per disposition for a specific supplier and a specific product;
- production return rates, global and per product;
- repair rates, global and per product for each operation;
- scrap rates, global and per product for each operation;
- suppliers returns rates, global, per supplier and per supplier for each product;
- corrective actions, including the number of corrective actions per product and per area, the success rates and the average number of days in retard in the corrective actions effectively implemented.

For the maintenance department, the required indicators were:

- Time analysis, including MTBF, MTTR and MTA, the operators time by intervention, man-power time by intervention and equipment, testing time per equipment and intervention, equipment operational time, man-power yield and lubrication cost (expressed in terms of intervention time for this particular operation);
- Costs analysis, including dissection by intervention, equipment, man-power per equipment, man-power per intervention, materials per equipment and per intervention, spares per equipment and per intervention, total maintenance cost, subcontracts per equipment and per intervention, corrective and preventive maintenance and quality of maintenance actions per equipment.

3.3 Legacy Systems Integration

At Growela, system integration was made at the database level. Growela had a MRP system (Gestical, by Softel) and a LIM system (GCI, also by Softel). However, these systems were built over a standard commercial database server (Informix Online) which was fully supported by all applications.

The data collection system was integrated with a special application which made the interface between the shop-floor terminals and the database.

4. CONCLUSION

The overall conclusion is that a TdB application is useful within the context of new manufacturing management concepts. The prime goal was the construction of an application with the following requirements:

- Usefulness.

  Performance indicators are related, at the various management levels, with company departments. Therefore, the TdB must be useful for people working in those areas, both in its semantic content and in the way information is shown to the users.
- Simplicity.

  It must be easily accessed by everyone, that is, it should not require deep knowledge on any subject. The information should be shown in such a way that it can be easily perceived and understood.
- Flexibility.

  As concepts and strategic goals evolve, so must performance indicators do it also. It should be easy to define new indicators and adapt the existing ones.
- Completeness.

  Regarding performance evaluation, the defined set of indicators should be enough to report and document current shop-floor status. This set of indicators must be made available in the TdB application, so that only one interface is presented to the shop-floor.
- MES integration.

  Since MES are the primary data source, a very well integrated software solution must be available in order to extract the maximum performance from the system. Nevertheless, the connection requires a deep study on new software development approaches and, specially, in new data sources.

5. BIBLIOGRAPHY

[1] Beatriz Ribeiro, Paula Silva, João José Ferreira, José Mendonça; "Analysis of PCM tools and standards and specification of PCM tool set"; RIC deliverable D.3.2.1; 95/10/30

[2] João Neta, Rui Picas, Beatriz Ribeiro, José Mendonça; "New concept of data encapsulation and manufacturing data servers; Server functionality specification according to client application requirements"; RIC deliverable D.4.3.1/2; 95/10/30

[3] João Neta, Rui Picas, Beatriz Ribeiro, José Mendonça; "Specification of Distributed Heterogeneous System Architecture - Final Specification"; RIC deliverable D.4.1.2; 95/10/30

[4] Brian Maskell; "Performance Measurement for World Class Manufacturing"; Productivity Press

[5] João Neta; "Quadros de Bordo Dinâmicos"; Msc paper under the Industrial Automation subject; 95/07/31

[6] Eve Chapello, Marie-Hélène Delmond; "Les tableau-de-bord, outils de introduction du changement"; Revue Française de Gestion

[7] "MES to COMMS integration: declaration of interdependenmce"; AMR Report; 92/08/31
[8] "Cleaning up the MES: positioning manufacturer's needs with software vendors capabilities"; AMR Report; 92/01/31
[9] Jim Esch; "A fine MES"; BYTE; 95/12/31
[10] Robert Orfali, Dan Harkey, Jeri Edwards; "Client-Server computing"; BYTE; 95/04/30
[11] "MES in process manufacturing"; CIME review
[12] Brian Smith; "Three-tiers for client-server"; Data-based advisory; 96/10/31

PART EIGHTEEN

Optimization and Analysis

44

Visual Modeling Interfaces for the Cutting Stock Problem

L. Lessa Lorenzoni, F. J. Negreiros Gomes,
Universidade Federal do Espírito Santo
Av. Fernando Ferrari, s/n - CEP 29060-900, Vitória-ES - Brasil
email: [lessa, negreiro]@inf.ufes.br

Abstract

This work describes the design and development of visual modeling interfaces for Decision Support Systems - DSS in the context of Computer Integrated Manufacturing - CIM, that are based in optimization models and algorithms. The DSS can show the model to the user and then he can form clear mental images of the structure and functions of the model. As a case study, we present an application, of Visual Interactive Modeling to the cutting stock problem in the shoe industry.

Keywords

Visual modeling interactive, Modeling by Example, Cutting Stock.

1 INTRODUCTION

Optimization problems arise in several fields of application. In the literature, the attention has been concentrated in the development of models and algorithms. However, concerning the input/output aspects, little effort has been presented to turn the modeling process into a comfortable task to the end user.

Specifically, in the Computer Integrated Manufacturing – CIM there is a great need of systems optimization tools, but the lack of appropriated user interfaces does not have encouraged their utilization in this area.

In this way, we have been involved in the application of visual interface concepts such as Visual Interactive Modeling – VIM, Modeling-by-Example – MbE and Objects with Dynamic Behaviour – ODB, in the design and development of intelligent interfaces for optimization problem in the context of CIM.

This paper is structured as follows. In the section 2, we present the cutting stock problem and its characterization in the shoe industry. Section 3 explains the methodology we use in the development of the graphical interface. Section 4 presents the design of the graphical interface to the cutting of pieces in the shoe industry.

2 THE IRREGULAR CUTTING STOCK PROBLEM

2.1 Problem Characterization

The irregular cutting stock problem, illustred by Figure 1, Provedel (1995), is settled when we desire to cut a set of moulds over a plate with the objective of minimizing the material waste involved in the cutting process.

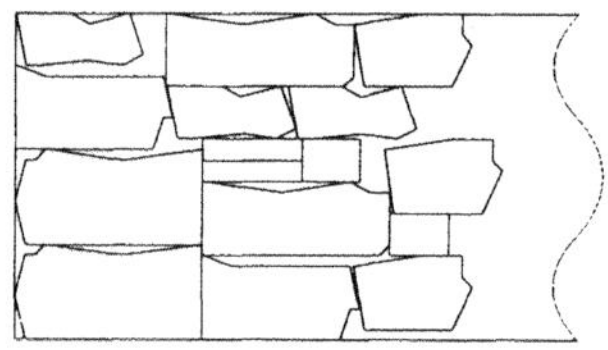

Figure 1: Cutting Stock Problem.

The pieces representation is done through circunscribed irregular polygons and the stock plate is a retangle with enough area to held the whole demand.

It should be avoided the pieces overlay, but it is allowed the rotation of the pieces over the plate. In some cases, the allocation of certain kind of pieces is allowed only into specific regions of the plate.

2.2 The Cutting Stock Problem in the Shoe Industry

The shoe industries use the techinical modeling approach as a manufacturing metodology. Basically, this method consists to create a shoe model or to duplicate (reproduce) the model of an existing shoe, to build the moulds, to determine the quantity of raw material to be used, to decide the moulds layout on the plate and to execute the moulds cutting process.

We note that there is a high level of raw material waste, with respect to the techinical modeling process mainly because two factors:

- it is a systematic approach;
- it uses an empirical process to decide the moulds layout.

With the objective to optimize the process of mould layout construction, Provedel (1995) presented various combinatorial models and heuristic methods.

3 METHODOLOGY

Aiming the systematization of the modeling process and solution of the mould allocation problem, we propose the development and implementation of a graphical visualization engine that supports the visual representation of objects as well as the active interaction with the user. In this way, we adopt three paradigms in the system design, which are described in the following sections.

3.1 Visual Interactive Modeling - VIM

Visual Interactive Modeling is the combination of friendly interactive interfaces, computer generated visual screens to represent the model state, and mathematical and symbolic problem models within Decision Support Systems - DSS, Bell (1991).

The visualization is an important characteristic in a modeling environment. It could be crucial to help a DSS user with respect to:

- to acquire new perspectives of problem structures through the generation of different views of the decision circunstances;
- to explore his own abilities in order he can recognize alternatives and significant strategies during the problem solution process.

Within a system that follows the VIM paradigm, the user of such system can increasily build, complete (fill up) and test model. This makes the man-machine interaction easier and concrete, and the modeling process stimulates the user learning concerning non-structured situations.

3.2 Modeling by Example - MbE

The Modeling by Example methodology is based on interpretative-computational intelligence. It allows the decision makers to acess the knowledge, that is stored in the DSS, through a dynamic example exchange, Angehrn (1991).

The main idea that supports the interactive consulting activity is to let the system to communicate its knowledge of the problem by a concrete example exchange.

The MbE, different from the expert systems where the main objective is to emphasize the autonomous capabilities of the system to solve problems, explores Artificial Intelligence techniques, Nilsson (1980), to improve system abilities in a high level communication with the user. The goal of the MbE is to give support to the decision maker creativity and intuition without attempting to substitute his judgement. It supports the user in the flexible application of complex methods into a VIM environment. So, the users do not need to interrupt their own decision process or to reformulate their models because the change from a descriptive to a normative mode.

3.3 Objects with Dynamic Behaviour

Taking into account that the visual aspect is a dominant characteristic of such systems, the user interface is thought as to play an important role.

In this context, depending on the type of user, the same object could have different views. We use the concept of objects with active behaviour to represent an object with various views, Camarinha-Matos and Pinheiro-Pita (1993). For example, the developed interface is configurable in accordance with the type of the user (Client, Planner, Designer and Administrator).

4 GRAPHICAL VISULAIZATION ENGINE

The graphical visualization engine, illustrated in Figure 2, allows the user to describe and visualize his problem in a graphical form instead of a mathematical form and it is divided in three modules: Model Specification, Example Exchange and Solution Visualization, which we describe bellow.

Model Specification: Allows the user to build an instance of the problem, where he specifies the plate and moulds geometry, the constraints and the parameters, by means of a visual language.

Example Exchange: In this module the user could interact dynamically with the system, evaluating the various instances of the solutions, and he can propose partial solutions and new specifications according to his own view of the problem context and experience.

Solution Visualization: Allows the model visualization and also exhibits a report with informations such as the executed algorithm, the raw material waste and processing time to generate the solution.

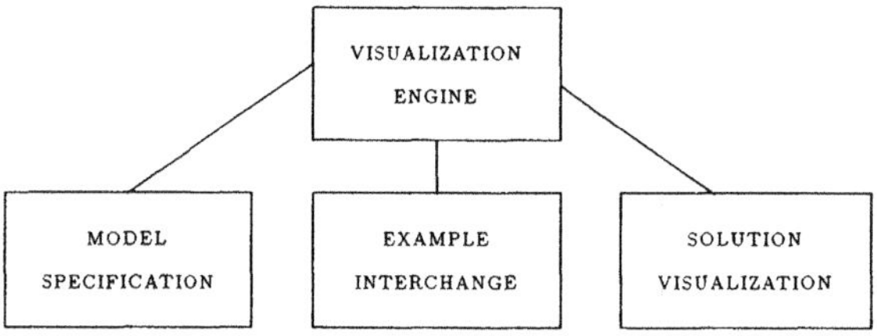

Figure 2: Graphical Visualization Engine

4.1 Description of the Graphical Interface

We follow the works of Alvarenga *et al.* (1993) and Negreiros *al.* (1993), where they describe respectively, the DSS project in a generic way and the user interface subsystem, in a more specific way.

The system combines the funcionality of the spreadsheets and the optimization algorithms together a graphical presentation of the results. The interface project is shown in Figure 3.

The interface is divided in the following parts:

1. Menu Actions: exhibits the methods that can be applied to the selected object.
2. Work Area: it is reserved to the selection, interaction and presentation of problem solution.
3. Object Graphic Menu: it is composed by visual objects (icons) which could be chosen by the user to execute a desired action.
4. Status Line: it is an area reserved to the exhibition of warning, error and help messages.

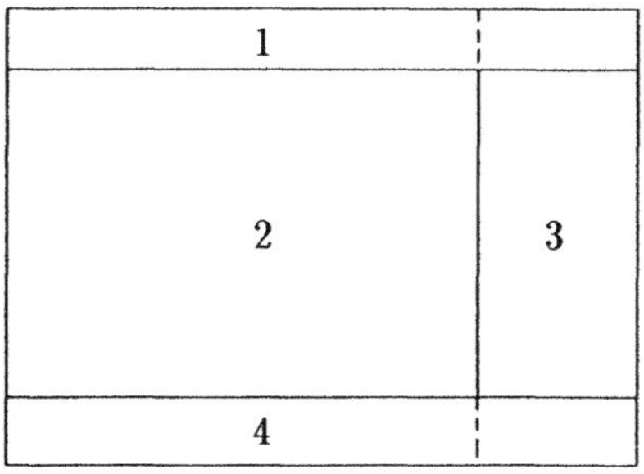

Figure 3: Graphical Interface for Cutting Stock Problem.

The interaction method is done through an Graphical User Interface - GUI that supports windows, graphic objects, mouse, menus and browsers. The interface system for the optimization models is sufficiently generic to support the model management functions, as follows:

- Model construction.
- Knowledge base interface.
- Model storage, retrieval and maintenance.
- Model and algorithms linkage.
- Model solution.

The interface configures itself in accordance to the type of user and/or in accordance to the problem to be solved. For example, in the case of the rectangular cutting stock problem, the moudls are represented by their height and width. By other side, in the case of irregular moulds, they are represented through a list of the polygon vertices that circumscribe each of them.

The DSS is constituted by three basic constructions, called modeling primitives:

- Objects: concrete or abstract entities (plate, moulds) with a set of attributes (type, area, shape, cutting direction and location).
- Relations: objects that describe functional or structural dependence among other objects (neighborhood, cutting regions).
- Constraints: restrictions on the object attribute values. For example, piece 1 must not be cut in area B or the permitted orientations of piece 2.

These primitives are ready to use by user in a interactive graphic environment.

Figure 4 shows a GUI to the cutting system with respect to model building. It exhibits a screen in which the user specifies the geometry of each mould to be cut as well as various parameters and restrictions. For example, the system asks the user to input the demand for the last inserted mould (the white one in Figure 4) and the user informs that it is 4 units.

After finishing the model building process, the user identifies it and then he is able to solve the problem, choosing the appropriated action. Figure 5 presents the solution of the proposed problem. At this point, the user can interact with the system adjusting the proposed solution in accordance to his own knowledge and experience as it is shown in the Figure 6.

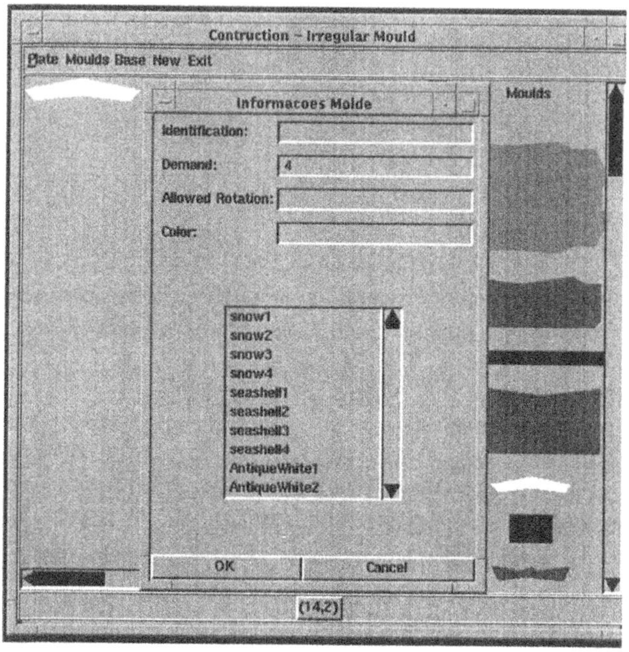

Figure 4: Model Construction - Irregular Mould.

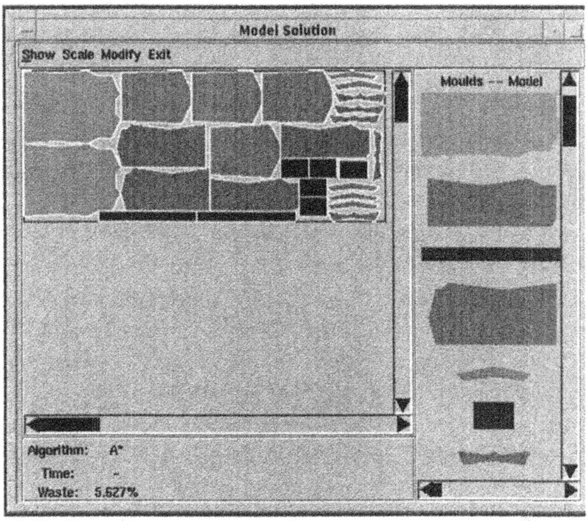

Figure 5: Solution.

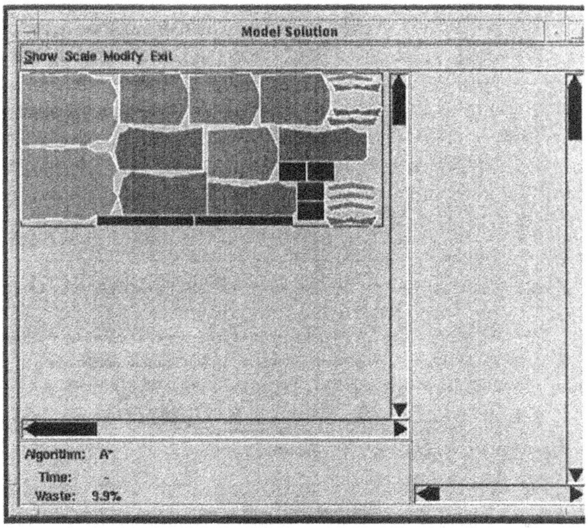

Figure 6: Modified Solution.

A detailed description of the cutting system interface can be found in Lorenzoni (1996).

5 CONCLUSIONS

The implementation of the cutting system was done using an UNIX machine with the X-Window Graphical Library together the Tk/Tcl programming language, Ousterhout (1993). The system is one of the instances of the general layout problem, Alvarenga *et al.* (1996), which is a common application in CIM.

The system provides the users with a modeling environment to the solution of complex problems and the man-machine interaction is done through a visual language. The interface provides an active support which allows the interaction between models and users. This is implemented through a set of active agents which autonomously fire appropriated answers, such as moulds overlay, exceeded demand, forbiden cutting regions and also supports the process of decision making. Because the system is based on active cooperation between man and computer it should be thought as a balance automation system, so it can contribute to a sustainable industrial production.

6 REFERENCES

Alvarenga, A.G., Provedel, A., Negreiros, F., Daza, V.P., Sastrón, F., Arnalte, S., *Integration of a irregular cutting stock system into CIM. Part I - Information flows*, Proceedings of the ECLA-CIM'93, Lisbon, Nov 22-26 (1993).

Alvarenga, A.G., Provedel, A., Negreiros, F., Ahonen, H., Lorenzoni, L., Daza, V.P, Pinheiro-Pita, H.J., Camarinha-Matos, L.M., *Multipurpose Layout Planner for Cutting Stock Problems: Implementation Issues*, Balanced Automation Systems - Architectures and Design Methods, IFIP, Chapman & Hall, (1996).

Angehrn, A. A., Modeling by Example: A Link between Users Models and Methods in DSS, *European Journal of Operational Research*, Vol. 55, (1991).

Bell, P.C., *Visual Interactive Modelling: the past, the present and the prospects*, European Journal of Operational Reserch, Vol. 54, pp. 274-286, (1991).

Camarinha-Matos, L.M., Pinheiro-Pita, H.J., *Comportamento de objectos activos na interface gráfica do sistema CIM-CASE*, 4as Jornadas de PPPAC, Lisboa, (1993).

Lorenzoni, L.L., *Visual interfaces for the optimization of industrial cutting in the context of CIM*, MSc Dissertation, PPGEE, UFES, (1996) (in portuguese).

Negreiros, F., Alvarenga. A.G., Lorenzoni, L.L., Pinheiro-Pita, H.J., Camarinha-Matos, L.M., *Objects dynamic behavior for graphical interfaces in VIM - An application case study*, Proceedings of the ECLA-CIM'93, Lisbon, Nov 22-26 (1993).

Nilsson, N.J., *Principles of Artificial Intelligence*, Tioga Publishing Co., (1980).

Ousterhout, J.K., *Tcl and Tk Toolkit*, Computer Science Division, Department of Electrical Engineering and Computer Sciences, University of California, Berkeley, CA 94720, (1993).

Provedel, A., *An optimization system for industrial cutting and its integration in the context of CIM.*, MSc Dissertation, PPGEE, UFES, (1995) (in portuguese).

45

Developing Finite Element Simulation of Sheet Metal Stamping Processes - a contribution to Virtual Manufacturing

A.D. Santos [1,2], P. Montel [1], J.F. Duarte [1,2], A.Barata da Rocha [1,2]

[1] *INEGI/CETECOP - Institute of Mech. Engineering and Industrial Management*
Rua do Barroco, 174/214, S.Mamede Infesta, Portugal
[2] *DEMEGI - Department of Mechanical Engineering of the University of Porto*
Rua dos Bragas, 4099 Porto Codex, Portugal

Abstract

Finite Element Modeling (FEM) can be a valuable addition to computer integrated manufacturing (CIM) along with other kinds of analysis and tools such as computer aided design (CAD), manufacturing (CAM) and engineering (CAE).

A FEM program for analyzing sheet stamping operations has been developed to simulate sheet metal stamping processes. Most of these processes involve axi-symmetric, plane strain or plane stress analysis for which a two-dimensional finite element simulation code can give important and necessary directions to the best compromise of stamping variables to be used, before tool manufacturing and parts production. Major advantages of such analysis are fast answers and possible use of PC's instead of workstations.

This paper describes the use of a static explicit finite element simulation code and its implementation to work and be used on a Pentium based PC. Comparison of simulated results and experimental ones is presented, as well as comparison of CPU time efficiency for PC and workstation.

Keywords

FEM, CAD, CAE, CIM

1. INTRODUCTION

Simulation of sheet metal stamping processes by using Finite Element Method has been proving to be a useful and reliable aid to help tool designers to make right decisions before tool manufacturing and trial production, thus saving time and cost [2]. However, some difficulties must be overcome in order to obtain such goals. Concerning sheet metal forming processes the main problems to overcome in simulation include an efficient approach to describe tool geometry [Santos, 1992], the treatment of tool-work contact [Makinouchi, 1990(1)] in order to deal with the highly deformation dependent contact problems with friction, and also the robustness of the code. From these variables the robustness of the code plays an important role, since sheet forming simulation is prone to instabilities due to the non-linearity characteristic of these processes [Saracibar, 1990]. Many elasto-plastic finite element codes employ the implicit time integration scheme, which has the advantage of large time increments and satisfies rigorously the equilibrium at the end of time step [Makinouchi, 1990(2)]. However sometimes calculations stop due to lack of convergence. On the contrary the explicit integration scheme restricts the time increment to a very small size in order to maintain out of balance forces within admissible tolerances and this scheme has the important advantage of a stable and robust solution. The lack of convergence for implicit analysis is more evident when 3-dimensional forming problems are simulated [Rebelo, 1992], due to additional instabilities. The robustness of the explicit analysis is the main reason why the code herein presented uses such formulation.

Another important concern when developing a code is its user friendliness and its compromise with speed of calculation. The existent growing development of fast processors for personal computers is beginning to make possible the use of PC's to simulate sheet metal forming with the advantages of using common, non-expensive computers with user friendly interfaces.

This paper presents the implementation of 2-dimensional code to be used in a personal computer including the development and the use of standard PC tools for pre and post-processing. Comparison of simulation and experimental results are performed to test the accuracy and reliability of the explicit code and comparison of CPU time is performed between PC and workstation in order to test the efficiency and cost effectiveness of this solution.

2. FEM CODE

The simulation results were obtained by using ITAS code. This is a Japanese FEM code developed by Makinouchi and co-workers [Makinouchi, 1990(1)].

2.1 Variational principle

The updated Lagrangian rate formulation is the base of the incremental elasto-plastic finite element code. The rate form of equilibrium equations and boundary conditions are described by the principle of virtual velocity, as it is proposed by McMeeking and Rice [Mc Meeking, 1975], on the form:

$$\int_V \left\{ \left(\overset{\circ}{\sigma}_{ij} - 2 \cdot \sigma_{ik} \cdot D_{kj} \right) \cdot \delta D_{ij} + \sigma_{jk} \cdot L_{ik} \cdot \delta L_{ij} \right\} \cdot dV = \int_{S_t} \dot{t}_i \cdot \delta v_i \cdot dS \tag{1}$$

Here, $\delta \underset{\sim}{v}$ is the virtual velocity field satisfying the boundary condition $\delta \underset{\sim}{v} = 0$ on S_t. V and S denote the region-occupied, at time t, by body and its boundary, respectively. S_t is the part of the boundary S, on which the rate of the nominal traction vector, $\underset{\sim}{\dot{t}}$ is prescribed. $\underset{\sim}{\sigma}$ denotes the Cauchy stress tensor, $\underset{\sim}{L}$ is the gradient of the velocity field ($\underset{\sim}{L} = d\underset{\sim}{v} / d\underset{\sim}{x}$), $\underset{\sim}{D}$ and $\underset{\sim}{W}$ are respectively, the symmetric and anti-symmetric parts of $\underset{\sim}{L}$, and $\overset{\circ}{\sigma}_{ij} = \dot{\sigma}_{ij} - W_{ik} \cdot \sigma_{kj} + \sigma_{ik} \cdot W_{kj}$ represents the Jaumann derivative of $\underset{\sim}{\sigma}$.

2.2 Constitutive equation

The small strain isotropic linear elasticity and the large deformation rate-independent work-hardening plasticity is assumed. In order to deal with anisotropy of sheet metal, Hill's quadratic yield function and the associative flow law are used. Following Cao and Teodosiu [Cao, 1989], the constitutive equation of the present model is written in the form:

$$\overset{\circ}{\sigma}_{ij} = C^{ep}_{ijkl} \cdot D_{kl} = C^{ep}_{ijkl} \cdot L_{kl} \tag{2}$$

where the constitutive matrix C^{ep}_{ijkl} is:

$$C^{ep}_{ijkl} = \lambda \delta_{ij} \delta_{kl} + \mu(\delta_{ik}\delta_{jl} + \delta_{il}\delta_{jk}) - \alpha . f_0 Z_{ij} Z_{kl} \tag{3}$$

In this equation λ and μ are the Lamé elastic constants, δ_{ij} is Kronecker's delta and α is a constant; $\alpha = 1$ for plastic loading and $\alpha = 0$ for elastic state and unloading. f_0 and Z_{ij} are expressed by:

$$f_0 = \frac{2\mu}{Z_{ij} Z_{kl} + Z_{ij} Z_{kl} + H_0 / (2\mu)} \tag{4}$$

$$Z_{ij} = M_{ijrs} \sigma'_{rs} \tag{5}$$

Here H_0 is the work-hardening modulus, σ'_{ij} is the stress deviator and M_{ijkl} denote Hill's anisotropic parameter, assuming that sheet remains orthotropic during deformation, while the orthotropic axis are subjected to a time-dependent rotation R_{ij}. The explicit form of M_{ijkl} is written as:

$$\bar{\sigma}^2 = M^0_{ijrs} \cdot \sigma_{ij} \cdot \sigma_{rs}$$

$$= F(\sigma_{22} - \sigma_{33})^2 + G(\sigma_{33} - \sigma_{11})^2 + H(\sigma_{11} - \sigma_{22})^2 + 2L\sigma_{23}^2 + 2M\sigma_{13}^2 + 2N\sigma_{12}^2 \qquad (6)$$

where F, G, H, L, M, N are material parameters; assuming that their value is kept constant during deformation, thus the evolution of M_{ijkl} is given by:

$$M_{ijrs} = R_{im} \cdot R_{jn} \cdot R_{rp} \cdot R_{sq} \cdot M^0_{mnpq} \qquad (7)$$

2.3 Finite Element Discretization

Equation (2) corresponds to an inviscid behaviour, since it is insensitive to the choice of the time scale. Thus, the time increment can be replaced by the increment of any other monotonously increasing parameter, e.g. an imposed displacement. Performing a standard finite element discretization, namely dividing the sheet into finite elements and introducing the expression of the incremental displacements in terms of the shape functions into the incremental forms of (1) and (2) yields a system of algebraic equations of matrix form:

$$[K] \cdot \{du\} = \{df\} \qquad (8)$$

where $\{du\}$ is the vector of nodal displacements increment and $\{df\}$ the vector of nodal forces increment, corresponding to a trial increment dw of the loading parameter, while $[K]$ is the symmetric elasto-plastic tangent stiffness matrix [Makinouchi, 1985].

ITAS code uses an explicit approach to obtain the solution of Eq.(8). The stiffness matrix $[K]$ is described at time t and it is regarded constant within the time increment Δt. The so-called *Rmin* method [Yamada, 1968; Makinouchi, 1990(2)] is used to impose limitation to the time step in such a way that no significant changes in the stiffness matrix occur during the increment. This method assures the accuracy of the explicit integration scheme and it is used to choose the size of the increment. This size is calculated assuming that, during one increment:

(I) - a finite element can not change from elastic to plastic state and vice versa

(II) - the largest value of the incremental principal strains attains a limiting value of $\Delta\varepsilon_{max}$=0.002

(III)- the largest absolute value of the incremental rotations attains a limiting value of $\Delta\theta_{max}$=0.5°

(IV) - a free node will not come into contact

(V) - a contacting node will not get free

(VI) - a sticking node will not slide

(VII)- a sliding node will not stick.

The last 4 conditions are concerned with boundary conditions (contact and sliding states) and they assure that, during one incremental step, boundary conditions are kept unchanged.

The procedure to apply for Rmin strategy and calculation of the increment size involves the following steps:

- at the beginning of the increment a fictitious incremental tool displacement Δa is prescribed;
- the stiffness equation is solved for a fictitious solution Δu, correspondent to the previous tool displacement Δa;

- for each of the previous conditions I~VII, a ratio **r** is found, between the obtained solution Δu and a "real" solution that validates each condition; the minimum of all these ratio **r** will be the value of Rmin ;
- the obtained solution Δu is weighted by Rmin coefficient (0<Rmin<1), the size of the increment will be Rmin* Δu and this is the value used to update the configuration of the sheet, the total displacements, the total (Cauchy) stresses and the yield limit of each element.

2.4 Finite Elements

The four node bilinear isoparametric element is the basic element used. This element can be used with different stress integration methods: the full integration, the reduced integration, the selective reduced integration and the hour glass control stabilized matrix method [11]. From these integration schemes the third is the one used in the simulations herein presented, since it offers a good compromise between performance and results.

2.5 Description of Tool Geometry

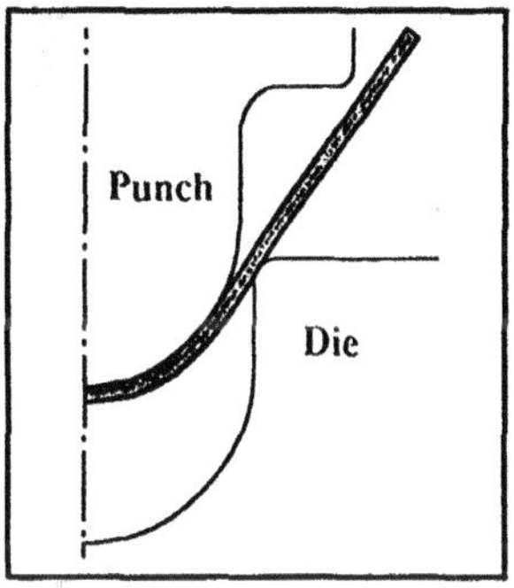

Figure 1 - Function approach

Several tool descriptions can be used to define tool geometry. They include the Point Data approach, the Mesh approach and the Parametric approach [Santos, 1993]. For 2-dimensional problems the function approach is suitable due to its easy generation and complete potential for tool description. In this approach the geometry of tool surface is expressed by a combination of subdomains which are represented by elementary functions of the form:

$$f(x,a,c)=0 \quad \text{in domain D} \tag{10}$$

In this equation **a** and **c** are constants which define shape and position of the curve. For 2D situation straight lines and arcs are used and explicit form of Eq.(10) are given by:

$$y=c.x+a \qquad for \quad x_A \le x \le x_B \tag{11}$$

$$(x-a_1)^2+(y-a_2)^2=c \qquad for \quad \theta_A \le \theta \le \theta_B \tag{12}$$

where θ is the angle for the arc of circle.
Motion of the tool is given by an incremental rigid motion Δa as

$$f(x,a+\Delta a,c)=0 \tag{13}$$

Tool is modeled by considering its boundaries as being rigid.

2.6 Pre and Post-processing

In order to use a PC to run a simulation code for sheet metal forming it is important also to have or develop pre and post processing capabilities that can be used on the same computer. In our case we are using an widespread CAD software (AutoCad) for pre-processing the shape of tools to be used in simulation. Such data is written into a file in DXF format and read by the code at the same time as the sheet data, material data and other variables (Figure 2). Concerning post-processing, AutoCad is used to output sheet geometry as well as contacting nodes and its force vectors (Figure 2). To output stress and strain distribution we are using a visualization code developed at our Department of Mechanical Engineering of Porto University.

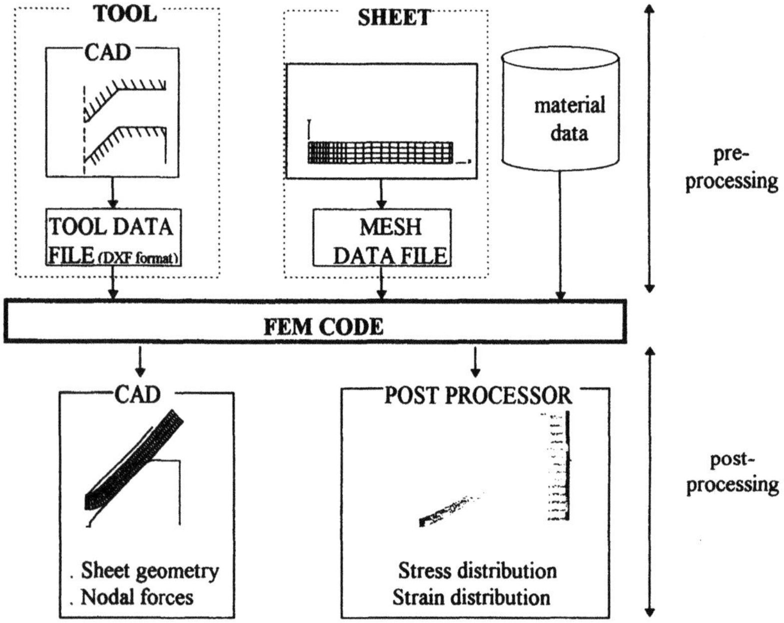

Figure 2 - pre/post processing organization.

3. RESULTS

In order to test the accuracy and reliability of the code, two test examples have been simulated and comparison with experiments have been performed. The tests were a V-bending example and a sheet bending example with a cylindrical punch.

3.1 V-bending

This example uses a standard tool from a press brake machine. Tool geometry as well as comparison of simulated and experimental geometries are presented in Figure 3 for three stages of the process (10, 15 and 20mm punch displacement). Material is mild steel with 3mm thickness, and sheet is discretized by a mesh of 90 elements (only 1/2 of the geometry is used for calculation, due to symmetry). Punch reaches the 20mm displacement after 690 steps. As seen, there is a good agreement between simulated and experimental deformed geometry.

3.2 Sheet bending with a cylindrical punch

This test uses the same standard tool but now with a cylindrical punch. Sheet characteristics are the same as in the previous example. Calculated and experimental geometries are presented in Figure 4 for the same stages as the V-bending process. The deformed geometry obtained by simulation agrees quite well with experiment also for this example.

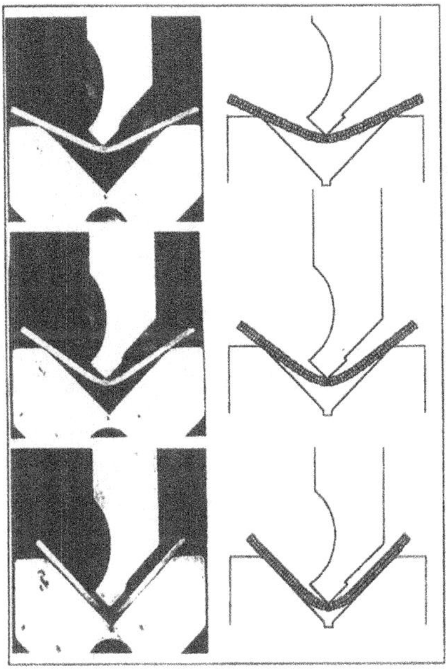

Figure 3 - compared geometries for V-bending

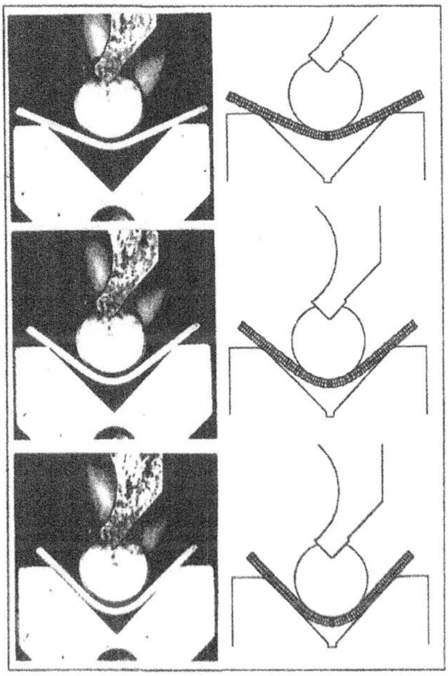

Figure 4 - compared geometries for cylindrical punch bending.

3.3 CPU time comparison

Time comparison between PC with a Intel Pentium processor running at 90MHz and a HP workstation model 720 was performed to test the efficiency of this implementation. Figure 5 shows comparison time results for calculation of the V-bending process example, with and without time optimization during code compilation.

To reach the 20mm punch displacement, maximum time reported as 100% in Figure 5 is 6mn 14s. This time corresponds to workstation CPU time without optimization. As it can be seen the PC computation time is very close to the workstation computation time (with optimization), which shows the great potential of using a PC to perform this kind of simulation.

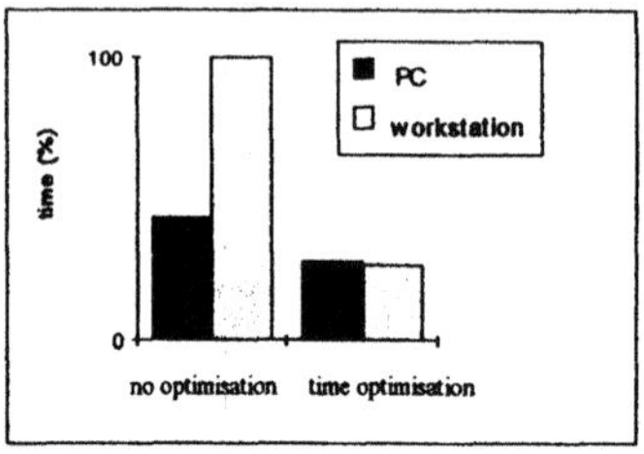

Figure 5 - CPU time comparison

4. CONCLUSIONS

An integrated system for analysis of 2D sheet metal forming processes has been presented. Until recently such system could only work by using workstations, specially due to the processing time needed for the FE code. It is shown that with actual processing power of PC's this task can be accomplished by using this kind of computers and in order to work efficiently it has been developed and presented pre and post-processing capabilities.

The most important part of this system is the FE code. The explicit formulation of the presented code gives the necessary robustness to overcome all non-linearities and difficulties of this kind of processes. Two simulation examples have been presented and comparison between simulation and experiment has shown a good agreement of the results.

It is expected that this type of implementation and its further development can contribute for the user friendliness and efficiency of Finite Element Simulation, which is nowadays a proved "tool" to be used in the design of sheet metal components, allowing us to predict forming defects that may appear during production, thus eliminating or reducing the trial and error stage after tool fabrication and before parts production.

5 ACKNOWLEDGMENTS

The authors wish to thank the financial support provided by Junta Nacional de Investigação Científica e Tecnológica of Portugal (JNICT), through project PBIC/TPR/2557/95. Also they would like to express their gratitude to Dr.A.Makinouchi and Sheet Forming Simulation Research Group of Japan for the availability of ITAS code and the opportunities and fruitful discussions during the stay at Riken Institute of Japan, of one of the authors. The authors are also grateful to Mr.Hipólito Reis, Mr.Fernando Junqueira and Mr.António J.Gomes for their help during experiments.

6 REFERENCES

Cao H.L., Teodosiu C. (1989) "Finite element calculation of springback effects and residual stresses after 2D deep drawing", Computational Plasticity'89, pp.959-971.

Duarte J.F., Santos A.D., Barata da Rocha A. (1995) "FE Simulation as an Aid to Design Sheet Metal Parts", SheMet'95-3. International Conference on Sheet Metal, June 26-28, University of Central England, Birmingham.

Makinouchi A.(1985) "Elasto-plastic stress analysis of bending and forming of sheet metal", Comp. Model. Sheet Metal Form. Proc., Eds. Wang N.M. & Tang S.C., Warrendale, pp. 161-176.

Makinouchi A., Shirataki Y., Liu S.D., Nagai Y. (1990)(1) "Generalization of Tool-Work contact conditions for elasto-plastic analysis in Forming Processes", 3rd ICTP, pp.1161, Kyoto, Japan.

Makinouchi A. (1990)(2) "Contact descriptions in finite element simulation of metal forming processes, Computer aided innovation of new materials, CAMSE'90, Conference Proceedings pp.597-602, August 28-31 1990, Tokyo, Japan.

Mc Meeking R.M., Rice J.R. (1975) "Finite element formulation for problems of large elastic-plastic deformation"- International Journal of Solids Structures, pp.601-616, vol.11.

Rebelo N., Nagtegaal J.C., Taylor L.M., Passman R. (1992) "Comparison of implicit and explicit finite element methods in the simulation of metal forming processes", Numiform'92, Conference Proceedings pp 99-108, September 14-18 1992, Valbonne, France.

Santos A., Makinouchi A. (1992) "Comparison of different approaches to describe tool geometry in FE simulation of 3D sheet stamping processes"- Computational mechanics 92 - theory and application, Conference Proceedings, pp.171, Vol.1, December 17-22, 1992, Hong Kong.

Santos A., Makinouchi A. (1993) "Contact strategies to deal with different tool descriptions in static explicit FEM of 3D sheet metal forming simulation", Numisheet'93, Conference Proceedings pp 261-270, August 31-September 2 1993, Isehara, Japan.

Saracibar C. (1990) "Analisis por el metodo de los elementos finitos de procesos de conformado de laminas metalicas", Dissertation, Politechnique University of Catalunya, Barcelona, Spain.

Takizawa H., Makinouchi A., Santos A., Mori N. (1991) "Simulation of 3-D sheet bending processes", FE-simulation of 3-D sheet metal forming processes in automotive industry, VDI berichte 894, Conference Proceedings pp 167-184, May 14-16 1991, Zurich.

Yamada Y., Yoshimura N., Sakurai T. (1968) International Journal of Mechanical Science, 10, 343-354.

7 BIOGRAPHY

Abel D. dos Santos is Assistant Professor of Faculty of Engineering of Porto University and responsible for Section of Materials and Technological Processes of the Department of Mechanical Engineering and Industrial Management. He is also Assistant Researcher at INEGI - Institute of Mechanical Engineering and Industrial Management - where it is mainly involved with Sheet Metal Forming Simulation.

He has been graduated at the Department of Mechanical Engineering of Porto University and obtained his Msc. degree in Structural Mechanics at the same University. His PhD degree has been obtained at Tokyo University, Japan, being supervised by Prof. T.Nakagawa of Tokyo University and A.Makinouchi of Materials Fabrication Laboratory of Riken Institute. His PhD work has been mainly concerned with development of a Finite Element code for Sheet Metal Forming Simulation intended to be applied mainly in the automobile industry.

INDEX OF CONTRIBUTORS

KEYWORD INDEX

GPSR Compliance

The European Union's (EU) General Product Safety Regulation (GPSR) is a set of rules that requires consumer products to be safe and our obligations to ensure this.

If you have any concerns about our products, you can contact us on ProductSafety@springernature.com

In case Publisher is established outside the EU, the EU authorized representative is:

Springer Nature Customer Service Center GmbH
Europaplatz 3
69115 Heidelberg, Germany

Batch number: 10755248

Printed by Printforce, the Netherlands